Containing the Atom

Containing the Atom

*Nuclear Regulation in a
Changing Environment,
1963–1971*

J. Samuel Walker

UNIVERSITY OF CALIFORNIA PRESS
Berkeley · Los Angeles · Oxford

Published 1992 by the
University of California Press
Berkeley and Los Angeles, California

University of California Press
Oxford, England

Prepared by the Nuclear Regulatory Commission; work made for hire.

Library of Congress Cataloging-in-Publication Data

Walker, J. Samuel.
    Containing the atom : nuclear regulation in a changing
environment, 1963–1971 / J. Samuel Walker.
        p.   cm.
    "Prepared by the Nuclear Regulatory Commission"—T.p. verso.
    Companion vol. to: Controlling the atom : the beginnings of
nuclear regulation, 1946–1962 / George T. Mazuzan and J. Samuel
Walker.
    Includes bibliographical references and index.
    ISBN 0-520-07913-2 (alk. paper)
    1. Nuclear energy—Law and legislation—United States—History.
2. Nuclear industry—Law and legislation—United States—History.
3. U.S. Atomic Energy Commission—History. I. Mazuzan, George T.
Controlling the atom. II. U.S. Nuclear Regulatory Commission.
III. Title.
KF2138.W35   1992
353.0087'22—dc20   91-46446
                                                                      CIP

Printed in the United States of America

1 2 3 4 5 6 7 8 9

The paper used in this publication meets the minimum requirements of American
National Standard for Information Sciences–Permanence of Paper for Printed Library
Materials, ANSI Z39.48-1984 ∞

Contents

Illustrations

Preface

This book is the second in a series of volumes on the history of nuclear regulation sponsored by the United States Nuclear Regulatory Commission. The first volume, *Controlling the Atom: The Beginnings of Nuclear Regulation, 1946–1962*, published by the University of California Press in 1984, examined the formative years of the efforts to promote the safety of the fledging nuclear power industry. This volume picks up the story at about the time that the nuclear industry experienced an unprecedented and unanticipated boom. The so-called "great bandwagon market" made an enormous impact on approaches to nuclear regulation. So too did the growth of environmentalism as a major public concern and political issue, which occurred at virtually the same time as the nuclear boom. This book investigates the interaction between those contemporaneous forces and the development of policies designed to ensure both the expansion and the safety of nuclear power.

Like *Controlling the Atom*, this sequel focuses on the United States Atomic Energy Commission (AEC), the agency that before its disbandment in 1975 was primarily responsible for nuclear power safety. Regulation was but one of the three major statutory functions that the AEC carried out. It devoted the bulk of its attention and resources to the other two: development and testing of nuclear weapons and promotion of the commercial use of nuclear power. This book is an effort to explain the AEC's regulatory positions and policies. It is not a full account of the AEC's activities, though it does attempt to show how the agency's other responsibilities affected its regulatory programs.

The period that this book covers is necessarily untidy in its boundaries. The end of Glenn T. Seaborg's tenure as chairman of the AEC was, in many ways, a logical place to end the volume. But some issues of significance that arose during his chairmanship culminated after he left the AEC, and they have received limited attention here. Given the length of this volume, readers might be surprised to learn that major subjects have not been included, but such are the exigencies of historical scholarship. The most prominent issue that I cover only in passing is radioactive waste disposal. The reason is that the AEC's attempt to settle the question by building a waste depository in Kansas came to a head only after the period examined in this book. The entire issue will be covered in the next volume in a way that will describe the background, outcome, and aftermath of the Kansas project. So will other matters of interest and importance that had their origins in the period between 1963 and 1971 but became more prominent later.

Given the complexity of the subject of nuclear regulation and the imposing amount of documentary material in a variety of collections, this book is not an exhaustive treatment of the topic it covers. The book centers on policy-making considerations and decisions at the highest levels of the AEC and its regulatory staff. It proposes answers to questions that I found to be most interesting and important. What were the primary concerns of the AEC in framing its regulatory programs? How did the AEC's statutory conflict of interest between its responsibilities to encourage the use of nuclear power and to ensure the safety of nuclear power affect its regulatory positions? What was the AEC's response to a changing technological and political environment? How was the agency perceived by outside constituencies, including Congress, the nuclear industry, and the general public, and how accurate were those perceptions? What influence did both supporters and critics of nuclear power and the AEC have on regulatory programs? Why did the overwhelming acceptance of nuclear power in the early 1960s give way to growing skepticism by the end of the decade? To what extent was the AEC successful in achieving its goals of promoting nuclear power, governing nuclear safety, and maintaining public confidence? Scholars with different interests are likely to focus on other questions that can increase our knowledge of the history of nuclear power regulation. Some might even (perish the thought!) take exception to my conclusions. But I am convinced that the questions I address are essential for a full understanding of the subject of nuclear safety, with all its implications for developments affecting technology and science, politics, public health, and energy supplies.

The Nuclear Regulatory Commission (NRC), which was established when the AEC was abolished, placed no restrictions, implicitly or explicitly, on me in the course of writing this book. I had complete access to all relevant documents and complete independence in deciding on the structure, approach, direction, and conclusions of this volume. Indeed, the NRC has provided an exceedingly pleasant setting in which to conduct historical scholarship. It provided the resources I needed but did not interfere with how I used them. From the highest levels of the agency on down, I enjoyed consistent support for and appreciation of what I was doing. I wish to emphasize that this book embodies the results of my own professional training and judgment, and I, not the Nuclear Regulatory Commission, bear full responsibility for its contents. The book does not represent an official policy position of the NRC.

I received very able assistance in the preparation of this volume from people both inside and outside of the NRC. My greatest debt is to my former colleague and the coauthor of *Controlling the Atom*, George T. Mazuzan. Perhaps most importantly, he hired me to be his associate at the NRC. Before he left the agency to become the historian of the National Science Foundation, he drafted three chapters that are a part of this volume. He wrote chapter four, which remains largely intact (much of it was previously published as "'Very Risky Business': A Power Reactor for New York City," *Technology and Culture*, 27 (April 1986): 262–284). Dr. Mazuzan also drafted chapters two and three, and even after substantial revision, he deserves credit (or blame) as a major contributor. After leaving the NRC, he has remained a source of advice and encouragement that I always appreciate and sometimes badly need.

I have benefitted immeasurably from the comments of dozens of readers who critiqued draft chapters of this book. I am particularly grateful to those who read the entire manuscript and offered constructive criticism: Roger M. Anders, Joseph J. Fouchard, Victor Gilinsky, George T. Mazuzan, Leo E. Slaggie, Roger R. Trask, and Allan M. Winkler. The other scholars, former AEC officials, and NRC employees who read selected chapters are too numerous to name, but I profited greatly from their efforts.

A legion of NRC staff members gave me invaluable assistance by retrieving seemingly ancient records, locating printed materials, obtaining books and articles on a timely basis, and providing a wide variety of administrative services. To avoid doubling the size of this book, I cannot name them all. But I do thank them all. A few NRC employees merit special mention. John H. Austin explained nuclear power technology in terms that even a lapsed diplomatic historian could understand. Leo E. Slaggie and Steven F. Crockett performed the same service on an even more perplexing topic, the legal aspects of nuclear regulation. My friends and colleagues in the Office of the Secretary of the NRC were marvelously supportive of my project and tried earnestly, some would say with mixed success, to help me come up with a title for this book.

I was fortunate to receive the help of three talented young historians in conducting the research for this volume. Kevin R. Hardwick and Mark A. Vargas, then graduate students at the University of Maryland, carried out a variety of research tasks as summer interns with gratifying efficiency and enthusiasm. They also developed a grasp of the subject of nuclear regulation so quickly that they challenged my thinking in engaging, if unsettling, ways. Anne L. Foster, then of History Associates, Inc., undertook several complex research projects and performed them with remarkable skill, especially in light of her own primary interest in diplomatic history. The work of all three greatly speeded my progress on this volume.

I am grateful to the many professional historians and archivists who provided assistance to me in the course of my research and writing. The Department of Energy houses many records of vital importance on the history of nuclear regulation, and its historical office treated me with courtesy and helpfulness that went far beyond the call of duty. I am especially indebted to Roger M. Anders, not only for sharing his knowledge of the records but also for sharing insights from his own scholarly work on nuclear issues. I am also very appreciative of the camaraderie, cooperation, and professionalism of former chief historian Jack M. Holl, his successor B. Franklin Cooling, and staff historians Terrence R. Fehner and Francis G. Gosling.

The staff members of the National Archives with whom I dealt were nearly all gracious and helpful. I am especially grateful to James Hastings, Lee Johnson, Michael McReynolds, Rodney Ross, Sarah Stone, and Sharon Thibodeau. For responding to sometimes desperate phone calls and/or making research productive and enjoyable, I am also indebted to David C. Humphrey of the Lyndon B. Johnson Library; Dwight M. Miller of the Herbert Hoover Library; Lana Beckett of the University of Southern California archives; Christopher M. Beam and

Lois Griffiths of the Edmund S. Muskie Archives at Bates College in Lewiston, Maine; James E. Carothers and Stephen Wofford of Lawrence Livermore National Laboratory; and Richard A. Baker and Donald A. Ritchie of the U. S. Senate Historical Office.

Finally, I owe a special word of thanks to Elizabeth Knoll, my editor at the University of California Press. She mastered the intricacies of the government contract process, the prelude to the press' own standard review process, with apparent ease, or at least without complaint. This in itself sets her apart from most editors, or indeed, from most citizens. More importantly, she provided constructive, thoughtful, and exceedingly useful criticism of my manuscript in a friendly and encouraging way. She failed only in her persistent efforts to get me to shorten the manuscript. For the length of the book, as well as for any errors that remain in spite of the contributions of so many individuals, I take full responsibility.

J. Samuel Walker
Rockville, Maryland
October 21, 1991

CHAPTER I

The Context of Nuclear Regulation, 1946–1962

During the 1960s, two contradictory assessments of nuclear power technology competed for public attention and support. They reflected the concurrent growth, on the one hand, of the use of nuclear power for electrical generation, and, on the other hand, of concern about the environmental and public health hazards of the technology. Those who strongly favored the expansion of nuclear power presented a seductive image of a beneficial technology that satisfied both energy and environmental claims. While acknowledging that nuclear power production posed a risk to public health, they insisted that the risk was small and that the advantages of the technology far exceeded its disadvantages. They argued that nuclear energy, properly channeled, could play a major role in improving standards of living throughout the world. Glenn T. Seaborg, chairman of the U. S. Atomic Energy Commission, and William R. Corliss, a free-lance technical writer, articulated the long-standing convictions of nuclear proponents in a book they published in 1971: "Nuclear power will soon be so cheap and so abundant that it will greatly accelerate the development of the hungry, poor parts of the world. If energy is cheap and abundant, so will be food, water, clean air, and all the amenities of what we call civilization." This vision of nuclear power had first appeared immediately after World War II and prevailed, at least in popular attitudes, virtually unchallenged into the 1960s.[1]

A more alarming view of nuclear power began to appear in the early 1960s and within a short time became increasingly familiar. It depicted the technology as a grave danger to public health and a serious threat to the environment. A growing chorus of critics suggested that the risks of nuclear power were severe enough to outweigh its benefits. Free-lance authors Richard Curtis and Elizabeth Hogan expressed the widening concern in a book they published in 1969. If the "concept of nuclear plants as clean, safe, reliable 'good neighbors' is a myth," as they argued in their book, "then a single major accident may cost the lives of tens and perhaps hundreds of thousands of Americans, affect the health of millions more, and cause property damage and loss in the billions of dollars." They added: "Even if such an event is forestalled, the gradual accumulation of immense stores of radioactive material and the inevitable release of a measure of its radiation represent a long-term threat not only to the population of our country but to life everywhere on the planet."[2]

At the center of the emerging controversy over nuclear power stood the Atomic Energy Commission (AEC). Although individuals from a wide variety of fields and institutions, including scientific research centers, engineering firms, law offices, health agencies, manufacturing companies, utilities, government organizations at the local, state, and federal levels, environmental groups, and both mass and trade media, participated in atomic energy affairs in important ways, the role of the AEC in nuclear development and safety was singularly vital. In the Atomic Energy Act of 1954, Congress assigned the agency responsibility for both promoting and regulating the nuclear power industry. As a result, the AEC performed a variety of sometimes conflicting functions; it was at once a promoter, a subsidizer, a supplier, a publicist, an educator, a research sponsor, a guarantor of safety, an inspector, an enforcer, and an arbiter. It was also a target of sharp criticism by both sides of the nuclear power debate.

The AEC was created by the Atomic Energy Act of 1946, which Congress passed after months of acrimonious debate over the administration and control of the fearful new technology that had destroyed the Japanese cities of Hiroshima and Nagasaki. The law, passed as postwar disputes with the Soviet Union were intensifying into the cold war, emphasized the military applications of atomic energy. The principal functions it assigned the AEC were the production of "fissionable materials" that fueled nuclear bombs, and the development and testing of new weapons. It required that the AEC own all fissionable materials and the plants that made components for atomic bombs. The 1946 act encouraged the AEC to investigate the civilian uses of nuclear technology, but this clearly was a secondary goal. The preoccupation with the military applications of the atom and the tight government monopoly of the technology ensured that progress in exploring the potential of peaceful nuclear energy would be, at best, uncertain.

The 1946 act also established the organizational framework of the AEC. The agency was headed by five commissioners, appointed by the president for five-year terms and confirmed by the Senate. Congress later designated the chairman as the "official spokesman" of the AEC, but otherwise, the commissioners exercised equal responsibility and authority. They made formal decisions by a majority vote. The commissioners focused on matters of policy and priorities while another official, the general manager, directed the agency staff and day-to-day operations. The staff communicated with and reported to the Commission through the general manager.[3]

In January 1947, when the AEC officially took over the functions of the Manhattan Project, which had built the first atomic bombs during World War II, it continued the practice of using contractors to perform its major tasks. The number of people who worked directly for the AEC throughout its existence was a small percentage of the number who worked for contractors that it hired. In 1952, for example, the AEC employed about 6700 people while its contractors employed about 135,000; by 1969 the figures had changed little, to about 7200 and 125,000 respectively. Contractors operated the many installations that the AEC owned to turn out materials and components for nuclear weapons and to conduct scientific research on atomic energy. They included the huge complexes that produced enriched uranium, plutonium, and tritium for nuclear bombs: the Hanford Works in eastern Washington (operated by the General Electric Company), Oak Ridge in Tennessee (operated by the Union Carbide and Carbon Chemicals Company), and Savannah River in South Carolina (operated by the DuPont Company).

Contractors also ran scientific laboratories for the AEC. Two major facilities operated by the University of California specialized in research on and design of nuclear weapons: Los Alamos Scientific Laboratory in New Mexico and Livermore (later Lawrence Livermore) Radiation Laboratory in California. National laboratories at Oak Ridge and Argonne, Illinois, conducted research on both military and civilian uses of nuclear energy; the AEC established another in Brookhaven, New York devoted exclusively to peaceful applications. Union Carbide operated Oak Ridge National Laboratory and consortiums of universities operated the Argonne and Brookhaven centers. In addition, the AEC hired contractors to run other installations scattered around the United States that helped to design, develop, fabricate, and test components for nuclear weapons, and to a lesser extent, carry out research on other aspects of atomic energy. *Time* magazine vividly described the magnitude of the AEC's "expanding empire" in January 1952: "The AEC controls a land area half again as big as Delaware—and is growing more rapidly than any great U. S. business ever did. Its investment in plant and equipment . . . makes it bigger than [the] General Motors Corp."[4]

To oversee the AEC's management of its vast operations and grave responsibilities, the 1946 Atomic Energy Act created a unique and powerful congressional committee, the Joint Committee on Atomic Energy. The Joint Committee consisted of eighteen members, nine from the Senate and nine from the House of Representatives. No more than five of the nine members from each chamber could come from the same political party. After 1954, the chairmanship of the committee alternated between the House and the Senate every session of Congress. The Joint Committee wielded enormous influence in Congress; it was one of the few committees established by statute rather than by the rules of the Senate and the House, and it was the only joint committee authorized to consider legislation and recommend it to Congress. Both the House and the Senate referred all proposed legislation on atomic energy to it, and the law gave it full jurisdiction over "all bills, resolutions, and other matters" relating to the AEC or atomic energy. Congress generally approved legislation or recommendations it received from the Joint Committee, particularly if the committee itself reached a consensus on an issue. In the immediate post-World War II period, as two informed observers put it, "the Joint Committee *was* Congress as far as atomic-energy matters were concerned." Even after atomic energy became a less awe-inspiring and esoteric subject than it initially seemed, the Joint Committee's decisions and judgment usually won the easy, if not automatic, endorsement of Congress.

The Joint Committee's authority over the AEC was at least as imposing as its influence in Congress. Statutory prescriptions and its own self-interest forced the AEC to foster and maintain the support of the committee. One key provision of the 1946 act instructed the AEC to "keep the Joint Committee fully and currently informed with respect to the Commission's activities"; the 1954 act went even further by changing the last phrase to "all of the Commission's activities." The Joint Committee strictly and aggressively enforced this requirement, sometimes in the face of complaints and resistance from the AEC. The provision was a potent instrument for airing criticism and influencing the policies of the AEC.

The Joint Committee's status as the AEC's sole oversight committee gave it additional leverage. If the committee agreed with the AEC, it had the power in Congress to be an effective and generally unchallenged ally. If, on the other hand, it opposed the AEC's position, the agency had nowhere else to go. The AEC could not use the time-honored practices of trying to play one committee against another or one house of Congress against the other. This was an especially vital consideration in the Joint Committee's control over the AEC's budget. Although the committee initially exercised little power over the AEC's funding, it gradually increased its role so that by the early 1960s it had assumed responsibility for authorizing the AEC's entire budget.[5]

The Joint Committee and the AEC shared the objectives of expanding the use of civilian nuclear power while ensuring its safety, but they did not always concur on the best means to accomplish their goals. The Joint Committee, for example, frequently grumbled about the AEC's licensing procedures, regulatory requirements, and other measures that it viewed as obstacles to the development of nuclear power. The AEC attempted to meet objections raised by the Joint Committee; only rarely did it defy the wishes of the committee during the 1960s—and then only with great reluctance. The agency and the committee existed in a kind of symbiotic relationship; they depended on one another to fulfill their ambitions. After the mid-1960s, when public opposition to nuclear power became more common and more vocal, critics often cited the lack of careful oversight by the Joint Committee as a major flaw in atomic energy programs. The effect, ironically, was to draw the committee and the AEC even closer together in defense of their mutual interests.

The most visible and influential members of the Joint Committee during the 1960s were its alternating chairmen, Congressman Chet Holifield and Senator John O. Pastore, and its ranking minority-party member, Congressman Craig Hosmer. Holifield, who was a charter member of the Joint Committee, was so active and powerful that he earned the nickname "Mr. Atomic Energy." Before his election to Congress, he had operated his own menswear shop in suburban Los Angeles; he turned his attention to politics after the great depression of the 1930s hurt his business so badly that he lost his home to foreclosure. When the economy improved he paid his debts and revived his business, but he was irreversibly hooked on politics. In 1942, he was elected to Congress as a Democrat from a newly created district in California.

1. Joint Committee on Atomic Energy hearing, 1965. Committee members seated at table, left to right: Henry M. Jackson, John O. Pastore, Chet Holifield, Melvin Price, Wayne N. Aspinall, Craig Hosmer, and (around corner) William H. Bates. Committee executive director John T. Conway is seated slightly behind the table between Pastore and Holifield. (National Archives 434-SF-19-12)

Holifield never graduated from high school or received formal scientific training, but he took a keen interest in atomic energy after Hiroshima. In 1946, he gained a seat on the newly established Joint Committee; he held it until his retirement from Congress in 1974.

Holifield was always a strong advocate of nuclear power development. During the 1960s he emphasized that the technology provided the means both to meet rapidly rising demands for electricity and to protect the environment from air pollution, an issue of increasing public concern. Nuclear power, he argued, was vital to the production of "abundant, economical, and reliable electric power" in a way that would "minimize the effects on our environment." For that reason, he was highly critical of AEC policies or regulatory procedures that he thought retarded the growth of the industry. He was even more exasperated by nuclear critics whom he found unreasonable and ill-informed; he denounced their "scare tactics" and "emotionalism." Holifield was outspoken in urging action to promote the use of the nuclear power, ensure its safety, and win its public acceptance. He was less inclined to address questions about setting priorities when those goals conflicted or reconciling the contradictions inherent in the promotion and regulation of the nuclear industry.[6]

Pastore's views on nuclear issues did not differ in any significant way from those of Holifield, though his style of leadership was more relaxed and less confrontational. His approach was partly a function of his personality and partly a function of the fact that, like most of the senators on the Joint Committee, he devoted less time than the House members to its activities. Pastore, who received a law degree from Northeastern University in 1931, served in several offices in his home state of Rhode Island, including the governorship, before winning election to the U. S. Senate in 1950. He was appointed to the Joint Committee two years later.[7]

Both Holifield and Pastore were liberal Democrats and supporters of the Great Society programs of President Lyndon B. Johnson. Craig Hosmer, by contrast, was a conservative Republican from southern California. When Richard M. Nixon succeeded Johnson in the White House, Hosmer was the president's congressman; he represented the district that included Nixon's home in San Clemente. He held a law degree from the University of Southern California, and following service in World War II, worked for a short time as a lawyer at the AEC's Los Alamos laboratory. He won election to Congress in 1952 and became a member of the Joint Committee six years later. Despite his political differences with Holifield and Pastore, Hosmer shared their outlook on the need for and advantages of nuclear power. He and Holifield, in fact, were fast friends and mutual admirers.

Hosmer expressed his opinions in colorful and often humorous terms. He lampooned what he viewed as exaggerated fears of nuclear power, for example, in a speech he gave in 1969: "Not long ago I was accosted by a man distressed over a nuclear power plant under construction in his locality. He was concerned about being thermally radiated and atomically irradiated, but he wasn't quite sure which alleged hazard would do him in first. However, he was certain that both the flora and the fauna in his area would get it before he did." Hosmer was philosophical about his position as the perpetual ranking minority member of the Joint Committee. "It gives me some heckling privileges not shared by others," he remarked, "and a constant opportunity to play Devil's Advocate." It was a role that Hosmer played skillfully, enabling him to attract more attention and wield greater influence than his minority-party status might normally have allowed.[8]

The government monopoly over atomic technology that the 1946 Atomic Energy Act established effectively, though not purposefully, discouraged the development of civilian nuclear power. By the early 1950s, however, the AEC, several utilities, and the Joint Committee were seeking ways to foster the growth of the industry. They were motivated by several considerations. One was concern for the long-term availability of energy resources. The President's Materials Policy Commission, for example, reported in 1952 that shortages of fossil fuel might become serious as early as 1975 and urged that nuclear power be developed to help meet future requirements. Of more immediate significance, at least to government officials, was the fear that if the United States did not actively promote atomic power it would fall behind other nations. This would be a major setback to its international prestige and scientific preeminence in the field of atomic energy. It would also run the risk of allowing the Soviet Union to score a propaganda and technological victory in the cold war. The determination to assert American leadership in nuclear power technology was intensified by an impulse to show that the atom could provide benefits that were as dramatic as its destructive power. President Dwight D. Eisenhower articulated that goal when he told the United Nations in December 1953 that "this greatest of destructive forces can be developed into a great boon, for the benefit of all mankind."[9]

With those considerations in mind, the Joint Committee drafted legislation to supersede the 1946 act. Congress approved and Eisenhower signed the new law in August 1954. It redefined the AEC's atomic energy program by ending the government's exclusive control of technical information and making the development of a commercial nuclear industry an urgent national goal. The measure directed the AEC "to encourage widespread participation in the development and utilization of atomic energy for peaceful purposes." At the same time, it instructed the agency to prepare regulations that would protect the public from the radiation hazards of nuclear power. Thus, the 1954 act broadened the AEC's mandate to include three major responsibilities: to continue its production and testing of nuclear weapons, to promote the private use of atomic energy for peaceful applications, and to ensure the safety of commercial nuclear power.[10]

The AEC promptly took action to carry out the objectives of the 1954 act. To encourage commercial development of nuclear power it created a "power demonstration reactor program." The agency offered to perform research and development on power reactors in its national laboratories, to subsidize additional research undertaken by industry under fixed-sum contracts, and to waive for seven years the established fuel-use charges for the loan of fissionable materials (which the government would continue to own). For their part, private utilities and nuclear vendors would supply the capital for design and construction of plants and pay operating costs other than fuel charges. The purpose of the demonstration program was to stimulate private participation and investment in evaluating the technical and economic feasibility of different reactor designs. At that time, no single reactor type had emerged as the most promising of the several that had been proposed.

The response of private industry to the AEC's inducements was cautious. Although experiments with AEC-owned reactors had shown the technical feasibility of using nuclear fission to produce electricity, many scientific and engineering questions remained to be answered. Despite the subsidies that the AEC offered, the capital and operating costs of atomic power were certain to be much higher than those of fossil fuel plants, at least in the early stages of development. The prospects of realizing short term profits were dim; an American Management Association symposium concluded in 1957: "The atomic industry has not been—and is not likely to be for a decade—attractive as far as quick profits are concerned."

In addition, nuclear power presented unfamiliar and disconcerting risks. Based on experience with government test reactors, the AEC and industrial leaders regarded the chances of a disastrous nuclear accident as remote. But they did not dismiss the possibility entirely. As an executive of the General Electric Company told the Joint Committee in 1954: "No matter how careful anyone in the atomic energy business may try to be, it is possible that accidents may occur." Many utilities were interested in exploring the potential of nuclear power, but, mindful of both its costs and its hazards, few were willing to press ahead until more information and experience were available. The AEC was gratified, and rather surprised, that by August 1955 five power companies—either as individual utilities or as consortiums—announced plans to build nuclear plants. Two decided to proceed without government assistance and three others submitted proposals for projects under the AEC's power demonstration program.[11]

The Joint Committee was less impressed with the response of private industry to the 1954 act and the AEC's incentives. The Democratic majority on the committee favored a larger government role in accelerating nuclear development. This conflicted with the position of the AEC, which wanted private industry to take the lead in achieving commercially competitive nuclear power. In 1956, Holifield and Senator Albert Gore introduced legislation that would direct the AEC to construct six nuclear plants, each of a different design, to "advance the art of generation of electrical energy from nuclear energy at the maximum possible rate." The Gore-Holifield bill provoked a major clash between Democrats on the Joint Committee, who claimed that it was necessary to maintain American leadership in the race for nuclear power, and the AEC, which insisted that private industry was best able to expedite progress. AEC chairman Lewis L. Strauss declared that "we have a civilian power program that is presently accomplishing far more than we had reason to expect in 1954." The Gore-Holifield bill was narrowly defeated on the floor of the House by a coalition of Democrats from coal-producing areas and Republicans. Nevertheless, the views it embodied and the impatience of the Joint Committee for rapid development of nuclear power placed a great deal of pressure on the AEC to show that its reactor programs were producing results.[12]

In that atmosphere, the AEC set up its regulatory program. The fundamental goal in drafting regulations was to protect public health and safety without imposing excessive requirements that would discourage private investment in nuclear technology. Within a short time after passage of the 1954 act, the AEC organized a small regulatory staff; its assignment was to prepare rules rigorous enough to assure safety but flexible enough to allow for rapid changes in nuclear technology. The staff promptly established procedures for licensing power reactors and drafted regulations on key issues, including qualifications for reactor operators and the distribution and safeguarding of fissionable materials.[13]

At the same time, the regulatory staff drafted standards for protection against the hazards of radiation, which was of vital importance for public health and the future of the nuclear industry. The principal danger of nuclear power to the public was that an accident in a plant would allow the escape of massive amounts of the radioactive materials that were a product of nuclear fission. This would present a serious threat to anyone exposed to the air or water they contaminated. The expanding use of atomic energy for peaceful purposes also underscored an occupational health issue—the level of radiation to which workers could be exposed in their jobs. Scientists and physicians had long recognized that high levels of radiation could cause severe health problems, ranging from loss of hair and skin irritations to sterility and cancer; in extreme cases they could be fatal. But the experts were much less certain about the consequences of exposure to small amounts of radiation. This was a question of concern primarily to scientists and health professionals until the mid-1950s, when the general public became alarmed by the possible dangers from radiation in "fallout" from nuclear bomb tests.

A spirited debate over the hazards of fallout in the late 1950s and early 1960s focused on the risks of exposure to low levels of radiation. It moved the question from the rarified realms of scientific and medical discourse to the front page, and for the first time, the public became aware of and worried about the problem. The fallout controversy was a major influence on the AEC's radiation protection standards. The agency based its regulations on the judgment of scientific groups that recommended acceptable exposure limits, called the "maximum permissible dose." As a result of growing public concern as well as the expanding use of atomic energy, those groups sharply reduced their suggested permissible doses for occupational and public exposure. The AEC adopted the lower limits as a part of its regulations.[14]

The AEC's regulations applied to its licensees but not necessarily to its own operations or those of its contractors. The 1954 act specifically excluded the AEC's weapons production facilities from compliance with the licensing requirements that private companies would be expected to meet. The AEC adopted a policy that declared its intention to make certain that its "reactor facilities are designed, constructed, operated, and maintained in a manner that protects the general public, government and contractor personnel, and public and private property against exposure to radiation from reactor operations and other potential health and safety hazards." It also stipulated that the AEC, at its discretion, might depart from existing standards "when overriding national security considerations dictate." The director of regulation could provide advice and recommendations but did not exercise jurisdiction over the safety of AEC-owned reactors.[15]

By the time that Eisenhower left office in 1961, the AEC's efforts to promote commercial nuclear development had shown some encouraging results. One government-owned and two privately owned reactors were operating and twelve other units were in various stages of planning or construction. It was an impressive beginning for an industry that hardly existed before the 1954 Atomic Energy Act. But neither the AEC nor the Joint Committee was satisfied. The reactors in operation or on order were generally small experimental models that were far from being commercially competitive or technologically proven. Interest in further development among utilities seemed to be flagging and the future of the industry appeared to be precarious. The Joint Committee blamed the situation largely on the AEC and hoped that the new administration of John F. Kennedy and a new AEC chairman would place greater emphasis on and achieve greater success in promoting commercial power.[16]

Kennedy's selection as chairman of the AEC was Glenn Seaborg, who was, in many ways, an inspired choice. He was a renowned scientist and an experienced university administrator who had served on several important AEC and presidential advisory committees since the end of World War II. Seaborg was a registered Democrat but not active in partisan politics. He took cautious and noncontroversial positions on the most divisive nuclear issues of the time—whether a moratorium on atmospheric testing of nuclear bombs should continue to be observed and whether the government should assume a greater role in encouraging the commercial development of nuclear power. In Seaborg's confirmation hearings before the Joint Committee, he expressed support for accelerating the nuclear power program but withheld comment on how it should be done, and he offered no definite opinions on the issue of resuming nuclear testing. His nomination easily won the approval of the Senate and he took office on 1 March 1961.

Seaborg rose to the high levels of government from humble beginnings. He was born in 1912 in the small mining town of Ishpeming, Michigan, where his father was a machinist. His family moved to Cali-

2.   Glenn T. Seaborg (National Archives 434-P-DPI991-5)

fornia when he was a boy. He attended the University of California at
Los Angeles, where he majored in chemistry and made Phi Beta Kappa.
He then moved on to the University of California at Berkeley, where he
received his Ph.D. in chemistry. A short time later, in 1940, he made his
mark in scientific circles, along with two of his colleagues, by first
identifying and extracting minute amounts of the element plutonium.
During the war, he headed a group of Manhattan Project scientists who
devised methods to chemically extract plutonium in amounts large
enough to be used in a bomb. The first nuclear weapon ever exploded,
in a test at Alamogordo, New Mexico, and the bomb that destroyed

Nagasaki were fueled by plutonium. Despite his key role in the construction of atomic bombs, Seaborg expressed reservations about their use against Japan. He was one of a handful of scientists who signed a petition to Secretary of War Henry L. Stimson in 1945 that urged that the bomb be demonstrated to observers at an isolated site. In that way, Japan might be persuaded to surrender before the bomb was unleashed on its cities.

After the war, Seaborg returned to Berkeley, where he shared in the discovery of eight new elements. In 1951, he and his colleague Edwin M. McMillan received the most coveted of scientific awards, the Nobel Prize, for their pioneering achievements. Seaborg served for several years as the director of nuclear chemical research at Berkeley's Lawrence Radiation Laboratory, and in 1958, was named chancellor of the university. From that position he moved to the AEC.[17]

Seaborg was calm, deliberate, and thoughtful. He was a conciliator who sought to find solutions acceptable to opposing sides on an issue; he had a rare ability to mediate controversial questions without losing his composure or making enemies. Seaborg could be aloof and detached; he kept his own counsel and followed his own instincts more than he relied on the advice of others, an outgrowth, perhaps, of his long experience with atomic energy issues. This was a source of some irritation to the Joint Committee, which sometimes felt that he did not take its views seriously enough. Seaborg was meticulous to a fault—one example was the voluminous and detailed diary he kept, with entries ranging from accounts of meetings he attended to correspondence with family members. He was the devoted father of six children and participated eagerly and regularly with them in camping, hiking, golf, bowling, school work, and other activities. When the Seaborgs went on vacation, he recalled, "we were a pretty sight, the eight of us packed into our red station wagon with a luggage rack often packed full of equipment and food."

Seaborg was an avid sports fan who attended many professional and college events; he invariably recorded the scores of the games he saw in his diary. He described himself as an "ardent hiker and nature-lover"; one of his innovations as AEC chairman was to request that a modest hiking trail be carved out of woods around the agency's Germantown, Maryland headquarters. Inevitably, it came to be called the "Seaborg Trail." Seaborg was not known for a keen wit, though on occasion he displayed an engaging sense of humor. In 1970, he and his wife were awarded "his and her" certificates of appreciation from a nuclear industry group. The audience expected Seaborg to respond with, in the words of a trade publication, "one of his famed scholarly lectures." Instead, he opened with light-hearted remarks about his courtship and marriage, and then addressed the issue of population control, "about which," he said, "I am often questioned." He continued: "I can assure you that I am strongly in favor of population control. My wife, Helen, is strongly in favor of population control. My son, Peter, is strongly in favor of population control. My daughter, Lynn, is strongly in favor of population control. My son, David, is strongly in favor of population control. My sons, Stephen and Eric, and my daughter, Dianne, are all strongly in favor of population control."[18]

Seaborg was not a strong, aggressive, hands-on administrator of the AEC. He kept informed but was not inclined to get deeply involved in many of the issues with which the agency dealt. This was especially true of regulatory matters; he paid little detailed attention to them until they became sources of controversy and criticism of the AEC in the late 1960s. He was much more interested and active in emphasizing the advantages of nuclear power in speeches, conferences, and other public appearances. This was a part of his effort to promote the use of nuclear power, though there was an element of self-aggrandizement in it as well. He traveled so much to make speeches and to receive awards and honorary degrees that Holifield once suggested that he was neglecting more important duties. Seaborg spent a great deal of time on international correspondence, visits, and meetings. He participated in those activities not only to build closer ties with nuclear experts in other nations but also to accept the accolades his foreign colleagues showered on him. A staff member of the White House Office of Science and Technology observed a similar dual purpose in a proposal of Seaborg's to increase government spending on research for more new elements. In addition to the scientific merits of the argument, he suspected that "subconsciously, at least, someone would like to have an element named Seaborgium."

Seaborg displayed a passive style of leadership that sometimes frustrated those who sought a more active response to problems. When two scientists from Lawrence Livermore Radiation Laboratory began to sharply attack the AEC for its radiation protection standards in the late 1960s, for example, Seaborg resisted pressure to take punitive action against them. He argued that once the facts were known, the claims of the two dissenters would not stand up to scrutiny. In Seaborg's mind, the key to nuclear growth and acceptance was education, and he did his utmost to present the advantages of the technology to a broad segment of the public.[19]

After Seaborg took office, both the AEC and the Joint Committee remained troubled by the seeming stagnation of the nuclear industry. The Kennedy administration showed little interest in rapidly expanding the use of nuclear power; it was much more concerned with the exploration of outer space. Nevertheless, Seaborg, in an attempt to mobilize support for nuclear power, convinced Kennedy to request a report from the AEC on the status of and future prospects for the technology. The agency submitted its findings in November 1962. Not surprisingly, it claimed that nuclear power was essential to meet the long-term energy demands of the United States and that the AEC should play a major role in developing the advanced reactors that would be needed in the future. More surprisingly, the AEC's report also suggested that existing reactor models were "on the threshold of being competitive with conventional power in the highest fuel cost areas." It predicted that nuclear plants might provide up to 50 percent of the nation's electrical generating capacity by the year 2000. The document's analysis and projections were disputable, and indeed, were questioned by critics who doubted that nuclear power was as far advanced or as crucial for filling future energy requirements as the AEC maintained.[20]

Despite some uncertainty and controversy over the prospects for nuclear power, the environment in which the industry first emerged in the early 1960s was stable and supportive. Politically, public attitudes toward nuclear power were, by all indications, strongly favorable. Faith in the ability of science and technology to solve problems and improve standards of living seemed firmly entrenched. Confidence in the federal government as the promoter of national interests, guardian of personal welfare, and sponsor of scientific progress was equally prevalent. The United States easily filled its requirements for power with conventional fuels, and the extent and immediacy of the need for nuclear electricity were unclear. Nevertheless, the expansion of nuclear technology appeared essential for meeting growing energy demands. In the early 1960s, environmental concerns, which came to play a major role in public attitudes toward nuclear power, were only beginning to make an impression on the nation's political consciousness.

The technological as well as the political environment was relatively tranquil during the early years of nuclear power development. Few plants were operating or under review for licenses, the size of individual units was small, and the volume of business that the AEC's regulatory

staff handled was modest. Interest in nuclear power was largely restricted to a few experts in industry and government. They acknowledged the possibility that a nuclear accident could cause a public health
disaster, but they viewed the chances of such an occurrence as remote
enough to strain credibility. Furthermore, they believed that the potential benefits of nuclear power unquestionably exceeded the risks. Few
observers seriously challenged their assessment. Indeed, nuclear proponents, despite their attentiveness to concerns about safety, were even
more worried about the economic outlook for the technology. They
feared that the promise of nuclear power would be undercut by its high
capital costs, and that if the technology failed in the marketplace it
would never fulfill its potential. The AEC's 1962 report attempted to
address those concerns by asserting, with little supporting evidence, that
nuclear power was on the verge of economic competitiveness. Within a
short time after the report appeared, however, the nuclear industry
experienced a boom that made the AEC's predictions look more prescient and more astute than even the most optimistic observers anticipated in 1962.

CHAPTER II

The Nuclear Industry and the Bandwagon Market

During the late 1950s and early 1960s the use of nuclear power to generate electricity was a novel and developing technology. Since relatively few plants were operating, under construction, or on order, the scope of the AEC's regulatory functions such as reactor siting, licensing, and inspection was still limited. Its regulatory activities focused on writing rules and establishing procedures for the nuclear power industry to follow. During the later 1960s, however, the nation's utilities rapidly increased their orders for nuclear power stations, participating in what Philip Sporn, past president of the American Electric Power Service Corporation, described in 1967 as the "great bandwagon market." At the same time, the size of plants being built also grew dramatically. The sudden and unexpected arrival of commercially competitive nuclear power made an enormous impact on the regulatory programs of the AEC, and they, in turn, greatly influenced the course of nuclear power development.

The nuclear industry included two basic components: vendors that manufactured reactors and supporting systems, and utilities that purchased reactors and added them to their power supply. In addition, other industries, such as uranium mining and milling, fuel fabrication, and waste management, provided services essential to the use and growth of nuclear power. Four reactor vendors controlled the market during the 1960s: the Westinghouse Electric Corporation, the General Electric Company, the Babcock and Wilcox Company, and Combustion Engineering, Incorporated. Among the four, Westinghouse and General Electric quickly assumed leadership in selling their systems to utilities. Both were widely diversified multinational corporations that had long dominated the market for electrical equipment such as turbine-generators, switchgear, and transformers. After gaining experience with nuclear energy through contracts on government-sponsored projects in the post-World War II years, they drew on their imposing financial resources to move ahead in the development of nuclear power. Combustion Engineering and Babcock and Wilcox were established suppliers of fossil boilers and steam generators, but they trailed Westinghouse and General Electric in winning buyers for nuclear reactors.[1]

Westinghouse's coleadership in the nuclear market resulted to a significant extent from its early involvement in developing a nuclear reactor for submarine propulsion. In the summer of 1946, the Navy assigned Captain Hyman G. Rickover to study reactor technology for possible use in submarines. Rickover soon became convinced that "light-water" designs that researchers were investigating offered important advantages. Light-water models used ordinary water both as coolant and moderator. As a coolant the water kept the reactor from overheating; as a moderator it slowed the speed of the neutrons that caused nuclear "fission." The fission of the nuclei of elements used for nuclear fuel (uranium−235 in light-water reactors) generated the heat needed to produce power, and moderating the speed of neutrons increased the efficiency of the fission process. Alternate reactor designs used "heavy-water" (rich in deuterium, a heavy isotope of hydrogen), helium gas, or liquid sodium as a coolant, and heavy-water or graphite as a moderator.[2]

Rickover persuaded Gwilym A. Price, the president of Westinghouse, that his company should move into the field of atomic energy. Aware that General Electric was already acquiring nuclear experience by operating the AEC's plutonium reactors at the Hanford installation and an experimental reactor near Schenectady, New York, Price established an atomic power division to meet the challenge of Westinghouse's chief rival. By 1948, Westinghouse, with Rickover's backing, was running the Bettis Atomic Power Laboratory, located near the company's corporate headquarters in Pittsburgh, for the AEC. Closely supervised by Rickover, the Bettis facility worked on the Navy's program to develop a reactor for use in submarines. Those efforts came to fruition in 1954 when the first nuclear-powered submarine, the USS *Nautilus*, went to sea with a Westinghouse reactor.[3]

Although Westinghouse, the Navy, and the AEC focused their work at Bettis on naval propulsion, company officials recognized that their experience with naval reactors could eventually be applied to electrical generation. The *Nautilus* used a pressurized-light-water reactor that Westinghouse designed. In this model, water that circulated through the fuel was kept under pressure to prevent it from boiling. The heat from the water was transferred to a secondary loop through a steam generator, and the steam in the secondary loop drove turbines to create electric power. In the course of its work on pressurized reactors at Bettis, Westinghouse made a number of important contributions to nuclear technology, including the improvement of reactor fuels and the development of specialized equipment for reactors. Largely because of the success of Westinghouse's submarine program, the AEC selected the company, again under Rickover's direction, to design the first large-scale civilian nuclear power reactor at Shippingport, Pennsylvania. The pioneering plant used Westinghouse's pressurized-water reactor technology to produce electricity.[4]

Building on its experience, Westinghouse took another major step in the field of nuclear power when it contracted with the Yankee Atomic Electric Company, a consortium of New England utilities, to supply the reactor for the Yankee Nuclear Power Station at Rowe, Massachusetts. Rated at 140 electrical megawatts, the Yankee plant first began operating in August 1960 and achieved full power in June 1961. The cost of building the plant turned out to be 20 percent less than originally estimated. In addition, Westinghouse sold pressurized-water reactors for two larger plants in 1962. By that time, it had clearly established itself as one of the two foremost manufacturers of nuclear power reactors, and it could make a strong case for being *the* industry leader.[5]

Even before Westinghouse entered the nuclear power field, its primary competitor, General Electric, had begun to gain first-hand experience with the technology. In 1946, Harry A. Winne, the company's vice-president in charge of engineering, became fascinated with the prospects for the development of a nuclear industry. His interest was triggered by his membership on a panel of high-ranking government and industry officials that submitted a widely publicized report to Secretary of State James F. Byrnes on means to control nuclear weapons and to promote peaceful applications of atomic energy. When the officer in charge of the Manhattan Project, General Leslie R. Groves, sought a firm to operate the plutonium production reactors at Hanford the same year, General Electric, at Winne's prodding, negotiated a contract to assume the task. In exchange, the AEC agreed to provide a nuclear development center and test reactor that would be operated by the company and built near its Schenectady headquarters. Located in West Milton, New York, General Electric named the facility the Knolls Atomic Power Laboratory.[6]

By the mid-1950s, General Electric had embarked on a major effort to capitalize on its nuclear experience by pushing hard for the commercial use of nuclear power. Company chairman Ralph Cordiner emphasized his commitment to that goal by declaring that "the atom is the power of the future and power is the business of General Electric." The firm initiated its nuclear power operations in 1953 when it decided to focus on developing a new type of light-water reactor, the "boiling-water" model. It featured a completely different design than the pressurized-water reactor. The boiling-water reactor used a single loop, boiled the water that circulated through the fuel, and sent the steam directly to the turbines. Just as its basic design for producing steam departed greatly from that of a pressurized-water reactor, so did its auxiliary and safety systems. General Electric built its own experimental reactor to work on the boiling-water reactor concept, the Vallecitos Atomic Laboratory near Pleasanton, California. In August 1957 the AEC approved an operating license for the Vallecitos plant; it was the first ever issued to a privately owned reactor. General Electric, in designing the boiling-water reactor, drew heavily on research conducted at the AEC's Argonne National Laboratory, where scientists demonstrated that the single-loop system could operate without causing serious radioactive contamination of the steam turbine.[7]

In 1956, General Electric signed a contract with an old customer, the Commonwealth Edison Company of Chicago, to build a boiling-water reactor at a site near Morris, Illinois. The 180 electrical megawatt Dresden plant, completed in 1959, was the first privately owned commercial reactor in the United States to go on line. In addition, General Electric provided a small reactor (about forty-eight electrical megawatts) for the Pacific Gas and Electric Company's Humboldt Bay station on the coast of northern California. General Electric rapidly expanded its nuclear operations; by 1959 it had assigned more employees (14,000) to nuclear projects and committed more money ($20 million) to nuclear research and development than any other American company. This helped the company to make technical improvements in its boiling-water design and to challenge Westinghouse for the leadership of the nuclear manufacturing industry.[8]

The only threats to the dominance of Westinghouse and General Electric were two companies that were longtime vendors of boilers and steam generators for conventional power plants, Babcock and Wilcox and Combustion Engineering. Each managed to gain a small but significant share of the power reactor market. Babcock and Wilcox, which had provided boilers for electrical generating stations since 1881, acquired its initial nuclear experience as a supplier of components for naval reactors. It established its Atomic Energy Division in Lynchburg, Virginia in 1953, and, as its first major commercial project, manufactured the reactor for the Indian Point nuclear plant, which Consolidated Edison of New York opened in 1962.

Like Babcock and Wilcox, Combustion Engineering began its nuclear operations by furnishing components to the Navy. It moved into nuclear power after purchasing the General Nuclear Engineering Corporation, a business founded by reactor experts from Argonne National Laboratory in the mid-1950s. Combustion Engineering made its first sale of a commercial reactor in 1966 when the Consumers Power Company of Michigan accepted its bid for the Palisades nuclear plant. Both Babcock and Wilcox and Combustion Engineering built pressurized-water reactors, which left General Electric as the sole manufacturer of boiling-water models. Westinghouse and General Electric, despite inroads made by the other two vendors, continued to dominate the light-water reactor market. By 1968 they had received 77 percent of the orders; Babcock and Wilcox and Combustion Engineering divided the remainder.[9]

The nuclear components that vendors sold to utilities were the key parts of a series of interrelated systems in a power plant. In popular usage, the term "nuclear reactor" sometimes referred to an entire nuclear generating station and sometimes only to the nuclear portion of a facility. Among specialists, the terms "nuclear steam supply system" or "primary system" described the reactor and its supporting equipment, which included the pressure vessel (a huge container that housed the reactor fuel), pipes, pumps, heat transfer apparatus, and safety and control systems. Major nuclear vendors provided the primary system to utilities, including its design. Within the primary system were numerous components that the vendor purchased from subcontractors. Nevertheless, the vendors were responsible to their customers for the entire primary system.[10]

The nuclear steam supply system used fuel made in the first stages of what was called the "nuclear fuel cycle," including uranium mining and milling, chemical refining, and fuel fabrication. After the fuel was used in a power plant, it went through the final stages, often called the "back end," of the fuel cycle, including storage, transportation, and reprocessing of irradiated fuel, and management of radioactive wastes. Taken together, the fuel cycle was a critical consideration in the total operation and cost of a nuclear plant.

After uranium ore was removed from surface or underground mines, it was sent to mills that produced "yellowcake," a uranium oxide concentrate chemically extracted in a partially refined form. Many of the companies that engaged in uranium mining also mined other ores; uranium was only a part of their operations. Some companies mined uranium and left the milling to others, while a few of the largest firms, such as the Anaconda Company, the United Nuclear Corporation, and the Kerr-McGee Corporation, integrated mining and milling systems.

The yellowcake from uranium mills was chemically transformed into gaseous uranium hexafloride, which was sent to gaseous diffusion plants for further processing. The AEC built and owned the gaseous diffusion facilities. Because the plants were so costly, government construction amounted to an indirect subsidy to the nuclear industry. The AEC contracted with fuel fabricators to feed the uranium hexafloride through a series of diffusion stages to enrich its concentration of the fissionable isotope, uranium–235. Natural uranium contained only about 0.7 percent of the isotope uranium–235; the enrichment process increased the percentage to about 3 percent, the concentration needed to operate a light-water reactor. The enriched hexafloride gas was then converted to uranium oxide in the form of pellets. Fuel fabricators loaded the pellets into tubes, usually made of the metal zirconium, to form fuel rods for a reactor. The tubes, in turn, were welded and fabricated into clusters known as fuel elements. By 1967, the four steam supply system vendors had captured virtually all the fuel fabrication market for commercial reactor orders above 100 electrical megawatts in size. The United Nuclear Corporation was the only other company to gain a share of the market. Another half-dozen firms made fuel for smaller research and development reactors and for reactors other than light-water designs.[11]

Because of national security considerations, Congress stipulated in the 1946 Atomic Energy Act that the AEC would retain title to all fissionable materials. The less restrictive 1954 law allowed private companies licensed by the AEC to use, but not own, "special nuclear materials" (that is, plutonium, uranium–233, and uranium enriched in the isotope 233 or the isotope 235 that could serve as fuel for reactors or atomic bombs). The government retained ownership to guarantee an adequate supply of special nuclear materials for defense requirements. As nuclear power achieved greater commercial acceptance, the AEC sought to further reduce the role of the government in nuclear development. It proposed to the Joint Committee on Atomic Energy that private companies be permitted to own special nuclear materials. The committee conducted hearings in the summer of 1963, and, after further deliberation and hearings, an amendment to the 1954 act that carried out the AEC's recommendation passed in August 1964. There was no strong opposition to the measure in principle; the issue that caused the greatest difficulty was ensuring a smooth transition from government to private ownership.[12]

The amendment repealed mandatory government ownership of special nuclear materials, thus placing another key element of the technology in private hands. For an interim period, the AEC could sell or lease special nuclear materials in its own inventory, but by 30 June 1973, all previously leased materials would be converted to private ownership. After 1 January 1969, the AEC could furnish uranium enrichment services to private industry in installations that the government would continue to own. Called "toll enrichment," the AEC could deliver to either domestic or foreign purchasers a quantity of enriched uranium in return for a fee. This assured domestic and foreign companies reliable access to diffusion plants, which was of great importance to American nuclear vendors who were selling reactors both in the United States and abroad. The delay in beginning toll enrichment allowed the government time to reduce its inventory without disrupting the domestic uranium market.[13]

The first stages of the fuel cycle ended when the fuel rods were loaded into a reactor; the back end of the fuel cycle began when the fuel was exhausted, or "spent." The disposition of highly radioactive spent-fuel elements was an unresolved problem. The assumption of the AEC and other nuclear proponents was that the spent fuel would be chemically reprocessed to recover fissionable material left in the fuel and convert it to usable forms. The other radioactive wastes would be isolated for eventual long-term disposal. This would provide another source of nuclear fuel to the growing industry and also largely relieve owners of individual power plants of the burden of storing spent fuel. The first commercial reprocessing plant, located at West Valley, New York, about thirty miles from Buffalo, received a construction permit from the AEC in 1963 and began operating in 1966. It was plagued by financial problems and by technical deficiencies that exposed plant workers to excessive levels of radiation and released radioactive materials to streams that ran through the site. It shut down for improvements in 1972 but never opened again. Companies that planned other reprocessing facilities during the late 1960s were unable to complete operational plants.[14]

The failure of commercial reprocessing was a major blow to the AEC's strategy for dealing with radioactive waste from nuclear power plants. But even if reprocessing had proven to be more successful, disposal of the high-level wastes that remained after recovery of the usable elements in spent fuel was a problem that still had to be faced. In 1970, in response to the growth of the industry and increasing expressions of concern about the lack of a policy on high-level waste disposal from scientific authorities, members of Congress, and the press, the AEC announced that it would develop a permanent repository for nuclear wastes in an abandoned salt mine near Lyons, Kansas. Geologists and other experts regarded large salt cavities as the most promising geologic formations in which to place intensely radioactive wastes. The AEC aired its plans for the Lyons project without conducting thorough geologic and hydrologic investigations, and the suitability of the site was soon challenged by the state geologist of Kansas and other scientists. The uncertainties about the site generated a bitter debate between the AEC on the one side and members of Congress and state officials from Kansas on the other. It ended in 1972 in great embarrassment for the AEC when the reservations of those who opposed the Lyons location proved to be well-founded. The back end of the fuel cycle continued to resist easy technical solutions, and, as a result of the Lyons debacle and other problems, began to emerge as a major source of public concern.[15]

The primary system of a power reactor and the various functions performed in the fuel cycle constituted the uniquely nuclear systems required for the production of atomic power. A nuclear plant also used equipment like that in a conventional power station. The nonnuclear components included the turbine-generator and what was called the "balance-of-plant." The turbine-generator in a nuclear unit was much the same as that found in a fossil-fuel plant, although it was not identical. A light-water nuclear facility needed a larger turbine-generator that ran at a slower speed than one driven by conventional fuel. The turbine-generator was a massive piece of equipment. In a 1000 megawatt nuclear plant, for example, it weighed about 5000 tons and measured about 20 feet by 23 feet by 220 feet. By the early 1960s, General Electric and Westinghouse were the only domestic manufacturers of turbine-generators. Several foreign suppliers occasionally sold equipment in the

American market, but not enough to represent a serious challenge to General Electric and Westinghouse.[16]

The term "balance-of-plant" described those elements that were not a part of the nuclear or turbine-generator systems. They included such functions as site preparation, structural engineering, licensing administration, and conduct of start-up procedures, and such components as accessory electrical equipment, substation buildings, towers, and communication systems. Those services were delivered by architect-engineering companies and engineering-construction firms. The architect-engineering companies provided professional expertise on determining general specifications, conducting bid invitations, and performing balance-of-plant design. They did not, however, take responsibility for construction of plants. The "engineer-constructors" managed both architect-engineering and plant construction functions. The leading architect-engineers and engineer-constructors for nuclear plants were also among the most prominent firms in performing similar services for fossil-fuel facilities. In several cases in the mid-1960s, General Electric and Westinghouse entered contracts for nuclear units in which they agreed to provide the balance-of-plant through subcontracting; they assumed the responsibility for building the entire facility. To encourage utilities to "go nuclear," they offered so-called "turnkey" plants at attractive prices; the term derived from the promise that, theoretically, the purchasing utility would only have to turn a key to place the installation on line.[17]

One structure of particular importance in a nuclear plant was the containment building. Its purpose was to prevent uncontrolled release of radioactive materials into the environment in the event of an accident. All commercial light-water reactors had containment structures; steel pressure designs or various types of concrete shells were most commonly used. The design and construction of the containment building often was subcontracted by the firms that furnished the balance-of-plant equipment and services. Because of the structure's vital role in the safety of the plant, its designers and builders worked closely with the manufacturers of the primary system. The Chicago Bridge and Iron Company was the leading supplier of free-standing steel containment buildings. Engineering-construction firms generally built concrete containment structures themselves.[18]

Nuclear power vendors represented many segments of American industry, but a few companies largely controlled the business. The concentration of services, experience, and expertise in a relatively small number of firms had long characterized design and construction in the electric-power industry. The arrival of nuclear power on a large commercial scale in the 1960s was a new development in terms of technology but not in terms of market patterns. The four major vendors of light-water reactors quite naturally capitalized on the relationship they had already cultivated with the nation's utilities through years of supplying fossil-fuel plants and components. Those companies faced a new challenge, however, in trying to convince utility managers that nuclear technology should join conventional fuels in their mix of energy sources.[19]

During the first half of the twentieth century, the electric utility industry in the United States developed in an evolutionary way. The revolutionary technological breakthrough that paved the way for the modern power industry took place shortly after the turn of the century with the introduction of the steam turbine, which was far more efficient than the reciprocating steam engine it replaced. Over the following decades, the most significant advances were gradual improvements in two areas. One was reducing production costs by building larger generating plants; this allowed "economies of scale" by decreasing the number of sites, foundations, control rooms, auxiliary equipment, and other expenses a utility paid to complete an individual power plant. The second area of improvement was achieving greater "thermal efficiency" by applying higher steam temperatures and pressures to convert energy from fuel into electricity; this dramatically lowered the amount of fuel needed to produce a unit of power. Those refinements provided greater efficiency and reliability at a decreasing cost per unit for the growing number of commercial and residential customers.[20]

The success of evolutionary changes tended to instill a conservative management mentality in utility executives; they generally exhibited a strong aversion to risk. They were convinced that cautious business and engineering practices had supported the remarkable growth of the industry. From the early years of the century, the consumption of electricity increased by an average of 7 percent annually, which meant that demand roughly doubled every ten years.[21]

The electric utility industry in the United States had grown up and remained decentralized. At the beginning of 1967, more than 800 separate entities were operating. Some 200 investor-owned utilities controlled 75 percent of the nation's generating capacity. Federally owned facilities, such as the Tennessee Valley Authority, the Bonneville Power Administration, and projects run by the Army's Corps of Engineers, accounted for an additional 13 percent. The other 12 percent was managed by a variety of municipal utilities, cooperatives, and other small systems. Only a limited number of all the power companies and agencies were potential buyers of nuclear power plants. Small utilities, municipal systems, and cooperatives acting alone lacked access to the capital needed to build large generating stations. Nevertheless, the post-World War II growth of "power pools," in which several independent electrical systems invested jointly in plant construction and shared the power through interconnections, allowed smaller companies to benefit from the increasing size of generating stations. The systems that were a part of a power pool could invest in large plants, whether nuclear- or fossil-fueled, in which firms with greater financial resources owned the majority interest.[22]

Although pooling arrangements allowed for more widespread participation by smaller companies, the basic decisions on whether or not to order a nuclear power station still rested with the seventy largest utilities, which had the resources and the market to build giant-sized plants. In the 1950s, most of them responded cautiously to the AEC's attempt to promote rapid nuclear development because of the high costs of nuclear plants and uncertainties about reliability and safety. Many remained hesitant even when light-water technology showed promise of becoming commercially tenable in the early 1960s. The traditional conservatism of the utility industry made it wary of exercising the nuclear option. Utility executives did not regard nuclear power, even though it employed a new type of fuel, as a fundamental change in the methods of converting steam energy into electricity. This could be done more cheaply with other time-tested and abundant energy sources. Therefore, the utilities that responded to the AEC's early push for nuclear power development in the 1950s—Consolidated Edison, Commonwealth Edison, Detroit Edison, and Pacific Gas and Electric—were exceptions within the risk-averse industry.[23]

By the early 1960s, General Electric and Westinghouse, along with some utilities, had concluded that a second generation of nuclear plants that applied economies of scale could be competitive with fossil units in the northeastern states and California where fuel costs were high. Acting on that conviction, Westinghouse persuaded two utilities to build nuclear facilities that were larger than any ordered to that time. In December 1962, the Connecticut Yankee Atomic Power Company, a consortium of utilities, signed a contract with Westinghouse for a 550 electrical megawatt reactor at Haddam Neck. It was more than twice the size of any reactor in operation. The following month, Southern California Edison announced plans to build a 375 electrical megawatt

Westinghouse reactor on the California coast between Los Angeles and San Diego. Although neither plant was expected to be cost-competitive with fossil-fuel units in its region, the two orders helped stimulate greater interest in nuclear power.[24]

When a utility's management decided to purchase new generating equipment, whether nuclear or fossil, it analyzed a number of variables and weighed different options relating to its own resources and needs. One critical consideration that was common to all utilities was the projected future requirements for power. A company calculated the minimum demand for its power over a specified period of time, usually in increments of a year. This was called the "base load." Utilities tried to meet base-load requirements with large-capacity units that operated at full power over long periods of time. New fossil-fuel stations that were used to satisfy the base load more than doubled in size during the 1950s. It seemed apparent to vendors and utilities that if nuclear power units were to compete with conventional plants in filling base-load needs, they would have to achieve similar economies of scale.[25]

Utilities also had to prepare for times when demands on their systems were greater than the base load could handle. The industry called those periods the "peak load." Peak loads usually were of short duration and were met by small capacity generating units of twenty-five to fifty electrical megawatts. Those plants also provided standby power to ensure system reliability. Sometimes, peak load power was purchased from or exchanged with other utilities on interconnected grids. Between the base load and the peak load was another range known as the intermediate load; it covered fluctuations in daily power requirements. Utilities depended on plants operating at less than full capacity to meet intermediate load demands.[26]

Both suppliers and utilities assumed that nuclear plants would fill base-load requirements. This meant that a nuclear facility had to compete with fossil-fuel plants that were the most efficient in production and the least costly per unit of power output. A nuclear plant required significantly higher capital outlays to build, but it offered the advantage of lower fuel costs over the life of the facility than those anticipated for conventional energy sources. In the 1960s, a utility purchasing a new plant had to make its decision based on a comparison of the long-term costs of fossil-fuel technology with an extrapolated cost estimate for nuclear power. Since utility executives could not draw on past experience with large nuclear units, they took an economic risk if they elected to buy a nuclear plant.[27]

Spurred by Westinghouse's sales of large reactors to Connecticut Yankee and Southern California Edison, General Electric made a daring move in 1963 to increase its own business and to convince utilities that nuclear power had arrived as a safe, reliable, and cost-competitive alternative to fossil fuel. The company had not received an order for a nuclear plant since 1961, and that project, planned for the coast of northern California at Bodega Bay, was threatened by strong public opposition. General Electric targeted nuclear power as a potential growth area that it would aggressively seek to expand, even if it had to accept considerable risks. It engaged in a bidding war with Westinghouse and with conventional-plant suppliers for a contract to build a large unit for the Jersey Central Power and Light Company. *Time* magazine reported in July 1963 that General Electric and Westinghouse were "knocking heads" over the Jersey Central plant and that the contract was "almost a must" for General Electric. The company eventually won the bid by offering the first turnkey contract for the 515 megawatt facility, named Oyster Creek and located near Toms River, New Jersey. General Electric agreed to supply the entire plant, even to the extent of obtaining permits and licenses. It anticipated that it would lose money on its fixed-price contract of $66 million, but it hoped to show that the plant would produce electricity more cheaply than a conventional unit and help to stimulate the market for nuclear power.[28]

The Jersey Central contract opened the "turnkey era" of commercial nuclear power and came to symbolize the competitive debut of the technology. Albert Tergen, president of General Public Utilities, of which Jersey Central was a subsidiary, told AEC chairman Seaborg that General Electric's bid suggested that it was "no longer economic to build fossil fuel plants on the Eastern seaboard." The utility explained the basis for its decision in a detailed report, published in early 1964, that analyzed the financial projections for the Oyster Creek plant. It compared the costs of a nuclear plant with those of a coal-fired plant at the same site and with a coal-fired plant at a mine-mouth site in western Pennsylvania; the projections for the annual costs of nuclear-generated power appeared more favorable than those of either of the coal units. Oyster Creek, then, was the first case in which the estimated cost of nuclear power competed successfully with fossil fuel. And it did so without any direct subsidies from the AEC. Seaborg told President Johnson that Oyster Creek represented an "economic breakthrough" for nuclear electricity.[29]

Despite Seaborg's assessment, the extent to which the outcome of the

Oyster Creek bidding applied to other plants was a matter of uncertainty. Philip Sporn, who was a keen observer of trends in the utility industry as well as a former president of one of the nation's largest power companies, was asked by the Joint Committee on Atomic Energy for his evaluation of the Oyster Creek decision. He concluded that for a number of reasons it was not a major breakthrough in the economics of nuclear power. The bid, he argued, reflected conditions that would not always prevail, such as the intense competition between General Electric and Westinghouse over Oyster Creek and overly optimistic projections for the costs of nuclear power over the life of the plant. Sporn pointed out that General Electric's price for building the plant was based not on experience but on extrapolation. He noted, quite accurately, that the company's eagerness to gain a competitive advantage prompted it to take a considerable financial risk in the contract it offered.[30]

Although Oyster Creek foreshadowed the nuclear power boom, some time elapsed before orders mushroomed into what Sporn later called the "great bandwagon market." The coal industry waged an intense battle with nuclear power in the months following the Oyster Creek decision. In several cases in which utilities considered building a nuclear unit, coal companies substantially reduced their fuel charges. In other cases, coal suppliers made price concessions to existing plants in a utility's system, which cut the overall cost of power production. This forced General Electric and Westinghouse to bid not only against each other for nuclear contracts but also against coal interests that were deeply troubled by the challenge of nuclear power.[31]

The competition to sell reactors to the nation's utilities set off a fierce corporate battle. Westinghouse followed the example of General Electric in offering turnkey contracts, and the two companies sold a total of twelve plants in the three years after Oyster Creek. In June 1966, General Electric announced that it would end its turnkey program and revert to selling only nuclear steam supply systems. Westinghouse again followed its rival's lead. The turnkey plants were a financial blow for both of the companies. One General Electric official commented: "It's going to take a long time to restore to the treasury the demands we put on it to establish ourselves in the nuclear business." The losses that the two firms incurred on their turnkey contracts ran into the hundreds of millions of dollars. But the turnkey offers were instrumental in convincing several utilities in high fuel-cost areas that nuclear power was a sound investment.[32]

Other developments within the electric-power industry during the mid-1960s also encouraged nuclear power expansion. Since the early part of the decade utilities had greatly increased their pooling arrangements. This affected their long-range planning because it allowed them to build bigger facilities without fear of over-expansion. Extra or reserve power produced by large generating units could be sold to other companies on the interconnection. If utilities needed justification for building more reserve power, they received dramatic support in November 1965 when a failure in one part of an interconnected system plunged a large portion of the northeastern United States into darkness.[33]

The implications of interconnection worked in favor of nuclear vendors, who cited them as a major selling point for plant designs that featured economies of scale. By extrapolating on the expenses of and experience with smaller units, they were able to reduce the estimated cost in dollars per kilowatt of nuclear plants to a range competitive with fossil-fuel facilities of a comparable size. This helped overcome a prominent disadvantage that nuclear vendors faced—the heavy capital requirements for building their plants. Designs for nuclear plants leapfrogged from the 500 to the 800 to the 1000 electrical megawatt range even though operating experience was still limited to units in the range of 200 electrical megawatts or less. The practice of "design by extrapolation" had been used for fossil-fuel units by Westinghouse, General Electric, and other plant manufacturers since the early 1950s. Scaling up the size of plant components rather than waiting to learn from operating experience with smaller facilities allowed vendors to meet escalating demands for power and for larger generating stations. They took advantage of new and stronger construction materials and computer-assisted design techniques to increase the size of the plants they sold, and they built large margins of error into their calculations. In the 1950s and early 1960s, design by extrapolation appeared to be a successful approach, and it was natural that vendors extended it to nuclear units.[34]

In addition to the positive effects of turnkey contracts, system interconnections, and increasing unit size on the attractiveness of nuclear power, other considerations enhanced its appeal to utilities. One was growing national attention to air pollution. Passage of the Clean Air Act in late 1963, providing federal assistance to state and local governments to conduct research on air pollution, embodied the concern over the problem. Coal-fired electrical plants were major contributors to air pollution and were obvious targets for clean-up efforts. As the campaign to improve the environment gained strength, the electric-utility industry became more mindful of the cost of air-pollution control in fossil-fuel plants. They increasingly viewed nuclear power as a desirable alternative to paying the expenses of pollution abatement in coal-fired units.[35]

Even as coal plants were drawing criticism as sources of foul air, the costs of coal and the transportation of it slowly began rising. At the same time, the 1964 law that allowed private ownership of special nuclear materials lowered the long-range cost estimates for nuclear fuel. While the government held to a fixed price for enriched uranium of $8.00 a pound until 1970, utilities and fuel fabricators negotiated prices for delivery after that time as low as $4.50 a pound.[36]

By the mid-1960s, a series of economic trends and other developments had presented a favorable climate for the growth of nuclear power. The outlook was so promising that W. Kenneth Davis, the president of the Atomic Industrial Forum, a private organization that promoted industrial applications of nuclear energy, declared in November 1965: "This is now a developed, mature industry." His view seemed to be confirmed within a short time when the "great bandwagon market" for nuclear plant orders began to gather momentum. The Tennessee Valley Authority signaled the coming boom when it announced plans in 1966 to construct twin General Electric reactors at Browns Ferry, Alabama. Since TVA was located in a coal-producing area that could provide exceptionally low prices for coal, its decision made an even greater impact on the nuclear market than had Oyster Creek. The title of an article in *Fortune* magazine summarized the effect of TVA's action: "An Atomic Bomb in the Land of Coal." The size of the reactors, 1065 electrical megawatts each, was equally stunning. General Electric extrapolated from the size of much smaller operating units to project economies of scale that would compensate for the capital cost of the twin plants. To many utility analysts, Browns Ferry offered conclusive evidence that nuclear power had come of age. "Coal and atomic energy competed head on for the contract, on equal terms," the *Fortune* article noted, "and atomic energy won decisively."[37]

The boom market reached a peak during 1966 and 1967, exceeding, in the words of James Young, vice president of General Electric's Nuclear Energy Division, "even the most optimistic estimates." In 1965, the year before the market accelerated, Westinghouse and General Electric each sold three reactors domestically. One more nuclear unit, a high-temperature, gas-cooled reactor to be built at Fort St. Vrain, Colorado, was purchased from the General Atomics Division of General Dynamics. The total capacity of those seven plants was 17 percent of all of the facilities ordered during the year. In 1966, by contrast, utilities bought twenty nuclear units that made up 36 percent of the electric capacity committed. General Electric captured nine orders and Westinghouse six. Babcock and Wilcox and Combustion Engineering broke into the power reactor market with three and two orders respectively. Nuclear vendors fared even better the following year; they sold thirty-one units that represented 49 percent of the capacity ordered in 1967. Westinghouse led the field with thirteen orders, General Electric received eight, and Babcock and Wilcox and Combustion Engineering won five apiece. In 1968, the number of reactor orders dropped to seventeen, nine for General Electric, five for Westinghouse, and three for Babcock and Wilcox. The percentage of the capacity filled with nuclear orders, however, remained high at 47 percent.[38]

The bandwagon market orders were large plants that far exceeded the size of operating reactors. The Oyster Creek reactor, rated at 515 electrical megawatts, was ordered in 1963 and began commercial operation in 1969. During those six years, the AEC issued thirty-eight construction permits for plants that were larger than Oyster Creek, and twenty-eight of them were in the 800 to 1100 electrical megawatt range. The degree of extrapolation from small plants to mammoth ones was a matter of concern even to some strong nuclear advocates. The Joint Committee on Atomic Energy expressed concern that vendors had made a "far greater extrapolation in capacity than has ever been made in a step with conventional energy sources." As the first large units neared the operating stage, industry officials acknowledged some "opening-night jitters" about their performance. By the late 1960s, it was apparent that design by extrapolation with conventional plants was not as successful as anticipated earlier. "We hoped the new machines would run just like the old ones we're familiar with," complained a utility executive about his huge coal-burning stations. "They sure as hell don't." The AEC's regulatory staff increasingly focused on the safety problems that unproven designs for large nuclear plants might raise.[39]

Nevertheless, the prospects for nuclear power expansion remained auspicious. The AEC projected in 1967 that by 1980 the nuclear generating capacity of the United States would range from 80,000 to 110,000 electrical megawatts, more than double the estimate it had made for the same year in 1962. It restated, with less equivocation, its earlier prediction that nuclear plants would provide 50 percent of the nation's electrical power by the year 2000.[40]

The AEC's financial support had been a vital element in the initial development of nuclear technology. One observer remarked in 1963 that the industry was so dependent on government subsidies that, "every time the AEC burps, the industry excuses itself." The AEC sought to lessen that dependence during the 1960s, but its programs and policies continued to provide important financial assistance to the industry. Although Congress amended the law in 1964 to allow private ownership of special nuclear materials, the government operated all enrichment services and facilities. Although private firms carried out some radioactive waste management, the AEC performed the principal research and development on waste disposal, particularly on high-level wastes. The AEC discontinued the programs it had sponsored in the 1950s and early 1960s that provided subsidies for demonstration reactors, but it remained active in reactor research and development.[41]

Industry also benefitted from the government indemnity program embodied in the Price-Anderson Act. This act, passed in 1957, provided $560 million in liability insurance in the event of a catastrophic nuclear accident. Although plant owners paid fees for the coverage, the price and the availability of insurance underwritten by the government was a bargain. Private insurance companies provided only a small percentage of the total coverage.[42]

Utilities were further aided by an anachronistic AEC licensing policy. Although nuclear power achieved commercial acceptability by the mid-1960s, the AEC continued to license reactors as research facilities under section 104b of the Atomic Energy Act rather than under section 103, which was designed for commercial reactors. Technically, a reactor could not be issued a section 103 license until the AEC found that the technology had developed to the point of "practical value for industrial or commercial purposes." The agency concluded that it could not make such a finding without long-term operating experience that would prove "practical value." This placed the AEC in the awkward, and to some minds, ridiculous, position of proclaiming the arrival of nuclear power on the one hand while refusing to subject it to section 103 licensing requirements on the other.

The use of section 104b benefitted utilities, especially larger ones, because it exempted them from a prelicensing antitrust review that was mandatory under section 103. It avoided the possibility of disputes over antitrust rulings that could slow already complicated licensing procedures and extend the already lengthy lead time required to place a nuclear unit on line. After a campaign to enforce antitrust laws in nuclear licensing led by Senator George D. Aiken of Vermont, a member of the Joint Committee on Atomic Energy, Congress amended the 1954 Atomic Energy Act in December 1970 to require the Justice Department to review construction permit applications. The effect was to remove commercial reactors from the jurisdiction of section 104b and place them under section 103. If the Justice Department found evidence that antitrust laws might be violated, it would advise the AEC to hold a hearing on the issue. The new law applied to twenty construction permit applications under AEC consideration and retroactively to seven others already under construction.[43]

The onset of the bandwagon market marked the arrival of competitive commercial nuclear power. The growth of the number of plants ordered and the increasing size of individual units exceeded the most optimistic projections of the early 1960s and fulfilled the AEC's objective, as defined in the 1954 act, of encouraging the development of a nuclear power industry operated by private enterprise. At the same time, the rapid expansion of plants and the practice of design by extrapolation placed severe burdens on the AEC's regulatory staff. The unexpected emergence of the industry raised unanticipated safety issues that the AEC sought, with mixed results, to resolve. The growth of the industry also awakened alarm among the public about the environmental impact and the hazards of nuclear power, and this, too, created difficult problems for the regulatory side of the agency. The AEC's dual responsibilities for both promoting and regulating the nuclear industry created a dichotomy in its response to the reactor boom of the 1960s— elation over developmental progress and uneasiness about its implications for reactor safety.

CHAPTER III

The Regulatory Framework

Between 1954 and 1962 the AEC built a regulatory organization and developed a regulatory process to evaluate safety issues, publish regulations, review license applications, and verify that its rules and license conditions were observed. Agency officials hoped that the regulatory framework they established would be sufficiently rigorous to ensure that nuclear power was safe, but not so inflexible or stringent that it discouraged the growth of the technology. Striking a balance between those objectives was always a difficult task, and it became even more formidable after the surge in orders for reactors began in the mid-1960s and the size of plants under review dramatically increased. The AEC reexamined and revised its regulatory procedures to meet the new demands on its resources, and its licensing requirements and structure continued to evolve. Nevertheless, the regulatory process remained a chronic source of complaints from both supporters and critics of nuclear power.

Compared with the other major functional units within the AEC, the regulatory organization was a small component of the agency. Total AEC employment, excluding contractors, hovered around 7000 throughout the 1960s. The regulatory staff, despite a rapidly increasing work load, expanded slowly from 339 in 1964 to 540 in 1971. The size of the regulatory organization reflected the interests and priorities of the agency. Chairman Seaborg estimated in 1967 that the Commission spent only about one-sixth of its time on regulatory affairs, though that percentage probably grew in later years.

The regulatory staff was physically as well as functionally separated from the weapons-producing and developmental arms of the AEC. In 1963 it moved from agency headquarters in Germantown, Maryland to a building eighteen miles south in Bethesda, Maryland. As a result, the regulatory staff developed its own administrative system for mail distribution, travel, record-keeping, correspondence, and other chores. The direct cause of the move to Bethesda was shortage of space in Germantown, but the segregation of the regulatory unit was symbolic of the AEC's attempt to keep regulatory operations removed from promotional ones. The size and location of the regulatory staff also were indications of its status within the AEC. Other offices viewed the regulatory function as something of a backwater in comparison with more glamorous and exciting weapons and reactor development activities. They recognized the vital importance of reactor safety, but they also worried that an overzealous regulatory staff could conceivably interfere with their own programs. For their purposes it seemed advantageous to keep the regulatory organization isolated, limited in size, and overburdened with work.[1]

The head of the regulatory staff was Harold L. Price, who had served in that capacity since the AEC had established its regulatory organization in 1954. His title was director of regulation. A native of Luray, Virginia, Price received a bachelor of arts degree from the University of Virginia in 1928 and a law degree from the same school in 1931. He practiced law in Roanoke, Virginia for several years before taking a job with the U. S. Department of Agriculture. From there he moved to progressively more responsible legal positions in other government agencies. In 1947 he joined the Atomic Energy Commission as an assistant general counsel in the Oak Ridge Operations Office. After transferring to AEC headquarters in Washington in 1949, he became the agency's deputy general counsel two years later. When Congress passed the 1954 Atomic Energy Act, the AEC's general manager asked Price to set up the licensing program for privately owned reactors, and in 1955 named him the director of the Division of Civilian Application, which carried out regulatory duties. During subsequent staff reorganizations, he remained as chief of the regulatory unit. In 1961, in an effort to separate regulatory functions from other AEC activities, the Commission removed Price and his staff from the supervision of the general manager. After that time, he reported directly to the commissioners.

Price was an unassuming, cautious, no-nonsense bureaucrat. Although he often presented himself as an unsophisticated country lawyer,

3. Harold L. Price (Herbert Hoover Library)

he had a sharp intellect and a keen sense of bureaucratic politics. His job required that he balance the constant and sometimes conflicting demands of AEC commissioners and staff, Congress, the nuclear industry, and the public to ensure reactor safety without impeding the development of nuclear technology. Price, perhaps by necessity, was a workaholic with few outside interests or activities. He could be brusque and impatient; Congressman Hosmer, with tongue in cheek, once described him as "at least as sweet, kind and gentle" as Admiral Rickover, a man of legendary spleen. But Price earned the loyalty and respect of his staff, and despite his lack of formal training in science or engineering, he dealt knowledgeably with the policy aspects of technical issues. He approached regulatory problems in a business-like way and displayed what Commissioner James T. Ramey once described, in jest, as "a lot of common sense for a lawyer." Price was not inclined to undertake bold initiatives or to promote major policy departures. He believed that his role was to carry out the wishes of the Commission as efficiently and promptly as possible. "I work for the Commission and I work under their policy guidance," he once summarized his position in the agency during Joint Committee hearings. His ability and savvy enabled him to retain the confidence of the Commission through seventeen years of controversy over the AEC's regulatory programs and performance.[2]

In 1964, in response to a growing regulatory work load, Price reorganized his staff structure by creating five divisions out of the three that had previously existed. They were: the Division of Reactor Licensing, which evaluated license applications, judged reactor operators' qualifications, and maintained liaison with other AEC units involved in aspects of reactor safety; the Division of Safety Standards, which developed and assessed standards for reactor design, siting, construction, and operation, reviewed and recommended changes in radiation protection standards, and participated in safety research programs; the Division of Material Licensing, which issued licenses for medical, industrial, and agricultural uses of radioisotopes and other nuclear materials; the Division of State and Licensee Relations, which coordinated contacts with state governments; and the Division of Compliance, which inspected licensee activities to ensure that they conformed with AEC regulations. In 1967, Price again revised the regulatory structure. He divided the Division of Safety Standards into two new units, the Division of Reactor Standards and the Division of Radiation Protection Standards. He anticipated a rapidly increasing level of activity in the areas of reactor licensing, standards development, and inspection.[3]

The burden of the surge in reactor orders that began in the mid-1960s fell first on the Division of Reactor Licensing. The director of the division, Peter A. Morris, complained in November 1966 that his staff was already "swamped" and that it was likely to face a growing flood of applications. Morris, who held a Ph.D. in physics from the University of Virginia, had recently been appointed the division's director after holding a number of other jobs in the nuclear industry and the AEC. He believed that his experience enabled him to understand industry's needs and concerns as well as safety issues, and he emphasized the "importance of industry being able to continue to count on efficient and business-like handling of its license applications."[4]

The AEC handled applications in a process that, despite its goal of simplicity, had grown more elaborate over the years. The fundamental regulatory objective of the AEC after passage of the 1954 Atomic Energy Act was to ensure that public health was protected without imposing overly burdensome requirements that would impede industrial development. One commissioner, Willard F. Libby, articulated an opinion common among AEC officials when he remarked in 1955: "Our great hazard is that this great benefit to mankind will be killed aborning by unnecessary regulation." Distinguishing between essential and excessive regulation was a matter of judgment and experience, and the complexities of nuclear technology and politics gradually increased the complexity of the licensing process. The 1954 act laid out the basic framework for licensing plants by outlining a two-step procedure. An applicant that met the AEC's requirements would first receive a construction permit, and after the plant was built, an operating license. This arrangement allowed the construction of reactors to go forward before all the technical questions about their operation had been answered. Within the two-step licensing structure, the AEC established its own requirements and procedures.

The first phase of the licensing process as it existed in 1964 preceded the formal filing of an application. When a utility was considering the construction of a power reactor, it generally met informally with the regulatory staff to discuss possible sites for the prospective plant. At that point the staff advised the utility about its siting requirements on critical matters such as population density, seismology, meteorology, geology, and hydrology. It also outlined the information that should be included in a formal application. Once the utility decided to build the plant at a particular site, it submitted its application for a construction permit. The staff first conducted a preliminary review of the document to make certain that it contained the necessary information. If not, it returned the application for additional work. If so, it formally docketed the application and the review process officially began.

The major portion of the construction permit application consisted of the "preliminary hazards summary report." Once it received this document the staff of the Division of Reactor Licensing undertook a detailed evaluation. It analyzed the suitability of the site; the soundness of design methods and calculations; the adequacy of the proposed plant's containment building and safety systems for preventing the escape of radioactive materials in case of an accident; and plans for providing power to the plant, handling fuel, and disposing of radioactive wastes. The staff also assessed the technical and financial qualifications of the applicant to build and operate the plant safely. It usually found that the application was incomplete or unsatisfactory in some respects. In those cases, it held conferences in which the utility and the reactor manufacturer could address its questions, propose revisions, and resolve differences. The AEC did not expect the preliminary hazards summary report to provide complete design details of the plant's safety systems, but it insisted that the report offer "reasonable assurance" that the reactor could be built and operated without threatening public health.[5]

As soon as the Division of Reactor Licensing received an application for a construction permit, it sent copies to the Advisory Committee on Reactor Safeguards (ACRS) for a concurrent and independent evaluation. The ACRS was a body of reactor experts from outside the AEC who met regularly to review applications and discuss safety issues. It had initially been formed to advise the AEC on the the design and construction of government-owned reactors, but after the passage of the 1954 Atomic Energy Act, it played a major role in evaluating the safety of commercial plants. The ACRS had fifteen members, who usually met as a full committee once a month for a period of three days. Smaller groups convened more frequently in subcommittee sessions. The members of the ACRS were appointed by the Commission for four-year terms; most held full-time positions in universities, national laboratories, or other research institutions. Employees of the nuclear industry were precluded from membership because of conflict-of-interest considerations, though some retired industry officials served on the committee. The ACRS included well-known and respected authorities from a variety of technical fields related to nuclear energy.

In 1957, following a disagreement in which the ACRS took a more cautious position than the regulatory staff on issuing a construction permit for a technologically advanced fast-breeder reactor, Congress made the committee a statutory body, directed that it review every plant application, and required that its licensing recommendations be made public. The committee's review of construction permit applications proceeded simultaneously with that of the Division of Reactor Licensing, but it focused on novel features rather than simply duplicating the effort of the staff. As soon as the ACRS received an application it assigned it to a subcommittee for examination. The subcommittee discussed the proposal with the regulatory staff and the applicant to raise questions and concerns. It submitted its findings and evaluation to the full ACRS, which also received a safety analysis of the application from the regulatory staff. The entire committee then considered the application, which often included further meetings with representatives of the utility and reactor manufacturer and with the regulatory staff. Once the ACRS reached a consensus on its collective opinion, it sent a letter to the chairman of the AEC. The letter frequently contained suggestions for issues that remained to be resolved before the plant received an operating license. On occasion, members of the ACRS added dissenting views if they strongly disagreed with the position of their colleagues.

The ACRS set its own agenda, elected its own chairman, operated on its own schedule, and reached its own conclusions. It was not obligated to conform with the views of the regulatory staff or the wishes of the Commission. The members of the ACRS, like their counterparts in the AEC, supported the use and the growth of nuclear power, but they had much less of an institutional stake in the rapid expansion of the technology. In the early years of nuclear development, the committee commanded greater prestige than the regulatory staff, but as the staff increased its size and technical capabilities it enhanced its stature within the nuclear community. Indeed, as the staff gained greater parity with the ACRS, some nuclear proponents suggested that the role of the committee in the licensing process should be diminished. The regulatory staff and the ACRS did not always agree on safety questions, and each sometimes complained about the methods and performance of the other. The independent role of the ACRS exacted a price by causing greater complications, more potential for conflicts and delays, and general untidiness in the licensing process, but it provided a different perspective and a second opinion against which to measure the evaluations of the regulatory staff.[6]

At the same time that the staff of the Division of Reactor Licensing reviewed the application and waited for the analysis of the ACRS, it solicited the advice of other government agencies. As a matter of course, it sought the comments of the U. S. Geological Survey and the U. S. Coast and Geodetic Survey on geological and seismological aspects of the proposed site, the U. S. Fish and Wildlife Service on the potential radiological effects of the plant on marine and animal life, and the U. S. Weather Bureau on meteorological conditions. The staff also sent copies of the construction permit application to state and local government officials and placed it in the AEC's Public Document Room, where it was available for public examination.[7]

Once the regulatory staff completed its review of the application and received reports from other agencies and the ACRS, it prepared a safety evaluation for a public hearing before an Atomic Safety and Licensing Board appointed to consider the proposed reactor. The licensing boards were a recent innovation that had evolved from the Joint Committee's requirement, first established in a 1957 law and later revised, that the AEC hold a public hearing on every construction permit application. For a time, the AEC had used a single hearing examiner to conduct a hearing. This practice had created problems that arose principally from the fact that the hearing examiners were trained in administrative law rather than in reactor technology. In 1962, at the urging of the Joint Committee, the AEC established an Atomic Safety and Licensing Board panel. It initially consisted of eleven part-time members with strong technical qualifications from outside the AEC, three full-time AEC hearing examiners, and one part-time administrative law expert. In each licensing case, the board that conducted the public hearing was made up of a hearing examiner, who served as chairman, and two technical experts drawn from the panel. The Commission was responsible for selecting the three licensing board members for an individual proceeding.

The Atomic Safety and Licensing Boards were, to an important extent, another of several measures adopted by the AEC to separate its promotional and regulatory responsibilities. Within a short time after the 1954 Atomic Energy Act became law, informed observers expressed concern about the inherent conflict of interest in the AEC's mandate to encourage the development of the atomic industry and to regulate its safety. One possible solution was to create separate agencies, but in light of the limited knowledge of nuclear technology, the shortage of qualified nuclear scientists and engineers, and the likelihood of conflict between different agencies, neither the AEC nor the Joint Committee favored such an approach. The licensing boards were an effort to address the problem without establishing a separate regulatory agency. They were designed to help insulate regulatory decisions from undue influence by other AEC programs and to provide a public evaluation of the regulatory staff's judgment by a board made up of a majority of outsiders.

To further avoid the appearance or the substance of a conflict of interest in its licensing deliberations, the AEC also imposed restrictions on communications between the regulatory staff on the one hand and a licensing board and the Commission on the other. The AEC first set up a "separated staff" requirement, in which members of the regulatory staff, the general counsel's office, and others who had worked on a particular licensing case were prohibited from discussing it with the Commission or the licensing board except on the public record after the AEC issued a public notice of the hearing. They remained "separated" until the board's decision. This procedure later became subsumed in a more general *ex parte* rule. It prohibited private contacts between the Commission and a party, including the regulatory staff, involved in a proceeding; oral or written communications had to be placed in the public record.[8]

The AEC announced a hearing of a licensing board at least thirty days before it was scheduled to begin. A short time later, within about twenty days of the hearing, the AEC published the regulatory staff's evaluation of the proposed plant's safety and the letter report of the ACRS. They constituted the basic documentary evidence for the proceeding. The hearing could be contested or uncontested. In theory, it could consider conflicting views between the staff and the applicant over the merits of the plant proposal, but in fact no applicant went to a licensing board hearing if the staff opposed issuance of a construction permit. The staff and the applicant resolved their differences before reaching the hearing stage. Both, therefore, appeared before the board in support of the application. Other parties could challenge the application with oral or written testimony, either by participating fully in the proceedings as intervenors or by making "limited appearances" in which they simply submitted a statement of their position.

The purpose of the licensing board was to make certain that the staff and the applicant provided sufficient evidence to justify a construction permit and to judge the validity of their technical review. The depth of the review and the approach by which an individual board arrived at its decision varied according to the nature of the application and the membership of the board. The precise role of the licensing boards remained an open question: was their function primarily to certify the correctness of procedures and the completeness of applications or was it to conduct a *de novo* technical review to check on the assessments of the staff and the ACRS?

Once the licensing board handed down a decision, the fate of the application, at least within the AEC's licensing system, rested with the Commission. If the commissioners did not formally review the board's ruling, it became effective on a specified date. If they elected to review it, either on their own initiative or on petition from one of the parties in the proceeding, their decision completed the construction permit application process. On occasion, the Commission modified a licensing board finding by requesting additional information or ordering further review. When the commissioners issued or affirmed a favorable decision, the director of regulation issued the construction permit. Commission decisions could be appealed to federal courts; they were the last resort for anyone who objected to the AEC's action on an application.[9]

The construction permit allowed the applicant to build the plant, but operating it required another series of safety reviews. As construction of the plant proceeded, the applicant submitted a "final hazards summary report" that showed the completed design and operating plans. This document received another comprehensive review from the Division of Reactor Licensing and the ACRS. After they finished their evaluations, the applicant might face another public hearing, but, unlike the requirements for a construction permit, a hearing was not mandatory. At its discretion, the Commission could order a hearing if intervenors requested it or public interest seemed to warrant it. A ruling by a licensing board was again subject to Commission review; in the absence of a hearing the approval of the application became effective on a specified date. In many cases, the AEC granted a "provisional operating license" for a period of up to eighteen months that limited the power output of the reactor until its performance could be further evaluated.[10]

The complexity and redundancy that characterized the licensing process by 1964 were a source of concern to the nuclear industry. As one observer described the situation: "A utility that wants a permit to build a nuclear plant now is in about the same position as a rider starting a cross country race on a foggy night without ever having been over the course before. There is no one who can really give him much idea of what he will be up against. About all he can be sure of is that there are some unpleasant surprises ahead of him." Anticipating an increasing number of reactor applications in the near future, the AEC wanted to avoid "unpleasant surprises" and complications that would discourage the growth of the industry. In January 1965 the Commission decided to appoint a special panel to study its licensing procedures and recommend ways in which they might be simplified.[11]

The idea for the review originated with Commissioner Ramey, a veteran participant in atomic energy affairs and a strong advocate of nuclear power development. A graduate of Amherst College and the Columbia University School of Law, Ramey had begun a long career in government service in 1941 as a staff attorney with the Tennessee Valley Authority. His work impressed the chairman of the TVA, David E. Lilienthal, and in 1947, when Lilienthal became the first chairman of the AEC, Ramey also moved to the new agency. As a lawyer with the Chicago operations office, he drew on his experience in TVA to design contract arrangements between the AEC and private firms to carry out particular tasks. He also gained a great deal of knowledge about reactor development programs.

In 1956, the Joint Committee on Atomic Energy appointed Ramey as its staff director. Like members of the committee, he pushed for more rapid development of commercial nuclear power and complained that the AEC was moving too slowly. Ramey won the respect and admiration of members of the Joint Committee, and when a seat on the AEC opened in 1962 they pushed hard for his appointment to it. The White House was reluctant to agree because of Ramey's close ties to the Joint Committee and Commission members harbored similar reservations. In addition, some senior AEC officials remembered and resented Ramey's role in bitter clashes between the agency and the Joint Committee. But the lobbying of the Joint Committee, especially Holifield, secured Ramey's selection. He was reappointed by President Johnson in 1968, and when his term expired in 1973, he had served longer, almost eleven years, than any other member of the Commission.

Ramey's experience with atomic programs, along with his knowledge, energy, and commitment, made him an active and influential participant in a broad range of AEC activities. From the time he joined the Commission, he took greater sustained interest in regulatory affairs than any other commissioner. In the late 1960s and early 1970s, when opposition to nuclear power was becoming increasingly visible, he did more than any of his colleagues to reach out to nuclear critics in an effort to address their concerns and find common ground. His attentiveness to regulatory issues did not mean that he had curbed his enthusiasm for rapid nuclear development; one industry official described him in 1966 as "industry's best friend on the Commission." Rather, it suggested that he recognized more clearly than his fellow commissioners the intimate and inseparable relationship between safety questions and industry growth. He realized that a major nuclear accident would be a severe setback to nuclear progress, but he also worried that excessive regulation or public apprehension would have a similar effect. He guarded against actions that would impose what he viewed as unnecessary burdens on the industry or raise public fears.

Ramey was an outgoing personality whose legal and activist orientations occasionally caused conflict˜ with the more passive and detached demeanor of the scientifically minded Seaborg. He sometimes surprised his colleagues by showing unusual sensitivity to perceived slights; he once circulated a memorandum rebuking another Commission member for failing to check with him before deciding "to shoot from the hip" on regulatory matters at an industry meeting. Despite his prickly side, Ramey had a genial bearing and a good sense of humor that helped win many friends among the different groups connected with atomic energy. An avid golfer, he did not hesitate to sell the advantages of nuclear power during golf outings with utility executives. By late 1964, he was concerned that efforts to promote nuclear technology were threatened by a formidable and unpredictable licensing process.[12]

At Ramey's suggestion, the Commission agreed in January 1965 to form a Regulatory Review Panel to study licensing procedures. It appointed William Mitchell, a former general counsel of the AEC who was an attorney in private practice in Washington, to chair the committee. With the exception of Manson Benedict, a professor of nuclear engineering at the Massachusetts Institute of Technology, the committee members were either high-level employees of or consultants to nuclear utilities or vendors. The seven panelists began their work immediately. Over a period of five months they met with regulatory staff officials, the ACRS, licensing board panel members, commissioners, and Joint Committee staff representatives to discuss licensing requirements. After gathering information from those groups, the Mitchell panel convened for four days in Williamsburg, Virginia to draft its report. It submitted its recommendations to the AEC on 14 July 1965.[13]

The Mitchell panel gave a strong endorsement to the work done in the past by the regulatory staff and the ACRS, praising them both for "remarkable" performances. It affirmed that it had "every confidence that the public interest is being protected by the regulatory process as it now exists." Nevertheless, it emphasized that changes were needed if reactor orders increased as much as expected. The panel found that a utility had to plan on at least a year to receive a construction permit from the time it decided to build a nuclear plant at a particular site. The length of the licensing process, the panel argued, "penalizes the economic position of nuclear power." It urged a number of revisions in licensing procedures so that the time required could be reduced to about six months. This, it contended, "would be an important step in encouraging the growth of the nuclear industry."

The Mitchell panel's first recommendation was that the AEC emphasize that the regulatory staff offered "the public's primary protection in reactor safety matters [and] that its review of the safety of a reactor project is the most complete, thorough, and objective review conducted during the regulatory process." It argued that the "high caliber of the regulatory staff" and the nature of its review should be made clear to the public. Although the ACRS should retain a vital role in evaluating the work of the staff and in focusing on novel safety designs, it could not conduct a careful review of each of the increasing number of applications. The panel advised that the statute requiring an ACRS review of every construction permit and operating license application be modified. It thought that the ACRS should be relieved of the obligation to examine every plant proposal, and instead, should look at only the ones of particular interest or concern to it. The committee would decide for itself which applications it wished to review. This would ease the work load of the ACRS and allow it to "function as an advisory committee instead of a second regulatory staff."

The Mitchell panel directed most of its attention to the rules and the conduct of licensing board hearings. It urged that the proceedings be revised by, for example, restricting or excluding discussion of issues over which the AEC had no statutory jurisdiction, placing a limit on the time allowed a board to reach a decision, and appointing an alternate member of a board to step in if one of the original members could not sit through a prolonged hearing. The Mitchell panel also suggested that *ex parte* rules be relaxed so that the Commission could consult with the regulatory staff, though it stipulated that any facts or opinions presented to the commissioners by the staff should be included in the public record. The report stressed the panel's strong conviction that the licensing boards should be clearly instructed not to undertake *de novo* technical reviews of reactor applications. The "growing tendency" to perform a third independent evaluation, it argued, "makes no real contribution to reactor safety." The Mitchell panel contended that the proper functions of the boards were to determine whether the application was complete and the staff and ACRS had carried out an adequate review of it, provide a public forum for judging the application, and adjudicate opposing views in contested cases.

In addition to its recommendations for changes in licensing procedures, the Mitchell panel advised the AEC to take a number of actions that could alleviate the procedural difficulties facing applicants and shorten the time needed to obtain a construction permit. They included, for example, a more precise definition of what information the AEC expected in an application and clearer technical guidelines on performance requirements. The panel also called for better coordination between the regulatory staff and other AEC divisions in planning reactor safety research.[14]

The report of the Mitchell panel received warm praise in industry circles and the AEC acted promptly to consider and implement its recommendations. Moving with what the trade journal *Nucleonics Week* called "rare speed" and motivated in part by the panel's conclusions, the AEC undertook new efforts to clarify the information it sought in applications and to define licensing requirements. It also drafted new rules and legislation to carry out many of the panel's proposals to reform the licensing process. Some of them, such as the addition of an alternate licensing board member, were straightforward and caused no division of opinion. Others were less clear-cut. The AEC only partially followed the recommendation that licensing boards be barred from conducting a thorough technical evaluation. It made an explicit policy statement that in the case of an uncontested hearing, the licensing board would not perform a *de novo* technical review. But in the case of a contested hearing, it refrained from such an unambiguous directive. The regulatory staff expressed misgivings about a strict prohibition on the grounds that a licensing board should be free to explore any issue on which it wanted further information or clarification. The Joint Committee, which had pressed the AEC to add technically qualified members to the boards in 1962, seconded the regulatory staff's position. As a result of those arguments, the Commission decided to "discourage rather than prohibit" *de novo* reviews at the licensing board stage.[15]

The Commission also decided not to carry out fully the Mitchell panel's recommendation on relaxing *ex parte* rules. Harold Price had long sought ways to make possible greater communication between his staff and the Commission because he believed it would enable the commissioners to address safety issues more knowledgeably in their review of a licensing case. The Mitchell panel gave his position additional support. As a result, the commissioners elected to permit exchanges between the regulatory staff and themselves, and between the regulatory staff and the licensing boards, in uncontested hearings.

Contested hearings posed more of a problem. The Mitchell panel had focused on uncontested proceedings, and in April 1966, the Commission organized another committee, again chaired by William Mitchell, to investigate the procedural aspects of contested applications. The second Mitchell panel, which was comprised of three members, submitted its report in June 1967. It concluded that many of the AEC's existing practices to ensure a fair hearing for all parties were sound, though it recommended a few minor procedural amendments. On the *ex parte* issue, it strongly advised against easing the rule in ways that the first Mitchell panel had suggested. It argued that neither a licensing board nor the Commission should communicate with the regulatory staff in contested proceedings, even if they intended to place information about the discussions in the public record. The Commission accepted this recommendation. The findings and proposals of the Mitchell panels made a major impact on the AEC's licensing procedures by helping to bring about important changes. But they did not satisfy all the concerns about the licensing process or end the complaints about its complexity and uncertainty.[16]

The Mitchell panels focused their attention on ways to streamline the licensing process and ease the burdens it presented to applicants. This was a major concern to nuclear proponents in both the industry and government. At the same time, the conflict of interest in the AEC's authority both to promote and regulate nuclear power remained a lively issue. Opponents of nuclear plants often cited it and suggested that the AEC was an unreliable regulator because of its dual responsibilities. Therefore, agency officials continued to look for ways to counter the criticism and to show that the regulatory staff did not base its judgments on promotional considerations.

One possible solution to the problem that both the AEC and the Joint Committee occasionally entertained was the creation of separate agencies, but the AEC still thought that such a step would be premature. In 1963, Ramey predicted the establishment of an independent regulatory agency within a period of ten years. Three years later he thought that a separation might occur in five to ten years. Other AEC officials gave similar estimates; Seaborg and Price remarked in a meeting in 1970 that the AEC had been saying that the creation of a regulatory agency might be appropriate "in five or ten years" since 1955.[17]

The question of dividing the functions of the AEC received some attention when the Joint Committee held extensive hearings on the regulatory process in 1967. Both AEC and industry officials argued that experience with nuclear power technology was too limited to create a regulatory body that was divorced from research and development activities. This reasoning was unconvincing to nuclear critics, one of whom described the existing arrangement as "schizophrenia which is inflicted upon the AEC by law." Some members of the Joint Committee were also dubious about the position of the AEC and the industry on the separate agency issue. The most outspoken was Representative John B. Anderson of Illinois, who told a group of nuclear power professionals in 1968 that "suggestions that consideration of this [separation] question can be put off for another 5 or 10 years simply aren't realistic." He suggested that new technical information could be communicated to an independent regulatory agency without undue hardship. Joint Committee staff director John T. Conway stated publicly the same year that the committee might decide to hold hearings on the subject.[18]

In light of the growing interest in the separate agency question, the AEC undertook a review of it. General counsel Joseph F. Hennessey told the Commission in October 1968 that the establishment of a new agency would offer advantages, including "maximum objectivity and impartiality in reactor safety evaluation." But he indicated that the disadvantages would be greater, especially the burdens that an independent agency could place on the AEC's developmental and operational programs, the duplication of administrative functions, and the "obstacles to informal consultation." The AEC remained opposed to the creation of a separate regulatory agency, at least for the immediate future. But it recognized that its dual responsibilities undermined its regulatory credibility. Therefore, in late 1968 it began to look for alternatives that might ease the conflict-of-interest problem without requiring a complete separation of the regulatory staff from the AEC.[19]

With Ramey providing the impetus, the Commission decided to deal with the issue by establishing a new panel to rule on appeals of licensing board decisions. The AEC had considered the idea of an appeal panel previously, but never found it desirable. In October 1968, with the separate agency question becoming increasingly prominent, Ramey told his colleagues that "the time has come when we ought to move on this matter." The Commission decided to study the concept of an appeal panel that would assume at least some of its responsibilities for reviewing licensing board actions. The advantages, as outlined by Hennessey, would include alleviating demands on the Commission's time and, more importantly, defusing the conflict-of-interest issue by removing the Commission from involvement in the cases it delegated to the appeal panel. The primary disadvantage was that an independent appeal panel would diminish the Commission's policy-making authority. If the appeal panel made the final decision in a licensing case, it could act in ways that were inconsistent with existing policy or that set policy precedents. If, however, the Commission retained final authority, the appeal panel would simply add another layer to the regulatory process without doing much to resolve the conflict-of-interest question.[20]

After considerable discussion, the Commission decided that the advantages outweighed the disadvantages. In August 1969 it announced the establishment of the Atomic Safety and Licensing Board Appeal Panel. To reduce the possibility that the appeal panel could issue a decision that undercut its position on general regulatory issues, the Commission stipulated that, on its own motion, it could review a decision that conflicted with existing policy or dealt with an "important question of public policy." It did not attempt to define the precise meaning of this condition. The Commission would, at its discretion, assign some but not all cases under appeal to the new panel. The appeal panel would consist of the permanent chairman and vice-chairman of the licensing board panel and a third member chosen by the Commission. The permanent chairman and vice-chairman of the licensing board panel were positions that the AEC had established in 1966; they were responsible for appointing board members in individual proceedings and for performing other administrative duties.[21]

The nuclear industry found the creation of the appeal panel an unwelcome innovation. A committee of the Atomic Industrial Forum complained that "the effect of introducing the Appeal Board will be to add another level of review, thus prolonging the time required to complete licensing proceedings." The AEC, despite its reluctance to further complicate the licensing process, believed that creating the appeal board was a necessary and useful means to increase the credibility of its regulatory program. Its action did not, however, end criticism over the agency's built-in conflict of interest. Harold Price lamented in January 1971 that "as far as the public is concerned there is no distinction between actions by the regulatory [staff], [the] ACRS, the Commission or the ASLB's." All, he added, were "considered a part of the nuclear power promotional effort."[22]

Although the AEC's statutory conflict of interest and the idea of establishing a separate regulatory agency were recurrent issues throughout the 1960s and early 1970s, they never acquired the urgency of more pressing concerns. On a day-to-day basis, the primary administrative problem that faced the regulatory staff was keeping pace with the growing work load that the rapid development of the nuclear industry produced. The flood of reactor orders placed enormous pressure on the Division of Reactor Licensing to process them promptly. Utilities were greatly concerned about delays in licensing that set back their schedules and threatened their load requirements. They were distressed about licensing procedures that added to the lead time, about seven years in 1967, that they needed to plan, build, and begin operating a nuclear plant. One indicator of the trend was that the average time the Division of Reactor Licensing took to process a construction permit application had expanded to more than eighteen months by 1970.

The regulatory staff sympathized with the complaints of power companies, but its spokesmen pointed out that applications were not only more numerous but also more complex than those for smaller plants. While they regretted the delays that the "avalanche of applications" caused, they refused to accept the entire blame for the increased average time for issuing construction permits. Peter Morris told a packed ballroom at the annual meeting of the Atomic Industrial Forum in November 1967 that his staff could grant a construction permit in an uncontested case within seven months if it received a "perfect application." Unfortunately, he said, the AEC never saw such a proposal; as a rule the applications as originally submitted were incomplete or technically deficient. The growing frequency of contested applications further extended the licensing process by lengthening hearings in individual cases and by generally stretching the resources of the regulatory staff. By the late 1960s, intensified public concerns about the health and environmental effects of nuclear power made intervention in licensing proceedings by nuclear opponents more common.[23]

While the AEC continued to consider administrative changes to streamline licensing procedures, it sought to deal directly with the surge in reactor orders by increasing the size of the regulatory staff. The growth of the industry added to the work load of all the regulatory staff's divisions, but it placed the greatest burden, at least initially, on the Division of Reactor Licensing. In 1964, the division employed thirty-two engineers to review applications. The number more than doubled by the end of 1967, but this was not enough to keep up with the volume of business. As more plants received construction permits, the staff of the Division of Compliance also increased substantially in a vain attempt to stay abreast of the expanding work load.[24]

Beginning in 1968, the growth of the regulatory staff slowed to a virtual standstill. As President Johnson, and following him, President Nixon, sought to slash government expenditures and imposed ceilings on hiring, the AEC's budget faced unprecedented restraints. *Nucleonics Week* reported in December 1969 that the agency was "caught up in the tightest financial bind in its 23-year history." The budgetary pressures forced the AEC and its contractors to lay off hundreds of workers in operational and developmental programs. The regulatory staff avoided personnel cutbacks, but it was unable to enhance its size as rapidly as demands on its resources grew. Between 1965 and 1970, while the staff increased by about 50 percent, its licensing and inspection case load increased by about 600 percent. The agency requested relief from the Bureau of the Budget, and in 1969 Seaborg appealed personally to Nixon for funds to hire more regulatory personnel. Those efforts proved largely futile, however, and the shortage of qualified professionals continued to plague the regulatory staff.[25]

In 1971, after the AEC undertook a series of efforts, on the one hand, to streamline licensing procedures, and on the other hand, to deal with its inherent conflict of interest, the regulatory process remained a source of acute dissatisfaction to the nuclear industry, the Joint Committee, and nuclear critics. The agency's attempts to balance its responsibilities for encouraging the growth of nuclear power and for ensuring nuclear safety led to a complex regulatory system that generated complaints from both nuclear advocates and opponents, though for quite different reasons. Industry representatives objected strongly to what they viewed as unnecessary requirements or undue delays in licensing deliberations. A. Eugene Schubert, vice-president of the Nuclear Energy Division of General Electric, told Seaborg in 1968 that the effect of the AEC's approach to licensing "on cost and scheduling to both the [applicant] and the manufacturer was very serious."[26]

Earl Ewald, chairman of the board of the Northern States Power Company, offered a similar opinion from the perspective of a utility that was suffering through a lengthy and controversial licensing proceeding. "If the delays encountered in this licensing procedure are duplicated in connection with the other nuclear power plants scheduled for commercial service in the next few years," he wrote, "it can safely be asserted that the splendid promise of nuclear power will have had a very short life." Another utility executive was even more blunt, calling the licensing process "a modern day Spanish Inquisition" carried out by "AEC engineers, scientists, and consultants [who] have no serious economic discipline." Industry officials were highly critical of the ACRS for imposing what they viewed as excessive burdens that lengthened the licensing process. In 1968, Congressman Hosmer proposed in a speech to an industry group, "half in jest," that the ACRS be abolished. To his surprise and somewhat to his chagrin, he received a standing ovation.[27]

Hosmer and his colleagues on the Joint Committee were careful observers and frequent critics of the AEC's licensing process and its impact on the nuclear industry. They regularly held hearings on the subject and often voiced their displeasure. By the late 1960s, the Joint Committee was particularly troubled by the growing role of intervenors in regulatory proceedings and their ability to delay the issuance of licenses. Holifield told AEC officials in 1971 that "this process is running wild. . . . I think you are going a little bit far along the line of permissiveness in the field of intervention."[28]

Intervenors did not share the same view. They emphasized that the AEC's dual responsibilities prejudiced its licensing process and discouraged discussion of vital issues. One of the most prominent and thoughtful exponents of this position was Harold P. Green, a professor of law at George Washington University, former AEC attorney, and counsel for intervenors in several licensing cases. "Despite the statutory provisions for licensing reactors 'in a goldfish bowl' with public hearings and public disclosure of safety analyses," he wrote in 1968, "in actual practice the regulatory procedures tend to stifle public awareness and discussion of safety issues." He urged that the licensing process be revised to provide greater opportunity for the public to receive a clear understanding of the risks of nuclear power.[29]

The AEC continued to seek ways to improve the licensing process, but it obviously could not adopt reforms that would please all of its critics. It recognized that its efforts to ensure nuclear power safety without placing undue burdens on the industry would inevitably raise objections from partisans of one side of a regulatory issue. It was not an "easy task," Seaborg declared in reply to Earl Ewald's complaints, "to strike the fine balance which properly accommodates the goal of conducting reasonably expeditious hearings and at the same time safeguards the legitimate interests of the public in participating in the regulatory review process."[30] The AEC's regulatory process—complex, controversial, and imperfect as it was—provided the framework in which a variety of siting, safety, environmental, and public health issues were considered and debated between 1963 and 1971.

CHAPTER IV

Reactors Downtown?
The Debate over
Metropolitan Siting

From the earliest days of the civilian nuclear power program, the siting of reactors was a critical issue because of both safety and economic considerations. The principal hazard of nuclear power production was public exposure to radiation, and the AEC's siting policies were a key element in protecting public health from the consequences of an accident that released radioactivity into the environment. The siting of plants was also a vital factor in evaluating their financial benefits, and hence, in encouraging the growth of nuclear technology. The projected capital requirements for any central power station necessarily included transmission costs. If utilities were going to invest in nuclear power, they wanted to locate plants as close as possible to the population centers they served. Both nuclear vendors and utilities, therefore, strongly pushed for what came to be called metropolitan siting. The regulatory staff resisted the pressure, but its position was contested not only by the industry but also within the AEC. Eventually, the agency settled on an informal siting policy that was a compromise between the differing views.

The earliest government-owned reactors, built in the late 1940s and early 1950s, had been placed far from population centers. It was not long, however, before newer facilities were located closer to populated areas. Argonne National Laboratory built a small research reactor on its grounds near Chicago and General Electric constructed a "Submarine Intermediate Reactor" at West Milton, New York, some twenty-five to thirty miles from the tri-city area of Schenectady, Troy, and Albany. To compensate for the risks of siting reactors closer to populated regions, designers relied on, among other things, a containment building to mitigate the consequences of any projected accidents. The West Milton reactor, for example, was enclosed in a large steel containment structure and the Argonne reactor in a gas-tight concrete building. The first commercial reactor at Shippingport, Pennsylvania, completed in 1957, also featured a containment building. Except for a few experimental reactors at remote sites, two gas-cooled reactors, and a plutonium-producing plant at Hanford, all nuclear plants constructed in the United States after 1957 included containment structures as a compensation for closer proximity to population centers.[1]

The AEC, despite painstaking efforts, was unable to delineate the specific relationship between distance from population centers on the one hand and containment and additional safety features on the other. In 1959 the regulatory staff drafted a proposal in which siting would be evaluated by such factors as population density, meteorological and seismological conditions, hydrology and geology. Although it disavowed any intention to commit the agency to an inflexible position, it stressed distance rather than "engineered safeguards" (the term that was then commonly used, though it was later superseded by "engineered safety features"). Both industry officials and the Advisory Committee on Reactor Safeguards (ACRS) strongly objected to the staff's recommendations, arguing that they were too rigid and too heavily weighted toward site isolation. Eventually, the staff revised its original draft to give greater emphasis to engineered safeguards. Rather than publishing a formal regulation, in 1961 the AEC issued less binding "site criteria" for the interim guidance of industry. After long deliberations to try to devise generally applicable requirements, judging the suitability of a site remained a case-by-case determination, just as it had been with the first reactor applications.[2]

The regulatory staff's early drafts of its siting criteria contained a sample calculation for determining acceptable distances from population areas. Industry groups protested strongly that its inclusion would make the guidelines unduly rigid. The final version of the siting criteria deleted the calculation, which was then published separately as an AEC "Technical Information Document" titled "Calculation of Distance Factors for Power and Test Reactor Sites" (TID–14844). As a compromise between the staff's position and industry objections, the siting criteria referred to but did not include the technical information document, which showed the computations used to arrive at an acceptable distance and provided supplementary explanatory information.[3]

The 1962 site criteria affirmed that applicants were "free—and indeed encouraged—to demonstrate to the Commission the applicability and significance of considerations other than those set forth in the guides." The final document emphasized that it was "intended as an interim measure until the state of the art allows more definitive standards to be developed." The criteria, then, reflected the realities imposed by a developing industry and the rapidly changing nature of the technology. The AEC believed that the guidelines provided an ample margin of safety while at the same time allowing the industry to build commercial power reactors reasonably close to populated areas with the greatest demand for electricity.[4]

By placing increased confidence in engineered safety features and continuing to make site evaluations on a case-by-case basis, the AEC regulatory staff opened the way for utilities to request approval for plants in urban areas. In several instances, utilities attempted to demonstrate that sites in densely populated locations could be made suitable for power reactors by installing new engineered safeguards. This generated a great deal of concern, among both the public and the regulatory staff, and eventually led to an uneasy resolution of the metropolitan siting issue. The AEC, without formally adopting a rule, acted to prohibit construction of nuclear plants in heavily populated urban sites, though it did not define precisely where it drew the line of acceptability.

Even before the 1962 site criteria went into effect, the Consolidated Edison Company of New York (Con Edison) announced plans for a commercial reactor at a location that did not meet any of the AEC's distance guidelines. The site, therefore, would have to be judged strictly on the basis of the proposed plant's engineered safeguards. On 10 December 1962, Con Edison applied to the AEC for a construction permit to build a 1000 electrical megawatt facility along the East River in Queens, in the heart of New York City. The proposed Ravenswood nuclear station would sit about one-half mile north of the Queensboro Bridge that connected Queens with Manhattan. Central Park was approximately a mile-and-a-half west of the site and the United Nations complex was about the same distance southwest. The plant would use a Westinghouse pressurized-water reactor supplemented with two oil-fired superheaters. The Ravenswood facility would be the largest power reactor in the world. Nuclear plants then operating in the United States were much smaller; those with the greatest capacity were Con Edison's Indian Point plant

PLANT SITE

RAVENSWOOD NUCLEAR GENERATING UNIT "A"
AERIAL PHOTOGRAPH LOOKING NORTHEAST
FIG. B-11

4. Site of proposed Ravenswood nuclear plant in New York City. The Empire State Building is at the lower left; the United Nations is in the center along the East River. (AEC Docket 50-204)

with 275 electrical megawatts (163 from nuclear and 112 from an oil-fired superheater) and Commonwealth Edison's Dresden station with 202 electrical megawatts.[5]

Con Edison's interest in the Ravenswood site was based mainly on economic considerations. The company had determined that a 1000-megawatt nuclear plant, by taking advantage of economies of scale, would be competitive with a fossil-fueled facility at the outset and more economical than a conventional plant over its lifetime. Transmission costs were particularly vital in the selection of the site. They were expensive under any conditions, but Con Edison incurred much higher-than-usual charges in New York City. The company figured transmission costs in urban areas at $3 million per mile because it had to place many lines underground.[6]

Concerns over air pollution and availability of fuel also weighed heavily in Con Edison's decision to build a nuclear plant at the Ravenswood site. Air pollution from an oil-fired plant already under construction at the site would be aggravated by another fossil-fuel burner, but a nuclear facility would alleviate the problem. Con Edison's other major consideration was the possible interruption of coal supplies if it built a coal-fired plant. Strikes by coal miners and carriers had caused critical difficulties in the past, and cost and space requirements precluded the storage of large reserves on site. Utility officials saw the nuclear option at Ravenswood as a way to avoid those pitfalls.[7]

Con Edison's proposal gave credence to the AEC's report to the president on civilian nuclear power, issued and widely noted a month earlier, which argued that commercial light-water reactors were on the threshold of being competitive with conventional plants. But the Ravenswood site posed a delicate problem for the agency's regulatory staff. It had to judge the potential hazards of allowing construction of a nuclear plant in the heart of the nation's largest city against the benefits of the proposed facility for the AEC's reactor development program. The staff turned first for guidance to the recently published reactor site criteria. To aid in evaluating a site, the criteria required applicants for construction permits to make projections about certain factors: the radiation released from the nuclear core of the reactor during a hypothesized major accident, the effectiveness of safety features in preventing the escape of radioactive materials into the environment, and the meteorological and geological conditions pertinent to the site. On the basis of its assumptions and predictions the utility could then calculate the isolation distances spelled out in the site criteria: the "exclusion zone," "low population zone," and "population center distance." The criteria defined the exclusion area as the vicinity surrounding a reactor in which the licensee had the authority to control all activities. Normally, residence within the exclusion area was prohibited. The criteria provided that the exclusion area be of such size that an individual located at any point on its boundary for two hours following a postulated accident would not be exposed to more than a specified amount of radiation.[8]

The low population zone defined an area immediately surrounding the exclusion area that contained a density and total number of residents who, in the event of a serious accident, could be given appropriate protective measures (for example, evacuation). The site criteria did not specify a permissible population density or a total population for the zone because the situation could vary from case to case. Instead it determined that the low population zone should be an area in which an individual who was exposed to the radioactive cloud resulting from the postulated release of radioactivity and who was located on the zone's outer boundary during the entire period of the cloud's passage would not receive radiation exposure in excess of the same limits that applied for the exclusion zone. The criteria defined the third element, the population center distance, as the span from the reactor to the nearest boundary of a built-up area "containing more than about 25,000 residents." It stipulated that this should be at least one and one-third times the distance from the reactor to the outer boundary of the low population zone. The guidelines also stated that in applications for plants close to large cities a greater distance might be necessary. The criteria suggested that the AEC's technical information document (TID–14844), demonstrating the procedural method and sample calculation for a hypothetical reactor, be used as a beginning point for evaluation of any proposed site. The calculated distances could then be adjusted upward or downward depending on the physical characteristics of a site and the specific features of a reactor, such as the leak rate of containment or engineered safety features.[9]

Application of the sample calculation in TID–14844 to a hypothetical reactor with the equivalent power rating of the proposed Con Edison plant produced an exclusion distance of one mile, a low population zone distance of 16.5 miles and a population center distance of 21.5 miles. Ravenswood's actual proposed exclusion area was 675 by 550 feet. Within a five mile radius, a daytime population of five million people lived and worked. At night the number diminished to three million. Within a half mile radius, Con Edison estimated the population to be 28,000 during the day and 18,800 at night. The population statistics obviously precluded application of the calculated low population zone and population center distance. Engineered safeguards would have to compensate for proximity to a large population. It fell to the AEC's regulatory staff and the ACRS to evaluate Con Edison's reactor design and determine its suitability for the Ravenswood site.[10]

As a pioneering company in the nuclear power field, Con Edison appeared to be the type of utility that could make the Ravenswood facility a reality. The company's experience with the Indian Point reactor, its application for Ravenswood declared, had been "invaluable in aiding our determination to bring the benefits of atomic power to our customers at the earliest possible time." Harland C. Forbes, who had become chairman of the board in 1957, was an aggressive engineer who believed that nuclear energy would be an important part of Con Edison's mixture of fuel sources. Forbes and his senior staff harbored no apprehensions about the safety of the proposed plant. Shortly after the Ravenswood announcement, the chairman told a reporter that operating Indian Point had given the company confidence that a nuclear station could be built at Ravenswood, "or in Times Square for that matter," without hazard to employees or the community. He believed that the future of nuclear power as a source of electrical energy in large cities was at stake in the Ravenswood case. Forbes later suggested that Ravenswood raised a key question that the AEC faced in making decisions about siting: "Either these plants are safe to build or they're not," he said, "whether it would be 500 people affected or 5,000,000."[11]

Con Edison's plans for Ravenswood soon sparked intense opposition from local citizens. The utility's announcement of the Ravenswood application in December 1962 came at a time when all the city's newspapers were shut down by a strike. This might have delayed organization of opposition for a short time, but by January 1963, a group of residents neighboring the Ravenswood site was seeking more information about the company's plans. The Astoria-Long Island City Community Council, consisting of nearly all the civic and religious organizations in Queens, arranged a public information meeting that representatives of Con Edison, the AEC, the governor of New York, the mayor of the city, and the president of the borough of Queens planned to attend.[12]

Sleet and snow in New York on the night of 19 February 1963 did not prevent a crowd of some 250–300 people from filling the auditorium at St. Rita's Roman Catholic Church in Queens. The three hour session quickly became as stormy as the weather outside. Politely at first, the

audience listened to Con Edison spokesmen give details about the proposed plant, including the safety devices designed to ensure that the
reactor posed virtually no danger to public health. Robert Lowenstein,
director of the AEC's Division of Licensing and Regulation, stressed that
he could not discuss the merits of Con Edison's application, but he assured the crowd that the agency would evaluate it carefully. He testified
that neither the AEC staff nor the ACRS had reached any conclusions
about the proposal.[13]

The mood of the meeting changed dramatically after Robert C.
Beardsly, a biology professor at Manhattan College, challenged the
optimistic assessment of the plant's risks that Con Edison presented. He
offered his scientific opinion that there was no safe dose of radiation
and that there was ample reason to suspect that even the smallest
amounts were harmful. Beardsly was followed by Queens borough president Mario J. Cariello, who minced no words in denouncing the Ravenswood application: "I was opposed to this project, I am opposed, and I
will continue in that stand until convinced otherwise." The audience
roundly applauded. The chairman of the Astoria-Long Island City Community Council, who had called the meeting, later reported that "although there was a difference of opinion between the speakers of the
evening as to the dangerous effects of atomic radiation, the audience
showed no such disagreement, and unanimously applauded all opposition to the building of such a plant in our community."[14]

Grassroots feelings against the Ravenswood facility continued to
grow. A trickle of letters to the AEC early in 1963 increased in the spring
and summer. Most of them expressed the opinion that nuclear plants had
no place in a metropolitan area, though few correspondents opposed
nuclear power in other locations. Meanwhile ad hoc committees and
established groups in the local area rallied against the Ravenswood reactor. The February meeting led to the creation of the hundred-member
Committee Against Nuclear Power Plants in New York City, headed by
Irving Katz, a biochemist who lived in Ravenswood. Beardsly, the Manhattan College professor who had spoken against Con Edison at the
February meeting, helped organize the Scientists' Committee for Radiation Information, which issued several reports designed to explain nuclear energy to the public. Although initially the committee was officially
neutral toward the plant, it eventually took a position against the Ravenswood site. By late spring, another ad hoc group, the Committee for a Safe
New York, was urging citizens to tell their representatives in Congress
that the dividends of Con Edison stockholders "should not be raised at

the expense of the safety of millions of New Yorkers." A political organization that opposed the application, the Queens County New Frontier Regular Democratic Club, sponsored another community-wide information meeting on 25 April in Forest Hills. Although not as volatile as the crowd that attended the February meeting, the large gathering displayed unequivocal opposition to the plant. All the turmoil prompted an inquiry into the Con Edison project by the New York City Council. The Democratic majority leader, Eric J. Treulich, introduced and called hearings on a bill prohibiting industrial reactors in the city.[15]

Con Edison chairman Forbes was questioned about the expanding local resistance to Ravenswood when he testified before the Joint Committee on Atomic Energy on 3 April 1963. Forbes downplayed the protests by commenting that there had been little public interest either in favor or opposition. He told the committee that "there have been a few community meetings at which one or two people have raised some question about the genetic effects of radiation and so forth, some of which is rather silly. I think some of that has to be expected." While admitting that public opposition "could develop," he thought it unlikely. "It seems to me," Forbes declared, "that the public in general has reached the point where it has accepted nuclear plants as a matter of course as they would any other plants."[16]

The opposition that Forbes deprecated gained some national publicity the following day when David E. Lilienthal, the former chairman of the AEC (1947–1950), testified before the Joint Committee. He had been asked to elaborate on and explain a lecture he had given at Princeton University two months earlier. At that time, he had criticized the AEC's 1962 "Report to the President" for overstating the need for nuclear power and for understating the fact that, even if it were cost-competitive, it differed significantly from other forms of energy because of the hazards associated with it. In his testimony before the Joint Committee, Lilienthal remarked that if he lived in Queens, he would consider operation of a reactor at Ravenswood "very risky business." Recalling that isolation of reactors had been the policy during his tenure as AEC chairman, Lilienthal questioned whether enough safety features had been developed in the interim to justify metropolitan siting. He added that he "would not dream of living in the Borough of Queens if there were a large atomic power plant in that region." Lilienthal's statements elicited sharp rejoinders from several members of the Joint Committee. Senator Pastore accused him of being "unfair" and "making these statements rather loosely," and Congressman Holifield suggested that he was influenced by oil interests. But Lilienthal's comments and his stature gave opponents of Ravenswood an unexpected windfall of prominence and credibility.[17]

The increasing discord over Ravenswood became a political issue when Democratic majority leader Treulich introduced his bill banning industrial reactors in New York City. As a representative from Queens, Treulich stated that he would reserve judgment on the wisdom of his legislation until after hearings were held. But he emphasized that an inquiry into the matter was needed, and he hoped that his bill would ensure that the questions surrounding Ravenswood were thoroughly aired.[18]

The City Council hearing on the Treulich bill, held on 14 June 1963, featured a long list of speakers who expressed their differing opinions about Ravenswood. Demonstrators marched and distributed literature in the plaza outside City Hall. Inside the Council chamber, a packed crowd of 350 people listened as fifty-nine witnesses spoke on the bill during the seven-and-one-half-hour session. The speakers, including scientists, engineers, lawyers, and concerned citizens, represented the utility, the AEC, civic groups, and parents' organizations. Earl L. Griffith, senior vice president of Con Edison, outlined the utility's proposal for Ravenswood, backed by four company scientists and engineers. The AEC's Robert Lowenstein outlined the regulatory process the Con Edison application would have to pass before the agency granted a permit. He also read a statement from AEC chairman Glenn T. Seaborg that raised a serious question about the legality of the proposed city legislation. Seaborg suggested that the 1954 Atomic Energy Act gave the U.S. government exclusive regulatory authority to decide on the suitability of the Ravenswood site. State senator Seymour R. Thayler of Queens made one of the strongest protests against the plant. He told the city legislators that four out of five Queens residents opposed the plant. Despite all the safety precautions the engineers could build into the reactor, he argued, "the mind of man has not yet invented an accident-proof piece of mechanical equipment."[19]

Although the hearing was inconclusive, the presentations by Con Edison and the AEC greatly disturbed a careful observer and strong proponent of nuclear power. In a rare editorial, the weekly trade journal *Nucleonics Week* depicted the New York hearing as a missed opportunity for the industry. Noting that more and more attention would be focused on siting nuclear plants in urban centers, it argued that many people were asking intelligent questions that could not be "brushed

under the carpet." What was needed at the hearing, the editorial suggested, was an authoritative presentation for laymen on the "abc's" of
nuclear power. It criticized the AEC for sending a lawyer (Lowenstein)
instead of a technical expert to testify. Lowenstein had to "duck vital
technical questions" and therefore reinforced doubt in the minds of
some of the listeners. But in the editor's opinion, Con Edison's representatives were even less satisfactory. The paper criticized the utility's cava-
lier assumption that it did not have to "go all out on educating the City
Council and the public" and its failure "to give immediate and full
answers to some good questions."[20]

The *New York Times* echoed *Nucleonics Week*'s opinion. In an editorial on 26 August 1963, it complained that Con Edison had not presented a "compelling reason why the construction of a nuclear reactor
inside the city limits" was necessary. "What is still lacking," the editorial concluded, "is a clear statement of why there is no other place to put
the plant than [in] the heart of the metropolitan area."[21]

The issue continued to generate controversy and gain headlines in the
fall of 1963. Lilienthal's earlier questioning of the benefits of nuclear
power had irritated many nuclear proponents. Seaborg, for one, joined
the debate in early November in an address to Sigma Delta Chi, a
national fraternity of journalists. Without mentioning his predecessor
by name, he presented a point-by-point refutation of both Lilienthal's
Joint Committee testimony and his charges in earlier lectures. Seaborg
explained why utilities wanted to locate facilities in metropolitan areas
and pointed out that any application had to pass AEC regulatory review, which, he said, had been "ultra conservative with respect to
safety." He declared, in contrast to Lilienthal: "I would live next door
to the atom. I would not fear having my family residence within the
vicinity of a modern nuclear power plant built and operated under our
regulations and controls."[22]

Seaborg's talk was widely reported in the press and set the stage for
further controversy. At a joint meeting of the American Nuclear Society,
a professional society populated principally by engineers and physicists
with an interest in nuclear energy, and the Atomic Industrial Forum,
Lilienthal reiterated his criticism of the nuclear establishment. He again
questioned whether enough was known about the hazards to the public
to permit nuclear plants in congested areas. Furthermore, he charged
that Seaborg's recent comments at the Sigma Delta Chi convention had
in effect prejudged the Ravenswood application. Lilienthal's speech
stirred a sharp reaction. AEC commissioner Ramey, Emerson Jones,

consultant to the managers of the Consumers Power District of Nebraska, Robert E. Ginna, chairman of the Rochester Gas and Electric
Corporation, and Joseph Howland of the University of Rochester decided to meet with the press and rebut his views. They maintained that
Lilienthal's remarks were "entirely uncalled for" and vigorously defended the safety record of the nuclear industry. Later that day,
Lilienthal wrote privately in his journal that he thought the "truth
squad" tactic was a "stupid, reckless stunt." He predicted that the AEC
would hear more misgivings about its role in protecting the public from
the hazards of nuclear power plants.[23]

In the midst of the year-long public debate over the Ravenswood site,
the AEC regulatory staff and the ACRS quietly worked on their evaluation of the Con Edison application. Following its usual procedures, the
staff reviewed the preliminary hazards summary report on the reactor
that the utility submitted. It also met informally on several occasions
with Con Edison representatives. The ACRS, likewise following its standard format, established a Ravenswood subcommittee that proceeded
independently to evaluate the hazards report.[24]

Because of the dense population around the site, the proposed reactor's acceptability would have to be based on the AEC's evaluation of the
reliability of its engineered safeguards. The safety features had to compensate for the fact that the facility could not meet the minimum distance
standards suggested by the AEC's reactor site criteria. Without distance
to rely on as a safety factor, the staff was forced to consider the reactor
design and to project ways in which the engineered safeguards might fail
to perform properly. This procedure would determine whether the reactor could be constructed at the proposed site.

The most important safety feature in the plant design was the so-
called double containment. According to Con Edison's preliminary hazards report, the containment would withstand penetration even in the
worst conceivable accident (instantaneous release of the radioactive primary cooling system from the piping into the containment followed by
100 percent of the nuclear core melting). The structure was a reinforced
concrete "igloo," approximately 150 feet in diameter, 167 feet in
height, and seven feet thick, that rested on a solid concrete pad. The
outer shell consisted of five-and-one-half feet of reinforced concrete.
This provided the bulk of the shielding from radiation inside the reactor. On the inside of the outer concrete shell was a welded steel vessel or
membrane, fabricated from one-quarter inch carbon steel plate. Within
that shell, the designers placed another one-quarter inch welded steel

vessel that was separated from the outer one by a distance of two feet. The space between the two membranes would be filled with low density, pervious concrete. Called the "negative pressure zone," it would be maintained at a level below atmospheric pressure by means of a pump-back system (another engineered safety feature) that would pump air from this area into the interior of the containment. Thus any radioactivity that leaked into this space could be discharged back into the inner vessel. The designers also included in the containment plans a spray system to reduce pressure that might build up within the containment structure in the event of an accident.[25]

The Ravenswood plant would be a pressurized water reactor, with a basic design similar to other Westinghouse nuclear units. Within the primary system, the pressure vessel, which housed the core of the reactor, acted as an important safety barrier. Within the core itself, metal cladding around the fuel served as an additional safety feature. Westinghouse also designed two independent emergency safety injection systems to supply borated water that would maintain neutron-absorbing coolant to the nuclear core in the event of a major loss-of-coolant accident. The preliminary hazards report postulated that a major loss-of-coolant accident, defined as the complete severance of the largest pipe in the primary coolant system, was the worst credible accident that could happen to the plant.[26]

Con Edison's safety evaluation report emphasized that the integrity of the double containment structure would not be breached and radioactivity could not escape from the plant, even in case of the worst conceivable accident (meltdown of the core due to a complete loss of primary coolant, which nuclear experts did not consider credible). As far as Con Edison was concerned, this safety feature alone would more than offset the need for the distance factors the site criteria had outlined. Therefore the AEC's regulatory staff decided to evaluate the design of the containment first. By concentrating their effort on that system, they could make a preliminary judgment about whether the Ravenswood site would be suitable. The staff believed that unless the design features of containment proved to be sound, it would be premature to analyze the remainder of the proposed plant.[27]

The preliminary hazards report was Con Edison's initial evaluation of the proposed plant's safety, and the regulatory staff wanted more technical information before reaching its decision on the site. After several meetings with Con Edison officials, the staff formally requested additional data on 9 August 1963. Most of the questions the AEC posed in its request related to such technical items as how adequate the containment design was, how to measure radioactive leakage rates in the containment, how penetrations through the containment could be monitored, and how filter systems could remove radioactivity from the containment.[28]

While waiting for Con Edison's reply, the regulatory staff continued its evaluation of the data on hand. It also conferred with the Ravenswood subcommittee of the ACRS and with members of the Con Edison staff. Although it made no formal decisions as a result of the meetings, the AEC staff suggested both to the ACRS and to Con Edison that the Ravenswood site probably would prove to be unacceptable. For example, in September 1963, Edson G. Case, assistant director of the Division of Reactor Licensing and Regulation, reported to the ACRS subcommittee on a recent computer study that analyzed the available information on the Ravenswood reactor. Case was an engineer and a veteran of the nuclear navy whose straightforward style and willingness to articulate his views fully and frankly commanded the respect of his colleagues, even those who disagreed with him. The computer study calculated the potential hazards to the public of an accident in which radioactive materials were released. It considered, among other things, containment spray rates, leakage from containment, and the efficiency of safety injection and pump-back systems. Case told the subcommittee that while the study could not be used to show that the Con Edison application was acceptable, it might well show that reliance on engineered safety features alone at Ravenswood would be unacceptable.[29]

Subsequent runs of the computer program supported Case's assessment; they indicated that even if all the engineered safety features operated as planned, the facility as designed could not meet the radiation exposure limits established in the site criteria. Within a short time, the staff sent a draft report to the ACRS that described its reservations about the Ravenswood application. It found the proposal to be unacceptable because it was not convinced that in the event of a major accident, there would be no release of radiation, or "fission products," from containment. The staff's position was "based upon the uncertainties involved in determining the adequacy of the engineered safeguards in the event of a fission product release. . . . The problem of demonstrating a guaranteed essentially zero release of fission products in the event of such an accident . . . appears overwhelming." Other engineering approaches might prove to be workable, but they "would have to be evaluated on their own merits."[30]

The staff apparently conveyed its conclusion to Con Edison. In November, the utility formally responded to the agency's request of 9 August for supplemental information. In addition to submitting the analyses requested, Con Edison conceded that as a result of its several meetings with the regulatory staff, various aspects of the design had to be clarified and amplified. It believed that particular attention had to be given to "detailed studies of additional engineered safeguards not utilized to date in existing plants" and it planned to file a formal amendment to its preliminary hazards report. The company also hinted that it had not made a final decision on the Ravenswood plant.[31]

While the regulatory staff's reaction to the Ravenswood application was negative, the ACRS initially seemed ambivalent as it undertook its independent review. Members of the subcommittee assigned to the project approached Con Edison's application with an open mind, agreeing that its review should be based on a thorough evaluation of the engineered safeguards rather than making an "arbitrary determination that the plant is located unnecessarily close to a large center of population." After initial study, however, some members began to question whether adequate safety features could be built into the facility. In its first meeting with the utility in September 1963, the subcommittee chairman, Franklin Gifford, director of the Atmospheric Turbulence and Diffusion Laboratory at Oak Ridge, commented that the "lousy" site required undue emphasis on designed safety elements. Committee member William Ergen, a physicist at Oak Ridge National Laboratory, suggested that Con Edison consider alternative sites, but utility officials demurred because of the high costs of transmission. Subcommittee members later agreed among themselves that "many paths exist by which the double containment scheme might be by-passed" and that even a "very small release [of fission products] may be intolerable at this site." At its monthly meeting in November 1963, one ACRS member observed that if Ravenswood was to be rejected "because of the elementary state of the reactor art," Con Edison should be informed soon to avoid waste of time and money.[32]

The precise impact on Con Edison of the reservations expressed informally by the regulatory staff and the ACRS is difficult to assess. But they undoubtedly played a major, and perhaps a determining, role in Con Edison's decision, announced on 3 January 1964, to withdraw the application for Ravenswood. Con Edison chairman Forbes told Glenn Seaborg that the action was based on the utility's opportunity to purchase a large block of hydroelectric power from Canada on an "economically advantageous basis." Forbes stressed that his company still looked to nuclear energy to "supply the additional thermal power needs for our system in the years ahead" and that withdrawal of the application did not devalue the "role which nuclear energy will play in our service area."[33]

In 1963, Con Edison had begun negotiations with the British New-foundland Corporation for the purchase of Canadian electricity. At the time, utility spokesmen said that if agreement could be reached, Con Edison would postpone all new plant construction, including Ravens-wood. The *New York Times* reported in January 1964 that the estimated cost of the Canadian power delivered to the edge of Con Edison's service area in Westchester County varied from five to six mills per kilowatt hour. This compared favorably to Con Edison's projection that electricity from Ravenswood would run six-and-one-half to seven mills per kilowatt hour. The *Times* report did not take into account, however, the high cost of new transmission lines from the edge of the utility's service area into the heart of New York City. At the April 1963 Joint Committee hearing, Forbes had emphasized that particular expense as the main economic attraction for constructing Ravenswood. Con Edison officials who met with the ACRS had underscored the same consideration. Those earlier statements were not forgotten by committee members. In a brief discussion of the Ravenswood cancellation at a January 1964 meeting, one ACRS member remarked that Con Edison's explanation for abandoning its plans was "in variance with [its earlier] argument for the need for short transmission lines and hence lower costs in the New York area."[34]

Con Edison's withdrawal of the Ravenswood proposal in the face of opposition from the AEC regulatory staff left the question of metropolitan siting unresolved. The first time the full ACRS had discussed the application in October 1963, one member had voiced some doubt that Con Edison's stated reasons for pursuing the Ravenswood site were its primary concerns. He conjectured that the proposal "may be more of an attempt to see if approval for city locations could be obtained." If this were the case, the result remained inconclusive. The regulatory staff, unable to satisfy the distance factors suggested in the recently published reactor site criteria, harbored deep reservations about the Con Edison application. The questions the staff raised about the proposed engineered safety features indicated its sense, never formally articulated, that reactor technology was not yet advanced enough to permit metropolitan siting without some degree of isolation. Joseph Lieberman, assistant director of nuclear safety in the Division of Reactor Development, told a reporter shortly after Con Edison's retreat that although some manufacturers believed they could build reactors safe enough to be located in downtown areas, the regulatory staff "apparently had not reached the point where it feels they have."[35]

The cancellation of the Ravenswood proposal left open the possibility of metropolitan siting in the future, not only for Con Edison but for other utilities as well. Had Con Edison pursued the application and been rejected formally by the AEC, it might have settled the issue by effectively eliminating the chances of locating plants in downtown areas. By abandoning the Ravenswood application, Con Edison postponed a final decision.

The reliability of engineered safety features that might be incorporated into the design of reactors was a source of continuing concern to the AEC. Early in 1964, the Commission asked the ACRS to study and report on the issue. By November, the committee had concluded that properly engineered safety features permitted the "reduction of distances required for protection of the public, and that engineered safeguards of a selected type should make feasible the siting of power reactors at many locations not otherwise considered as suitable." The committee report discussed various safety features and offered different evaluations of them. It considered, for example, containment structures and pressure suppression systems in boiling-water reactors to be effective safety designs. The ACRS expressed doubts, however, about core sprays (a system that would provide cooling water to the core in a loss-of-coolant accident) and safety injection systems (separate high- and low-pressure systems that would inject additional cooling water to the nuclear core in a loss-of-coolant accident) because they "might not function . . . in the event of an accident." The regulatory staff undoubtedly influenced the ACRS report. It had informed the committee that many engineered safety features had not been thoroughly tested or proven to be reliable. In its report to the Commission, the ACRS pointed out that while some safety features were based on engineering principles supported by tests, "others require developmental and proof testing." It made no policy recommendations, but its findings strongly indicated that case-by-case review of proposed sites and reactors would have to continue.[36]

Meanwhile, several utilities applied for and received approval of plants that were close to metropolitan areas but met the agency's siting guidelines with a combination of distance and safety features. For example, the agency granted construction permits to the Southern California Edison Company for its San Onofre unit (375 megawatts electric) in

February 1964, to the Connecticut Yankee Power Company for its
Connecticut Yankee reactor (550 megawatts electric) in April 1964, and
to the Jersey Central Power and Light Company for its Oyster Creek
facility (515 megawatts electric) in December 1964.[37]

A year after the Ravenswood controversy, the metropolitan siting
issue again came to the forefront. On 8 January 1965, representatives of
the Boston Edison Company met with the commissioners and the regulatory staff and discussed several possible sites for the utility's first reactor. The company preferred its Edgar site in Weymouth, Massachusetts,
about nine miles from the center of Boston. It stressed the absence of
zoning and land acquisition problems and the availability of transmission lines as principal reasons for favoring the location. Except for
Ravenswood, the Boston Edgar site was surrounded by the highest
population density of any reactor proposal received by the AEC to that
time. Within one mile of the facility lived 6000 people, within five miles
250,000, and within ten miles a million. The 600 electrical megawatt
power level of the proposed boiling-water reactor placed it in the range
of the largest plants that had received construction permits. The application raised anew the issue of whether such reactors should be located in
populous metropolitan areas.[38]

Shortly after the January meeting, Harold Price gained the Commission's agreement that Boston Edison's proposal made it appropriate for
his staff to develop a discussion paper on the broad issue of metropolitan siting. The paper the regulatory staff prepared reemphasized its
cautious position on metropolitan siting. It noted that the 1962 site
criteria had paved the way for metropolitan reactor proposals by encouraging applicants to demonstrate that engineered safety features could be
substituted for distance. Citing the Boston Edison proposal as an example, the staff predicted that the utility would design a containment
structure and associated engineered safety features that would bring
"maximum credible exposures from accidents within the levels specified" in the AEC's regulations. This would raise the question of whether
the AEC was ready to accept sites in heavily populated areas before
acquiring experience with the reliability of advanced engineered safety
features at sites further removed from metropolitan centers. The regulatory staff informed the Commission that it believed that "power reactor
technology and experience with advanced safeguards have not yet
reached the stage where it can be considered prudent from a public
safety viewpoint to permit the construction and operation of nuclear
power plants in densely populated areas of major cities."

The regulatory staff cited two technical reasons for its view. First, experience was limited with advanced types of containment structures and other engineered safety features and their reliability was still uncertain. Second, while assuming that the probable consequences of a major reactor accident "would be no greater for a reactor in a city than for one 20 or 30 miles out," the consequences of smaller accidents that seemed more likely to occur troubled the staff. Its apparent willingness to approve designs for reactors in nonurban locations that it would find unacceptable in populated areas did not mean that it was less concerned about the health and safety of rural residents. Rather, it reflected the conviction that a reactor accident in a lightly populated region would be easier to cope with. Harland Forbes's statement that a reactor was either safe or it was not, no matter how many people it might affect, ignored the fact that an evacuation in the event of a serious accident would be much more feasible if a population in the hundreds or even in the thousands rather than in the millions had to be moved. Further, a small release of radiation could be dispersed and diluted in the atmosphere in the country without creating the problems or seriously threatening the population in the way that it would an urban area.[39]

The staff paper included a suggested draft policy statement on metropolitan siting. It acknowledged the commendable safety record of the industry and the economic rationale for placing reactors in populated areas. But it also cited the lack of experience with large reactors, and concluded that the "public interest can best be served by continuing to exclude large cities as permissible locations for nuclear power plants."[40]

The commissioners discussed the paper on 11 February 1965. Clifford Beck, deputy director of regulation, told them that enormous technical strides were being made in reactor design that, in some instances, raised complex safety considerations. At the same time, utilities and vendors were increasing their pressure on the regulatory staff to permit building reactors in cities. Operating experience with a group of "second generation" reactors such as Oyster Creek and Connecticut Yankee, Beck contended, was necessary to assess several new safety features in larger reactors. He estimated that adoption of the policy outlined in the staff paper would delay the location of large reactors in cities at least until the 1970s. The commissioners questioned the length of time the staff thought necessary to acquire sufficient experience. Although no precise period could be determined, the staff believed that about ten years would be a reasonable estimate. Ramey, however, countered that four to five years might provide enough experience and that the AEC should review the situation in two or three years.[41]

Sensing the Commission's opposition to the staff's proposed policy on metropolitan reactors, Harold Price explained that the draft statement was not meant to be definitive. But he insisted that the impending Boston Edison proposal demanded a clear policy on metropolitan siting to help the utility determine whether or not to pursue the application. Because of the lack of experience with engineered safety features, the staff's detailed technical review could result in rejection of the application. Price was hoping for a categorical policy that would avoid the need for case-by-case reviews that were so expensive and time-consuming for both a utility and the regulatory staff. Undoubtedly he remembered that the Ravenswood controversy had lasted for over a year before concluding. The Commission, however, did not accept Price's appeal. Instead, it asked him to prepare a detailed technical briefing on safety criteria related to metropolitan siting. Although discussion of the question continued, the meeting indicated that the AEC would not soon issue a more definitive policy on metropolitan siting.[42]

The bulk of discussion within the AEC on metropolitan siting took place between the regulatory staff and the ACRS. The committee had already scheduled a meeting in March 1965 to discuss the implications of siting policy for upcoming congressional hearings on extension of the Price-Anderson Act. Passed in 1957 for a period of ten years, the law provided government indemnity insurance for nuclear plant owners. The nuclear liability law did not expire until 1967, but the Joint Committee scheduled hearings far in advance of the deadline to allow ample time for Congress to act. The metropolitan siting question potentially posed a dilemma for the AEC, which favored extension of the law. If the agency determined that reactors were safe enough to be located in cities, it would appear to obviate the need for Price-Anderson's protection against claims arising from a nuclear accident. But if the AEC imposed a ban on metropolitan sites it might hamper industry development by indicating that power reactors were not demonstrably safe.[43]

After discussing the issue, the ACRS opposed the regulatory staff's proposal to prohibit metropolitan sites, despite Price's insistence that a more explicit policy had to be enunciated to answer recent industry inquiries. The AEC's Division of Reactor Development added its considerable weight to the position of the ACRS by criticizing a ban as "unrealistic." It argued that such a policy would have "an adverse effect on the over-all public image of the safety of nuclear reactors." The Reactor

Development staff suggested that improving margins of safety through research programs and the development of design codes and engineering standards for plant safety systems would be vastly preferable to a prohibition on urban reactors.

The ACRS decided to send a draft report to the commissioners as a basis for informal discussion with them. By forwarding a working paper instead of its customary formal public report, the committee hoped to keep the siting question an internal matter for the time being. The draft stressed that a "flexible position" should be maintained on metropolitan siting and applicants "should be encouraged to use imagination and to employ improved provisions for safety." It also urged the AEC to provide guidance on what was acceptable for reactor designers and manufacturers. The committee made its strongest recommendation on this issue: "It would seem prudent to operate in metropolitan areas only reactors of a proven type, which do not represent a large extrapolation in power, involving radical changes in design from reactors already in service." In other words, metropolitan reactors "should closely duplicate reactors with demonstrable and favorable operating experience."[44]

In a meeting with the committee in May 1965, the commissioners made clear that they wanted to avoid any definition of metropolitan siting "which might preclude reactors at such locations." Although they generally approved the committee's guidelines they wanted to place them in the most favorable light. Seaborg, for example, commented specifically on the committee's suggestion that only reactors of proven design should be located in cities. He hoped that the committee meant that an applicant could anticipate favorable operating experience from a reactor at a remote site, which would allow construction of metropolitan facilities to proceed concurrently. The Commission again took no action on the question of metropolitan siting, and opposed making any public policy statement at the time. Despite the Commission's rebuff to his arguments, Price continued to maintain that metropolitan reactors needed proven engineered safety features and operating experience before being acceptable to the AEC.[45]

Meanwhile, Boston Edison officials continued their planning for a nuclear plant at the Edgar site. The company chose General Electric as its supplier and conducted an extensive public information program in the Boston area. The utility president, Charles Avila, reported to Seaborg that no negative public reaction had been apparent in a series of public meetings; indeed, a number of persons had urged the utility to build the reactor in their community. Avila told Seaborg that his company was ready to proceed with preliminary technical discussions with the agency.[46]

Boston Edison and General Electric provided technical details about the facility to the regulatory staff and the ACRS. Both agency groups viewed the Edgar site in the same terms as Ravenswood, and in a series of meetings they conveyed their uneasiness to Boston Edison. Finally, in a January 1966 meeting, Richard Doan, then head of the Division of Reactor Licensing, bluntly outlined the staff's position to Boston Edison officials. A retired manager of the Atomic Energy Division of the Phillips Petroleum Company, Doan had joined Harold Price's staff as director of Reactor Licensing in mid-1964. Years of experience with Phillips, as well as membership on the ACRS, had instilled in him a cautious position on reactor safety.[47]

Doan reminded Boston Edison's representatives that when the AEC had prepared its site criteria in the early 1960s, reactors were still located some distance from cities. The metropolitan sites proposed since then were difficult to assess because new safety features had to compensate for the lack of isolation. Doan had personally reviewed the technical data on the proposed reactor and found no significant safety advances in its design. Consequently, he informed the utility officials that, in his opinion, the regulatory staff would not approve the Edgar site until more operating experience with engineered safety features had been gained. In passing, Doan noted that no one at the agency wanted to state officially that large reactors in cities were not acceptable. He emphasized, however, that the AEC would sanction metropolitan sites only if the industry made improvements in the design and application of engineered safety features.[48]

Doan's blunt appraisal of the Edgar site came as "a body blow" to Boston Edison's plans, according to an attending utility official. It undoubtedly figured significantly in Boston Edison's eventual decision to move its choice of sites to one that would meet the AEC's guidelines. The warning from Doan placed the utility on notice that it would face formidable obstacles if it submitted a formal application for the Edgar site. When the company finally made an application, it abandoned Edgar in favor of a location near Plymouth, Massachusetts that met the AEC's site criteria.[49]

Although the AEC had not formally prohibited metropolitan siting, industry officials chafed at the regulatory staff's unofficial position. At an American Nuclear Society symposium on metropolitan siting in March 1966, Clifford Beck made the AEC's clearest public statement to date on the issue. He announced that there was little chance that the agency would allow a large reactor in an urban center in the near future. Con Edison official W. Donham Crawford told the gathering that his company would press for approval of city sites anyway. It must do so, Crawford insisted, because of transmission costs. Joseph C. Rengel, general manager of Westinghouse's Atomic Power Division, indirectly chided Beck by complaining that too much emphasis had been placed on engineered safety features for urban sites. He declared that manufacturers built plants that were safe for any site and affirmed that "safety is no less important in a remotely located plant than in a city." This symposium was not the only forum in which industry expressed its concern; both vendors and utilities continued to criticize the regulatory staff's position on siting.[50]

After dropping its plans for a reactor at the Ravenswood site, Con Edison, in late November 1965, announced that it would construct a second reactor at its Indian Point site in Buchanan, New York. Indian Point II, a pressurized-water reactor like the existing plant, would generate 873 electrical megawatts, more than three times the capacity of the original unit, which came to be called Indian Point I. Although the earlier reactor had been constructed before the 1962 site guideline was formally adopted, it still met those criteria because it incorporated additional safety features that offset the high population density around the site. The AEC never considered the original Indian Point plant a metropolitan reactor despite the heavy population in its immediate vicinity and its proximity (twenty-four miles) to New York City. The regulatory staff informally labeled it a "suburban" facility. Indian Point II, however, had to be judged anew against the site criteria's calculated distances for the exclusion area, low population zone, and population center distance.[51]

Another critical issue substantially increased the complexity of the Indian Point proposal. By the time that they received the application, ACRS members had begun to raise questions about the integrity of the containment building in the event of a major accident. They worried increasingly about two eventualities that seemed improbable but still possible: a sudden, catastrophic pressure vessel failure or a loss-of-coolant accident in which emergency cooling systems did not function properly. As a result, the ACRS recommended to the AEC that "means be developed to ameliorate the consequences of a major pressure vessel rupture," and it urged the development of improved emergency cooling systems. Westinghouse, Con Edison's nuclear steam supply system de-

signer, responded by adding several engineered safety features to Indian
Point II. It planned, for example, two independent safety injection systems to flood the core with borated water in case of a pipe rupture. As a
passive backup to the emergency cooling systems, Westinghouse also
designed a new and controversial water-cooled stainless steel tank
named the "core catcher." Further, the containment design included a
new internal recirculation spray and an air recirculation system that
would provide long-term cooling during and after an accident.[52]

After lengthy deliberations over those new safety features, the AEC
granted a provisional construction permit to Con Edison for Indian
Point II. At the same time, however, the members of the regulatory staff
settled on an informal siting standard. Because of the growing concern
about the possibility of a breach of containment, the staff agreed that it
would oppose construction of any large reactor at a site more populous
than that surrounding Indian Point. Indian Point II, in other words,
became an unofficial standard of size and location against which future
reactor proposals would be judged.[53]

A related decision came out of the intramural agency discussion over
Indian Point II and metropolitan siting. Throughout 1966, both the regulatory staff and the ACRS intermittently considered a scheme of classify-
ing reactor sites. The categories included metropolitan, suburban, rural,
and remote locations. The classifications were highly subjective; all participants agreed that such categories had to be based on flexible criteria
because of the large number of variables that had to be weighed in any
application. Nevertheless, an informal consensus evolved from their deliberations. It reinforced the regulatory staff's emphasis on closer scrutiny of designs for any reactors that might be proposed for a metropolitan
or suburban site. In effect, it meant that reactors planned for metropolitan sites would not be approved, at least for some time, because of the
extra scrutiny given them.[54]

Not long after the regulatory staff finished its evaluation of the application for Indian Point II, it received a new one for another metropolitan site. Public Service Electric and Gas Company, a New Jersey utility,
announced in December 1966 that it planned to construct a 993 electrical megawatt pressurized-water reactor on a 140 acre site at Burlington,
New Jersey. The site was approximately eleven miles southwest of Tren-
ton and seventeen miles northeast of Philadelphia, with a density of
population that was measurably higher than Indian Point.[55]

The AEC applied its informal decision not to approve a site surrounded by a population higher than that at Indian Point. The regula-

tory staff bluntly informed Public Service officials that Burlington was a poor location, and the ACRS unanimously agreed that the site "was unsuitable for the proposed reactor." To avoid flatly rejecting the application, the ACRS, at the suggestion of Harold Price, told the utility that it did not see how it could endorse the existing proposal, which left an opening for Public Service to amend its application with a different site. Public Service took advantage of this option, and announced in January 1968 that it was now planning to build two large (each 1050 megawatts electric) reactors in sparsely populated Lower Alloways Creek Township, New Jersey.[56]

As the site review for the Burlington facility proceeded, the AEC received other applications for large power reactors. One application that bordered between a suburban and a metropolitan site was a proposal by Commonwealth Edison of Chicago to build twin units near Zion, Illinois. Because the site and reactor size were similar to Indian Point II, it became important as a benchmark for metropolitan siting. The AEC added Zion to Indian Point II as an informal standard against which to judge other urban locations. The two sites also were used by the ACRS in an attempt to establish formal criteria on metropolitan siting.

Commonwealth Edison filed its application for the plants in the summer of 1967. The utility planned to construct Westinghouse pressurized-water reactors, each with a net electrical output of 1040 megawatts. The Zion reactors were the largest proposed to date in a region of relatively high population density. About half-way between Chicago and Milwaukee, Zion was located on the west shore of Lake Michigan, approximately six miles north of downtown Waukegan, Illinois and eight miles south of Kenosha, Wisconsin. The community of Zion had a population of 14,106, Waukegan, 55,719, and Kenosha, 67,899. The site of 250 acres was bordered on the north by open marshland that contained a few scattered residences and on the south by a state park. Eight-tenths of a mile to the west, tracks of the Chicago and Northwest Railroad paralleled the site. The land within a four-mile radius was primarily residential with some light industry. A comparison of population distribution between Zion and Indian Point II showed striking similarities. Using a 1965 census estimate for Zion and a 1960 figure for the New York site, in a zero to one mile radius Zion had a cumulative population of 1,260, Indian Point, 1,080; at four miles, Zion had 38,855, Indian Point numbered 38,730; and at ten miles, Zion had 200,000 while Indian Point numbered 155,510.[57]

The regulatory staff's evaluation highlighted the similarity, not only in population distribution but also in the engineered safety features in the design of the Zion reactors. The principal problem that the staff identified was the provision for radioactive iodine removal by chemical containment sprays in the case of a major accident. The Zion design provided only for a containment spray system that Commonwealth Edison calculated would meet the siting guidelines for iodine exposure to the public. The regulatory staff, however, projected that if atmospheric conditions were less favorable than assumed by Commonwealth Edison, the potential iodine exposure might exceed the guidelines. Because of the population density around Zion, it decided to require an additional independent safety feature in the form of charcoal filters in the containment building such as the Indian Point II reactor had. Otherwise, the staff approved the reactor design and the site for the Zion plants, largely on the basis of its comparability to Indian Point II. The ACRS, however, called for additional safeguards because the size of the Zion units was larger than Indian Point II. After considerable negotiation, Commonwealth Edison agreed to conduct research on problems that concerned the ACRS and make changes if they turned out to be necessary.[58]

Metropolitan siting, an issue that commanded the attention of the AEC throughout the 1960s, defied easy resolution and persistent attempts to define specific criteria. Although the AEC's 1962 siting guidelines theoretically allowed urban plants, if they included enough engineered safety features to offset the lack of isolation, the regulatory staff was reluctant to approve metropolitan sites. After the Ravenswood experience, the staff tried to convince the Commission that a ban on metropolitan sites should be imposed. The reasons the staff cited—lack of operating experience with large reactors, paucity of research on relatively limited accidents that might have greater consequences at city sites than in remote locations, and the expense incurred by utilities when the AEC rejected an application—were not enough to convince the Commission. It wanted to retain as much flexibility as possible on regulatory issues at a time when applications for nuclear plants were rapidly increasing.

On this, as on a number of other questions, the regulatory staff lacked the clout to carry its arguments against considerations that weighed more heavily with the commissioners. The ACRS, the Division of Reactor Development, and the commissioners themselves opposed a strict prohibition on metropolitan siting, largely because of their concern that it would discourage the growth of nuclear power. Consequently, the regulatory staff had no choice but to use a case-by-case approach in judging applications. Eventually it developed an informal standard—the Indian Point and Zion population densities—that it used in evaluating other cases. This position was a compromise within the AEC's bureaucracy; it kept the option of metropolitan siting in the future open while, at least for the time being, establishing a benchmark that the regulatory staff could apply in its licensing decisions.

The Indian Point and Zion standard was, in effect, an addendum to the 1962 siting guidelines. According to those criteria, one of the key factors to be considered in evaluating sites was that "where unfavorable physical characteristics of the site exist, the proposed site may nevertheless be found to be acceptable if the design of the facility includes appropriate and adequate compensating engineering safeguards."[59] For the sites where the population did not exceed the unofficial standard set by Indian Point and Zion, the regulatory staff was willing to grant construction permits, assuming that it was satisfied with the other requirements of the application.

The AEC applied a flexible approach to metropolitan siting. It attempted to balance the designs of engineered reactor safety features against population densities and theoretical knowledge about the consequences of possible accidents, while leaving a wide margin for unknown circumstances. It also stressed the need for more operating experience before allowing power reactors in what it defined as metropolitan areas. Several utilities, despite their strong incentives for metropolitan reactors, changed sites once they discovered that the regulatory staff placed limits, however informal and ill-defined, on plants in heavily populated locations. Although the AEC received a few more applications for metropolitan reactors, by the end of the 1960s, the pressure for siting reactors close to downtown population centers had greatly eased. Utilities recognized the staff's position and few made further efforts to build nuclear units in metropolitan areas.

CHAPTER V

Reactors at Faults

The Controversy over Seismic Siting

At the same time that the AEC was deliberating over the issue of metropolitan siting, it was evaluating another difficult and contentious siting problem—the safety implications of locating power reactors near earthquake zones. Because of limited knowledge about earthquakes and little operating experience with nuclear plants, the question was fraught with scientific uncertainties that soon triggered spirited public debates. Like its consideration of metropolitan siting, the AEC, in making judgments about approving applications for proposed plants, was torn between its goal of encouraging utilities to build reactors and its goal of increasing public confidence in reactor safety and regulatory procedures. On an issue in which scientific views clashed and no definitive answers were available, the AEC's assessments, not surprisingly, were also divided and equivocal. Doubts expressed by some agency officials about the safety of two proposed plants on the California coast doomed their applications, while other reactors near fault zones received construction permits. Eventually, the AEC sought to formalize its policy by publishing a set of guidelines for siting plants in the vicinity of potential seismic activity.

The debate over nuclear reactors and seismic hazards first arose after the Pacific Gas and Electric Company (PG&E) of San Francisco, the nation's largest privately owned utility, announced plans to build a plant on the California coast near the tiny town of Bodega Bay. The fishing village of 350 people, about fifty miles northwest of San Francisco, sat on the edge of a scenic harbor formed by a curving, rocky peninsula, called Bodega Head, that jutted into the sea. Russian sailors had landed there in 1812, establishing a sealing station and a territorial claim for the czar. Much later, Bodega Bay had served as an isolated setting for moviemaker Alfred Hitchcock's film "The Birds." Bodega Head featured beaches, dunes, and cliffs that endowed it with extraordinary natural beauty and made it, potentially at least, an appealing recreation spot. The state of California and the county of Sonoma had taken preliminary action to reserve it for a park. In addition, the University of California had considered establishing a marine laboratory on the bay.[1]

Bodega Head also offered many attributes as a site for a power plant. The population in the area was small, and the surrounding countryside of rolling hills and dairy farms was well-suited for transmission lines. By placing the plant near the point of the peninsula, PG&E engineers could design it to draw cooling water from the calm bay and discharge it into the ocean. This would not only make it easier to build the cooling intake but would also allow the heated water from the plant to be cooled by ocean currents and breezes. The harbor would provide a convenient water transportation terminal. With those advantages in mind, the utility quietly began making plans to acquire land and construct a power station at Bodega Head. On 2 September 1958 it announced its intentions, though it had not yet definitely decided whether to place a nuclear or fossil-fuel plant at the site.[2]

PG&E's plans aroused strong protests from some residents of the area around Bodega Bay. The objections centered on the damage the proposed plant would inevitably inflict on the scenic beauty of the site, the harm it might cause the local fishing industry, and the ill-effects it could have on marine research. The most outspoken and colorful of the opponents was Rose Gaffney, a seventy-six-year-old grandmother who owned a large parcel of land adjacent to the proposed site, a part of which PG&E had condemned under eminent domain. Gaffney, who had long battled the county government on environmental issues, took an equally defiant stand against the utility. A power station on Bodega Head, she declared, would be a "crime against humanity." After PG&E confirmed rumors that it had decided to build a large nuclear plant (325 electrical megawatts) at the site in June 1961, the protests widened.[3]

Opponents of the plant made their case before the California Public Utilities Commission in hearings convened in May 1962. Over a period of eight days, highlighted by Rose Gaffney's lengthy slide presentation on the history and wildlife of Bodega Head, conservationists delivered their arguments in a chamber crowded with sympathetic listeners. But the Public Utilities Commission was unmoved; it dismissed the objections and granted PG&E the land-use permit it sought. This did not end the protests, but it shifted their focus to the AEC, which received a formal application for the Bodega plant in December 1962.[4]

The chief vehicle for the opposition was a group formed the previous spring, the Northern California Association to Preserve Bodega Head and Harbor. The membership of the organization was not large—by December 1963 it was about 800. Its influence, however, extended far beyond its own membership, largely because of the efforts of its executive secretary, David E. Pesonen. A 1960 graduate in biology from the University of California at Berkeley, Pesonen was the conservation editor of the Sierra Club and represented it at the hearings of the California Public Utilities Commission on the Bodega project. Although neither the Sierra Club nor the Northern California Association to Preserve Bodega Head and Harbor opposed nuclear power as a matter of policy, both objected to PG&E's plans for a reactor, or any other power plant, at Bodega Head. "We didn't know much about radiation at the beginning," Pesonen later commented. "Our concern was scenery." When Sierra Club leaders expressed reservations about Pesonen's confrontational tactics, however, he resigned from his position with the organization. He took a part-time job as a laboratory technician while working indefatigably to win publicity for his cause and to organize protests against construction of the Bodega plant.[5]

The Northern California Association continued to submit its complaints to state and local agencies, but it concentrated its efforts on the licensing process of the AEC. Since the AEC did not weigh aesthetic or land-use questions in its site evaluation, the association placed a much greater emphasis on safety matters than it had earlier. And the safety issue that soon took precedence in the growing controversy over Bodega Head was whether or not the proposed plant presented seismic hazards that made it unsuitable for nuclear power. The concern was that an earthquake would cause safety systems to fail and that the damaged plant would release large amounts of radioactivity into the environment.

Although it did not immediately emerge as the central issue in the dispute, the seismic question had not been overlooked by either the utility or its opponents in the initial debates over the Bodega site. The proposed reactor would stand within a few hundred feet of the western boundary of the San Andreas fault, the 400-mile long fault zone that had been responsible for the 1906 San Francisco earthquake. The fault zone, about a mile and a half wide at Bodega, ran through the bay between the headland on which the plant would be built and the shore. There were two ways that an earthquake could cause the safety systems of a nuclear plant to fail. It could produce "ground displacement," in which the earth moved along a fault at the site of the plant. Even in the absence of a fault at the site, an earthquake that occurred nearby could damage the plant by generating severe ground shaking.

PG&E hired several prominent authorities to study the site, and their findings were generally favorable. In the preliminary hazards summary report that the utility submitted to the AEC in December 1962, Don Tocher, a University of California at Berkeley seismologist, and William Quaide, a geologist in private practice, concluded that the site showed no evidence of major faults, which would have made it unsuitable for the proposed plant. They saw no indications that existing minor faults had moved in the preceding few thousand years and suggested that the quartz-diorite granite that underlay the site was "a much better foundation than any other geologic formation on Bodega Head." The 1906 earthquake had caused substantial ground displacement on Bodega Head, and although Tocher and Quaide submitted that such massive tremors were rare, they predicted that one or two earthquakes of a similar magnitude would occur near the site within a century. Therefore, they recommended that the proposed plant be designed to withstand ground shaking from a nearby earthquake of major proportions. They also cautioned that their conclusions might be revised as further investigation revealed more information about the site.[6]

Another PG&E consultant, George W. Housner, a professor at the California Institute of Technology who was regarded as one of the foremost experts in the world on earthquake engineering, reviewed Tocher's and Quaide's report and endorsed their findings. He emphasized that in the absence of active faults under the proposed reactor, the critical problem would be to make certain that the plant could survive severe ground shaking caused by an earthquake that was centered nearby. Housner advised that the plant be built to withstand motion from a tremor of the magnitude of the 1906 earthquake, or 8.2 on the Richter scale. He expressed confidence that it could be designed to meet that objective.[7]

Opponents of the proposed plant took a much less sanguine view of earthquake risks at Bodega Head. David Pesonen charged in May 1963 that PG&E had misrepresented seismic conditions at the plant site.

Asserting that PG&E had misstated the distance of the site from the boundary of the San Andreas fault, he countered the utility's measurements by claiming that the site was only 1000 feet away rather than a quarter of a mile (1320 feet). The difference was of some importance because the AEC's siting criteria advised that a reactor should be at least a quarter of a mile from an active fault. Pesonen further alleged that PG&E had edited the conclusions of its consultants to make them appear more favorable in its hazards report to the AEC.[8]

PG&E dismissed Pesonen's allegations of procedural misdeeds, but its position faced a more serious challenge from the findings of a scientific expert consulted by the Northern California Association. In late 1962 Pesonen sent aerial photographs of the area of the proposed reactor to Pierre Saint-Amand, a seismologist at the U. S. Naval Ordance Test Station in China Lake, California. In April 1963 Saint-Amand and a colleague from China Lake went to Bodega Head and conducted a two-day inspection of the site. By that time PG&E had begun excavation, which it could do without a construction permit from the AEC, and Saint-Amand had the opportunity to view the open pit. He was very disturbed by what he saw. For one thing, the foundation in places was not solid rock but broken rock or alluvium. He contended that this would not only provide a poor base for the weight of the plant but also could exacerbate the effects of an earthquake by moving at a rate different than the bedrock. Saint-Amand was even more distressed by another observation—on one side of the hole he detected what he described as "a major fault zone." Although he did not submit his final report to the Northern California Association for several months, he concluded in a letter to the office of the U. S. Secretary of the Interior: "This one feature alone would cause me to recommend against construction of any major structure in the immediate vicinity." PG&E denied that the foundations of the site were unstable or that an active fault line ran through it.[9]

The burden of judging the conflicting geological claims fell on the AEC. The agency had included seismic considerations in its siting criteria by stipulating that a nuclear plant should not be built within a quarter of a mile of an active fault. Earlier drafts had cited the proper distance as between a quarter mile and a half mile, but to eliminate ambiguity in favor of easing the siting of plants, the AEC had adopted the lower figure in the final version of the criteria. The imprecision of the AEC's guideline, and the rather casual manner by which the regulatory staff arrived at it, reflected the fact that earthquake geology was an

inexact science. Geologists and structural engineers agreed that buildings should not be located directly over active faults, because there was
no certain way to design a structure to withstand the sudden and unpredictable movement of an earthquake. But other related questions stirred
dissension. There was no consensus, for example, on how close to a
fault it was safe to build, whether a structure could be placed on a fault
line that had apparently not moved for thousands of years, or how the
risk of a "possible" fault should be judged. Geology, at least as far as
understanding earthquakes was concerned, provided only contestable
hypotheses rather than immutable truths. Its practitioners examined the
existing evidence and drew conclusions, but, like historians, they could
offer no assurance that professional colleagues would reach the same
conclusions or that accurately reconstructing the past was an unambiguous guide to future behavior.[10]

The AEC was not well-equipped to evaluate the competing positions
on the seismic suitability of the Bodega site. It had no seismologists or
geologists on its staff. To make up for this lack of expertise, the AEC
hired two well-known authorities as consultants to conduct their own
investigation. In addition to its technical deficiency, the AEC's ability to
review PG&E's application impartially was suspect. The Bodega project
came at a time when progress in the agency's reactor development
program was sluggish and future prospects looked problematical. Although several plants were being built, the AEC had received very few
applications for new ones. This was a cause for concern within the AEC
but even more so on the Joint Committee on Atomic Energy. Committee
members complained bitterly about the "inadequacy" of the AEC's
civilian power program. In response, agency officials argued that their
efforts to encourage the use of nuclear power were beginning to produce
results, and one of the primary examples they cited was the proposed
Bodega plant.[11]

Opponents of the plant maintained that the AEC's eagerness to see it
operate would preclude an open-minded hearing on licensing it. Pesonen, for example, wrote in 1962: "Obviously the individual citizen
cannot look for protection to the AEC. . . . The AEC is firmly committed to construction of the Bodega reactor." That conviction received
some supporting evidence in the spring of 1963, when both the regulatory staff and the ACRS gave preliminary approval to PG&E's application. In both cases, however, the endorsements depended on further
exploration of the seismic hazards. The staff concluded that the difference between a quarter of a mile and the actual distance of the site from

the boundary of the San Andreas fault zone was of minor significance, but it added: "Information developed during excavation should be carefully evaluated for evidence of faults, the presence of which would require a reassessment of the suitability of the location proposed."[12]

Rather than waiting for the AEC's final evaluation, Pesonen's organization and other plant opponents turned to an agency they thought would be more sympathetic to their appeals, the U. S. Department of the Interior. They succeeded in winning the notice and support of high-level officials in the department. Speaking in Sacramento in February 1963, Undersecretary of the Interior James K. Carr warned against the destruction of California's natural beauty and cited the Bodega project as a disturbing example. Secretary of the Interior Stewart L. Udall stepped into the Bodega controversy in an even more direct and visible way. Udall was at that time writing a book, *The Quiet Crisis*, that was published later in 1963 and became a national best-seller. It was an eloquent appeal for greater attention to and action against environmental depredation. Udall did not take an antinuclear position; to the contrary, he proposed that by "allay[ing] our fears of fuel shortage once and for all," the work of atomic scientists was "the supreme conservation achievement of this century." But he also cautioned that "only prompt action will save prime park, forest, and shore line and other recreation lands before they are preempted for other uses." He viewed Bodega Head as a place that needed saving from such perils.[13]

On 18 February 1963 Udall wrote to Kermit Gordon, director of the Bureau of the Budget, to ask that procedures be established to allow Interior to review reactor applications to make certain that they complied with "the conservation efforts of this Department." He mentioned the Bodega site as one of particular concern. This led to a meeting between high-level staff members of Interior and the AEC. Representatives of both agencies concurred on the need for consultation and for strengthening existing lines of communication. The AEC already solicited the advice of the Fish and Wildlife Service, a part of the Department of the Interior, and its participants agreed that seeking the department's opinions on other issues would be useful. Harold Price pointed out, however, that the AEC lacked statutory jurisdiction over some matters of possible interest to Interior, such as the aesthetic impact of a nuclear plant. An Interior spokesman offered to draw up a draft memorandum of understanding that would specify how the coordination on questions of mutual concern would be carried out.[14]

Without waiting for an interagency agreement to be drafted, Udall announced his reservations about PG&E's plans for the Bodega plant. Like the Northern California Association to Preserve Bodega Head and Harbor, he focused not on land-use or aesthetic aspects but on the safety problems that the AEC was considering, particularly the seismic hazards. In a letter to Seaborg on 20 May 1963 he declared that the proximity of the reactor site to the San Andreas fault was "reason for grave concern." Udall suggested that a "very thorough investigation" of seismic conditions should be conducted, and he offered the services of experts from the Geological Survey, a part of his department. At the same time that he sent his letter to Seaborg he issued it to the press. The AEC was annoyed with Udall's press release, and pointed out in its reply that it was already consulting with the staff of the Geological Survey in its evaluation of the Bodega application. But it pledged to undertake further discussions with the Survey on "what further geological studies may be appropriate to assure that the geological conditions of the site are fully understood." Joint Committee member Craig Hosmer was more outspoken in his reaction to Udall's letter. He called it "specious and phony" and, noting that Udall aired his concerns about Bodega shortly after David Lilienthal criticized plans for the Ravenswood plant, complained that "Lilienthal and Udall are double-teaming nuclear power."[15]

Udall's letter enhanced the credibility and fired the enthusiasm of opponents of the Bodega station. It received prominent attention in northern California newspapers. At the request of the Northern California Association to Preserve Bodega Head and Harbor, opponents of the plant sent a flurry of letters to Udall to express support for his action. On Memorial Day, between 250 and 350 people gathered on Bodega Head to protest against PG&E's plans. Renowned jazz trumpeter Lu Watters, a local resident, came out of retirement to play for the rally. The highlight of the event was the release of 1500 balloons with warnings about radioactive fallout. "This balloon could represent a radioactive molecule of strontium 90 or iodine 131," each read. "PG&E hopes to build a nuclear plant at this spot, close to the world's biggest active earthquake fault. Tell your local newspaper where you found this balloon."[16]

As bumper stickers proliferated and radio stations played anti-PG&E ditties, a public information assistant in the AEC's San Francisco office observed that the campaign "by the Pesonen group . . . is unprecedented for its intensity in this area." It was also increasingly effective. The opponents of the Bodega plant received a boost when both the governor and the lieutenant governor of California announced support for their position. Governor Edmund G. (Pat) Brown told a press conference: "I don't like to see Bodega Head with a steam plant located out there in that beautiful place." Lieutenant Governor Glenn M. Anderson, in a letter to Seaborg, objected to the plant for aesthetic reasons but also emphasized safety concerns. "Confronted with the mounting conflict of professional opinion, the question of safety looms very large indeed," he observed. "I strongly urge that the Atomic Energy Commission withhold issuing a permit for operation of this unit at Bodega Head."[17]

Meanwhile, the AEC was seeking further information on seismic risks at the Bodega site. The regulatory staff might have hoped for expert appraisals that were prompt and clear-cut, but instead it got more delay and ambiguity. The consultants it hired did not agree on the severity of the earthquake hazards. After reviewing PG&E's reports, Nathan M. Newmark, a professor of civil engineering at the University of Illinois, concluded that it was "entirely feasible to design the proposed reactor to resist the maximum credible earthquake shock at the site." University of Washington seismologist Frank Neumann, however, contended that the utility was "employing every possible technical device to underestimate probable earthquake forces."[18]

The differing analyses of the AEC's consultants placed even greater importance on the field inspections carried out by the Geological Survey. But things were not proceeding smoothly on that front either. After receiving Udall's letter of 20 May, the AEC checked with Survey officials about what the Interior Secretary thought should be done that was not already under way. They conceded that Udall's letter had been out of date by the time he sent it and that his suggestion that the AEC consult the Geological Survey "did not involve anything in addition to the investigation" previously arranged. The Survey also informed the AEC that, contrary to its stated intentions, it could not submit its findings within two weeks after Udall's letter. The regulatory staff appealed to the Survey to conclude its work and internal reviews as soon as possible, but it took several months to complete even a preliminary report.[19]

The Geological Survey assigned two staff geologists in its Menlo Park, California office, Manuel Bonilla and Julius Schlocker, to conduct a detailed study of the site. They scrutinized the rock formations in the area, particularly those revealed in the ever-deepening hole (eventually reaching seventy-three feet below ground) that PG&E was digging. By 23 August 1963, Bonilla and Schlocker had found no evidence of recent faulting and had made no observations that seriously undermined the conclusions of PG&E's consultants. In the early stages of their investigation, the Geological Survey decided to send its senior seismologist, Jerry P. Eaton, to assess the damage that ground shaking from an earthquake might cause to a reactor at Bodega Head. Eaton made a one-day inspection of the site, but, as the AEC grew increasingly impatient, postponed writing his analysis of it.[20]

While the AEC and PG&E waited for a report from the Geological Survey, the seismic hazards of the site burst into the headlines. Pierre Saint-Amand, the seismologist who had inspected the area the previous April, formally submitted his findings to the Northern California Association to Preserve Bodega Head and Harbor. The organization promptly called a press conference on 29 August to air his conclusions. Saint-Amand described the location as "very poor" and added: "A worse foundation situation would be difficult to envision." He maintained that "one spectacular fault" was visible in the excavation. Saint-Amand's announcement was front-page news in major newspapers in the San Francisco Bay region and the feature story on television and radio stations.[21]

A few days later the AEC received information that gave credence to Saint-Amand's general conclusions, if not to his specific findings. While inspecting the wall of the excavation, Julius Schlocker of the Geological Survey and Don Tocher, PG&E's consultant, discovered what was clearly a fault line. Upon further examination, they saw that the fault extended into the bedrock at the site, about forty feet below ground. This suggested that the fault, at least in a geological time frame, was of relatively recent origin. It remained open to question, however, whether the fault should be considered active or inactive.[22]

PG&E and its supporters suffered yet another shock less than two weeks after the detection of the fault line in the hole where the reactor would be placed. On 25 September 1963 the Department of the Interior submitted the long-awaited preliminary report of the Geological Survey to the AEC. Based on the investigation as of 6 June, Schlocker's and Bonilla's conclusions were generally favorable to construction of the plant. But those findings, of course, had been superseded by the recent discovery. To make matters worse from PG&E's perspective, seismologist Eaton took a strongly negative position. Although he had found no unequivocal evidence that ruled out the site, he argued that existing uncertainties were enough to disqualify it. "Because we cannot prove that the worst situation will not prevail at the site," he wrote, "we must recognize that it might."[23]

5. Excavation for proposed reactor at Bodega Head, California. The harbor is immediately in front of the construction area; the mainland is in the background. (AEC Docket 50-205)

The AEC was surprised and PG&E was shaken by Eaton's report. Because he prepared it hastily after putting it aside throughout the summer, some of his colleagues in the Geological Survey were skeptical about it. But it created a sensation when the AEC released it to the press, along with the news that a fault line had been found in the site excavation. Once again seismic conditions at Bodega Head made headlines, including eight-column banners in some newspapers. PG&E denied that Eaton's views or the newly discovered fault line necessarily meant that the site was unsuitable. It described the fault as "a minor offset" that probably had not moved for tens of thousands of years.[24]

The new findings and conflicting expert opinions about the acceptability of the site made the controversy over the Bodega plant even more acrimonious. Pesonen called on the AEC to save PG&E from its "increasingly embarrassing mistake," and Adolph J. Ackerman, a consulting engineer from Madison, Wisconsin, wondered how "the management and directors of a distinguished company [could] justify abandoning their traditional standards of responsibility?" Supporters of the project struck back; an extreme example was an article in a Stockton newspaper that contended that the Northern California Association to Preserve Bodega Head and Harbor was a communist front organization.

In that atmosphere, the AEC still faced the problem of weighing divergent views of leading authorities on the seismic risks of the Bodega site. The situation had become increasingly unsettled. On the one hand, Saint-Amand and Eaton sharply criticized PG&E's plans, while on the other hand, the utility strongly reaffirmed its position. The AEC's own consultants were divided, and the extent to which Eaton spoke for the technical staff of the Geological Survey was unclear. Confronted with those muddled realities, the AEC did the only thing that seemed appropriate—it procrastinated and hoped for more certain guidance. The agency postponed any action until it received another report from the Geological Survey, and Seaborg announced: "We will take as long as required to get all the facts."[25]

The AEC received the Geological Survey's report, prepared by Schlocker and Bonilla, in January 1964. It did not categorically answer the key questions about the seismic hazards of the site. The two geologists were unable to determine whether the fault in the excavation should be considered active or inactive, or roughly when its last movement had occurred. They believed it was "an important zone of weakness" but could not be sure that "it would move in preference to other faults" on the headland. In the event of a major tremor on the San

Andreas fault, they thought it likely that the granite bedrock would rupture somewhere on Bodega Head.[26]

Schlocker's and Bonilla's findings were inconclusive but hardly reassuring. PG&E responded by insisting that the plant could be designed and built to operate safely. Its consultants reiterated that the foundations of the site were stable, the faults were minor, and the likelihood of significant ground movement was remote. The company received some unexpected and unsolicited support from two prominent geologists from the University of California at Berkeley, Garniss H. Curtis and Jack F. Evernden, who undertook their own investigation of the site. Although they both were initially "emotionally biased against use of the site," they changed their minds after their field inspection. They concluded: "We can see no reasonable objection to the site. . . . The evidence in the field makes it clear to us that no major displacement has ever taken place within the site area and none is to be expected in the near future."[27]

The new reports did little to ease the AEC's decision on licensing the Bodega plant. As PG&E grew increasingly disgruntled with the delays in the review process, the regulatory staff and the ACRS proceeded with their evaluations. Several staff members met with Schlocker and Bonilla, who took a somewhat more optimistic view of the hazards at the site than they had expressed in their written summary. Schlocker indicated that he thought that the chances of a ground rupture on Bodega Head in the event of an earthquake were "*very* low," but that it was a possibility. He added that he regarded the probability of movement along the fault in the excavation as even lower. Both Bonilla and Schlocker agreed with most other experts that an earthquake was most likely to occur near the line of the 1906 break, but they maintained that a major displacement could happen elsewhere. In short, they believed that an earthquake that would strike the proposed plant directly was a slight but nevertheless real peril. While the regulatory staff prepared its analysis, PG&E worked on design modifications that would accommodate a large displacement without impairing the reactor's containment system.[28]

In March 1964, as the AEC continued its deliberations, the controversy over the Bodega reactor was punctuated by an unexpected and disquieting event. A massive earthquake devastated southern Alaska; its magnitude, measuring 8.6 on the Richter scale, was the largest ever recorded on the North American continent. It was so great that it laid waste to the streets of Anchorage, reshaped the shoreline of several ports, and created huge tsunamis with enough force to toss fishing vessels inland.

Opponents of the Bodega plant saw the Alaskan earthquake as a harbinger of what could happen in California. "It is difficult to believe that PG&E can still seriously insist that 'engineering safeguards' will protect against earthquake damage," commented Pesonen. "Most of the buildings in Anchorage were 'earthquake proof' too." In a poignant scene at the U. S. Capitol, California Senator Clair Engle rose to introduce a resolution that urged the AEC to delay licensing the Bodega plant until it had "reasonable assurance [about] the geologic adequacy and seismic safety" of the site. He was attempting to make his first speech on the floor since undergoing brain surgery, but was unable to utter a word. After a few moments Senator Pat McNamara introduced the measure for him and aides helped him leave the Senate chamber.[29]

The regulatory staff, meanwhile, was preparing its evaluation of PG&E's application for submission to the ACRS, which was doing its own independent review. The staff was mindful of the implications of the Alaskan earthquake but more concerned about the judgments presented by Schlocker and Bonilla for the Geological Survey, particularly their suggestion that an earthquake could conceivably cause ground movement of one to three feet at or near the reactor site. After careful consideration, it concluded that PG&E had not shown that it could design the plant to withstand ground displacement of the magnitude that the Survey experts thought possible. Therefore, the staff told the ACRS: "Despite the fact that the risk of a large differential ground movement on Bodega Head is low, we do not believe that unproven design measures should be depended upon to solve this problem." The staff's opinion remained open to reconsideration, and it requested further information from the utility on a number of matters, including how vital components in the proposed plant would fare in the event of a major displacement and what action would be taken to protect against large tsunamis. PG&E engineers acknowledged that responding to the staff's questions would require substantial effort.[30]

The ACRS was still weighing the evidence and trying to arrive at its position. It was under no compulsion to agree with the regulatory staff, although the AEC preferred to speak with a single voice to reduce the chances of further controversy. The committee met with the staff and the utility in May 1964 but reached no decision. It took its own tour of the site the following month and held more meetings with PG&E's engineers and consultants. Even as the utility grew increasingly exasperated and discouraged, it attempted to answer the questions posed by the regulatory staff. It came up with a novel design to accommodate up to

three feet of ground movement in an earthquake. PG&E proposed placing the reactor on a base of special sand and allowing three feet of
clearance between the reactor's walls and the side of the hole in which it
would sit.[31]

In the end, it convinced only one of the two groups judging its plans.
The ACRS, based on its own review and the opinions of a structural
engineer it hired as a consultant, endorsed the utility's application in
October 1964. While acknowledging that details of the plant design still
needed to be completed, the committee accepted PG&E's arguments that
it had provided adequately for the consequences of ground displacement
or tsunamis. However, the regulatory staff, to the surprise of the ACRS,
still found PG&E's presentations unpersuasive. Once again, it decided
that the doubts raised by the Geological Survey were more compelling
than the assurances supplied by the utility. Reluctantly, it concluded that
although PG&E's design offered adequate protection from severe ground
shaking caused by a nearby earthquake, it was not convinced that the
plant could survive a sudden displacement of up to three feet in any
direction, which Survey experts thought could happen. The decisive consideration for the staff was that there was no way to test the design to
support PG&E's claims. "We do not believe," it wrote, "that a large
nuclear power reactor should be the subject of a pioneering construction
effort based on unverified engineering principles." Therefore, in the opinion of the staff, PG&E had not given "reasonable assurance" that the
plant could be operated without undue risk to the public. It regretted its
disagreement with the judgment of the ACRS, but it submitted that in
light of the conflicting views offered by leading authorities, "this is a kind
of case . . . on which reasonable men may differ." And in its view,
"Bodega Head is not a suitable location for the proposed nuclear power
plant at the present state of our knowledge."[32]

The breach between the regulatory staff and the ACRS surprised
even AEC insiders who were not directly involved in the regulatory
proceedings. "This is," noted Chairman Seaborg, "going to be a very
difficult decision for the Commissioners." On 27 October 1964 the
AEC released the text of both reports to the public. Opponents of the
plant were astonished and gratified; they claimed a major victory.
PG&E officials were stunned and company engineers were angry, but
they quickly decided to withdraw their application. One reason was
that Governor Brown delivered what *Nucleonics Week* called the *"coup
de grace"* at a press conference. Asked about his response to the AEC's
statement, he declared: "This nuclear danger is so great that you can't

take any chance whatsoever. And my immediate reaction is to say to PG&E: 'Let's go someplace else.' "

A second reason for PG&E's retreat was that it avoided the prospect of facing confrontational public hearings and a licensing board or Commission decision that might well, in light of the regulatory staff's decision, reject the application. On 30 October the company issued a statement that reaffirmed its own confidence in the site but added that it did not want "to build a plant with any substantial doubt existing as to public safety." It also emphasized its commitment to building nuclear plants at other locations. Six years after the initial announcement of its plans to place a power station on Bodega Head and after spending a total of $4 million on the project, PG&E was left with a great deal of frustration and a seventy-three-feet-deep hole in the ground.[33]

Plant opponents were obviously pleased with the outcome of the Bodega controversy. Pesonen sent a letter of apology to Harold Price "for any harsh words in the past." Praising the staff's "careful and sober assessment," he declared: "The entire performance was in the best tradition of the public trust." Price, still smarting from Pesonen's past attacks, replied with a one-sentence acknowledgment of receipt of the letter. But Pesonen's triumph was marred by uncertainty about the ultimate disposition of Bodega Head. Even after PG&E offered to lease its holdings to Sonoma County for recreational purposes, he warned supporters of his campaign that the utility might try again to build a reactor at the site after the opposition dispersed.[34]

Most proponents of the Bodega project and nuclear power in general did not view PG&E's abandonment of its plans as a major setback for the development of the technology. They stressed that the AEC's regulatory staff had recommended against the proposed plant for reasons that applied specifically to the site and suggested that the long-term prospects for nuclear power would not be adversely affected. *Nucleonics Week* went so far as to argue that the "decision . . . on the Bodega plant may prove to be a boon to the nuclear power industry." It reasoned that the regulatory staff's willingness to take a position inimical to the industry's wishes would enhance the agency's credibility and "should give a substantial boost to AEC's stock in the public mind." Some commentators, however, were not so optimistic. The harshest criticism came in an editorial in *Electrical World*. In contrast to *Nucleonics Week*, it contended that by publicizing an internal difference of opinion in "this sorry affair," the AEC "damaged public confidence in its competence."[35]

The controversy over the Bodega reactor highlighted a number of difficulties that the AEC faced in evaluating plant applications. It illustrated the problem of building reactors, especially ones of substantially larger size than those in operation, with little experience to draw on. Like the Ravenswood proposal, which the AEC was considering at the same time, the Bodega application required the staff to assess a controversial site, novel design features, and conflicting expert views. In the case of Bodega Bay, the burden of judging those issues was compounded by differing opinions and limited knowledge about earthquake risks. With those uncertainties, the divergence between the ACRS and the regulatory staff was understandable, though still something of a shock both inside and outside the agency. The ACRS found the plans and prescriptions of PG&E and its consultants, who were among the leading authorities in the world, convincing; the regulatory staff was more impressed with the reservations voiced by the Geological Survey. Neither could feel that its position was unassailable. The disagreement increased the stature of the regulatory staff and gave it greater parity with the ACRS as an independent review panel. It was viewed with more respect by Bodega opponents and with more suspicion by utilities and AEC offices that were pushing for expanded use of nuclear power.

The AEC was caught between its dual objectives of encouraging the development of nuclear power and ensuring the safety of the technology. In the cases of both Ravenswood and Bodega, it determined that the risks of the sites outweighed the benefits. But they were not easy decisions. Critics of the agency were justified in their complaints that it was inclined to grant construction permits to applicants. The AEC was under a great deal of pressure from its patrons in Congress to accelerate the nuclear power program and its own leadership was strongly committed to promoting the same goal. Yet the critics underestimated the willingness of the AEC to put developmental objectives aside when the safety of a proposed plant seemed doubtful. Every plant application carried an element of uncertainty, but in varying degrees, and Bodega crossed the nebulous line where the staff's concerns overrode its disposition to issue a construction permit.

Anxious as the AEC was to encourage utilities to build nuclear plants, it was also eager to win public confidence in its regulatory procedures. For that reason it took the protests of plant opponents and the views of the Department of the Interior seriously. Both suggested that there was not a "reasonable assurance" that the Bodega plant could operate safely, and the regulatory staff eventually accepted that argument. It was particularly responsive to the findings of the Geological

Survey; by contrast it found the protests of the Northern California Association to Preserve Bodega Head and Harbor annoying if not galling. But the one could not be separated from the other; Udall was at least partly influenced by the appeals of the association, and the Survey was at least partly influenced by the views of its superior. The regulatory staff was sensitive to public opinion, and on an issue on which the experts were hopelessly divided, the objections to the plant that opponents marshalled so effectively played an important role in the outcome, though one that was impossible to measure precisely.

The Bodega case did not resolve the policy questions it raised about seismic siting. Despite conflicting opinions about how the proceeding ended, informed observers agreed that placing reactors near fault zones remained an open and divisive question that seemed likely to elude clear or prompt resolution. Even as the controversy over the Bodega application played itself out, many of the same issues were being debated over a proposal for a reactor on the coastline of southern California.

In December 1962 the Los Angeles Department of Water and Power (LADWP), a public utility, announced that it had decided to build a 490 electrical megawatt nuclear plant near the community of Malibu. The reactor would be located on a 300-acre tract of land about thirty miles northwest of downtown Los Angeles and a few miles north of Sunset Boulevard and Muscle Beach. It would nestle in Corral Canyon beneath the rising heights of the Santa Monica Mountains. Before making its plans public, LADWP had requested a preliminary review from the AEC and the ACRS about the acceptability of the site. Both expressed concern about the population density in the surrounding area; it did not meet the agency's site criteria guidelines. They also raised questions about how the proposed plant would deal with the problem of earthquake hazards. But they offered tentative approval of the site, contingent upon the design of suitable engineered safeguards to protect against the dangers of earthquakes or other forces that might breach containment and expose the public to radiation injury.[36]

Perhaps mindful of the growing opposition to PG&E's Bodega plant and impressed with the way another utility, the Consumers Public Power District of Nebraska, had won support for a municipally owned reactor, LADWP undertook a major campaign to explain its plans for Malibu and to allay public fears. High-level utility executives contacted their counterparts in nearly forty city, county, and state agencies to inform them about the project. Company officials met with residents of the area around the proposed plant and commissioned a widely distributed pamphlet to provide details about it. LADWP also circulated the favorable views of prominent authorities. Smith Griswold, chief of the Los Angeles County Air Pollution Control District, for example, emphasized that the Malibu unit would produce electricity without increasing air pollution, a consideration that was especially appealing in a region plagued with smog. George W. Housner of Cal Tech, who had also consulted for PG&E, pointed out that the Malibu site was forty miles from the San Andreas fault and maintained that fault zones that were closer were inactive.[37]

LADWP's efforts deferred but did not prevent the growth of organized opposition to Malibu. Although at first there was little adverse reaction to the utility's plans, by August 1963 a group calling itself the Malibu Citizens for Conservation had formed to lobby against the proposal. The leadership and impetus in the Malibu protests came largely from property owners who worried not only about safety issues but also about the effect of the plant on real estate values. Their anxieties intensified when they learned that in the future LADWP might place as many as four reactors at the Malibu site. At a local planning commission meeting in November, residents of Malibu turned out in force to air their complaints. They included attorneys for comedian Bob Hope, who owned land next to the LADWP tract, singer Frankie Laine, and actress Angela Lansbury. Lansbury declared that the possibility of a nuclear plant in the area made her "hair stand on end," and added: "The two words 'atomic energy' are the most horror-packed words in the English language."[38]

A short time later, a tragedy in Los Angeles fueled the apprehensions of Malibu critics. As a result of ground shifting unrelated to seismic conditions, a reservoir maintained by LADWP burst open, releasing 250 million gallons of water, destroying many homes, and killing five people. One plant opponent expressed a common view when she raised questions about the utility's application for an AEC construction permit in a letter to California Senator Thomas H. Kuchel. "May I point out that this is the same company which stated that the Baldwin Hills reservoir was perfectly safe in a residential district," she wrote. "Ask those who lost their homes, and contact the five who were killed about the reliability of the L.A. DWP engineers." Those kinds of fears were further heightened by the Alaskan earthquake of April 1964. By then, the Malibu proposal had become a lively local political issue. In response to the growing protests spearheaded by the Malibu Citizens for Conservation, the Los Angeles County Board of Supervisors overruled an earlier zoning decision that had allowed LADWP to proceed with its plans for the Malibu unit. The board's ruling was of dubious legality, and it was soon contested by the City of Los Angeles. The local disputes underscored how complicated and controversial the licensing of the plant had become.[39]

In that atmosphere, the seismic conditions at the Malibu site emerged as a major source of contention. The AEC requested both the Geological Survey and the U. S. Coast and Geodetic Survey, a part of the Department of Commerce, to examine the geological and seismological characteristics of the area. The Coast and Geodetic Survey had also offered its views on Bodega Head, but it had not played a major role in those proceedings. As the two agencies undertook their studies, two private geologists who lived near the proposed site reached their own conclusions. Frank A. Morgan, a resident of Malibu and for many years the chief petroleum geologist for the Richfield Oil Company, argued that a "broad zone of extreme faulting" ran through the site. He suggested that the foundations were unstable and that substantial ground movement had occurred recently. He received support from another petroleum geologist, Thomas L. Bailey of nearby Ventura, who claimed that the "Corral Canyon site is extremely unstable and about as unfavorable as can be found in this region."[40]

Other experts, however, took a much more favorable position. LADWP's consultants conceded that there was a fault zone close to the plant site, but submitted that it was a minor offset that in the worst of circumstances would not cause a tremor of more than six on the Richter scale. They found no evidence of recent activity and contended that ground displacement would not be a problem at the site. They saw the primary threat to the reactor as ground shaking, or in their term, acceleration, from a distant earthquake. But they believed that the plant could be designed to withstand this possibility. Their views won the general endorsement of the two federal agencies that examined seismic conditions at the site. The Coast and Geodetic Survey focused on the dangers of ground acceleration and tsunamis; it recommended that provisions be made to protect against ground shaking and seismic sea waves of up to fifty feet above sea level. The Geological Survey investigators, Robert F. Yerkes and Carl M. Wentworth, were unwilling to describe the Malibu fault as either active or inactive because they thought that such terminology was meaningless, but they concluded that "the probability of ground displacement at Corral Canyon in the next 50 years is very low."[41]

Based on the information the AEC received from other agencies and its own consultants, the regulatory staff and the ACRS decided in favor of the Malibu application. Their final approval depended, however, upon the completion of a design that they believed would provide adequate protection against ground acceleration, landslides, and tsunamis, as well as reactor accidents. They felt comfortable that the containment and engineered safeguards were suitable only after a series of meetings with the utility and improvements in the design of the plant. This took until February 1965, and by that time, the seismic issue had been rekindled.[42]

The previous month the attorneys for the Marblehead Land Company of Malibu had submitted a report to the AEC that sharply challenged the evaluations of the LADWP and Geological Survey. The author was Barclay Kamb, a professor of geology and geophysics at Cal Tech, whom the land company had asked to perform a detailed study of the site. He agreed with the data and findings of the Survey investigators, but he strongly disputed their conclusions. He contended that their assessment of the chances of ground displacement contradicted the geologic evidence they presented. Kamb believed that earthquakes that caused major ground displacement at or near the site were not only possible but likely. He further dissented from the opinions of the Survey and LADWP by arguing that a tremor of a magnitude greater than six on the Richter scale was a "significant probability" in the Malibu fault zone. He also warned that a large earthquake as far away as the San Andreas fault could produce massive landslides at Malibu. "In relation to the possible range of exposure to fault hazards in southern California, the Corral Canyon site ranks among the more hazardous possible," Kamb declared. "It is well to remember that inattention to geological fact has repeatedly caused serious consequences in the Los Angeles area, of which the recent Baldwin Hill reservoir disaster is a striking example."[43]

Kamb's report did not reach the AEC in time to influence the regulatory staff's final hazards analysis of the Malibu plant. But it commanded a great deal of attention when the Atomic Safety and Licensing Board, in the next step in the licensing process after the favorable recommendations of the ACRS and the regulatory staff, held public hearings on the Malibu proposal. Four intervenors—the County of Los Angeles, the Malibu Citizens for Conservation, the Marblehead Land Company, and Bob Hope—formally opposed the application when the proceedings opened in Santa Monica in March 1965. The licensing panel was made up of chairman Samuel Jensch, chief hearing examiner of the AEC, Hood Worthington, a retired scientist and executive from the E. I.

DuPont Company, and Lawrence Quarles, dean of the School of Engineering and Applied Science at the University of Virginia. In a room packed with an overflow audience, they listened to the supporters and opponents of the plant present their cases. After six days of testimony, the board asked LADWP to expand its excavation work in search of evidence that might resolve the differences between the experts who offered their views. The hearings would reconvene after the new investigations were completed.[44]

The utility dug a new trench five feet deep and 250 feet long at the site, but it fed rather than ended the controversy among experts. The trench revealed a displacement, which plant opponents found ominous but which LADWP called insignificant. The Geological Survey investigators prepared a lengthy report that reaffirmed their earlier findings. They continued to assert that the chances of a major earthquake or ground displacement were remote. When the licensing board hearings resumed in July 1965, so did the debate over the seismic hazards at Malibu. The sessions stretched into the fall of 1965 and eventually consumed forty-one days, a record length for a licensing board proceeding. After listening to more than sixty witnesses during that time, the board seemed most impressed with the views of Barclay Kamb, who gave a three-and-a-half hour lecture on earthquake geology, and by LADWP's admission that the Malibu plant was not specifically designed to withstand ground displacement.[45]

One other issue, involving the Geological Survey's evaluation of the site, caused a stir during the hearings. When the regulatory staff had originally received the report of Yerkes and Wentworth, it understood that the Survey believed the hazards of the Malibu site to be substantially less than those at Bodega Head. But it was concerned that the language used in both cases was similar, and urged that the Survey replace the words "very low" with "negligible" to describe the probability of ground displacement at Malibu. Survey officials agreed, though the investigators themselves had reservations because they did not want to imply that the possibility should be completely ignored. The first public report of the Survey employed the term "negligible," while a later one added that this meant "the sense of very low." Lawyers for the intervenors grilled the regulatory staff on this matter and later suggested that the change of wording it prompted was an "act of irresponsibility" that raised "questions about the competence of the staff, its objectivity, and the extent to which it is vulnerable to pressures toward promoting the development of nuclear power."[46]

During the licensing board's deliberations following the hearings, the issue took on considerable importance. In asking written questions of the utility the panel seemed to adopt the interpretation that even if the hazards were "very low," they were not "negligible." One LADWP official was so disturbed by the board's language that he told members of the regulatory staff that the application was a "dead duck." If the board required the utility to redesign the plant to accommodate ground displacement, he thought the project would "go down the drain." The official's fears proved to be prescient. On 14 July 1966 the licensing board issued its decision. It denied the assertions of LADWP and the regulatory staff that the chances of ground displacement from an earthquake were low enough to be disregarded. It found that "the probability of faulting and permanent ground displacement is high enough so that we cannot conclude that there is reasonable assurance that no undue risk is involved." The board ruled in favor of a construction permit, but only after the utility modified its design to provide adequate protection against displacement.[47]

The licensing board's attempt to strike a balance between the competing claims on the seismic hazards at Malibu was frustrating and disappointing to all the parties involved in the case. One LADWP official groaned, "The agony isn't over yet." But the utility thought it could "live with" the decision and decided to pursue the application by revising the plant design. The intervenors complained that the board assumed that the plant could be built to withstand ground displacement and called for a flat rejection of the application. The regulatory staff took sharp exception to the licensing panel's ruling. Citing estimates that surface displacement had not occurred around the Malibu site for at least 10,000 years, it declared: "To deny Corral Canyon as a location for the proposed nuclear power plant . . . would, in the judgment of the regulatory staff, represent an unwarranted, extreme viewpoint which would not be consistent with the standards applied in other areas of nuclear power plant design." It reiterated that it found the chances of ground rupture to be so low that they "could be disregarded," though this was a more optimistic interpretation of the seismic risks than the Geological Survey investigators supported. The staff criticized the board for demanding "nothing less than absolute assurance that permanent ground displacement will not occur."[48]

For completely contradictory reasons, both the intervenors and the regulatory staff appealed the licensing board's action to the Commission. This placed the staff in the extraordinarily awkward position of asking the Commission to overrule a decision that LADWP had accepted. In March 1967, the commissioners denied the intervenors' petition and also rejected the staff's assessment of the earthquake hazards at the site. They based their ruling largely on their awareness, as Ramey told the Joint Committee, that earthquake science was "only in the early stages of its development" and that in geological time, 10,000 years was "not very long." In addition, the commissioners might have been more conscious than the staff of the political costs of approving the application. Moreover, since the boom market for reactors was well under way, they had little reason to be concerned about the effect of their judgment on the reactor development program. In any event, they upheld the licensing board's ruling and remanded the case to it. They instructed the board to conduct further proceedings to determine the suitability of the utility's new designs. LADWP continued work on plant modifications for three more years, but at a halting pace. The AEC had begun to prepare seismic site criteria, and the utility elected to wait for their completion before drawing up final plans. In June 1970, however, LADWP terminated its contract with Westinghouse and effectively killed the project.[49]

The Malibu case was in many ways a replay of the Bodega Head proceedings. Both applications foundered after citizen groups raised doubts about seismic hazards. Each generated major scientific controversies after experts who shared those reservations challenged the positions of professional colleagues who found the sites and plant designs suitable. In both cases, the critics cited enough evidence to convince some AEC officials that a license should not be issued, at least until more information was available. The regulatory staff relied heavily on the opinions of the Geological Survey in both proceedings. Although it went overboard in placing the findings of Yerkes and Wentworth on Malibu in the most favorable light, it insisted that it supported that site after rejecting Bodega Head because the Survey found the risks to be much less severe. And in both cases, the AEC, rather in spite of itself, was responsible for the termination of the application. The major difference between the rulings was that the regulatory staff that opposed Bodega thought the Malibu location to be acceptable, only to be overridden by the licensing panel and the Commission. Despite its commitment to rapid development of the nuclear industry, the AEC was not a bureaucratic monolith and its licensing decisions, at least in cases where a part of the staff had serious misgivings, were not foregone conclusions.

The importance of public objections in promoting debate over the

seismic risks of proposed plants and influencing the AEC was apparent
from the relative ease with which an application that did not face major
protests received approval. This occurred in the case of the San Onofre
unit, located about forty miles south of Los Angeles and fifty miles
north of San Diego. It was a 375 electrical megawatt facility built jointly
by the Southern California Edison Company, which financed 80 percent
of the costs, and the San Diego Gas and Electric Company, which
financed the other 20 percent. The plant was located on the beach
within the boundaries of the Camp Pendleton Marine Corps Base,
about two miles from the town of San Clemente.[50]

The proposed plant aroused virtually no opposition after it was announced in January 1963. Two individuals objected to its construction
during licensing board hearings, but they did not raise the seismic issue.
The U. S. Coast and Geodetic Survey and the Geological Survey examined the site and found little evidence of seismic activity. The utilities
agreed to design the plant to withstand ground acceleration greater than
any that had occurred in the past, and the question of surface displacement was not raised by intervenors or the AEC. Descriptions of the
seismic conditions at the site were similar to those at Malibu, and the
San Onofre proceedings had ended by the time that the chances of
ground displacement became a major issue there. The muted response
to the San Onofre plant was attributable to the fact that it was removed
from residential areas and distant from known major fault zones, and
although it occupied a scenic spot, the Marines had long kept it closed
to the public. At the same time that bitter debates over Bodega Bay and
Malibu were making headlines, San Onofre received a construction
permit with hardly a ripple.[51]

The controversies over Bodega and particularly over Malibu, highlighting the ambiguities in judging seismic risks and the disagreements
among leading authorities, spurred the AEC to develop seismic siting
criteria. The agency decided to provide direction to utilities on the
safeguards against earthquake hazards that it would expect, though,
like the general siting criteria, it offered flexible guidelines rather than
rigid requirements. As early as February 1964, while both the Bodega
and Malibu battles were still being waged, the Advisory Committee on
Reactor Safeguards expressed concern to the AEC that the seismic issues
that had arisen might cause "more conservatism than is necessary" in
the design of nuclear plants. A year later, in the wake of Bodega's
demise and the opening of licensing board hearings on Malibu, the
ACRS urged that, in light of the lack of consensus among experts on

seismic siting, the AEC undertake an effort to "provide a basis for determining earthquake protection requirements in a general way, if possible."[52]

In November 1966 the regulatory staff submitted a preliminary draft of seismic siting criteria for the consideration of the ACRS. It stated the AEC's basic policy at the outset: "A reactor should not be located within an active earthquake zone, nor at distances less than about 1/4 mile of the accepted boundaries of such a fault zone." That much was clear and undisputed, but on more debatable issues the staff also offered some specific guidelines. It conceded that existing knowledge did not make possible a precise distinction between an active and an inactive earthquake zone. It declared, however, that for the purposes of reactor siting, a fault could be viewed as inactive if it had not moved for at least 10,000 years. The same held true for "subsidiary, secondary, or sympathetic faults tributary to active fault zones." The paper further proposed that a nuclear plant be designed to withstand "without impairment of function" both ground shaking and displacement from the largest earthquake that could be anticipated in the area. All reactors should be built so that ground displacement of a few inches would have no effect, and in cases where displacement of up to five feet could occur, the AEC would evaluate the adequacy of the design. If more severe displacements seemed possible, the site would not be appropriate.[53]

The regulatory staff's draft was brief and straightforward. It got longer and more complicated after review by and a series of discussions with other agencies and the ACRS. In response to comments it received, the staff expanded and refined the definition of an active fault. A fault would be considered to be active if it demonstrated one or more of several conditions, including signs of "historic movement based on instrumental measurements," observations or reports of seismic activity of a magnitude of 4.0 or greater, indications that it had moved once in 35,000 years or twice or more in 500,000 years, or evidence of having caused surface rupture. In addition, the new draft listed the minimum distances that a reactor could be located from a fault zone according to whether it was designed to withstand ground displacement and according to the magnitude of the largest earthquake that could be expected in the region. In general, the paper emphasized that the seismicity of the entire vicinity and not just the plant site itself had to be taken into account. The draft made clear that its guidelines were necessarily imprecise and that although the staff intended them to be conservative, "some risk, however small, must be implicitly accepted."[54]

As it did in preparing all its regulations, the regulatory staff attempted to provide an ample margin of safety in its seismic criteria without imposing excessive requirements that would discourage utilities from building nuclear plants. Its success in achieving that goal was questioned by the AEC's Division of Reactor Development and Technology, headed by Milton Shaw. Shaw was a hard-driving veteran of Admiral Rickover's staff who was committed to encouraging the rapid growth of the nuclear industry. *Nucleonics Week* once described him as "probably without peer in convincing someone that nuclear power is to be embraced with little or no reservation." Shaw and members of his division were concerned that the draft seismic criteria would effectively rule out many sites and place unreasonably costly burdens on utilities.[55]

The Division of Reactor Development and Technology was also uneasy about the potential impact of the guidelines on reactor projects under consideration, especially a combination power and sea-water desalting plant that the AEC and the Department of the Interior were interested in supporting. For a time, the Bolsa Island reactor, to be placed on a man-made island just off the coast near Los Angeles, aroused great enthusiasm from President Johnson and Secretary Udall as well as the AEC, and Shaw did not want to see it undermined by the regulatory staff's seismic criteria. In addition, Pacific Gas and Electric was planning to build a 1060 electrical megawatt nuclear plant on the California coast at Diablo Canyon, near the town of San Luis Obispo. Although the focus of concern about earthquake hazards was on the west coast, fault zones to which the criteria would apply also existed in the eastern part of the country. The regulatory staff and the ACRS had already investigated seismic conditions around the Connecticut Yankee reactor in Haddam Neck, Connecticut, and fault zones around other eastern projects seemed likely to raise safety questions.[56]

After lengthy discussions with the Division of Reactor Development and Technology, the regulatory staff completed a revised version of the seismic criteria in January 1969. The new draft was more equivocal in defining seismic hazards than earlier ones. It introduced a new term to the consideration of seismic conditions by identifying geologic structures that were "capable" of causing surface displacement. They were fundamentally the same as those listed as characteristics of active faults in previous drafts, but the new version added that the existence of one of them did not necessarily mean that a structure was capable of producing displacement, and therefore, was unacceptable. It also eliminated the clearest and least controversial policy statement in the previous papers, declaring that a plant could not be built over an active fault. Although the provisions of the new draft made that highly unlikely, it did not include a specific prohibition. The regulatory staff's revisions satisfied Shaw's division, but they were greeted cooly by a group of industry representatives, who complained that the guidelines were too vaguely worded, too expensive to meet, and too likely to provide opportunities for intervenors. The AEC made further revisions, largely to clarify definitions and terminology, and issued the guidelines for public comment in November 1971. With some new sections on designing plants to withstand possible ground acceleration and rupture, the seismic criteria were finally added to the regulations in 1973.[57]

While the AEC was preparing the seismic siting guidelines, it was evaluating a few applications in which earthquake hazards were an important consideration. The Bolsa Island project was abandoned as too expensive, but PG&E's Diablo Canyon application stirred some opposition over seismic conditions. Inspections of the site revealed no evidence of major or recent faulting, and the AEC granted a construction permit for one unit in 1968 and a second in 1970. A controversy over seismic hazards at Diablo Canyon later developed after a fault zone of substantial proportions was discovered about two miles offshore. A similar situation occurred at San Onofre, where Southern California Edison and San Diego Gas and Electric planned to construct two more reactors of 1100 electrical megawatts each. Following an earthquake in Los Angeles in 1971, concern arose about a long-dormant fault within a mile of the site and about newly found faults offshore. The AEC suggested that the utilities design the plants to resist much greater ground acceleration than the smaller unit already in operation, but they protested, claiming that such a requirement was unnecessary and inordinately expensive. Eventually the utilities and the regulatory staff compromised on this issue, which led intervenors to question the safety of the design. The Atomic Safety and Licensing Board ruled that the design was adequate and the plants received construction permits in 1973.[58]

Seismic siting issues throughout the 1960s were complicated and inevitably controversial. They combined uncertainties about reactor safety with unknowns about earthquake behavior. The public was faced with the specter of an earthquake disaster being compounded in unfathomable dimensions by a nuclear plant catastrophe. The AEC attempted to guard against such an occurrence by imposing what it regarded as strict standards of siting and design where seismic hazards existed. But its judgment and the seismic criteria it prepared could offer no absolute assurances, given the differing views among leading experts in the fields of seismology and geology and unresolved questions about reactor safety and engineering. Despite the AEC's rejection of the applications for the Bodega Head and Malibu plants, it was willing to grant construction permits in other cases where it found the hazards less severe. The AEC was subjected to criticism in either case. On the one hand, plant opponents insisted that the AEC overemphasized its developmental functions at the expense of its safety responsibilities. On the other hand, utilities and vendors complained that the AEC tended to be overly cautious in its evaluations. With the state of scientific knowledge at the time, there was no sure way to resolve differences of opinion. The agency attempted to provide guidance to utilities on its position on seismic siting, but the criteria also generated protests and debate. As long as the experts on the probability and risks of seismic siting continued to disagree, the only thing that was certain about locating, or not locating, nuclear plants in the vicinity of fault zones was that it would continue to provoke controversy.

CHAPTER VI

Dilemma over Disasters

The Extension of Indemnity Legislation

The growth of the nuclear power industry would have proceeded much more slowly, if at all, without a government-sponsored indemnity program that insured nuclear plant owners against the consequences of a severe accident. Nuclear proponents in both industry and government frequently and publicly acknowledged that a catastrophic accident was a possibility, no matter what precautions they took or safety systems they installed. But they insisted that the possibility, though real, was remote. The problem they faced was that they could not be sure how remote the chances were or how much personal injury and property damage a plant disaster would cause. The estimates that experts offered were necessarily imprecise, and efforts to quantify both the probability and the consequences of an accident foundered because of the lack of data and experience. Those questions, which had been examined before Congress passed indemnity legislation in 1957, resurfaced in the mid-1960s as it considered extending the existing law.

The original law, sponsored by Senator Clinton P. Anderson and Representative Melvin Price and passed as amendments to the Atomic Energy Act of 1954, was intended to remove a major impediment to the development of nuclear power as well as to ensure that victims of a severe accident would receive compensation. Even before Congress opened nuclear technology to commercial applications by passing the 1954 act, Francis K. McCune, general manager of the Atomic Products Division of General Electric, had warned that progress in the industrial use of atomic energy would require insurance against a nuclear catastrophe underwritten by the government. After the new atomic law went into effect, other industry representatives presented the same argument. They pointed out that they could never provide absolute assurances that a major accident would not occur, and that a nuclear plant disaster could cause deaths, injuries, and damages that far exceeded the resources available to private insurance companies. Therefore, the growth of the nuclear industry depended on some kind of government program to augment the coverage that private underwriters could offer.

The AEC, the Joint Committee on Atomic Energy, and industry groups, all committed to the expansion of the use of civilian nuclear energy, investigated approaches to the insurance problem that might be adopted. Eventually, after much discussion and deliberation, they decided that the best alternative was a government indemnity program that would provide liability insurance to nuclear plant owners. Private underwriters made available up to $60 million of insurance against property damage and third-party claims—an amount that far exceeded previous coverage for any hazard. But it still was not enough to handle the potential costs of a severe nuclear accident. Therefore, the Joint Committee drafted legislation that would provide an additional $500 million in indemnity coverage. The dollar amount was a rather arbitrary figure suggested by James T. Ramey, then staff director of the Joint Committee, and accepted by Anderson. They sought to strike a balance between those who urged unlimited liability, including the AEC and industry representatives, and members of Congress who opposed giving plant owners a "blank check."

Before taking action on indemnity legislation, the Joint Committee asked the AEC to estimate the probability and the consequences of a major reactor accident. The agency, in turn, requested that scientists at Brookhaven National Laboratory, a research institution on Long Island that specialized in projects related to the peaceful applications of nuclear energy, conduct the investigation. Brookhaven was run by a consortium of nine universities (Cornell, Columbia, Harvard, Johns Hopkins, MIT, Pennsylvania, Princeton, Rochester, and Yale) under contract to the AEC. After several months of study, Brookhaven submitted its report, "Theoretical Possibilities and Consequences of Major Accidents in Large Nuclear Power Plants," to the AEC in March 1957. The document became better known by the shorthand label the AEC assigned to it: WASH–740.

The most prominent theme that WASH–740 emphasized was uncertainty. The investigators agreed that the probability of a catastrophic accident was "exceedingly low," but they added that the lack of operating experience with reactors did not provide a "dependable statistical basis" to estimate how low. The best they could offer was an educated guess that the chances of a major accident ranged from one in one hundred thousand to one in a billion per year for each operating reactor. Projecting the consequences of such an accident was equally problematical. The Brookhaven scientists considered three different cases. In the worst example they cited, they assumed that a large amount of radioactivity escaped into the environment under highly unfavorable conditions. In very unlikely but theoretically possible circumstances, they speculated that an accident could cause up to 3400 deaths, 43,000 injuries, and $7 billion in property damage in areas outside the plant.

After receiving the WASH–740 report from Brookhaven, the AEC submitted it to the Joint Committee, where its impact on the consideration of indemnity legislation was slight. The Price-Anderson amendments, featuring the $500 million government indemnity, passed the committee and the Congress with little debate. It stipulated that AEC licensees would be assessed an annual fee to help support the insurance program. In the event of an accident that caused more damage than the $560 million provided by the government and private companies, claimants would have to appeal to Congress for additional compensation.[1]

Price-Anderson authorized the AEC to provide indemnity protection to plants it licensed within a period of ten years. Unless Congress extended the law, the AEC could not offer the indemnity program to owners of reactors licensed after 1 August 1967. As enacted, the measure left some ambiguities about the scope of its coverage. One of them arose in 1963 when the Jersey Central Power and Light Company asked how the expiration date of the law would apply to a plant that had received a construction permit but not an operating license. The utility was preparing to award a contract for the construction of the Oyster Creek generating station, and it anticipated that construction would not be completed until 1968. The company informed the AEC that although its attorneys believed that Price-Anderson covered a facility that was granted a construction permit before the termination date, they also thought that the law could be interpreted to apply only to plants that had been issued operating licenses by 1 August 1967. Citing "the apparently unanimous agreement" among nuclear vendors that government indemnity insurance was "indispensable" to any reactor owner, the utility wanted to make certain that Oyster Creek would be covered even if Congress refused to extend Price-Anderson. It requested that the law be clarified to include beyond question any projects that received construction permits before the expiration date.[2]

In response to Jersey Central's inquiry, the AEC reexamined the provisions and legislative history of Price-Anderson. General counsel Joseph F. Hennessey reported that in his opinion the law applied to a plant that was issued a construction permit before 1 August 1967, even if it had not yet received an operating license. The Joint Committee reached the same conclusion. But to eliminate the ambiguity that concerned Jersey Central and other utilities, the AEC suggested legislative action, either in the form of a brief clarifying amendment or an extension of Price-Anderson beyond 1967. To deal with the immediate problem, it favored an amendment. At the same time the agency staff began to study the long-term issues of whether Price-Anderson should be extended, and if so, what changes should be made in the existing law.[3]

The addition of a clarifying amendment appeared to be a relatively simple solution to the problem raised by Jersey Central. To the annoyance of the AEC and the Joint Committee, however, it became more complicated after the intervention of supporters of coal interests. In response to the declining fortunes of the coal industry, a coalition of bituminous coal producers, railroads, mining equipment manufacturers, electric utilities, and the United Mine Workers of America had joined forces to form the National Coal Policy Conference (NCPC) in 1959. The purpose of the new organization was to promote the use of bituminous coal and the welfare of the industry. One approach it adopted was to protest what it viewed as favored treatment for competing industries, including nuclear power, by the federal government.[4]

In early 1963, the NCPC and other representatives of coal interests embarked on what *Nucleonics Week* called "no-holds-barred opposition to [the] AEC's civilian nuclear power program." In hearings before the Joint Committee, they denied that there was any pressing need to develop atomic power and called for an end to federal subsidies intended to encourage the growth of the nuclear industry. NCPC president Joseph E. Moody accentuated those points a short time later, declaring that "it is difficult to understand how Congress and the AEC can continue to try to justify spending hundreds of millions of dollars in subsidies to force-feed atomic energy power plants which are not needed." Nuclear spokesmen struck back. Chauncey Starr, president of Atomics International, for example, cited the hazards of air pollution from coal and asserted that an "atomic power plant may be 150,000 times safer than the routine operation of a fossil fuel plant."[5]

As the dispute escalated, the coal lobby added Price-Anderson to its list of complaints about nuclear power. After the AEC proposed its amendment to resolve Jersey Central's concerns about the expiration of the indemnity law, coal interests seized the opportunity to voice their objections. They argued not only that Price-Anderson was an unwarranted and improper subsidy for the nuclear industry, but also tweaked nuclear proponents by questioning why it was necessary. In March 1964, Congressman John P. Saylor, who represented bituminous coal regions in western Pennsylvania, introduced his own amendment to Price-Anderson. In contrast to the AEC's measure, his bill provided that a nuclear plant would not be covered by the indemnity law unless it received an operating license by 1 August 1967. "The atomic energy industry insists that atomic powerplants are safe and . . . the Atomic Energy Commission supports this claim," he declared. "Under the circumstances, there is no reason for the tax-paying public . . . to be forced to underwrite insurance for a commercial venture."[6]

Saylor's bill and his arguments nettled both the AEC and the Joint Committee. Chet Holifield suggested that "the Government would be reneging on its promises" and that "it would be a definite moral breach of faith . . . if this particular bill were passed." But Holifield, who had opposed the original Price-Anderson legislation, exhibited some lingering doubts about the indemnity program. In a hearing on the AEC's proposal to clarify the law to cover any plant that was issued a construction permit by 1 August 1967, he sought assurances from Commissioner Ramey that it was still needed to promote nuclear development and protect the public. Holifield also urged the AEC to review the question of whether Price-Anderson should be extended so it could be fully addressed before the 1967 deadline. Then, quite casually, he asked: "You will make an effort, then, to update the Brookhaven Study on reactor hazards which was used for the 1957 amendment between now and the time you come forward asking for any extension?" Ramey replied: "Yes, sir." The AEC's clarifying amendment to Price-Anderson received congressional approval within a short time. But the effort to secure its passage produced some unintended long-term results: tensions between coal and nuclear interests increased and the AEC committed itself to an update of the WASH–740 report.[7]

Shortly after the Joint Committee hearing, the AEC acted on its pledge to Holifield by making arrangements for a new study of the theoretical probability and consequences of a reactor accident. It followed the same procedures it had used in the preparation of the original WASH–740. Within the AEC, the regulatory staff assumed responsibility for completion of the report, and it, in turn, requested that Brookhaven National Laboratory undertake the technical evaluation and write a "semi-final draft." The laboratory would appoint a staff of "principal contributors" to conduct the study, and the regulatory staff would organize a steering committee of AEC officials to consult with the Brookhaven scientists. The two groups would then collaborate on the final version of the report.[8]

The agency official who was assigned the leading role in the WASH–740 update was Clifford K. Beck, deputy director of regulation. He was the logical choice for the task, not only because he had chaired the AEC's steering committee for the original report but also because he was the highest ranking member of the regulatory staff with the necessary technical qualifications. The oldest of eleven children, Beck grew up in the Piedmont region of North Carolina and worked his way through tiny Catawba College. He taught school for several years to help his siblings attend college, and then went to graduate school, earning a Ph.D. in physics from the University of North Carolina in 1942. He conducted research on the enrichment of uranium isotopes for the Manhattan Project during World War II, and after the war worked at the Oak Ridge Gaseous Diffusion Plant on the safe handling and storage of radioactive materials. In 1949 he moved to North Carolina State College as head of the department of physics. He persuaded the AEC to license the construction of the nation's first university research reactor, and after it began operating in 1953, his department awarded the first Ph.D.'s in the new discipline of nuclear engineering. Beck's achievements at North Carolina State won him wide recognition, but in 1955 he left to join the AEC's recently established regulatory staff.

Beck was the regulatory division's leader on scientific and technical issues, both in explaining them to the Commission and the Joint Committee and in serving as a liaison between policymakers and the technical staff. His role complemented that of Harold Price, who regularly discussed the legal and policy aspects of safety issues with the Commission and the Joint Committee but who generally deferred to Beck on technical matters. As chair of the steering committee for the WASH–740 revision, he brought a solid technical background, wide experience, and an international reputation as a pioneer in the field of nuclear engineering.[9]

6.   Clifford K. Beck (National Archives 434-PON-20-1)

In July 1964 Brookhaven agreed to undertake the technical reassessment of WASH–740 and promised to try to complete its report within three months. Kenneth Downes, a mechanical engineer who had served as Brookhaven's project director on the 1957 study, received the same assignment on the update. He enlisted a group of ten scientists to work on it, while Beck recruited a steering committee of the same size from various AEC divisions. The organization of the study went smoothly, but it quickly hit snags over the more troublesome matters of its ground rules, assumptions, and purposes. Within a short time, the inherent difficulties of conducting the review and divergent opinions on how to resolve them embroiled the Brookhaven and AEC experts in a series of disagreements.[10]

Some of the quandaries and complexities of preparing the report were apparent to members of the AEC's steering committee when they first met on 5 August. They agreed that "a maximum degree of objectivity" in the final report was desirable. They qualified that conclusion, however, by adding that it "must avoid the twin pitfalls of over-pessimism, which might produce great difficulties in gaining public acceptance of nuclear energy, and under-pessimism, which might appear to be a 'white-wash' of the problem." At the outset, then, the committee was keenly aware of the political implications of the study, even if it was not certain about what the findings would be. Some members thought that the revised WASH–740 would produce less disquieting results than the original. There appeared to be reason for optimism on this point. Since 1957, much more knowledge about and experience with reactors had accumulated from both commercial and experimental plants. Further, the development of engineered safety features and improved methods of containment could reduce the consequences of projected accidents.[11]

Other changes since 1957, however, could make the effects of an accident more severe. The most apparent difference was the greater size of plants. The original WASH–740 had based its estimates on a unit of 500 thermal megawatts, which was the equivalent of 100 to 200 electrical megawatts. Three operating commercial plants already exceeded that size, and several under construction or being planned were even larger. For example, Oyster Creek was rated at 515 electrical megawatts, Connecticut Yankee at 550, and San Onofre at 375. Larger plants would contain a correspondingly larger amount of radioactivity, the "fission product inventory," that could escape into the environment in an accident. Thus, the outcome of the revised WASH–740 would depend heavily on how its authors weighed those countervailing considerations and on how they calculated the probability of an accident.[12]

Within a short time after the Brookhaven team began working on the update, its assumptions and procedures stirred the concern of the AEC's steering committee. Citing their lack of expertise and the paucity of reliable data, the Brookhaven scientists declined to make any estimates of the probability of reactor accidents. They also decided not to consider the mitigating effects of engineered safety features in their study. As Downes reminded the steering committee, their assignment was, among other things, to calculate the consequences of the worst imaginable accident, which he defined as "the most pessimistic . . . that cannot be shown false." He added that strictly following those assumptions would produce "very horrible results" because of the larger size of nuclear plants. One such estimate was that an accident could cause as many as 45,000 deaths, and Brookhaven soon decided not to include death and injury figures because they were so problematical and certain to elicit a highly emotional reaction.

AEC officials complained that if Brookhaven refused to estimate probability or to consider the engineered safety features in newer reactors, there was no point in doing the new report. When they suggested that a number of recent experimental findings and technical improvements would provide a more optimistic assessment, Downes responded that the mitigating effects of those factors were either unproven, insignificant, or potentially defective. If containment worked as designed, for example, radiation would not escape into the environment. But if the containment shell were breached, which was entirely conceivable, the results of a major accident could be very serious. Another of the Brookhaven scientists emphasized that "unless some mechanism can be found to make their assumptions impossible, the numbers look pretty bad." When AEC officials probed for ways to get better results, David Okrent, who represented the ACRS on the steering committee, grumbled that his colleagues seemed "unhappy with the catastrophic results and . . . secretly hope some other group will supply optimistic probabilities." Beck agreed that "the results of the report could not be ignored just because they were not pleasant."[13]

Despite Beck's statement, meetings on the WASH–740 update underscored the differences between the Brookhaven scientists, who drew their conclusions from the available data without regard for policy implications, and the AEC staff members, who grew increasingly worried about the political impact of the new report. Stanley A. Szawlewicz, chief of the research and development branch of the AEC's Division of Reactor Development, expressed grave concern about the impact of publishing the update if Brookhaven's assumptions were used. His division was primarily responsible for fostering the development of civilian nuclear power and he feared that the revised WASH–740 would impede the growth of the nuclear industry. He argued that nuclear opponents would emphasize the worst-case aspects of the report in ways that would be difficult to refute. "No matter what statements are made on the incredibility of the upper limit accidents," Szawlewicz wrote to colleagues in his division, "those would be ignored by reactor siting antagonists, including the coal lobby." He suggested that the ground rules be changed from a study of the upper limit of damages from a theoretically possible but unlikely accident to a study of "reasonable accidents that are still considered incredible by most standards."[14]

Szawlewicz's complaints were echoed by others. William B. Cottrell, coordinator of the Nuclear Safety Program at Oak Ridge National Laboratory, told Downes that he was "singularly disappointed" that the Brookhaven study did not include information from experiments on reactor safety that he and his colleagues had conducted. He submitted that Oak Ridge findings on the release of fission products from a reactor would significantly reduce the consequences of an accident as projected by Brookhaven. He added his view that "your present study will be subject to much misunderstanding and misinterpretation and will have a net result that will be quite detrimental to the exploitation of the potential benefits of nuclear science." Representatives of the AEC's Division of Reactor Development and nuclear safety experts from Oak Ridge and the National Reactor Testing Laboratory in Idaho informally agreed that their best options for limiting the damage that the Brookhaven report could cause were to persuade the steering committee to guide the assumptions behind the study in a more "reasonable direction," and to seek additional data that might qualify the severity of Brookhaven's estimates.[15]

It became increasingly obvious that the update of the WASH–740 report, undertaken in response to what seemed to be an offhand question from Chet Holifield, had placed the AEC in an awkward dilemma. On the one hand, even though Brookhaven decided not to estimate the number of casualties, its projections of damages from an accident could provide support to the foes of nuclear power. Controversies over the licensing of Ravenswood, Bodega Bay, and Malibu had made the AEC acutely aware of the power and the concerns of an aroused public. Although opponents of those plants had rarely mentioned the findings of the original WASH–740, agency officials feared that the new study would offer compelling and explosive material for antinuclear spokesmen. They were particularly worried that the coal lobby would take advantage of the update in its campaign against the growth of the nuclear industry. The upper-limit projections of damages from an accident that Brookhaven would produce were certain to be alarming, and the coal lobby could be expected to publicize and emphasize them.

On the other hand, a worst-case estimate, or something like it, was required to advance the case for the extension of Price-Anderson. If the Brookhaven investigators took engineered safety features into account or allowed for the containment structure functioning as designed, the consequences of an accident beyond the boundaries of the plant were likely to be so low that the need for Price-Anderson would be doubtful. This too would lend credence to the position of the coal lobby. In that way, it could threaten the continued existence of a law that the AEC and other nuclear proponents regarded as essential for the development and welfare of the nuclear industry.

The Brookhaven investigators and the AEC's steering committee considered the dilemma raised by the WASH–740 revision at a meeting on 16 December 1964. Kenneth Downes outlined the model used for the worst-case analysis. It assumed that in a 1000 electrical megawatt reactor the core would melt and drop to the bottom of the pressure vessel, where the temperature would rise to more than 800 degrees centigrade within three hours. It further assumed that engineered safeguards would fail to perform and that the containment structure would be breached by a large hole. Under those conditions, Downes pointed out, an accident could spread radioactivity over an area the size of the state of Pennsylvania. He knew of no data, including the results of recent experiments, that substantially reduced the potential consequences. This voided one of the suggestions of the AEC's Division of Reactor Development for viewing the effects of the upper-limit accident in a more favorable light. The other suggestion—changing the ground rules of the study by estimating the effects of a somewhat less than worst-case catastrophe—proved equally unproductive. The idea never received serious consideration by the steering committee or the Brookhaven scientists, perhaps because the concept was inherently contradictory or because it would be an obvious departure from the original WASH–740.

Clifford Beck took the lead in suggesting ways to address the problems created by the WASH–740 revision during the 16 December conference. He told his colleagues that since the damages cited in the original report would be higher in the update and since an estimate of the effects of a worst-case accident was necessary for an extension of Price-Anderson, the key question was how to present the new information. This was especially important because the Brookhaven experts indicated that the upper-limit consequences would be forty to one hundred times worse than the 1957 study showed. Beck reemphasized the need to balance the dreadful effects of a catastrophic accident with the extremely low likelihood of it occurring. He promised to push work on accident probabilities, but he also acknowledged that the results would necessarily be imprecise and ill-suited to countering effectively the damage estimates.

The other option that Beck proposed to mitigate the impact of the new study was to obscure its findings. He recommended that Brookhaven prepare two reports. The first would be a summary of fifteen to twenty pages that stated its conclusions in general, qualitative terms, omitting the numerical estimates of the consequences of the worst-case accident. It would make clear that the results would be worse than the original WASH–740 showed without specifying how much worse. It would also refer to a second study, which would be published several months later and include detailed data, calculations, and explanations without relating them to the liability issue. In response to questions, Beck maintained that since this approach would indicate that the effects of a major accident would be considerably more serious than those estimated in the 1957 study, it would not harm the chances that Price-Anderson would be extended. "If it is reported that there is no reason to lower the consequences indicated by WASH–740," he said, "then Price-Anderson will probably be continued." Beck denied that he wanted to hide anything; he declared that both reports should be made publicly available. But his suggestion offered a way out of the WASH–740 dilemma. It could satisfy the Joint Committee and smooth the passage of an extension of the indemnity law without focusing attention on the frightening estimates of the potential consequences of a nuclear accident. Although some of the AEC and Brookhaven staff members seemed uneasy with Beck's proposal, they agreed to pursue it.[16]

While Brookhaven worked on a draft report on the consequences of a major accident to present to the steering committee, Beck received an analysis of the probability of such an occurrence. He had made arrangements with the Planning Research Corporation, which was developing methods of judging probability for other AEC projects, to perform the task. As he had anticipated, the company's conclusions provided little solace for those who were hoping that the probability estimates would offset Brookhaven's damage assessments. Basing their analysis on available data from a total of 1500 reactor-years of operation of various types of nuclear plants in the United States, the firm's experts determined that they were 95 percent certain that the probability of a catastrophe was, at most, one in five hundred per reactor-year of operation. They believed that the chances of a severe accident were actually smaller, but the data base was too narrow to offer a more precise or reassuring figure. If the probability of an accident were as high as one in five hundred, they declared, "there would be serious reservations about the safety of nuclear reactors." Therefore, until more data became available, estimates of probability would continue to depend upon the judgment of experts. In that regard, the situation had changed little since the preparation of the original WASH–740.[17]

After receiving the report from the Planning Research Corporation, Beck drafted a discussion of accident probability that he planned to include in the WASH–740 update along with Brookhaven's account of consequences. He acknowledged that the impressionistic estimates in the 1957 report could not be reliably quantified in the revision because of limited data. But he submitted that the chances of a catastrophic accident had been significantly reduced since that time because of enhanced understanding of reactor behavior, new engineered safeguards, and greater operating experience. He cited the estimates of the Planning Research Corporation, and added that its "quasi-probabilities," developed from the "expert judgment of knowledgeable persons," indicated that the probability of a reactor disaster was perhaps as low as one in a billion per reactor-year. But Beck also noted that hundreds of minor incidents had occurred in plants over the previous seven years, and warned that in several cases cracks and corrosion in pressure vessels, control rods, pipes, and valves could have led to serious accidents. He concluded that while "there is basis for confident belief that the likelihood of reactor accidents of sufficient severity to endanger the public is exceedingly low," there remained "the possibility that such accidents might occur."[18]

By late January 1965 Brookhaven had completed its draft on the upper-limit consequences of an accident. It itemized its assumptions and described the conditions under which a catastrophe could occur, but it carefully avoided making any numerical estimates of damages. Instead, it offered a summary statement that was concise, circumspect, and understated to the point of being misleading: "Given the occurrence of a reactor accident which is at least theoretically possible, we have found no reason to believe that the extent of damages would be any less than those estimated in WASH–740; conceivably the damages could be substantially greater." The Brookhaven investigators were more specific in the estimates that they planned to publish later in a separate document. They projected that an accident could contaminate with significant levels of radioactivity an area of 10,000 to 100,000 square kilometers and cause damages of $17 billion.[19]

After the sections on both the probability and the consequences of an accident had been drafted, Beck sent copies to members of the Atomic Industrial Forum's Committee on Reactor Safety to review. He did so largely because Brookhaven and AEC officials, particularly in the Division of Reactor Development, wanted to solicit the views of industry experts, who they hoped might offer some information or perspectives that would soften the potential impact of the report. The Forum had already been informed about the plans to update WASH–740; Commissioner John G. Palfrey had mentioned the new study in a speech to the organization's annual conference in December 1964. Beck discussed the drafts that he and Brookhaven had prepared at a meeting with the Forum's reactor safety committee the following month. He asked for guidance on two fundamental issues: whether the drafts overlooked important data that might modify their findings, and whether the results of the update should be published in detail or as "abbreviated summary statements."[20]

Predictably, the committee members were concerned about the study's conclusions. They posed questions about the basis for the estimates of probability and voiced objections to the assumptions used in the assessment of consequences. Beck, Downes, and others involved in the preparation of the report explained their ground rules and procedures exhaustively, if not always convincingly, to their audience. Although the discussion centered on technical matters, some of the Forum representatives focused on the larger issues the new study raised. Committee chairman Harold E. Vann asked, for example, whether one should conclude from the results that more reactors should not be built. Downes replied no, that the Brookhaven investigators believed instead that their findings showed that "complete reliance must be placed on engineered safeguards." Addressing the question of how to present the report to the public, the Forum committee suggested that the entire study be summarized in a few bland phrases stating that the effects of an accident in which engineered safety features failed completely would be "considerably worse" than the original WASH–740 had hypothesized. Beck did not think that such an approach was possible because too many people knew that Brookhaven had undertaken an extensive and detailed review.[21]

Industry spokesmen persisted in their appeals that the results of the study be withheld from the public. Reporting to Beck on the views of his committee shortly after its meeting on the WASH–740 revision, Vann recommended that the ground rules, assumptions, calculations, and results of the drafts be evaluated by the ACRS, the AEC regulatory staff, or contractors conducting reactor safety programs. "We question what purpose will be served by the report if the assumptions depart so far from reality as to make the analytical results meaningless," he wrote. While the drafts were being reappraised by other authorities, the AEC could submit a brief statement to the Joint Committee disclosing that the update of the 1957 study was not yet complete, and perhaps add that preliminary work indicated that the consequences of a catastrophic accident "appear to be somewhat greater." Many delegates to a meeting of the American Nuclear Society in February 1965 expressed even stronger objections to releasing the WASH–740 revision. They urged, according to one account, that it "be abandoned and not published, on the ground that the original report was quoted more often out of context than in, and thus did untold harm."[22]

The complaints of the industry reinforced the reluctance of Brookhaven and the AEC to publish their quantitative estimates of the upper-limit consequences of an accident. On 17 March 1965, with the drafts on probability and consequences largely completed and the deadline for submitting the update to the Joint Committee drawing near, Beck outlined the situation to the Commission. This came as unwelcome and perplexing news to the commissioners, who had not been advised previously about the progress of the revision or the problems it was creating. They did not want to release Brookhaven's worst-case projections, which they feared would seriously retard the growth of nuclear power. Yet they recognized that the WASH–740 update was tied to the extension of Price-Anderson, the expiration of which could be equally detrimental for the future of the technology. They also realized that some people outside the AEC and Brookhaven knew of the existence of the draft reports. With those considerations in mind, the Commission was inclined to accept Vann's suggestion that in place of a detailed report it send a brief letter to the Joint Committee. Beck prepared a draft letter that the AEC might submit to Holifield. It emphasized that a severe accident was "highly improbable" and said little about the potential consequences: "Calculations show that the upper limits in damages from this hypothetical sequence of circumstances would not be less, and under some circumstances could be substantially more, than the upper limits of the maximum consequence accident reported in the 1957 study."[23]

Although the commissioners seemed favorably disposed to accepting the suggestion of the Atomic Industrial Forum and limiting their report of Brookhaven's findings to a short letter, one member, John G. Palfrey, had second thoughts. After thinking more about the options available to the AEC, he decided that he disagreed with Vann. He urged that the WASH–740 revision be published. Palfrey, who had joined the Commission in 1962, was a veteran observer of and participant in atomic energy issues. His experience in the field began shortly after World War II with two years of research on legal and political aspects of the new technology at Princeton's Institute for Advanced Studies. He worked on the AEC's legal staff from 1947 to 1950, and then became a member of Columbia University's law faculty.

Palfrey was well-acquainted with the history and the legal complexities of the Price-Anderson Act. He had served as codirector of a committee that wrote an influential report for the Atomic Industrial Forum in 1956 on the need for a government indemnity program. Nine years later, he departed from the views of the Forum on the issue of how to handle the WASH–740 update. Palfrey placed it in the context of contemporaneous debates over metropolitan siting and the extension of Price-Anderson. He told his colleagues on 25 March 1965 that by considering licensing reactors in urban areas, the Commission was assuming that engineered safeguards were reliable. At the same time it was contending that Price-Anderson was still necessary. "This is tantamount," Palfrey argued, "to saying the risk of catastrophe is small enough to impose on large populations, but not small enough for the reactor manufacturers or operators to assume." He added: "If, on top of that, we defer, recast, or avoid publishing the Brookhaven reports, we will look still worse."[24]

The Commission equivocated. On 21 April Beck informed the steering committee that the Commission was weighing two alternatives for reporting to the Joint Committee. One was a "Short Form," consisting of a brief letter from Brookhaven to the AEC outlining their findings in general terms and the letter that Beck had drafted from the AEC to the Joint Committee. The second option was a "Long Form," which would include the full text of the reports drafted by Beck on the probability of accidents and by Brookhaven on the consequences. Neither the "Short Form" nor the "Long Form," however, would contain Brookhaven's quantitative projections of the damages from an upper-limit accident.[25]

As the Joint Committee's hearings on the extension of Price-Anderson drew near, the Commission decided against using either the "Short Form" or the "Long Form." It elected instead to send a two-page letter to the Joint Committee that provided even less information than the "Short Form." Before sending the letter, Seaborg discussed its contents with Holifield and showed him a draft. After receiving Holifield's endorsement, Seaborg sent a slightly revised version of Beck's proposed letter to the Joint Committee on 18 June. The AEC chairman stressed the low probability of a catastrophic accident, but added that "we cannot say . . . that the likelihood is nonexistent." He affirmed that the consequences of such an accident could be "substantially more" than the original WASH–740 had shown, and concluded: "Thus in our opinion, the answers to your two questions—that the likelihood of major accidents is still more remote, but the consequences could be greater—do not decrease but rather accentuate the need for Price-Anderson extension."[26]

The Commission acted to prevent information about Brookhaven's and Beck's draft reports from becoming public. Ramey learned that reporters for *Nucleonics* "had trapped" Brookhaven scientists into admitting the existence of their study. He called editors at the journal and persuaded them "not to publish anything regarding the report." The Commission also agreed to approach industry groups that had asked to review the completed document and explain to them that no final report would be published.[27]

The two main proponents of publishing at least the general results of the WASH–740 revision, Beck and Palfrey, went along with the Commission's decision not to release it or even acknowledge its existence. The letter that Seaborg sent to the Joint Committee deliberately obscured the work that had gone into preparing the draft reports, but it accurately, if delicately, summarized the conclusions that the task force had reached. Brookhaven and the AEC had made a conscientious attempt to improve on the 1957 study, but their effort, like the original WASH–740, had foundered because of a lack of data and operating experience. The update used more sophisticated computer analyses that provided some new perspectives on fission-product release and atmospheric dispersion of radioactivity following an accident, but they did not substantially change the estimates of the 1957 report. The new study affirmed the obvious fact that if safety systems failed completely, the effects of an accident in a large reactor would be more serious than in a small one. It did not, however, provide any major methodological or informational breakthroughs. By refusing to release the new report, the AEC did not deprive scientists or engineers of vital new insights about reactor safety. Therefore, the benefits of publication seemed small compared to the costs—handing nuclear critics an issue they would surely exploit.

In the end, Beck and Palfrey accepted the view that even an abbreviated version of the WASH–740 update should not be released, doubt-

lessly with some mixed emotions. Once they reached that conclusion, it
was a short and logical step to participating in the effort to obfuscate
the detailed findings and outcome of the study. It would be difficult if
not impossible for the AEC to avoid releasing the findings of the draft
reports if their existence became widely known. Beck explained to members of the steering committee, who had not been involved in the deliberations over what to do with the draft reports, that the "final handling" of them required "a long series of complicated maneuvers and
negotiations . . . to arrive at a final written document which would be
acceptable to all responsible parties." He told them that the update had
"not in fact, been completed," which was technically accurate but less
than a candid and full accounting.[28]

Palfrey, who clearly recognized the potential costs of covering up
Brookhaven's worst-case estimates, gave misleading answers to an inquiry about the WASH–740 revision. In August 1965, David Pesonen,
well-known to the AEC for leading the campaign against the Bodega
Bay plant, asked him for a copy of the report, which Palfrey had said
was under way in his speech to the Atomic Industrial Forum the previous December. Palfrey replied with a copy of Seaborg's 18 June letter to
the Joint Committee, but Pesonen was not satisfied. He reminded Palfrey that in the December address he had mentioned that a new study
was being prepared. Pesonen wondered when it would be available to
the public. He thought that Seaborg's letter to Holifield suggested that
the update had been completed, and he added: "I get the impression
that your letter to me is not entirely responsive to my original request."
Palfrey answered that "there was a review of the 1957 study," but that
"no new report is in existence or contemplated."[29]

Pesonen was not convinced. In October 1965 he published an article
in The Nation that accused the AEC of suppressing the WASH–740
update. He reviewed the findings of the 1957 report and stressed the
objections of industry representatives to publishing the results of the
new one. Pesonon concluded: "For public relations reasons, as well as
for legislative success in extending Price-Anderson, the AEC appears to
have abandoned or suppressed the updated report—a major research
project of potentially widespread public importance." The AEC responded to inquiries stirred by the article by restating what Palfrey had
told Pesonen. It did not cite the extent of the Brookhaven investigation
or the findings it produced.[30]

The WASH–740 revision began in the summer of 1964 as an effort
to discover or to project the probability and the consequences of a

severe reactor accident through the best scientific methods available. As work on the study continued and hopes that it would produce favorable results vanished, the AEC became progressively less forthright about its status and conclusions. It ended a year later with a bland and understated report. The AEC's concern about the impact of publishing Brookhaven's worst-case estimates was understandable and its decision to suppress them was probably inevitable. But it was also unfortunate. The AEC had sound reasons for fearing that opponents of nuclear power would exploit the figures to alarm the public. The agency compounded the problem, however, by consulting with the nuclear industry while refusing to inform the public fully about the findings of the new study and by dissembling about its very existence. Representatives of the nuclear industry and the AEC's Division of Reactor Development strongly protested plans to publish the WASH–740 revision, even in a watered-down form. Advocates of a fuller and franker presentation eventually agreed that the disadvantages of following their recommendations were greater than the possible benefits. As Beck told the steering committee, the final letter was sent to the Joint Committee because it was acceptable to all interested parties.

The AEC's decision was inept as well as ill-advised. Beck, Palfrey, and perhaps others were aware that the effort to revise WASH–740 was no secret. It was well-known within the nuclear community and by the nuclear opponents the AEC was so worried about. The AEC might have considered this problem largely resolved by an informal arrangement it made with the coal lobby in the spring of 1965. They reached an understanding in which agency representatives would refrain from attacking coal-fired plants for causing air pollution if coal interests would refrain from attacking nuclear power on safety issues. The unwritten agreement reflected a perspective expressed by Palfrey in May 1965: "I think the interests of no one will be served if coal and atomic energy were to go at each others' throats—one crying pollution of the air, the other government subsidy and risk of atomic catastrophe." Although the coal interests continued to contest Price-Anderson as an unwarranted subsidy to the nuclear industry, they promised to suspend their objections on the grounds of safety. Whether the AEC consummated the arrangement specifically with the WASH–740 revision in mind is unclear. But at the least, it smoothed the decision not to report Brookhaven's worst-case projections by removing the AEC's primary concern that the coal lobby would ask embarrassing questions about the existence and/or the findings of the update.[31]

Yet by refusing to air the worst-case estimates, the AEC handed other nuclear opponents an emotional issue by default. The scientific evidence was necessarily imprecise and inconclusive, and it would have been difficult for the AEC to explain the deficiencies in its data and the assumptions used to arrive at its upper-limit accident consequences to the public. The alternative, however, was that it might be forced to do so under much more awkward circumstances if the report became public knowledge. It is unclear how extensively such considerations were weighed as the Commission sought a way out of its dilemma, but it seems apparent that satisfying short-term needs was the only course of action that could win a bureaucratic consensus within the AEC. As happened on other occasions throughout the history of the AEC's civilian nuclear programs, the agency's commitment to nuclear development compromised the integrity of its regulatory program.

At the same time that the AEC was laboring, and agonizing, over the WASH–740 update, it was reviewing the question of whether Price-Anderson should be extended beyond its 1967 expiration date. In March 1965 the agency submitted a lengthy report on the subject to the Joint Committee on Atomic Energy. It concluded, not surprisingly, that the continuation of Price-Anderson coverage was still needed to encourage industrial participation in atomic energy activities. The AEC advised the Joint Committee that industry representatives with whom it had consulted unanimously expressed the same judgment. Although it maintained that experience with the act since 1957 had not revealed any major flaws, it identified a few areas that required clarification or revision.

The most glaring shortcoming involved the procedures that victims of an accident would have to follow to make claims. Under the Price-Anderson act, claimants would have to use existing state tort laws to prove negligence or fault before they received compensation. The AEC suggested that this would place an unwieldy and expensive burden on accident victims. It would also create confusion, uncertainty, and public outrage that "would greatly magnify the setback to the Nation's nuclear power program." The AEC submitted that several methods were available to redress the problem and it urged that "a precise rule of law at the Federal level be established." It further recommended that other questions, including the application of statutes of limitation to injuries from nuclear accidents, the "discovery period" in which a person had to submit a claim for injuries or damages, and liability coverage for accidents occurring in the transportation of nuclear materials, receive due consideration by the Joint Committee. It did not believe, however, that any of those items was so urgent that it had to be addressed before Price-Anderson was extended.[32]

The Joint Committee held three days of hearings on Price-Anderson in June 1965. The sessions followed predictable patterns. The AEC reiterated its appeal for extension of the law and for clarification of the issues it cited in its report. Representatives of nuclear utilities and vendors unanimously stressed that Price-Anderson was essential for the continued growth and vitality of the industry. The only major subjects of dispute were the amount of insurance that private underwriters were willing to offer and the objections to the law that the coal lobby raised.[33]

Both the AEC and the Joint Committee had expressed hope that the amount of private insurance available to reactor owners could be increased in light of the experience gained and the safety record exhibited by the nuclear industry since 1957. Holifield was particularly insistent on this point. In a speech he had delivered several months earlier he had argued that even though private insurance companies were not yet able to assume full liability coverage for nuclear power, they should be willing to increase their share from $60 million to $100 million, allowing the government's liability to be reduced by the same amount. "Ultimately," Holifield declared, "I would hope that we could completely dispense with the need for governmental indemnity and rely entirely upon the private insurance market." To his disappointment, the two private nuclear insurance pools reported that they would augment their coverage by only $14 million. They explained that their reserve fund was insufficient to handle a higher amount, or in their parlance, to provide "an adequate spread of risk against possible catastrophic incidents." Holifield grilled insurance pool executives about their policies at length, and was partly placated only by their pledge to try to reach the $100 million mark within ten years.[34]

The coal lobby generally adhered to its agreement not to attack nuclear power on safety issues. Representatives of the National Coal Policy Conference and the other major trade association, the National Coal Association, mentioned the WASH–740 revision only once in passing, and other than an occasional query about why Price-Anderson was necessary if nuclear power were safe, focused their criticism on two issues. One was that the law provided inadequate protection to the public because of what they termed its "no-recourse provision." Brice O'Brien, general counsel of the National Coal Association, contended that Price-Anderson deprived citizens of their right to sue manufacturers and owners of nuclear plants for damages in the event of a

serious accident, and suggested that, therefore, the law was unconstitu-
tional. He called for the elimination of "no-recourse" by making companies liable for damages that exceeded the $500 million of govern-
ment insurance. The AEC responded that in the unlikely event of a
catastrophic accident, victims would have a recourse. The legislative
history of Price-Anderson made clear that they could appeal to Congress for additional funds if necessary. The AEC also argued that the
limitation of liability included in the Price-Anderson act was entirely
constitutional.[35]

The other objection that coal representatives emphasized was that
Price-Anderson granted a "massive subsidy" to the nuclear industry.
They complained that even in the absence of a major accident that
would draw funds from the U. S. Treasury, nuclear plant owners received liability insurance from the government that was much less expensive than what private insurance, if available, would cost. NCPC president Joseph Moody urged that the fee that the government collected be
raised substantially, adding that "some system must be devised whereby
the Nation's taxpayers are not held in escrow to nourish the nuclear
power industry."

Joint Committee members did not deny that the relatively small fee
assessed plant owners for Price-Anderson protection was a subsidy, but
they maintained that it was a legitimate and justifiable means to foster
the development of a new source of energy. They pointed out that coal
and other fuel industries also received subsidies from the federal government. Having made their points, the coal spokesmen harbored no illusions that they would carry much weight in the deliberations over extending Price-Anderson. Moody declared, to laughter in the hearing
room, that "appearing before this committee is one of the more enjoyable things that happens to me." He continued, to more laughter: "I will
say that sometimes I think it is also one of the most ineffective things I
do." The Joint Committee unanimously recommended a ten-year extension of Price-Anderson in August 1965 and Congress overwhelmingly
approved the measure a short time later.[36]

In addition to providing indemnity protection for another ten years,
the new law revised the 1957 version by specifying that the government
share of the $560 million coverage would be reduced by the amount
that private insurance increased. The Joint Committee reiterated its goal
that eventually "complete reliance could be placed upon the private
insurance market to provide insurance coverage for [the nuclear] indus-

try." It took no action on the question of procedures for making claims in the event of an accident. The committee announced, however, that it intended to address the issue "as early as practicable" and to make certain that an accident victim "will not be subjected to a series of substantive and procedural hurdles which would prevent the speedy satisfaction of a legitimate claim." In November 1965 it submitted a list of questions to the AEC regarding the possible legal and procedural complications of the problem.[37]

The AEC had been considering for some time different approaches to ensure that victims of an accident received compensation promptly if an accident occurred. As written, the Price-Anderson act depended upon the application of state tort laws to provide compensation to the public after a nuclear accident. This meant that a claimant might be required not only to prove damage or injury as a result of the accident but also negligence on the part of the reactor owner or some other party. Even if proof of negligence was not necessary, a claimant might have to overcome a number of other legal arguments that defendants might use. The result could be to frustrate the purpose of Price-Anderson by denying damages to accident victims. As David F. Cavers, a professor of law at Harvard University and a leading authority on the legal aspects of atomic energy, warned in an article published in 1964: "If, through failure to prove fault where that is required or through the interposition of defenses where it is not, no person could be held legally liable for a nuclear incident, then none of the hundreds of millions of dollars that the Price-Anderson Act has provided would be available to compensate the injured public." Even if claimants won their case, they could be subjected to years of litigation and delay.

Cavers and other experts suggested several ways to deal with the problem. One was to impose, as a condition of Price-Anderson protection, a requirement of "absolute liability" on plant owners. With liability assigned to the licensee, a claimant would not have to prove negligence or fault under state tort laws to recover damages under Price-Anderson. This option could not be carried out, however, without legislative action by the states, and they had shown no interest in adopting laws on liability for nuclear accidents. Another alternative was to pass a federal tort law that placed absolute liability on plant owners. But this raised delicate issues of federal-state relations, especially the power of the federal government to assume responsibilities normally handled by states. A third possible approach was to require that nuclear plant licensees agree, in order to receive Price-Anderson coverage, to a "waiver of defenses." This simply bypassed the issue of liability by stipulating that plant owners would not use proof of negligence or other defenses available to them under tort law. It would avoid many of the legal entanglements that could frustrate the intentions of the Price-Anderson law.[38]

In a series of meetings of representatives of the AEC, Joint Committee, nuclear industry, and insurance companies, it quickly became apparent that the "waiver of defenses" was the preferred approach to assuring prompt payment to accident victims. It offered a means to settle the problem without revising or overriding state tort laws. It also skirted the imposition of liability on plant owners, who were reluctant to accept the stigma or to endorse the legal precedent of federally mandated absolute liability. Claimants would not have to show negligence or fault to collect compensation, though they still would have to prove that a nuclear accident was the cause of the injury or damages for which they sought payment. The statute of limitations, which generally applied after three years, could not be used as a defense against claims until ten years after a nuclear accident.[39]

The major concern of the insurance pools about the waiver-of-defenses approach was that it would encourage "spurious claims" and "nuisance suits" from persons who contended that they had suffered radiation injury from the routine operation of a nuclear plant. Insurance company representatives urged that the AEC make it clear that the waiver of defenses would apply only to "sudden and identifiable" incidents. After further discussion with nuclear and insurance industry officials, the AEC developed the term "extraordinary nuclear occurrence" to indicate a threshold for claims under Price-Anderson. The concept of an extraordinary nuclear occurrence alleviated the anxieties of the insurance pools, but the definition of it remained an open and elusive question.

When the Joint Committee held hearings in July 1966 to consider amendments to Price-Anderson that would incorporate the waiver-of-defenses proposal, AEC general counsel Joseph Hennessey outlined what the AEC thought would constitute an extraordinary nuclear occurrence. It must be, he explained, an "event" in which something "identifiable" happened, which released radioactivity in amounts "substantially in excess" of AEC regulations, and which caused off-site property damage of more than $5 million. In response to questions, however, he suggested that even damages of less than $5 million could be classified as an extraordinary nuclear occurrence. When Congress passed the amendments to the indemnity law, it gave the AEC broad discretion to determine what it meant by an extraordinary occurrence.[40]

In May 1968, after further discussions with Joint Committee, Department of Justice, nuclear industry, and insurance representatives, the AEC published for public comment a draft regulation that defined "extraordinary nuclear occurrence." To qualify, and to determine whether a waiver of defense would take effect, an accident had to meet two conditions: it had to release a "substantial" amount of radioactivity "from its intended place of confinement," and it had to cause "substantial damages to persons offsite or property offsite." The AEC went into detail to explain what it meant. Under the first criterion, it decided that a "substantial" release of radiation had occurred if one or more persons was exposed to a projected whole body or bone marrow dose of 20 rem (a unit to measure radiation exposure), a thyroid or other organ dose of 30 rem, or a skin dose of 60 rem. Alternatively, a release would be "substantial" if a total of 100 square meters of property had been contaminated at certain specified levels for different kinds of radioactivity.

Under the second criterion, the AEC defined "substantial" as an accident that caused, within thirty days, the death or hospitalization of ten or more people, $2.5 million in damages to a single individual or $5 million in the aggregate, or $5,000 in damages to fifty or more individuals. The AEC emphasized that a determination of an "extraordinary nuclear occurrence" under its criteria did not imply a judgment on whether a claimant would receive damages; it simply meant that the waiver of defenses would go into effect. In cases that did not meet the conditions for an extraordinary occurrence, claimants could still proceed under tort law. After receiving comments on its draft regulation, the AEC made a few revisions. They included lowering the requirement in the second criterion for the number of people killed or hospitalized from ten to five. The new regulation became effective in November 1968.[41]

The issues of waiver-of-defense and extraordinary nuclear occurrence did not attract much attention. They were approaches to a problem that resisted easy resolution and that would arise only in the event of an accident that was regarded as highly improbable. They did not fully satisfy all of those involved in the negotiations and drafting of the regulations. Some critics thought the damage figures for declaring an extraordinary occurrence to be too low while others opposed the entire concept as burdensome to claimants. But, in response to the concerns of legal experts and the Joint Committee, the AEC worked out a solution that most authorities found acceptable and that provided important protection to the public. Despite any ambiguities in the new regulations, they were a definite improvement over the uncertainties and potential injustices that existed before they were formulated.

CHAPTER VII

Reactor Safety

Growing Concern over Larger Reactors

The scientists who worked on the revision of the WASH–740 report in 1964–1965 had considered and debated the probability and theoretical consequences of a major reactor accident. They conceived a worst-case accident that could not be dismissed as impossible but that they viewed as highly unlikely. As nuclear vendors designed and utilities placed orders for increasingly larger power reactors, however, safety experts became increasingly troubled that new problems might arise. The growth in the size of proposed plants was accompanied by a corresponding growth in concern about the possibility of severe accidents occurring in ways that the authors of the WASH–740 update had not envisioned. Although neither the AEC's regulatory staff nor the Advisory Committee on Reactor Safeguards viewed the complications that the new plants presented as beyond resolution, they found them serious enough to reexamine their assumptions about reactor safety and to make revisions in their regulatory requirements.

In judging the safety of proposed nuclear plants, the AEC staff and the ACRS had traditionally relied on two methods of protecting the public from the consequences of a reactor accident—remote siting and engineered safety features. After the early 1960s, remote siting received less emphasis; utilities campaigned, with some success, to place reactors near population centers and a few proposed, with less success, to build them at close-to-downtown urban sites. As the stress on isolated locations diminished, the dependence on properly functioning engineered safeguards necessarily grew, and the trend toward larger plants under-scored their indispensability. But even as the relative importance of engineered safety features increased, questions arose about their reliabil-ity in preventing a massive release of radioactivity into the environment as a result of a severe accident.

Engineered safety features served two basic purposes: first, to pre-vent an accident, and second, to limit the damages and the consequences if a reactor accident did occur. The function of the reactor in a nuclear plant was to generate heat and produce steam to drive turbines and create electrical power. The heat in a reactor came from the fission of the nuclei of uranium atoms. During normal operation, the core of a light-water reactor, which contained the uranium fuel pellets, was cooled by the circulation of water. The principal danger involved in the production of nuclear power was that if the circulation of coolant was cut off by some malfunction, the reactor could overheat, overwhelm its safety systems, and release radioactive materials, or "fission products," into the environment. This was the problem that occupied the attention of safety experts by the mid-1960s.

Although the design of and safety features in nuclear plants varied greatly, they all used a system of control rods to operate the reactor, manipulate the level of power, and prevent accidents. Control rods provided the first means of defense against the effects of an "excur-sion," a term that safety experts often applied to an unplanned increase in the rate of nuclear fission, and, as a consequence, of heat in the core. The control rods contained elements, such as boron or cadmium, that absorbed neutrons, the particles that caused fission by colliding with atomic nuclei. When the rods were inserted into the core they stopped the fission process. They automatically shut down, or "scrammed," the reactor, for example, if the power level rose above the designated set-tings or if excessive heat or pressure was present in the core. The reactor could also be scrammed manually by plant operators. The rapid inser-tion of the control rods in response to an indication of an operational problem was the primary, but not the only, means of preventing acci-dents. Nuclear plants also included other systems that were critical to avoiding the occurrence of accidents, such as sensitive and redundant instrumentation, emergency electrical power to run the essential plant equipment if the normal supply was interrupted, and backup equipment to go into operation in the event of a failure in a basic system.

If, despite all precautions, a reactor accident took place, nuclear plants were built to limit the consequences, both to the facility and ultimately to the population beyond its boundaries. Neither the industry nor the AEC dismissed the possibility that an accident could occur, and they sought to make certain that it did not turn into a catastrophe. The primary source of concern by the mid-1960s was an accident in which the supply of coolant to the core was lost through a break in the primary system. Under the worst-case circumstances, the core could overheat in a matter of seconds. Even the rapid insertion of control rods would not end the emergency. It would shut down the major source of heat by halting the fission process, but it would not stop the creation of "decay heat," which resulted from the spontaneous radioactive decay of fission products already in the core. Even at its highest point at the time of the shutdown, the decay heat would only amount to about 7 percent of the level before shutdown. But without adequate cooling, this would still be enough to cause serious damage to, and perhaps to melt, the core. If the core melted, it could set off a series of events that could allow radioactivity to escape into the environment.

In the event of a serious accident, an inherent feature common to all reactors would reduce the release of radioactive materials. The reactor fuel pellets and the metal tubes in which they were encased, called cladding, would retain significant amounts of the fission products created in the core. The extent of the protection the fuel and cladding would provide was uncertain, but it was, in itself, an insufficient barrier to the escape of radioactivity. To increase the margin of safety, reactor designers added a number of systems to guard against the effects of a loss-of-coolant accident. Those systems varied from plant to plant, but the decreased reliance on remote siting and the larger size of proposed power reactors in the early 1960s both magnified the significance and expanded the number of engineered safety features. The newer plants featured a series of separate systems to prevent the escape of radioactivity into the environment even after a serious accident. The importance of ensuring that they worked properly could hardly be overestimated. Commenting on the relationship between remote siting and engineered safeguards in 1964, Herbert Kouts, chairman of the ACRS, told Seaborg: "The protection of the public ultimately depends on a combination of engineered safeguards and adequate distances. Engineered safeguards which can justify decrease of the distances must be extraordinarily reliable and consistent with the best engineering practices as used for applications where failures can be catastrophic."

The engineered safeguards in nuclear plants differed in design and operation, but they served the same basic functions. A number of systems were placed in reactors to remove heat and reduce excessive pressure in the event of an accident. They included, for example, a passive core flooding system in pressurized-water reactors and core spray systems and a passive pressure suppression pool in boiling-water reactors. Both types of plants also used "safety injection" systems that would decrease heat levels quickly in a loss-of-coolant accident by shooting large volumes of water into the reactor vessel. Another system of filters, vents, scrubbers, and air circulators would collect and retain radioactive gases and particles released by an accident before they escaped from the plant. Those engineered safeguards, individually or collectively, would be effective if they worked according to design, but safety experts worried that they might be incapacitated by an insufficient supply of water, a loss of electrical power, or some other contingency during an accident.

The final line of defense was the containment building, an often dome-shaped structure made of steel and concrete that rose as high as twenty stories. In some cases it consisted of double steel walls covered by a massive layer of concrete designed to trap fission products. The containment building surrounded the reactor and associated steam-producing equipment as well as the safety systems. All commercial light-water reactors had containment buildings, though their design and complexity varied in different plants.[1]

Reactor safety experts were confident that in almost any situation the engineered safeguards built into a plant would protect the public from the effects of an accident. But they were troubled by the possibility that a chain of events could conceivably take place that would bypass or override all the safety systems. The authors of the WASH–740 update, for example, projected ways in which a series of safeguards might fail, and they envisioned a worst-case accident in which large amounts of radioactivity escaped into the environment because a door in the containment building was left open. The AEC's regulatory staff and the ACRS, in evaluating reactor applications, acted on their belief that plants they approved were safe, and that even in a severe accident the containment structure would almost certainly prevent the dispersion of harmful concentrations of radiation. But they also recognized that equipment failures and human errors that could lead to a serious accident could occur. "The fact is, no one is in a position to demonstrate that a reactor accident with consequent escape of fission products to the environment will never happen," Clifford Beck told the Joint Committee. "No one really expects such an accident, but no one is in a position to say with full certainty that it will not occur." The AEC tried to make certain that both the chances and the consequences of an accident were minimized; the conviction that existing plants were safe did not prevent the agency from striving to make the technology safer by seeking to learn more about it and to keep abreast of its continual evolution.[2]

The AEC based its decisions on safety issues and designs on operating experience, engineering judgment, and the results of experiments with test reactors. The experience with the first commercial reactors had been encouraging; they had inevitably encountered problems of a minor nature but had suffered no major accidents. Reactor engineers had learned a great deal from the operation of the early plants. The knowledge gained from experience was valuable, but it was of limited application to newer reactors. The size of approved and proposed plants was growing significantly and their designs were changing rapidly. The evaluation of applications for the newer reactors, like the older ones, necessarily required the careful exercise of engineering judgment. Some safety systems and devices simply could not be tested in ways that provided absolute assurance that they would work as planned. Engineered safety features that were installed to contain a severe accident, for example, were difficult and in some cases impossible to test fully or adequately. Nuclear engineers, therefore, proceeded on the same basis as their professional colleagues who worked on other kinds of structures—by learning as much as possible about plant materials and components, by building models to run simulated or small-scale experiments, and by using the information acquired to design a facility according to accepted engineering principles.

The Advisory Committee on Reactor Safeguards emphasized the need for reliance on sound engineering judgment in early 1965 while expressing irritation with the regulatory staff's opposition to a construction permit for the proposed Bodega Bay plant the previous fall. One major reason that the regulatory staff disapproved the Bodega application was the lack of a means to test PG&E's design for preventing earthquake damage. The ACRS argued that since tests could not be conducted on every new design and safety feature, such as those planned for the Bodega reactor, "we must place reliance in the laws of physics and in properly verified engineering principles."[3]

Despite their disagreement over the Bodega project, the ACRS and the regulatory staff agreed that the proper exercise of engineering judgment depended on gaining as much experimental data as possible. This was especially vital in light of the many questions about reactor behavior that remained unanswered. Consequently, safety research was an essential part of the regulatory program. Since the early 1950s, the AEC had sponsored hundreds of small-scale experiments at the National Reactor Testing Station near Arco, Idaho and at other sites. One especially important series of tests that had begun in 1955 was known as SPERT (Special Power Excursion Reactor Tests). It focused on the performance of reactor cores and their components under unstable conditions created by researchers, even to the point of the deliberate destruction of different types of cores. The SPERT program provided valuable information on reactor kinetics during abnormal situations and showed that many "inherent mechanisms" controlled the consequences of a nuclear excursion. But it offered little guidance on key issues of growing concern to reactor experts in the 1960s, particularly loss-of-coolant accidents.[4]

AEC officials were keenly aware of the problem. Stanley A. Szawlewicz, chief of the nuclear safety research and development branch of the agency's Division of Reactor Development, which ran research programs, declared in 1963: "The design of components for accident control or countermeasures . . . is often performed without full knowledge of the accident details for which the component is designed." He added that the AEC based its safety research projects on the conviction that the "causes, consequences and fears associated with major nuclear accidents can best be resolved by undertaking large-scale and engineering-type experiments to complement the basic research programs." The agency made plans to build new equipment to conduct tests on the effects of a loss-of-coolant accident. The most important project for those purposes would be the construction of the LOFT (Loss-of-Fluid Tests) reactor. It would be a fifty thermal megawatt pressurized-water reactor with a containment shell that would be used to test core spray systems, pressure suppression devices, filtration processes, and other engineered safety features. The culmination of the LOFT program would be a loss-of-coolant accident in which the core melted. The AEC hoped that the LOFT facility would be ready by 1965, and that it, along with other tests on core damage and fission product activity, would provide data to improve or to confirm the safety of power reactor designs.[5]

The ACRS reviewed the safety research program with careful, and increasingly disapproving, vigilance. It gradually became more outspoken in complaining that some of the studies undertaken at various laboratories were inadequate, inappropriate, and/or poorly designed. In December 1962, ACRS chairman Franklin A. Gifford told AEC general manager A. R. Leudecke that a number of findings from the SPERT series and the proposed loss-of-coolant tests would be useful. He suggested, however, that other experiments that the ACRS wanted performed would be more fruitful than some of those already carried out. Leudecke responded that some of the recommendations of the ACRS would be followed, but, in what became a familiar refrain, that budgetary constraints had postponed the pursuit of others.[6]

The ACRS remained dissatisfied with the direction and value of certain aspects of the AEC's research agenda. One member declared, in executive session with ACRS colleagues, that he thought that much of the safety research program was a "man-sized boondoggle." Others were less blunt but no less troubled. One source of concern was a series of tests on fission-product release conducted at Oak Ridge and Brookhaven national laboratories. They were small-scale tests run with fuel samples of no more than thirty grams that were intended to demonstrate the chemical and physical form that fission products exhibited and their path, mobility, and deposition when the fuel melted. ACRS members were not persuaded that the experiments proved anything significant about what might happen under accident conditions in an operating reactor. A second source of concern within the ACRS was that much of the AEC's safety research seemed only marginally applicable to existing needs. One member, Theos Thompson of the Massachusetts Institute of Technology, suggested that "the safety research program is approximately ten years behind the times." He urged that the committee do more to encourage the AEC to focus its research on potential accidents in the types of reactors that were being built for commercial operation.[7]

Thompson's colleagues on the ACRS agreed with his position. In August and again in November of 1963 committee chairman David B. Hall addressed letters to general manager Leudecke that were restrained in their wording but clear in their meaning. Hall emphasized the critical importance of making certain that engineered safeguards worked properly, particularly at a time when they were "increasingly used to justify sites that would otherwise be unacceptable." He called on the AEC to build equipment and design research projects for a wide variety of tests. They included investigations of fission-product release under accident conditions, pressure suppression mechanisms, and containment systems. Hall viewed the LOFT reactor as an essential part of the experiments the ACRS was proposing, especially since it would provide the means to judge the accuracy, reliability, and completeness of ongoing small-scale tests in other facilities. He cautioned that even LOFT would not be enough in itself to answer all outstanding questions about reactor safety, but concluded that "the Reactor Safety Research Program promises to be of great significance toward establishing how far engineered safeguards may be relied on in easing reactor site problems."[8]

The ACRS received a rather defensive response, drafted by the Division of Reactor Development and signed by assistant general manager Dwight Ink, to its comments on the research program. Ink suggested that the laboratory scale experiments on fission-product release had been more useful than the ACRS seemed to allow, but he made clear that the AEC had decided to expand the scope of its safety research substantially. He outlined plans for additional research, including the construction of a "Nuclear Safety Pilot Plant" at Oak Ridge to test fission-product behavior under simulated accident conditions, and "Containment Systems Experiments" at Hanford to study the effectiveness of different engineered safeguards in a simulated loss-of-coolant accident. The AEC hoped to start the pilot plant experiments in early 1964 and the containment programs by April 1965.

Ink also called the attention of the ACRS to a new study of piping designs, materials, and failures, which was undertaken because a large pipe break could lead to a serious accident. Finally, he cited a series of experiments on structural materials used to build nuclear plants. They were designed to test, among other things, how the strength, ductility, and other properties of metals in a reactor were affected by constant exposure to high levels of radiation. Ink agreed with the ACRS on the importance of safety research, but he, and through him, the Division of Reactor Development, insisted that the research program was making good progress in carrying out its objectives.[9]

Despite its appreciation of the additional effort and resources that the AEC was devoting to safety research, the ACRS was less certain about the value of the new projects. Its reactor safety research subcommittee expressed concern that the fission-product release tests in the nuclear safety pilot plant and the containment experiments would be, at best, of limited usefulness in understanding the effects of reactor accidents. By October 1964, the pilot plant had run its initial tests on fission-product behavior, and the subcommittee found the results to be interesting but ambiguous. Its members agreed that the significance of the program depended on how well it simulated a real accident, and that would not be known until LOFT began performing. They were even less sanguine about the prospects for the containment experiments, which were scheduled to begin in May 1965. Since those tests would operate with only simulated fission products, the subcommittee believed that they had a "poor capability . . . to serve [their] intended purpose." The Hanford researchers who would conduct the tests were more optimistic about obtaining useful data once correlations were drawn between simulated and actual conditions, but the subcommittee was unconvinced.

The uncertain applicability of the pilot plant and containment systems tests placed an even greater premium on LOFT. It would be a functional reactor; experiments with it would demonstrate how actual fission products behaved and plant components worked in a severe accident. But the ACRS subcommittee on safety research divided in its opinion of how useful even LOFT would be. Theos Thompson, for one, worried that "a few LOFT tests would be applied 'across the board' rather than as specific tests under specific sets of conditions." While the subcommittee reserved judgment on how helpful LOFT would prove to be, it favored prompt completion of the facility and conduct of the tests. Despite the importance of LOFT for enhancing the knowledge of reactor experts on the distribution of fission products in an accident and the effectiveness of engineered safeguards, the construction schedule for the plant slipped. In January 1964, the Division of Reactor Development reported that the reactor would not be completed until July 1966. After an initial series of nonnuclear preparatory tests, the culminating loss-of-coolant experiment would be conducted by the winter of 1968. In this test, researchers would study the results of a large (eighteen inch) pipe break in which they delayed the insertion of control rods, cut off the normal circulation of cooling water, and excluded the use of core sprays.[10]

The concerns that the ACRS aired about the AEC's research program were echoed by another of the AEC's outside advisory groups, the General Advisory Committee (GAC). The GAC had been established by the Atomic Energy Act of 1946 to provide assistance to the AEC on a broad range of scientific and technical issues. Its members, who were appointed by the president, ranked among the best-known scientists and engineers in the nation. During the early years of the AEC, when the agency had only a small technical staff, the GAC had exercised great influence. In later periods the weight of its opinions diminished somewhat, but it continued to command prestige and respect within both the scientific community and the AEC. The GAC only rarely commented on regulatory issues, but in March 1965 the Commission asked it to review the safety research program. In response the GAC established a reactors subcommittee that included Manson Benedict (chairman), head of the nuclear engineering department at the Massachusetts Institute of Technology, Lawrence R. Hafstad, vice-president of the research laboratories of the General Motors Corporation and a former director of the AEC's Division of Reactor Development, and William Webster, vice-president of the Connecticut Yankee Atomic Electric Company. The subcommittee drew on the services of Herbert Kouts, then chairing the ACRS reactor safety research subcommittee, as a consultant.

After holding discussions with officials of the Division of Reactor Development, the regulatory staff, and the ACRS, and visiting the research facilities at the National Reactor Testing Station in Idaho, the GAC subcommittee reached conclusions that generally supported the position of the ACRS. In a July 1965 report to the Commission, it advised that the reactor safety research program was "useful, necessary, and not wasteful of funds or personnel." It added that the tests would provide some, but not all, of the information needed to evaluate reactor applications and to judge the operation of engineered safeguards. It called for the addition of other research projects, some shift of emphasis in existing ones, and better coordination between various offices within the AEC. The GAC subcommittee expressed the same reservations about the pilot plant and containment experiments that the ACRS had cited, and it cautioned that despite LOFT's promise, the "AEC should be prepared to resist the temptation to regard a single experiment as if it answered all questions about fuel meltdown and fission product escape." The subcommittee also voiced acute dissatisfaction with the schedule for the LOFT program, which had slipped farther. The meltdown test was not expected to take place until April 1969. The subcommittee lamented that "this is so late that it cannot be useful in resolving siting problems of reactors until the early 1970s," and it urged that "every effort . . . be made to accelerate the schedule."[11]

Like the ACRS and the GAC reactors subcommittee, the regulatory staff objected to some aspects of how safety research was handled. Its complaints centered on organizational rather than technical issues, especially its relationship with the Division of Reactor Development, which was primarily responsible for directing the AEC's research program. Although the regulatory staff offered its views and received information on the status of safety research through many informal contacts with the Division of Reactor Development, it believed that involvement on a more regular and systematic basis was both desirable and necessary. It considered the research program to be, in Beck's words, "generally a good one," but it shared the concerns of the ACRS and the GAC that tests to date had not provided information that was directly applicable to large reactors.

Beck sought to ensure that the regulatory staff had access to the plans for and the results of research projects and a prominent role in designing and setting priorities for safety experiments. He found the existing situation "somewhat unsatisfactory," which he attributed to the physical separation of the regulatory staff (in Bethesda, Maryland) from the rest of the AEC, the "sharp expansion" of the research program, and the Division of Reactor Development's failure to appreciate fully the procedures and requirements of the regulatory staff. For a time, a liaison committee representing both staffs tried to resolve the problem, but it was unsuccessful, largely because effective coordination appeared to be a "full-time job." The regulatory staff remained uneasy about its position; it recognized that despite a tradition of informal cooperation with Reactor Development, the priorities of the two groups sometimes differed. As its name indicated, the main function of the Division of Reactor Development was to investigate the feasibility of different designs and applications of nuclear reactors, including those for merchant ships, submarines, rocket propulsion, and other purposes. By 1964, the division was making preliminary plans to accelerate development of fast breeder reactors, which created "fissile material" that could be used to produce more nuclear fuel than the reactor consumed. Reactor safety research was an important part of the Division of Reactor Development's responsibilities, but not its principal task or its primary concern.[12]

The GAC reactors subcommittee was sharply critical of the limited coordination between the regulatory and reactor development staffs in its July 1965 report on safety research to the Commission. Citing a "serious lack of interaction between the organizations responsible for the conduct of research and the regulatory staff," it declared: "The regulatory staff should participate more actively in planning the experimental program and should utilize the experimental results more fully and more promptly than it now does." The subcommittee urged the creation of a senior level internal group that would meet regularly and frequently to plan and evaluate the research program. The same month, the Mitchell panel, which the Commission had established to study regulatory procedures, included an identical appeal among its recommendations for regulatory reform. This was not a coincidence; Manson Benedict, the chairman of the GAC subcommittee was also a member of the Mitchell panel. He kept the panel informed about the findings of the GAC review.[13]

By the summer of 1965, three advisory panels, with the tacit support of the regulatory staff, were urging the Commission to make changes in research agendas and procedures. Although they did not convey a sense of urgency in their appeals, they were insistent that the AEC act promptly to meet the problems they identified. In response to the reports of the Mitchell panel and the General Advisory Committee, the Commission established another internal committee, the Steering Committee on Reactor Safety Research, to coordinate the research program. It was charged with responsibility for making recommendations to the general manager on research priorities, evaluating existing and proposed projects, and developing procedures for rapid dissemination of information obtained from tests both within and outside of the AEC. The membership of the steering committee included Milton Shaw, who had recently become director of the Division of Reactor Development, high-level members of his staff, leading regulatory officials, and representatives of other divisions with an interest in the research program. The Commission appointed John Swartout, assistant general manager for reactors, as chairman of the new committee and Clifford Beck as vice-chairman.[14]

The steering committee met fourteen times between August and November of 1965 and conducted broad-ranging discussions of the status of and prospects for reactor safety research. Swartout reported that the commissioners had indicated that their primary concern was to prove the safety of reactors so that nuclear units could be placed in cities. Representatives of both the reactor development and regulatory divisions thought this objective to be unattainable in the near term and misplaced as a rationale for research. Marvin M. Mann, assistant director of regulation, observed that "there is adequate justification for a program to increase safety, even if there is no guarantee of achieving any specific goal." The members of the steering committee agreed on the need for a wide range of tests, but they were less certain of which should take priority. This was a question of immediate importance, because the research budget was insufficient to carry out all of the recommendations of the ACRS and the GAC. Expenditures for safety research had steadily grown, from about $24 million in fiscal year 1963 to an estimated $35.4 million in fiscal year 1966. The increased funding reflected higher allotments for existing projects as well as the addition of new ones. But it was still not enough to pay for all of the proposed tests; Shaw remarked that "budget limitations prevent the AEC from doing everything that appears useful."[15]

The steering committee members agreed that setting priorities for research projects was essential, but they failed to act on their conclusion. They viewed the LOFT program as particularly important, a position that was consistent with the rise in the LOFT budget from $1.2 million in fiscal year 1964 to an estimated $5.2 million in fiscal year 1966. But they were uncertain about how the tests would be conducted and what preliminary information was needed to make them as useful as possible. It was clear that the value of the LOFT results would depend on running a series of other experiments that would help in the design and interpretation of the final meltdown test. Those supporting experiments, which would require another two years, were a part of the reason that the meltdown test was not expected to take place until early 1969.[16]

The discussions among regulatory and reactor development officials who served on the steering committee revealed no major sources of disagreement. But they did not confront potentially tough choices and divisive questions that could have caused conflicts of opinion, such as the priority of research projects, the allocation of resources, or the expenditure of funds for safety testing. The steering committee neglected to fulfill two of the key functions it had been established to perform: determining an order of priorities for research projects and achieving a working agreement on the relative roles of the regulatory and reactor development staffs in the direction of the research program. In a progress report sent to the Commission in December 1965, Swartout promised that the committee would deal with those issues in the future, but he did not explain why it had not yet done so. Part of the reason was that the the steering committee and the organizations represented on it did not view the questions that safety researchers would investigate as urgent matters. Agency officials regarded the research agenda as important but they did not think it was pressing. They were convinced that operating power reactors were safe and that careful evaluation of individual applications would ensure that new plants were equally safe. There were still relatively few applications for construction permits; the bandwagon market had not yet gotten under way. Therefore, the effort to design LOFT properly and to derive the maximum benefit from the meltdown test took precedence over how soon it was conducted, and deciding on priorities for other projects unrelated to LOFT seemed to be something that could wait.[17]

The steering committee failed to set priorities for the research program and to settle organizational questions about its direction not only because of a lack of urgency but also because of a lack of leadership on the part of the commissioners. The Commission responded to the concerns of its advisory committees about the research program by forming the steering committee, but subsequently took little interest in its activities. Like the staff, it did not regard the completion of research projects as vital to safety in the short run. The commissioners did not insist that the steering committee promptly perform the tasks that they had assigned, and they seemed to regard safety research more as a means to encourage urban siting than to discover and apply new information about reactor design and safety. The budget for safety research was increasing and important work was being done, but on matters of emphasis, purpose, and timeliness it continued to drift.

At about the same time that the steering committee was reviewing the status of safety research, the ACRS was raising new questions that cast doubts on prevailing assumptions about the safety of existing and proposed plants. Its principal concern was that the newer and larger reactors might undergo an accident that set off forces powerful enough to breach containment and discharge fission products into the environment. Safety experts had always recognized that radioactivity might escape from the plant through an open door in the containment shell or some other happenstance, but they had been confident that otherwise the containment structure would be strong enough to withstand the effects of any accident. In mid-1965, however, the ACRS began to waver in that conviction because of its growing uneasiness about what might happen if components of a plant's reactor pressure vessel failed.

The pressure vessel was a container made of steel, three to ten inches thick, that held the core of the reactor, including fuel assemblies, control rods, and related equipment as well as the coolant. It also supported the pipes that fed water into the core and the mechanisms that drove the control rods. A major rupture of the pressure vessel could trigger catastrophic consequences by disabling safety systems and by causing a complete loss of coolant. The question of the integrity of pressure vessels in nuclear plants had been raised earlier. In 1955 the American Society of Mechanical Engineers (ASME), which had prepared industrial codes on steam boilers and pressure vessels since 1915, established a special committee to study the problems peculiar to nuclear vessels. In 1963 it published a code on the construction of nuclear vessels that covered standards and specifications for materials, welding requirements, stress analyses, and inspection. At that time, reactor experts regarded a gross failure of a vessel as beyond credibility and therefore, they did not design safeguards that took it into account.[18]

Within a short time after the ASME code appeared, however, some authorities began to wonder if it was adequate to prevent a pressure vessel rupture in newer plants. Former ACRS chairman C. Rogers Mc-Cullough, speaking to a symposium on nuclear accidents in April 1965, cautioned that in "the evaluation of safety of reactors our basic under-standing of the problems has not kept pace with advances in reactor design and size." He cited pressure vessel integrity as a matter that required attention. One reason that the safety record of pressure vessels had been so good over the previous fifty years, he maintained, was that they had been inspected on a regular basis and repaired if flaws were detected. But he was disturbed about potential problems with nuclear vessels because they had not been subjected to periodic examinations and because the effects of radiation on the durability of their steel walls was uncertain. McCullough pointed out that research on the ways in which radiation could alter the properties of metals had only recently begun. Some recent tests in the United States and Great Britain had indicated that under certain conditions pressure vessels might be suscep-tible to rapid breaks at reactor operating temperatures.[19]

The ACRS was also troubled by the possibility of pressure vessel failure. At about the same time that McCullough aired his views, its subcommittee on reactor pressure vessels discussed the issue with indus-try representatives. Although the information the subcommittee re-ceived on experience with pressure vessels was generally encouraging, some nagging uncertainties about reactor vessels remained. In addition to the questions that McCullough had raised about regular inspection of vessels in operating plants and the effects of radiation on steel, members voiced concern about methods of fabricating pressure vessels and the safety implications of the thicker walls used for larger reactors. They projected that a catastrophic vessel failure could occur if the bolts that anchored the top, or head, of the vessel came loose, perhaps becoming missiles that could blast through containment or even allowing the head itself to be hurled through the containment structure. A more plausible and equally disquieting hazard was that the vessel walls would break, opening a large fracture and causing a loss of coolant from the core.[20]

The reliability of pressure vessels emerged as a major issue when the Commonwealth Edison Company applied for a construction permit to build a second unit on the site of its Dresden plant, located in a rural

7. Fabrication of light-water reactor pressure vessel at Babcock and Wilcox plant in Indiana. (National Archives 434-SF-29-20)

area about fourteen miles from Joliet, Illinois and forty miles from Chicago. Dresden had been the first privately owned commercial reactor to operate, having received its operating license in 1959 and gone on line the following year. In early 1965 the utility announced plans to add the second unit. At 715 electrical megawatts, it would be much larger than Dresden I; indeed, it would be considerably larger than any other plant licensed to that time.[21]

Even though the Dresden site was distant from major urban areas, several members of the ACRS regarded it as a possible prototype for metropolitan plants. Therefore, they focused their attention on the potential for pressure vessel failure. Committee member David Okrent, a nuclear physicist at Argonne National Laboratory, took the lead in pressing the issue. At a meeting with officials from Commonwealth Edison and General Electric, which would build the reactor, he quizzed them about the consequences of a pressure vessel rupture. His questions spurred lengthy discussions about whether or not a vessel failure should be considered a credible occurrence and if so, what the consequences might be. One of the General Electric representatives suggested that although containment could withstand a partial break in the vessel it could not hold up in the event of "a complete break." If a major fracture seemed credible, the ramifications were ominous. It would indicate that under certain conditions the containment structure might not be sufficient to prevent the dispersion of large amounts of radioactivity into the environment.[22]

ACRS members divided in their opinions about whether a massive pressure vessel rupture could take place and about the implications of the issue for granting a construction permit to the Dresden II plant and to future power reactor proposals. After long deliberations the committee decided on a dual approach that was satisfactory to everyone; it would approve the Dresden II application and also send a letter to the Commission outlining its concern about accidents arising from pressure vessel breaks. Although it viewed the likelihood of a massive failure as remote, it could not dismiss the possibility entirely. The ACRS found the Dresden II proposal acceptable because the site was removed from population centers, but it did not want its endorsement of the application to be cited as a precedent for locating similar reactors in populous areas. Okrent and others insisted that nuclear vendors and utilities be placed on notice that they should address outstanding questions about pressure vessel reliability before seeking approval for plants closer to cities. Even for reactors in remote sites, Nunzio J. Palladino of Pennsylvania State University argued, much more information about the performance of pressure vessels in larger units was needed if "failures were deemed credible."[23]

On 24 November 1965, ACRS chairman W. D. Manly sent a carefully worded letter on pressure vessels to Seaborg. He acknowledged that in the committee's view, the probability of a major failure that led to a breach of containment was "very low." Nevertheless, he told Seaborg that the ACRS believed that it was both possible and prudent to take action to ensure greater reliability of vessels and to reduce the consequences if a rupture occurred. Manly urged that the AEC and the nuclear industry devote greater attention to improving methods of inspecting vessels during fabrication and during operation. He also called for increased emphasis on problems such as stress analysis and the "propagation of flaws" in vessel walls. To keep the effects of a vessel failure to a minimum, Manly recommended that plants be designed to prevent an "internally generated missile" from penetrating containment, to ensure adequate core cooling in the event of a vessel break, and to trap most fission products inside the reactor even if containment was breached. He stressed that the ACRS did not consider pressure vessel failure to present an undue risk to public health and safety in areas that were not heavily populated, but added that addressing the measures the letter cited would be advisable if large reactors were to be located close to population centers.[24]

The ACRS tried to present its views in a way that would not raise unwarranted alarm, but its letter elicited a sullen response from the AEC and an openly hostile one from industry spokesmen. The AEC took the unprecedented step of attaching a statement of its own when it released the ACRS letter publicly. It emphasized that it was already committed to an "augmented and accelerated" safety research program that would consider new developments and promote reactor safety. It promised that the recommendations of the ACRS on pressure vessels would receive the "prompt attention" of the Steering Committee on Reactor Safety Research. While research was conducted on safety issues, the AEC statement declared, "the adequacy of safety provisions in each reactor will continue to be established by thorough and detailed analysis and evaluation on a case-by-case basis."[25]

Industry response to the ACRS letter was much less restrained. Several spokesmen complained that the ACRS seemed to be raising theoretical problems that reflected a trend in its evaluations toward academic questions. "They should have to prove the justification of their question," commented one anonymous industry representative. "We always have to prove—they can just think, opine." Another added: "Is this just thinking up things to protect against? If so where is it going to stop? Why bring it up now? Do they know something we don't know?" Industry officials acknowledged that a vessel rupture would be disastrous, but they denied that it could happen. A spokesman for Babcock and Wilcox, one of the companies that produced reactor vessels, declared: "I believe the possibility of massive failure of reactor vessels has been reduced to zero—and the more we build the better they will be." A number of observers criticized the ACRS for its lack of confidence in engineering judgment, using arguments similar to those that the ACRS had advanced itself in chiding the regulatory staff in the wake of the Bodega Bay controversy. "I think it's wise . . . that the ACRS recognize its responsibility for safety and not permit anyone to build anything that isn't safe," one person said. "The question is—you can analyze things forever and never get anything built . . . I wonder if it isn't coming to a ridiculous point."[26]

A few weeks after the ACRS sent its letter, the commissioners and regulatory staff officials discussed the pressure vessel issue with the committee. Asked about the response of the ACRS to the comments of industry officials, Manly replied that although some of them were thoughtful, others demonstrated "a general ignorance of the ACRS procedures and intent." He added that the committee believed that action to carry out its recommendations was imperative. The ACRS hoped that its letter would encourage the industry to pay more attention to quality assurance in the fabrication of pressure vessels and to improve methods of inspecting vessels in operating reactors. A longer-term goal was to catalyze design changes that would provide safeguards against the consequences of a vessel rupture. The ACRS thought that design innovations were especially important for reactors located in populated areas, but regulatory staff officials pointed out that distinctions between plants on safety requirements could not be drawn. Any measures to protect against pressure vessel failure that were prescribed for a metropolitan plant would have to apply to all plants.

The commissioners expressed concern about the impact of the ACRS's views on applications already under consideration and on the efforts of the nuclear industry to keep the costs of power reactors competitive with coal plants. The ACRS members did not see any need for a moratorium on licensing; they believed that applications should still be judged on an individual basis. They recognized that carrying out their recommendations on pressure vessels would increase the costs of reactors, but they did not think that the extra expense of improved inspection and surveillance methods would be prohibitive. The costs of developing and implementing new designs would remain unknown until more research was performed.[27]

The ACRS did not insist that its recommendations on designs be carried out immediately; in early 1966 it approved of granting construction permits to two plants in relatively isolated locations without requesting new design features. But it pressed for prompt implementation of improved inspection and surveillance of pressure vessels. When the ACRS informally urged the nuclear industry to tighten its procedures in those regards, it received little encouragement, which was hardly surprising in light of the industry's belief that pressure vessel failure was not a significant problem. The ACRS also approached the American Society of Mechanical Engineers' committee on nuclear vessels, which agreed to reexamine its 1963 code. Because of uncertainty about how long it might take ASME to act and what the results of its study might be, the ACRS called on the regulatory staff to impose more stringent standards than those in the existing code for the fabrication of vessels. Finally, it recommended a requirement that pressure vessels be built so that their entire surface could be inspected by either visual or ultasonic means during operation and that examinations be scheduled on a periodic basis.[28]

The regulatory staff followed the advice of the ACRS. It held discussions with industry representatives about the question of inspectability. Although it ultimately decided not to publish specific criteria on inspection of in-service pressure vessels, it placed industry on notice that applications would be scrutinized to make certain that plants were designed to allow careful examination for cracks and flaws. On the matter of quality assurance in the construction of vessels, the regulatory staff took the unusual step of preparing and issuing detailed criteria intended to upgrade the ASME code. The AEC normally published broad rather then specific instructions or guidelines and as a general rule deferred to the judgment of independent groups of outside experts, such as ASME or the National Committee on Radiation Protection. But in this case the regulatory staff believed that the ACRS had identified an issue of vital importance that demanded prompt attention.[29]

Working in conjunction with the ACRS, the Division of Reactor Development, and Oak Ridge National Laboratory, the regulatory staff prepared a new set of criteria to supplement the ASME code. It provided detailed instructions on the design, fabrication, inspection, and testing of pressure vessels and directed that the reactor owner employ a qualified representative to make certain that the work was performed properly. The AEC emphasized that the new criteria, issued for public comment in August 1967, were guidelines rather than firm requirements. The nuclear industry greeted the AEC's action with criticism and exasperation. One reason was, as the magazine *Nuclear Industry* put it: "Whether the criteria are, at the moment, guidance or requirements, it seems almost certain that, in one form or another, they will soon become requirements."[30]

A group of experts appointed by the Atomic Industrial Forum's Reactor Safety Steering Committee complained that the AEC's criteria were inappropriate and in some cases impossible to fulfill "on the basis of current technology." It argued that if the AEC thought the existing code was inadequate, "it should refer such matters to the appropriate code-writing group wherein the expertise and experience reside." Spokesmen for ASME made the same point, suggesting that the AEC criteria were overly detailed and overly prescriptive. One commented that the AEC had made some constructive recommendations but that they "were not broad enough to be put into a code." Nevertheless, ASME eventually adopted a number of the measures included in the AEC's criteria. Meanwhile, the agency used the criteria in judging reactor proposals and insisted that applicants specify how they would put the guidelines into practice.[31]

By July 1968 the ACRS was satisfied that major advances in the quality and reliability of pressure vessels had been achieved. This was apparent in long discussions over approving a construction permit for Commonwealth Edison's Zion plants. Despite the relatively high population in the area of the proposed reactors, most members of the ACRS agreed that the utility had adequately addressed the problem of pressure vessel failure. The committee drafted a letter to Seaborg that affirmed: "The ACRS is pleased to note that substantial progress has been made in improving the design, fabrication, and inspection processes for reactor pressure vessels." It added that more research was needed on some questions, such as finding better methods for remote inspection of vessels, evaluating the possibility of embrittlement of vessel walls from exposure to radiation, and assessing how increasing the thickness of vessel walls in larger plants (to ten to fourteen inches) might affect the properties and behavior of the steel used in them. In each case, the AEC and/or industry was conducting research.

The ACRS found less evidence of progress on the other issue that it had raised in its November 1965 letter to Seaborg—designing safety features that would protect against a massive pressure vessel failure if it occurred. The draft letter pointed out that although the ACRS thought that current designs were acceptable for plants already approved, it continued to believe that "substantial additional measures must be taken for sites of substantially higher population densities." The ACRS, for unexplained reasons, never sent this letter to Seaborg, but Okrent underscored the same point in comments he added to the ACRS's report on the Zion application. While he did not oppose a construction permit for Zion and agreed that important steps to resolve questions about vessel integrity had been taken, he urged that the AEC require "additional conservatism" in the design and construction of the Zion plants. Okrent reiterated his conviction, which was shared by other ACRS members, that what was acceptable for rural or even suburban sites was not suitable for areas that were more densely populated, at least until adequate means to cope with pressure vessel failure were available.[32]

Although concern over pressure vessel failure did not disappear within the AEC and the ACRS, it had eased considerably by the time that they reviewed the Zion application. The uncertainty about the reliability of vessels was the first issue in which a breach of containment from the effects of an accident seemed to be credible to reactor experts. But it was soon superseded by new misgivings about the performance of containment under other accident conditions. Instead of focusing on pressure vessel breaks, the ACRS and the AEC began to devote more attention to the possible consequences of a core meltdown caused by a loss-of-coolant accident.

Reactor experts had long recognized that a core melt was a plausible, if unlikely, occurrence. A massive loss of coolant could happen, for example, if a large pipe that fed cooling water into the core broke. If the plant's emergency cooling systems also failed, the build-up of decay heat could cause the core to melt. In older and smaller reactors, the experts were confident that even under the worst conditions—an accident in which a loss of coolant melted the core and the mass of fuel, in turn, melted through the pressure vessel—the containment structure would prevent a major release of radioactivity into the environment. As proposed plants increased significantly in size, however, they began to worry that a core melt could lead to a breach of containment. This became more of a concern not only because of the greater decay heat that larger plants would produce but also because vendors did not add to the size of containment buildings in corresponding proportions to the size of reactors. The effectiveness of containment in controlling rising heat and pressure was partly a function of its volume. The margin of safety in newer plants was reduced when the volume of containment relative to the power density of the reactor was decreased.

The greatest source of concern about a loss-of-coolant accident in large reactors was that the molten fuel would melt through not only the pressure vessel but also through the thick layer of concrete at the foundation of the containment building. The fuel would then continue on its downward path into the ground. This scenario became known as the "China syndrome," because the melted core would presumably be heading through the earth toward China. Other possible dangers of a core meltdown were that the molten fuel would breach containment by reacting with water to cause a steam explosion or by releasing elements that could combine to cause a chemical explosion. The precise effects of a large core melt were uncertain, but it was clear that the results could be disastrous. The ACRS and the regulatory staff regarded the chances of such an accident as low; they believed it would occur only if the emergency core cooling system (ECCS), made up of redundant equipment that would rapidly feed water into the core, failed to function properly. But they acknowledged the possibility that the ECCS might not work as designed. Without containment as a fail-safe final line of defense against any conceivable accident, they turned their attention to other means of guarding against the China syndrome. Both groups sought the same end, but they disagreed sharply on the urgency of the problem and the best way to resolve it.[33]

The China syndrome first arose as a critical issue in the deliberations of the ACRS during the summer of 1966 over applications for construction permits by Commonwealth Edison for a third Dresden reactor and by Consolidated Edison for a second Indian Point plant. Dresden III would be identical to the Dresden II unit already under construction, which led ACRS member Theos Thompson to a prematurely optimistic conclusion: "It would appear that the Committee, at long last, will finally have a very easy time with a major power reactor." The prospects for evaluating the Indian Point II proposal did not seem to be as certain because, at 916 electrical megawatts, the new plant would be three times as large as the existing one. But the difficulties of reviewing the application appeared to be simplified by the fact that it would occupy a site that the AEC already had found suitable. Both the Dresden III and the Indian Point II applications proceeded routinely until June

1966, when representatives of Commonwealth Edison and General Electric met with the ACRS. After committee members raised questions about the effects of a core melt on the proposed reactor, it soon became apparent that neither the utility nor the vendor had satisfactory answers. Therefore, the ACRS decided to delay action on both Dresden III and Indian Point II until it received more information on the ways in which plant designers would deal with the hazards of core melting. Despite the AEC's previous approval of a plant identical to Dresden III and of the site of Indian Point II, concern about the China syndrome inserted a new and complex problem into the licensing process.[34]

The ACRS held lengthy sessions with the utilities and vendors for both proposed plants to address the core melt issue. General Electric advanced no technical solutions for preventing a meltdown if the core coolant was lost and the ECCS failed; it placed its reliance on the effectiveness of emergency cooling. In response to the qualms of the ACRS, it agreed to improve the ECCS for both Dresden II and Dresden III by adding a core flooding system to the independent core spray and injection systems already planned. Westinghouse, the vendor for Indian Point II, was also unable to suggest any sure ways to cope with a fuel meltdown. It proposed to install a water-cooled "core catcher" to intercept and solidify molten fuel that melted through the pressure vessel. But it did not view the core catcher as a proven or primary safeguard in the event of a core melt. Like General Electric, the safety of its design depended not on contending with a core melt once it occurred but in preventing it from happening in the first place. In the case of Indian Point II, emergency cooling would be provided by two independent core-flooding systems.[35]

ACRS members divided in their opinions about the degree to which the revised Dresden III and Indian Point II applications met their concerns about the China syndrome. After lengthy discussions, they agreed on a course of action similar to that taken the previous year in addressing the pressure vessel issue—they would endorse the construction of the two plants with improved emergency cooling systems but also send a letter to the AEC outlining their misgivings about a fuel meltdown. By mid-August 1966, Okrent, the chairman of the ACRS at the time, Harry O. Monson, a nuclear engineer at Argonne National Laboratory, and Palladino had prepared a strongly worded draft for submission to Seaborg. It acknowledged that the probability of a meltdown and an ECCS failure was slight, but urged the AEC to take action promptly on two fronts. One was to seek ways to ensure that a reactor could withstand a core melt even without an effective ECCS and the other was to improve the reliability of emergency core cooling. It called for "an evolutionary process of design and a vigorous program of research" so that "a high state of development" could be achieved within two years. The draft reiterated the committee's view that existing designs were adequate for plants in remote locations, but not for units that might be placed in populous regions. "The Committee believes it prudent to provide still greater protection of the public," it declared, "particularly for reactor sites closer to population centers." Although Indian Point was clearly not a remote site, the ACRS deemed the application for the second unit to be acceptable if an enhanced ECCS was installed.[36]

When the ACRS had sent its letter to Seaborg on pressure vessel failure in November 1965, the AEC complained that it had received no prior notification about the committee's concerns. As a result, the AEC had attached its own statement when it released the ACRS letter. This time the ACRS decided to show a copy of its draft to Harold Price. Price, in turn, forwarded it to the Commission. The AEC was cool to the recommendations of the letter and worried about the impact of publishing them. In a meeting with the ACRS on 8 September 1966 the commissioners requested that the letter be withheld. The AEC did not believe that the problems cited by the ACRS were urgent; the regulatory staff had approved the Dresden III and Indian Point II proposals with few reservations. It thought that an accident in which the core melted and the ECCS failed was highly unlikely and that the best way to ensure that it did not occur was to devote greater attention to the reliability of emergency cooling. Furthermore, it considered the sites of the two plants to be satisfactory because they met the AEC's siting guidelines and because reactors were already located on them. The regulatory staff had concluded that sites closer to metropolitan areas than Indian Point were unacceptable, at least in the foreseeable future, so the appeals of the ACRS for additional requirements for plants in more populous areas seemed moot. Although it acknowledged that the ACRS had posed legitimate questions about core melting, it did not regard them as issues that demanded immediate resolution.[37]

In their 8 September meeting with the ACRS, the commissioners and staff representatives emphasized that the agency was already sponsoring research on the consequences of a core melt. They were deeply concerned that issuing the committee's draft letter "might lead to misunderstanding by the public." Since some public opposition to Indian Point II had developed, the ACRS's letter might appear at an especially awkward time. Seaborg feared that the letter could have a harmful impact on the industry and suggested that the problems it cited should receive further study before being aired publicly. He proposed that the AEC establish a task force to examine the safety implications of core melting and the adequacy of emergency cooling. He asked that the ACRS delay sending its letter until after the task force completed its report.[38]

The ACRS agreed to the AEC's request, but not before some of its members registered vocal protests. The dissenters complained that their approval of the Indian Point II and Dresden III applications had been linked to sending a letter to the AEC on core melting. After long deliberation, committee members decided to withhold the letter because they thought that the task force was a good idea and they hoped that the draft letter had impressed the AEC with the seriousness of their concern. They also worried that hasty action might undermine the credibility of the ACRS with the industry by lending support to critics who charged that it was capricious, overly cautious, or even antagonistic to reactor development. Although the ACRS elected not to transmit its detailed letter on core melting to the AEC, it communicated its general concern about the issue. In a letter that Okrent addressed to Seaborg in October 1966, the ACRS listed the research projects to which it believed the AEC should be devoting the greatest attention. The first two items it mentioned were methods to cope with the effects of a core melt and the performance of emergency core cooling systems.[39]

The ACRS's decision to go along with the AEC's request not to send the draft letter on core melting neither ended nor concealed some fundamental organizational and philosophical differences between the committee and the regulatory staff. Both the pressure vessel and the core melt issues highlighted an inherent distinction in their roles in the regulatory process. The ACRS was an advisory body with no formal authority in the licensing of reactors. Its primary function was not to issue licenses but to offer its collective opinion on applications and safety matters. This encouraged it to challenge prevailing assumptions, project plausible accident scenarios, and extrapolate from available data in a rather free-wheeling manner while deliberating over plant proposals. Indeed, one of its primary duties was to ask the "academic" questions that so annoyed the industry on the pressure vessel issue.

The regulatory staff performed a similar function in reviewing applications, but its inquiry was more prone to be tempered by its formal responsibility for determining whether proposals for construction permits and operating licenses should be approved. The questions that the

ACRS raised about pressure vessels, core melts, and other complex problems made the task of the regulatory staff more difficult and spawned occasional grumbling that the committee was overstepping its responsibilities by commenting on plant design, inspection, and other matters. Such issues could delay licensing, fuel public concern, and force the staff to reexamine issues that it had not fully considered. In some cases, such as the inspection and surveillance of pressure vessels, the staff acted promptly on the recommendations of the ACRS. In other cases, such as design innovations to guard against vessel failure or core melt, the regulatory staff largely ignored the advice of the ACRS. And in still other cases, such as the reliability of emergency cooling, it proceeded more deliberately than the ACRS urged.[40]

The regulatory staff's inaction or delay in carrying out some of the recommendations of the ACRS was partly a reflection of the fact that it, to a much greater extent than the committee, had to balance safety improvements against the AEC's commitment to developing nuclear power. It had to weigh carefully not only the views of the ACRS but also the developmental interests of the AEC. The Commission, the Division of Reactor Development, and other AEC units were ever wary of what they regarded as excessive regulatory zeal that might impede progress in the nuclear field. Although they did not interfere directly in regulatory proceedings, their position inevitably influenced the thinking of regulatory staff officials, who often found themselves situated uncomfortably between the questions and notes of caution sounded by the ACRS and the desire to encourage atomic development on the part of the commissioners and of other staff divisions. Some of the recommendations of the ACRS, if followed, would add significantly to the costs of nuclear power, and this could hurt the industry's ability to compete with coal. They could also cause licensing delays that would undermine the growth of nuclear power. This was an especially vital issue for the AEC in 1966, because by that time the boom in reactor orders was well under way. The AEC feared that the bandwagon market could be disrupted by protracted licensing procedures. The surge in reactor orders intensified the pressure on the regulatory staff to judge applications within a reasonable time, a task that the complex questions of the ACRS encumbered.

Above all, the regulatory staff diverged from the ACRS in its assessment of the urgency of many of the safety issues the committee cited. This was less a result of organizational functions or promotional concerns than a result of differing emphasis on the relative importance of engineering safety features and siting. The ACRS was more flexible than the regulatory staff on the question of metropolitan siting. The committee opposed placing reactors with existing safety features in urban areas but remained open-minded about metropolitan siting in the future. It drew a distinction between what was required for a plant located in a relatively remote area and what would be necessary for one located in or near a city. Therefore, it focused greater attention than the regulatory staff on the adequacy of engineered safeguards.

The staff, on the other hand, placed greater emphasis on siting. It was firmer in its opposition to metropolitan siting for the near and perhaps the distant future. Clifford Beck declared in August of 1966 that he did not believe that reactors would be suitable for urban sites until questions about pressure vessel failure and core melts had been answered, codes and standards had been developed, safety systems had been improved, and more operating experience had been gained.[41] Those obviously were long-term objectives. The staff did not accept the position of the ACRS that stricter safety requirements could be imposed on urban reactors than on rural ones. It insisted that any safety features that the AEC prescribed for a metropolitan plant would have to apply to all units. Therefore, it was reluctant to carry out recommendations that the ACRS made for urban sites unless they seemed necessary for all locations. In the case of pressure vessel inspection and surveillance, the staff agreed that immediate action should be taken; in the other cases it did not. The extent of its response to the ACRS on core melting and ECCS would await the outcome of the task force assigned to study the issue.

In October 1966, at about the same time that the ACRS sent its letter to Seaborg on the need for research on core melting, ECCS, and other matters, an accident at the Enrico Fermi Atomic Power Plant in Lagoona Beach, Michigan punctuated its concern. Although the accident did not turn out to be a serious threat to public safety, the fact that it occurred at all added some substance to the largely theoretical projections of both the ACRS and the regulatory staff about the possibility that a loss of coolant could cause a reactor's fuel to melt. The Fermi reactor had a troubled history spanning more than a decade. In 1956, it had been the third private commercial reactor to receive a construction permit from the AEC. The issuance of the permit had generated controversy because of the advanced technology that the plant would use. It would be a fast breeder reactor built by the Power Reactor Development Company, a consortium of utilities led by the Detroit Edison Company. Fast breeders were significantly more complex than the light-water reactors that other utilities were constructing. Their great appeal was that they would create, or breed, a larger quantity of "fissile material" that could be used for reactor fuel than the amount of fuel they consumed. The drawback was that their design and operation required a higher level of technological sophistication than light-water models. The Power Reactor Development Company viewed the Fermi plant as a demonstration reactor that would show the way toward the wide application of breeder technology.

At a time when even light-water technology was in an embryonic stage, the application to build a breeder reactor stirred considerable uneasiness. The ACRS had expressed grave reservations about approving the construction of breeder reactors until more experimental evidence on their design and operation was available. The AEC, however, had issued the construction permit anyway. This, in turn, prompted a suit by three labor unions that was not resolved until a ruling by the U. S. Supreme Court in 1961 upheld the AEC's position. Meanwhile, construction of the Fermi plant had continued. It was a small unit with a capacity of about sixty-five electrical megawatts, and unlike light-water reactors, would use liquid sodium as a coolant. Liquid sodium had the advantages of a high boiling point and excellent heat-transfer properties; its primary disadvantage was that it could burn or explode if exposed to air or water. After lengthy construction, design, and testing delays, the Fermi plant received its operating license in December 1965. It still took time to get the plant running continuously because of a gradual increase in power levels and because of various malfunctions and adjustments. Finally, in August 1966 the plant operated for more than fifty consecutive hours and for the first time produced electricity that was transmitted over commercial powerlines. Some uncertainties about the performance of the plant remained, however, and it was shut down for a time to analyze problems with fuel elements and other equipment.[42]

On 5 October 1966, as the reactor's power level was slowly being increased as a part of a test program, the accident occurred. The senior operator of the reactor first became aware that something was amiss when he noticed that the control rods were in an incorrect position. Upon further checking, he found abnormally high temperatures in two fuel rods. While he was trying to locate the cause of this condition, radiation alarms sounded in the reactor building. The operator announced a "Class I" radiation emergency, the least serious of four categories of emergency situations at the Fermi plant, and scrammed the reactor. The cause of the accident was unknown and the extent of the damage was uncertain. Plant technicians and an AEC inspector quickly determined that fuel melting had taken place. This was an unsettling finding, but it was balanced by evidence that the consequences of the accident were limited. Higher-than-normal radiation levels were localized and no liquid sodium coolant had leaked. The control rods moved and functioned properly, indicating that a massive melting of the fuel had not occurred. Extensive melting would have interfered with the operation of the rods. The liquid sodium coolant was circulating freely, which would guard against further core melting.

Within a few days, reactor experts concluded that two of the plant's 103 fuel assemblies had partially melted. It took much longer to find out what had made the fuel elements melt. It was impossible to look at the damaged core directly, and only after months of probing, inspecting, and remote viewing did the cause of the accident become known. It turned out that two pie-shaped pieces of zirconium, which had been installed to help guard against the effects of a loss-of-coolant accident by separating and dispersing molten fuel, had broken loose. They had blocked the flow of the liquid sodium coolant to the two affected fuel assemblies, and in that way had triggered the accident. Remote removal of the zirconium segments took several additional months, and was finally completed more than two years after the accident. The Fermi accident was a major embarrassment to the Power Reactor Development Company, especially when it received criticism for failing to show the zirconium sheets on its construction drawings and for responding with insufficient attention and promptness to preliminary indications that something was wrong with the reactor.[43]

The Fermi accident did not have a major impact on the deliberations of the ACRS and the regulatory staff over the consequences of core melting. The plant was entirely different in design and much smaller in size than the large light-water reactors they were considering for construction permits in 1966. The accident, therefore, had limited applicability to the regulatory process. It was, nevertheless, a reminder that unanticipated flaws in design or operation could cause an accident in which the flow of coolant was obstructed and that, therefore, the proper functioning of emergency core cooling was an indispensable part of ensuring reactor safety. The consequences of a core melt and the reliability of ECCS were the focus of concern for both the ACRS and the regulatory staff by the fall of 1966, and the manner in which they dealt with those issues awaited the findings of the special task force that the AEC established.

CHAPTER VIII

Accident Prevention

The Emergency Cooling Imbroglio

By the fall of 1966, the AEC had shifted the focus of its reactor safety goals from mitigating the effects of accidents to preventing accidents. The redirection in the emphasis of the regulatory program was evolutionary and incremental; the AEC did not make a formal or abrupt change of policy. Nevertheless, uncertainties about the integrity of containment during a core-melting accident imposed a modified approach to reactor safety. Previously, the agency had relied on the containment building as the final independent line of defense to keep fission products from escaping into the environment. Even if a serious accident took place, the damage it caused would be restricted to the plant. The AEC had always sought to guard against such occurrences, but at the same time it had recognized that the possibility of a major accident could not be dismissed.

Once it became apparent that under some circumstances the containment building might not hold, the key to protecting the public from a large release of radiation was to prevent accidents severe enough to threaten containment from occurring. Reactor experts thought that a breach of containment was most likely to happen as a result of a loss-of-coolant accident, which meant that in the existing state of the technology public safety depended heavily on the proper design and functioning of emergency core cooling systems (ECCS). Their purpose was to prevent core melting that could lead to the failure of containment. The problem was that both experimental work and experience with emergency cooling were very limited. The AEC remained confident that reactors were safe, but it lacked conclusive evidence to verify its judgment. Therefore, finding a way to test and to provide empirical support for the reliability of emergency cooling became the central concern of the AEC's safety research program.

Representatives of the nuclear industry expressed concern about the impact of the new emphasis on ECCS. Robert E. Richards of General Electric told regulatory staff officials in November 1966 that the "recent escalation of safeguard requirements" arising from the Dresden III application had disturbed the industry because of the possibility of "contractual confusion, delays, [and] added cost and uncertainties." He explained that utilities were particularly anxious about slippage in their construction and licensing schedules while vendors were most troubled about the increased costs of new equipment. A short time later Richards complained that rather than designing power plants, nuclear vendors were being forced to design safety systems. James F. Young, vice-president and general manager of General Electric's nuclear energy division, made similar points in a letter to Chairman Seaborg. He accepted the need to improve the ECCS in larger plants, but he also suggested that the enhanced capacity for preventing accidents might make possible "compensating adjustments" in safety features installed to reduce the consequences of accidents.[1]

Regulatory staff officials understood the concerns voiced by industry and made it clear that neither they nor the ACRS wanted to halt the construction of nuclear plants. But other than listening sympathetically, the staff did not act on industry's complaints. In response to Young's appeal for relief, Clifford Beck told the Commission that improving emergency cooling was no reason to downgrade accident mitigation features. Until the probability of a major accident could be regarded as negligible, he explained, plants needed both kinds of safeguards.[2]

While it awaited data from pending tests on core melting and ECCS, the regulatory staff believed that its position was sufficiently conservative to continue granting construction permits to plants with emergency cooling systems comparable to those of Dresden III or Indian Point II. At the suggestion of ACRS chairman David Okrent, the staff advised the owners of older operating plants, some of which had "virtually no" ECCS, "to assign top priority" to analyzing the adequacy of their systems to withstand a loss-of-coolant accident. In addition, owners of reactors under construction, after discussions with the regulatory staff, agreed to upgrade the emergency cooling capacities of their plants.[3]

Meanwhile, the task force that the AEC had established in October 1966 to study the issue of emergency cooling began its deliberations. The AEC had requested that the task force consider and make recommendations on several vital subjects: providing "additional assurance" that emergency cooling systems would effectively prevent substantial fuel melting, projecting the behavior of a molten mass of fuel and its interactions with other materials in the containment building, and assessing the prospects for design innovations that could cope with a major meltdown. Harold Price appointed William K. Ergen, a reactor safety expert and former ACRS member from Oak Ridge National Laboratory as chairman of the task force. The rest of the group came either from industry (six members) or from national laboratories and nonprofit research institutions funded by the AEC (five members). The regulatory staff and the Division of Reactor Development and Technology (as it had been renamed) sent representatives to meetings of the task force. In response to the appeals of the ACRS for prompt action on the core melting and ECCS questions, the AEC asked that the Ergen committee complete its review and submit its report within about two months.[4]

The task force started on its assignment immediately and held a series of lengthy meetings. Committee members agreed that in the event of a loss-of-coolant accident and a failure of ECCS, it was likely that containment would be breached. They still viewed containment as a "substantial safeguards system," but not one that could be regarded as a virtually inviolable, independent barrier to the escape of radioactivity. The task force divided over the implications of those conclusions, particularly on the degree to which existing knowledge established the reliability of ECCS. The investigation placed the task force in a delicate position because of the questions its findings inevitably posed about the safety of operating plants and existing designs. Since empirical information on the effects of core melting and the functioning of ECCS under accident conditions was limited, the committee could not reach unambiguous conclusions. Industry representatives were especially anxious to avoid casting doubts on the safety of reactors that were on line or approved for construction.

Ergen told the ACRS in December 1966 that the composition of the task force raised the "possibility that its final report would be too optimistic." Partly as a result of the complexity of the issues with which it was dealing and partly as a result of differences within the task force, the drafting of a report took much longer than the two-month deadline the AEC had requested. Even after the task force had completed a draft report, some members remained dissatisfied. Ergen and a minority of his colleagues on the committee thought that "the conclusions in many cases represent judgments rather than solid fact regarding the performance of cores [and] emergency core cooling systems." For a time they considered issuing a dissenting statement but decided against it after the other members agreed to make some revisions in the text.[5]

The AEC requested further changes in the report. Seaborg noted drily that the findings of the task force would "cause some excitement." Commissioner Ramey worried that certain sections of the document, if read alone, would generate unwarranted alarm. At the urging of the regulatory staff, the committee placed additional emphasis on informing readers that reactors were carefully designed, that severe accidents were unlikely, and that discussions of accident scenarios in the report should not be considered out of context.[6]

Even with its caveats, the Ergen report, published in October 1967, represented a milestone in the evolution of reactor regulation. It was the first public statement, at least from official sources, that acknowledged that containment could conceivably be breached following a loss-of-coolant accident. Although it did not highlight this information, it explained the problem in clear terms: "If emergency core-cooling systems do not function and meltdown of a substantial part of an irradiated core occurs, the current state of knowledge regarding the sequence of events and the consequences of the meltdown is insufficient to conclude with certainty that integrity of containments of present designs, with their cooling systems, will be maintained." The report included a detailed description of what might happen if the largest pipe that carried cooling water to the core completely ruptured. Reactor safety experts regarded such a "double-ended" break of the largest pipe, in which the coolant would rapidly gush from the severed pipe, as the worst-case loss-of-coolant accident. In the absence of an effective ECCS that would cover the heated core with cooling water, a meltdown of the core seemed likely to lead to penetration of containment within a time period ranging from a few hours to a few days.

If containment could no longer be viewed as an inviolable barrier to the escape of radioactivity, the key to public protection from the consequences of an accident was the ECCS. The Ergen report emphasized that "sufficient reliance" could be placed on existing emergency cooling systems to prevent a meltdown. It described potential problems with both core spray and core flooding systems, but concluded that small-scale, laboratory experiments conducted to date showed that, if properly designed, both were satisfactory methods of emergency core cooling. It cautioned that the ECCS installed in a particular plant should be subjected to careful evaluation to make certain that it reflected conservative engineering judgment. Even as it affirmed the soundness of existing approaches, the report cited the need for further research on the operation of emergency cooling systems to provide "additional assurance" that they would prevent a meltdown.

The task force urged action on two fronts to improve reactor safety. One was expanded research, not only on the functioning of ECCS but also on the behavior of molten fuel if emergency cooling failed. The Ergen report gave little attention to the question that the ACRS had emphasized—whether design improvements could enable a reactor to withstand a loss-of-coolant accident even without effective emergency cooling. The task force suggested only a "small-scale, tempered effort" in this regard, at least until more basic knowledge about the problems of dealing with a molten core had been acquired. The other recommendation of the Ergen committee for improving reactor safety was that all components of the primary system, which included the pressure vessel, pipes, pumps, valves, and other equipment used to cool the reactor, should meet standards as high as those of the pressure vessel. It pointed out that a number of small leaks in the primary systems of operating plants had occurred because of poor design or faulty welds. The report suggested that imposing more stringent requirements was "the best way of providing further assurance on the inherent reliability of the primary system." Even if its advice were followed, the task force did not claim that absolute safety could be guaranteed: "We can not prove the impossibility of severe accidents; at best we can try to reduce their credibility to a very low level."[7]

Despite the Ergen report's affirmation of the safety of existing reactor designs, the information it contained was unavoidably disquieting. It downplayed the uncertainties about the effectiveness of ECCS, but a careful reading made clear that research on and knowledge of the functioning of emergency cooling systems were limited. It specifically acknowledged that a breach of containment was possible after a meltdown accident. The analysis it provided of reactor accidents was more unsettling than that of the ill-fated WASH–740 update. The drafts of the WASH–740 revision had used a worst-case scenario in which containment was lost through happenstance, such as an open door, and not through the forces unleashed by an accident.

In contrast to its treatment of the WASH–740 revision, the AEC published the Ergen report. But the agency carefully avoided publicizing it. The regulatory staff drafted several different press releases, which ranged in detail from a sketchy announcement that the AEC had received a report on emergency cooling to a lengthy summary of the findings of the task force. The Commission decided on the least informative approach. The press release that accompanied the publication of the Ergen report merely noted that the task force had completed its work and that its study would be "of great value to the Commission." It gave no indication of the report's substance. The AEC also withdrew the "WASH" report number that had been assigned to the report, which would increase the difficulty, or at least the inconvenience, of obtaining a copy.[8]

The regulatory staff generally agreed with the conclusions of the Ergen report, but it found that in some respects the task force had been unduly optimistic. This was particularly apparent in its assessment of the report's discussion of emergency cooling. The regulatory staff concurred that core spray and core flooding were "satisfactory approaches" to emergency cooling. But in light of the vital role of ECCS in leaving "little or no doubt that these systems will function . . . should a loss-of-coolant accident occur," the staff took issue with the task force. It declared in a report to the ACRS: "Existing test data fall short of covering these areas adequately." The staff review of the Ergen report emphasized that the loss-of-fluid-tests (LOFT) and supporting experiments should "provide a generally satisfactory approach to the problem of verifying the adequacy of currently proposed systems."[9]

The ACRS, whose draft letter on core melting and ECCS had prompted the AEC to form the task force, was more critical than the regulatory staff of the Ergen report. Some members complained that the task force had skirted the issues of primary concern to the ACRS. Instead of suggesting ways in which emergency cooling could be improved and new designs could enable a reactor to cope with a meltdown, it had emphasized its confidence in the adequacy of existing ECCS designs. In this manner, it seemed to suggest that "present arrangements are entirely satisfactory," a conclusion that the ACRS disputed. In February 1968 the committee sent its comments on the Ergen report to Seaborg. After applauding the task force for performing "a valuable service," it also expressed disappointment that the report had provided little information on "design modifications or new design concepts." The ACRS acknowledged the "extremely low probability" of a loss-of-coolant accident in which the ECCS failed, but it reiterated its previous advice that the AEC sponsor a "vigorous" research program to gain a better understanding of a large core meltdown and the performance of emergency cooling systems.[10]

At about the same time, the ACRS and the staff were considering possible causes of a breach of containment to which the Ergen task force had given little attention. One involved the effects of a loss-of-coolant accident on zirconium, the metal used to fabricate fuel cladding in most reactors. If zirconium reacted chemically with steam, which was possible at elevated temperatures during a loss-of-coolant accident, the "metal-water reaction" would generate additional heat and intensify the problem of cooling the core. It would also give off hydrogen, which could burn or explode. Questions about the effects of a metal-water reaction had arisen as early as 1964; the best solution to the problem was an effective ECCS. The Ergen task force had concluded that fuel melting would not occur at temperatures below the melting point of zirconium (3600 degrees Fahrenheit), but it had failed to weigh the consequences of a metal-water reaction, which could occur at much lower temperatures.

A metal-water reaction was a source of concern not only because of the heat and hydrogen it produced but also because it could make the zirconium brittle. This, in turn, could make the fuel cladding shatter and the core collapse. Such a loss of "core geometry" would greatly diminish or eliminate altogether the possibility of cooling the core sufficiently to prevent a meltdown. Joseph A. Lieberman, who ran safety research programs for the Division of Reactor Development and Technology, told his colleagues in February 1968 that the chances of a metal-water reaction indicated "that the margin of safety for present designs is not large, at best, inasmuch as present systems are predicated on maintaining a well-defined geometry for post-accident heat removal."[11]

Another source of concern to the regulatory staff and the ACRS was the possibility of "thermal shock" to the pressure vessel following a loss-of-coolant accident. They were raising questions by mid-1967 about the effects of adding a large volume of cold water from the ECCS to the pressure vessel after a loss of coolant had heated its walls to abnormally high levels. Showering hot metal with cold water was likely to cause cracks in the pressure vessel, but experts differed in their views of whether or not the cracks would pierce the entire five- to twelve-inch thickness of the vessel wall. Cracks that completely penetrated the wall could lead to a disastrous accident by allowing the cooling water from the ECCS to flow out of the pressure vessel.[12]

Uncertainties about ECCS, metal-water reactions, and thermal shock underscored the importance of safety research, and the AEC hoped that its test programs would provide prompt and useful information. Under the direction of the Division of Reactor Development and Technology, the agency used contractors, including national laboratories, to carry out safety research; in 1967, twenty-five contractors were involved in more than fifty separate projects. Many of them were ongoing experiments, such as work on fission product behavior at Oak Ridge, containment integrity at Hanford, reactor kinetics at the National Reactor Testing Station in Idaho, and primary system failure at various sites. Others were newer programs established in response to recently identified safety issues, such as experiments on metal-water reactions at Argonne National Laboratory and on thermal shock at Oak Ridge.[13]

The centerpiece of the AEC's research agenda continued to be the LOFT program, which was conducted by the Phillips Petroleum Company at the National Reactor Testing Station under an AEC contract. A report on safety research projects that Phillips submitted to the AEC in January 1967 highlighted the significance of LOFT in achieving a more complete understanding of reactor behavior under accident conditions, including core melting in the event of a loss of coolant, metal-water reactions, fission-product release and transport, and containment integrity. Researchers in other projects compiled "analytical models" from their findings, and the calculations and computer codes they developed were essential prerequisites to designing LOFT. At the same time, LOFT would furnish the means to verify, refute, or reinterpret the conclusions derived from previous tests and to determine the validity of the analytical models.[14]

A short time after receiving the report from Phillips, the AEC, in response to its recognition of the increased importance of a properly functioning ECCS in larger reactors, shifted the primary objective of the LOFT program. Previously, the goal had been to investigate the progression and the effects of an "unperturbed" loss-of-coolant accident. In light of existing uncertainties about ECCS, the AEC decided to redirect the focus of LOFT to test the operation of emergency core cooling under accident conditions. LOFT would be conducted as a series of experiments, starting with small-scale nonnuclear tests and culminating in tests of how effectively the ECCS in an operating reactor responded to a loss of coolant.[15]

While construction of the LOFT reactor was proceeding, the Phillips Petroleum Company was conducting preliminary tests. The first phase began in 1965 with small-scale experiments to study the effects of a "blowdown," a term used to describe a rapid loss of coolant from the primary system. The tests employed a primitive model of the primary system of a pressurized water reactor with a "core" that was heated electrically to simulate nuclear decay heat. Researchers referred to them as the "semiscale blowdown tests." Their purpose was to provide a better understanding of the forces that a loss-of-coolant accident would produce and the conditions under which safety systems would have to perform. In September 1967, Phillips expected to complete this stage of the LOFT program by February 1969.

The company's researchers planned to move immediately to the next phase, which would use the same equipment to explore the functioning of ECCS. Under varying core conditions, they would inject emergency coolant at different rates and locations. They hoped that this procedure would expand their empirical knowledge of the ability of the ECCS to cool a heated core, and conversely, the impact of the forces of a blowdown on the effectiveness of ECCS. They also anticipated that the semiscale ECCS tests would yield important data on emergency injection rates, pressure, and timing. Phillips scheduled this stage of the LOFT program for completion in July 1969.

Once the small-scale experiments were concluded, the LOFT program would proceed to large-scale "integral tests." Phillips's investigators would run them in a fifty thermal megawatt pressurized-water reactor, and for the first time, would evaluate the performance of ECCS in an actual loss-of-coolant accident. They planned to conduct the integral tests in three stages. During the first phase, they would carry out experiments without a nuclear core in order to better assess the results of the small-scale tests, judge the validity of analytical models for predicting the effects of a loss of coolant, and gain further data on ECCS operation.

During the second phase of the integral tests, researchers would, for the first time, test the functioning of ECCS in a nuclear core. After allowing the reactor to run for at least twenty hours, they would cause a pipe to rupture and activate the ECCS. The purpose was to appraise the effectiveness of ECCS under design conditions in an operating reactor and, once again, to check the accuracy of analytical models. The final stage of the LOFT program would test the performance of "degraded" emergency cooling systems. The objective was to determine how well the ECCS would work if its "delivery rate" of coolant was slower than projected. This would not only enable researchers to evaluate existing designs and models but could also demonstrate whether a loss-of-coolant accident might cause "unexpected phenomena." Phillips planned to conduct the initial nuclear-core tests between May and December of 1971 and the final phase between January and September of 1972.[16]

The status of and plans for reactor safety research elicited different responses from the three groups whose programs or institutional interests were most involved—nuclear vendors, the ACRS, and the regulatory staff. Industry representatives did not believe the research agenda was particularly worthwhile and questioned whether much of it was even necessary. When the AEC circulated the January 1967 Phillips report on reactor safety projects for comment, it received a highly critical appraisal from a task force drawn from the Atomic Industrial Forum's Committee on Reactor Safety. Theodore Rockwell, chairman of the task force, told Milton Shaw, director of the Division of Reactor Development and Technology, that safety research had "not significantly served either the industry or the AEC Regulatory Staff in their review and evaluation of license applications." He complained that the research program placed too much emphasis on hypothetical and low-probability accidents, that it was too slow and too diffuse to be of much value in designing plants, and that it did not produce results that the regulatory staff and ACRS found sufficient "to satisfy their concerns."

Rockwell urged that the AEC establish a way to apply the findings of research projects to the licensing process, to make clear whether existing data on a given problem were adequate to license a plant, and if not, to spell out what additional information was needed. Objecting to excessive "research for its own sake," he appealed for programs that addressed a specific problem within a fixed time frame. And he recommended that more emphasis be placed on tests that would help in the prevention of "realistic" accidents. "Too much emphasis is currently being focused," Rockwell wrote, "on studying the cause and course of accidents which have only a remote probability of occurrence."[17]

Other industry spokesmen echoed the same sentiments. At a meeting with members of the ACRS and the regulatory staff in February 1968, representatives of reactor manufacturers maintained that they had "sufficient information for current designs" and that those designs were "adequately safe without further major R&D." If the AEC disagreed, they challenged it to provide information about precisely what constituted adequate safety. Industry officials viewed research as a useful means of confirming the conservatism of their designs and of increasing knowledge about reactor technology, but not as essential to the safe operation of the existing generation of nuclear plants. A representative of Combustion Engineering suggested that evidence from tests on core melting was not needed to place reactors at metropolitan sites and predicted that plants would be "located in popular areas before any R&D results [were] available." Theodore Rockwell reiterated his conviction that "many of our problems exist because of the unreal world we have built." He added: "We can't simulate instantaneous double-ended [pipe] breaks because things don't break that way."[18]

The position of the ACRS on safety research sharply contrasted with that of industry. While acknowledging that a fuel meltdown seemed unlikely, its members believed that efforts to resolve outstanding issues promptly were imperative. Although operating experience, experimental data, and analytical models provided a basis for confidence in reactor designs, the ACRS insisted that much remained to be done to confirm the safety of existing plants. It rejected the view that sufficient information was available to place reactors in metropolitan locations. As Raymond F. Fraley, executive secretary of the ACRS, pointed out, reactor safety systems had never been fully tested under the conditions of a major accident. "It is difficult to say just how much margin, if any, exists in the safeguards being proposed for today's nuclear plants," he wrote in March 1967. "Hopefully there is considerable margin but this has yet to be proven." The ACRS was anxious to ensure that empirical evidence be developed to demonstrate the degree of conservatism in the plants that were being proposed. Rather than criticizing safety research plans as excessive, ethereal, or largely unnecessary, it called for an expansion of the program. It sought to obtain more data and greater assurance about the safety of the designs of plants under construction before their owners applied for operating licenses.[19]

The regulatory staff stood between the positions of the industry and the ACRS. Like the industry, it was satisfied that designs for the plants in operation or under construction were fundamentally sound. Peter A. Morris, director of the Division of Reactor Licensing, declared in 1968: "The central problem of reactor safety is to define the degree of conservatism necessary for public acceptance of reactor applications." But regulatory staff officials were more troubled than industry spokesmen about the possibility that a major accident might occur and the uncertainties of what might happen if it did. They thought that research programs were essential to learn more about the technology and to increase margins of safety. Like the ACRS, they rejected industry's contention that enough information was available to locate plants at metropolitan sites. The regulatory staff continued to oppose metropolitan siting, at least until it had more evidence that the safety systems of newer reactors were reliable. As Edson G. Case, director of the Division of Reactor Standards, observed in 1967: "All of the data on existing plants *can* be extrapolated to the larger plants now being proposed, but the real question is in the absence of operational or other experience, how much confidence can one place in these extrapolations."[20]

The regulatory staff also took issue with the industry's position on the usefulness of the AEC's safety research projects. It denied that there was a meaningful distinction between research that applied to "realistic situations" and the agenda the AEC was following. Since it could not rule out the kind of serious accidents being investigated, the staff found the industry's complaints about overemphasis on hypothetical accidents unpersuasive. It also disagreed with the charge that the research program had not produced results that were helpful in the licensing process. Case told the Joint Committee in 1967 that experiments on metal-water reactions, fission-product activity, pipe ruptures, and blowdown forces had provided valuable information to designers and regulators. He admitted that "it would be advantageous to have the final results of the LOFT program available now," but he affirmed that "we are profiting already from the analytical program supporting the LOFT program." The regulatory staff did not believe that it should hold up construction permits or operating licenses until outstanding questions were fully addressed; Clifford Beck remarked that the AEC had never "been able to answer all safety questions," in part because "we have assumed pessimistic conditions that we felt would not occur." But, like the ACRS, the regulatory staff still sought to gain as much information as possible as quickly as possible.[21]

The value and the timeliness of the research program for the regulatory staff depended heavily on the cooperation of the Division of Reactor Development and Technology. Despite informal contacts and the establishment of the Steering Committee on Reactor Safety Research in 1965, differences over funding, priorities, and communication continued to arise. Like the ACRS, the regulatory staff pressed for a study of design improvements that could enable a reactor to withstand a fuel meltdown even if the ECCS did not work effectively. But Shaw and his staff ignored the appeals for such an investigation. The regulatory staff also expressed frustration that it did not always receive information from Reactor Development on the costs of or the schedule for research projects. Peter Morris expressed his exasperation at one point by protesting: "We feel an urgent requirement to have [this] information in order that R&D funds may be allocated with proper attention to priority to get the most useful results."[22]

George M. Kavanagh, the AEC's assistant general manager for reactors and the man to whom Shaw reported, explained to the Joint Committee in 1967 that the perspectives of the regulatory staff and Reactor Development were inherently divergent. "The research and development man is characteristically disappointed that the regulatory man still has questions after complex and costly experimentation and analysis," he said. "The regulatory man is characteristically disappointed that the research and development man does not find an answer that he can completely accept and that the research man is so long about the tests and the experiments that he does do." Nevertheless, Kavanagh thought that the AEC's research program was "operating effectively" because both the regulatory staff and Reactor Development were committed to ensuring safety.[23]

Despite Kavanagh's assessment, the differing perspectives and priorities of the two staffs were accentuated by a series of delays in the LOFT project. All phases of the LOFT program suffered schedule setbacks and design difficulties, particularly progress in the construction of the reactor in which the integral tests would be conducted. By early 1970, the Division of Reactor Development and Technology was estimating that the culminating nuclear experiments in the LOFT facility would not begin until late 1974. One important reason for the problems with LOFT was deficient management on the part of the Phillips Petroleum Company. Shaw complained in July 1968 that "previously recognized management and engineering inadequacies of the Phillips's organization" had become "even more serious." He attributed the company's performance, described in one AEC report as "far short of being adequate," to a general lack of corporate concern about or support for its contractual obligations for nuclear safety research. As a result, Phillips failed to recruit the engineering talent it needed or to provide other resources essential for the success of the LOFT project. Shaw concluded that "the chances of Phillips's taking effective actions, either to meet their current commitments or for the long term, are minimal."[24]

Phillips responded to the AEC's expressions of dissatisfaction by promising to reevaluate its performance on the LOFT project, but it denied that serious problems existed. By December of 1968, Shaw's prognosis was as gloomy as ever. While acknowledging that the company had increased its corporate involvement to an extent, he was not impressed with the results. "The objectives and plans associated with LOFT," he told his superiors, "are still not being adequately developed and commitments are not being met in a timely manner." A short time later, Phillips reached an agreement in which the Idaho Nuclear Corporation assumed the major responsibility for LOFT and other research programs. The managing partner and majority owner of the Idaho Nuclear Corporation was the Aerojet-General Corporation, a subsidiary of the General Tire and Rubber Company. Phillips became a minority stockholder in the Idaho Nuclear Corporation. The AEC was encouraged by this development, but lamented the lack of progress on LOFT at a time when reactor safety research was taking on increasing urgency.[25]

Shortcomings in the management of safety research at the Idaho test site presented serious problems under any circumstances, but they had been compounded when the AEC shifted the focus of the LOFT program in 1967 from a study of an unimpeded loss-of-coolant accident to a more complicated investigation of ECCS operation. This required not only that plans for experiments be reassessed and modified but also that the test reactor be redesigned. While some of the preparations completed before the redirection of the LOFT program were still useful, the design of the new equipment virtually halted construction for two years. Both the regulatory staff and Reactor Development viewed the reorientation of LOFT's objectives as essential, but the change inevitably exacted a toll in the costs and in the timeliness of the project.[26]

Progress on the LOFT project was further retarded by the Division of Reactor Development and Technology's insistence that construction of the test reactor meet rigorous quality standards. In response to growing AEC concern about improving the quality of components used in nuclear plants, the division began to apply more stringent requirements to test reactors. It did so for two basic reasons. One was to lessen the risk that the results of reactor experiments would be invalidated or obscured by failures of components, materials, or systems that were not being investigated. Quality assurance increasingly seemed to be a prerequisite for obtaining clear and usable experimental data. A second reason that Reactor Development placed greater emphasis on quality assurance in test reactors was to gather information to help develop realistic standards. As Shaw pointed out, a nuclear plant needed to comply with as many as 5,000 separate engineering standards, but professional standards-setting groups had established only about 100 that applied specifically to reactors.

Scientists and engineers who worked on the LOFT project agreed on the importance of quality assurance in commercial reactors, but they questioned the applicability of stringent standards to the test program. Phillips Petroleum argued in June 1968 that LOFT was not being built for the same purposes as a commercial plant and that imposing the same quality standards would cause major delays without compensating benefits. The company suggested that strict application of quality standards would require "extensive redesign" and time-consuming consultations with the AEC and reactor vendors. But it would provide little information of value for commercial plants because the LOFT reactor by design would undergo a loss-of-coolant accident. Reactor Development rejected those appeals; Shaw told the ACRS that substandard equipment should not be allowed to jeopardize the LOFT program. Construction of the LOFT reactor proceeded according to quality standards prescribed by the Division of Reactor Development and Technology. The AEC's goal of improving quality assurance was unassailable, but applying it to LOFT levied a price both in further delays and higher costs.[27]

The effects of the conditions and requirements that held up progress and raised the costs of the LOFT project—management deficiencies, program redirection, and increased emphasis on quality assurance— were exacerbated by reduced funding. Beginning in the latter half of the 1960s, when costs for the Vietnam war and Great Society programs sharply escalated, the Bureau of the Budget searched diligently for ways to cut government spending in other areas. The drive for budget austerity initiated by Lyndon Johnson continued after Richard Nixon took office. In 1967, the Johnson administration ordered federal agencies to slash their expenditures of funds already allocated by ten percent. As a part of its effort to find other means of saving money, the Bureau of the Budget asked the AEC to explain its plans to increase spending for reactor safety research. Why should safety research costs continue to grow, the bureau wanted to know, when the AEC was issuing "substantial numbers" of licenses for commercial reactors that it had judged to be safe?[28]

The AEC vigorously defended the need for safety research on both existing light-water reactors and on more advanced fast breeder reactors. It pointed out that the construction permits it had granted were not licenses to operate, and added: "Many safety issues remain to be resolved or to be better understood before a substantial number of these plants will be able to be licensed for operation." The AEC suggested that its budget projections were the "minimum required" to obtain the information it sought on existing reactors, but it also predicted that the

8. LOFT reactor under construction at the National Reactor Testing Station in Idaho, 1969. (National Archives 434-SF-109-19)

costs of research on light-water models would rapidly decline within a few years once LOFT and other major programs were completed. By that time, the agency anticipated that the bulk of its research funds would be devoted to fast breeder technology, and it made the case for advanced reactors as irresistible as possible. "The successful introduction of these advanced reactors into the utility market," it declared, "will provide a virtually unlimited supply of low cost energy which can be used to elevate man's standard of living and, with proper attention, improve man's environment."[29]

The AEC's appeals were only partly successful. The Bureau of the Budget did not insist on cutting the AEC's safety research budget, but neither did it allow the significant increases the agency sought. As a result, the Division of Reactor Development and Technology was forced to allocate funds that fell far short of what it requested. Both the Commission and the Joint Committee were committed to rapid development of fast breeder reactors, which gave Shaw the task of balancing the funding restrictions imposed by the Bureau of the Budget against the demands of his superiors and the Joint Committee for progress on the breeder. Since important safety questions about breeders remained to be addressed, they consumed a growing portion of the AEC's safety research budget. Light-water reactor research suffered accordingly. Between 1968 and 1971 the AEC spent approximately $23 million annually on light-water research, down from the more than $30 million it had disbursed annually in the mid-1960s.[30]

The development of a commercial fast breeder reactor had been a major goal of the AEC for years. The technology was extremely appealing because it promised the means to create "fissile material," which could be made into nuclear fuel, in quantities greater than the fuel it consumed. The breeder converted the most common isotope of uranium, U–238, into a fissile isotope capable of sustaining a chain reaction, plutonium–239. This could be done by placing a "blanket" of uranium–238 around a core of uranium–235 or plutonium–239 so that it would absorb free neutrons released from the fuel. By capturing neutrons, followed by radioactive decay, the uranium–238 could be transformed into plutonium–239, which could then be extracted and used to make more fuel. The technology was complex, but if it could be mastered, it offered the means to produce virtually unlimited supplies of nuclear fuel from cheap and abundant uranium–238.[31]

The AEC maintained that, over the long term, development of the breeder reactor was essential to provide adequate electrical power. Fossil fuels would eventually be depleted, as would the relatively rare uranium−235 (only 0.7 percent of uranium found in nature) used in light-water reactors. In the report on atomic power's contributions to meeting future energy needs that it submitted to President Kennedy in 1962, the AEC emphasized the crucial role that the breeder would play. The Division of Reactor Development and Technology was responsible for the agency's breeder reactor program, and Shaw applied his forceful manner and formidable energies to pushing it ahead. "We have the potential," he declared in 1968, "to satisfy the energy needs of mankind for a very long time by the process of breeding."

Shaw took his cue from both the Commission and the Joint Committee, where support for breeder development was strong and tolerance for delays was limited. Joint Committee chairman Holifield called the breeder "indispensable" for meeting energy and environmental demands and expressed impatience with the AEC for not making better progress on its "highest priority civilian program." In 1970, Seaborg described development of the breeder as "a priority national goal" that represented "the most decisive single step that could be taken now toward assuring an essentially unlimited energy supply, free from problems of fuel resources and atmospheric contamination."[32]

By the late 1960s, plans for a prototype breeder reactor were well under way, and the need for safety testing of breeder technology was becoming more urgent. At the same time, the AEC's budget for safety research was tightening. In order to proceed with the breeder program, therefore, Shaw elected to reduce funds for light-water research. He recognized that a number of important questions about light-water safety remained outstanding, but when forced to allocate the money available to him, he cut the amounts for light-water research and increased them for breeder safety. In absolute terms, the level of spending for breeder safety research was still small—$5.2 million in fiscal year 1969 and $7 million in fiscal 1970. But it represented a significant percentage of the funds budgeted for all safety research. The extent to which the regulatory staff objected to Shaw's budget decisions is unclear, but it seems apparent that even strong and vocal protests would have been futile. Shaw would have reallocated the research budget only if the Commission had directed him to increase light-water spending, but it was committed to his campaign to develop the breeder and supported his division of funds.[33]

The result of budgetary constraints and the AEC's assignment of top priority to breeder development was to slow progress on light-water safety research. The LOFT reactor received allocations that climbed gradually, but it also incurred higher construction costs that arose from the application of strict quality standards and from inflation. Tight budgets were not the only and perhaps not even the major reason for delays in the construction of the LOFT reactor, but they were symptomatic of the failure of the Commission to focus on the importance of obtaining information promptly and to attach a high priority to light-water safety research. While the regulatory staff was seeking more experimental data on the performance of safety systems, especially ECCS, for use in its licensing decisions, neither the Commission nor the Division of Reactor Development and Technology viewed the light-water research program with a sense of urgency. If the impact of budget shortfalls on LOFT construction was ambiguous, the effect of them on other research programs was not. A number of tests had to be suspended or terminated, and the Idaho Nuclear Corporation was forced to lay off researchers. This not only caused further delays in tests to support LOFT and other studies but also raised questions in the minds of some researchers about whether sufficient resources were available to run experiments that were technically sound.[34]

The budget stringencies facing the AEC's water-reactor research program gave increased impetus to appeals to the nuclear industry to take over a larger share of the burden. The AEC and the Joint Committee had long urged the industry to perform more safety research on its own, and the Bureau of the Budget had requested that the agency find ways to shift safety programs, and their costs, to nuclear vendors. Reactor manufacturers were not inclined to respond favorably. They were already spending considerable amounts on designing safety systems for their plants, including conducting their own investigations of the functioning of emergency cooling, containment, control rods, and instrumentation. They were sharing the expenses of other projects, such as studies of pressure vessel reliability, pipe durability, and fission-product filtration, with the AEC. They were opposed to taking on further responsibility for safety research partly because of the additional costs and partly because they still did not believe it to be necessary. Reactor manufacturers insisted that their designs allowed an ample margin of safety, and they denied that the AEC programs were essential for safe operation of their plants. As an ACRS staff member noted after a meeting with representatives of reactor vendors: "It seemed that the vendors believe that, since the AEC/ACRS raised certain safety questions, the AEC should finance the effort to find answers to these questions."[35]

Reactor manufacturers made their views clear in a meeting with regulatory staff and Reactor Development officials in November 1969. They reiterated their long-standing complaints that the AEC's research program produced results too slowly to be of much benefit in the licensing process and that it overemphasized work on accidents that were highly unlikely. The industry representatives focused on the problems with LOFT to underscore their points. They were unhappy with the slippage in the schedule for the integral testing and expressed doubts that it would yield results in time to be useful. "A substantial segment of the industry," grumbled one executive, "feels r&d progress is not oriented toward timely resolution of safety problems." Industry officials also wondered about the applicability of the tests, even when they were run, to much larger commercial plants, and about the extent to which LOFT would resolve outstanding safety issues. AEC spokesmen recited the value of LOFT for increasing general knowledge of the repercussions of a loss-of-coolant accident and the effectiveness of emergency cooling, but industry representatives remained unconvinced. As one put it: "There is growing concern that the LOFT program is soaking up money and manpower and just not getting anywhere." Because of their reservations about the usefulness of LOFT and their belief that much of the AEC's safety research was superfluous, reactor vendors resisted appeals to help the agency carry out its program by increasing their expenditures on research.[36]

The ACRS also complained about the research program in general and LOFT in particular, but its criticism reflected considerations that were quite different than those of industry. It regretted that as a result of budget cuts, "many safety research activities have not been initiated, have been slowed, or have been terminated." The ACRS chided the AEC for failing to sponsor a major effort to resolve issues it had cited as key concerns for years, including the "course of events following partial or large scale core melting" and design improvements to handle a core melt without a functional ECCS.

Members of the ACRS divided in their opinions about the usefulness of LOFT. Some thought it would provide uniquely valuable information about a loss-of-coolant accident and the performance of ECCS; others believed its applicability to large reactors was limited. They agreed, however, that LOFT should not be pursued if it compromised the AEC's ability to undertake what the ACRS viewed as more pressing experiments. The ACRS concluded that if the AEC found that continuing with LOFT was useful, it should make "every reasonable effort" to complete the program "on an improved time scale." In short, the committee was disgruntled with the effects of the tightened research budget, the refusal of the AEC to respond promptly to its recommendations on research priorities, and long delays in receiving results from safety investigations. As one unidentified ACRS member told *Nucleonics Week* in June 1970: "Increasingly we will find ourselves having to deal with problems with less factual information on hand than we would like." He added: "Of course, we have had to do that in various areas heretofore—and it always raises a certain amount of storm and strife outside."[37]

The ACRS member's comments proved to be prescient—within a short time the lack of conclusive data on the performance of emergency cooling systems triggered a major controversy and a great deal of "storm and strife." During the summer of 1970, the findings and analyses of ECCS tests carried out at the Idaho testing site and other locations had aroused concern among researchers. Some experiments had indicated that in the event of a loss-of-coolant accident, temperatures in the core could rise higher than previously believed and the coolant from the ECCS might not reflood the core as rapidly as predicted. While those problems were being reexamined, a new series of tests on ECCS operation stirred even greater consternation. Run as a part of the small-scale LOFT supporting experiments, they produced unexpected results that were the immediate cause of the debate over the effectiveness of ECCS designs.[38]

In November and December of 1970, Idaho Nuclear Corporation researchers conducted four "semiscale" tests. They were performed with a core that was nine inches long, compared with a 144-inch core in a power reactor. Their purpose was to study the effectiveness of ECCS under accident conditions in a pressurized-water reactor, though their results were also applicable to some extent to boiling-water reactors. The experiments were run by heating the simulated core electrically, allowing the cooling water to escape, and then injecting the emergency coolant. To the surprise of the investigators, the high steam pressures that were created in the vessel by the loss of coolant blocked the flow of water from the ECCS. Without ever reaching the core, about 90 percent of the emergency coolant flowed out of the same break that had caused the loss of coolant in the first place. In February 1971 the Idaho Nuclear Corporation reported: "Preliminary analysis of these tests indicates little or no cooling by the emergency coolant."[39]

In many ways the semiscale tests were not accurate simulations of designs or conditions in power reactors. Not only the size, scale, and design but also the channels that directed the flow of coolant in the test model were markedly different than those of an actual reactor. Unlike commercial plants, which had from two to four loops for the flow of coolant, the test equipment used only a single loop. This increased the possibility that emergency coolant could bypass the core and flow out the break in the single loop rather than reaching the core through other paths. Nevertheless, the results of the tests were disquieting. They introduced a new element of uncertainty into assessing the effectiveness of ECCS. The outcome of the tests had not been anticipated and called into question the analytical methods used to predict performance. "Nobody at this time is able to predict definitely from the models whether this would or wouldn't happen in a real loss-of-coolant accident," commented one scientist who worked on the experiments. "Although we feel pretty sure we can flood the core, our difficulty is that we can't quantify the margin of safety we have." For that reason, the semiscale tests generated serious concern.[40]

The results of the experiments caught the regulatory staff unprepared. Harold Price told the ACRS in March 1971 that his staff was uncertain of where it stood on the issues raised by the tests. Their outcome created doubts within the staff that it could testify "with reasonable assurance" that the emergency cooling systems were adequate in several plants under review for construction permits or operating licenses. In early 1971, Price had established a special task force of senior regulatory staff members to review outstanding safety questions. In March, in response to the results of the semiscale experiments, he directed it to evaluate on a priority basis the methods used to predict the performance of emergency cooling systems and to draft a "white paper" within a month. Until he received a report from the task force, Price wanted to avoid any public discussion of the semiscale experiments and their ramifications.[41]

The unexpected uncertainties over the performance of ECCS came at an exceedingly awkward time for the AEC. The Joint Committee was rebuking the agency for what it viewed as unwarranted licensing delays and was planning to hold hearings on the AEC's regulatory process. Complaints from the committee about the length of the licensing process were nothing new, but they took on increased intensity in late 1970 and early 1971 as a growing number of applications clogged the system and a shortage of electrical generating capacity emerged as a major concern in some areas. Chairman John Pastore told Seaborg that he and his colleagues had "long been concerned with the undue delays" that were attributable to "certain administrative obstacles." At the same time, public opposition to nuclear plants was growing and intervenors were citing misgivings about ECCS in their arguments against the licensing of reactors. If the results of the semiscale experiments became public knowledge, it would add credibility to their position and probably cause more delays. The regulatory staff worried not only about the effects of the tests on five plant applications ready for a decision but about dozens of others that were further back in the pipeline. The semiscale tests were hardly conclusive, but they raised important questions that had to be addressed.[42]

Within a short time after undertaking its investigation of ECCS performance, the task force that Price had established encountered major complications. Its chairman, Stephen H. Hanauer, a former professor of nuclear engineering at the University of Tennessee, past chairman of the ACRS, and since 1970 the technical adviser to the director of regulation, soon discovered bureaucratic and financial problems that made it more difficult to deal with the technical ones. In a meeting with Andrew J. Pressesky, assistant director for nuclear safety in the Division of Reactor Technology and Development, Hanauer acknowledged that "we have some real safety problems." He complained bitterly that he had learned that because of funding shortages, the Idaho Nuclear Corporation was going to "fire some of [the] good people that can work [on the] problems." Pressesky confirmed that of the twelve researchers working on LOFT support at the Idaho site, eight would lose their jobs before the end of the fiscal year unless more money became available.

To make matters worse, Hanauer learned from Pressesky that the test model for the recent semiscale experiments had been dismantled in order to build a more sophisticated version. There was, Pressesky added, "no money to put [it] back together." An exasperated Hanauer wondered: "Where else are we in trouble and don't know it?" In response, Pressesky cited as "another problem area" the latest results of experiments on the possible effects of a loss-of-coolant accident on fuel cladding.[43]

Scientists from Oak Ridge National Laboratory who had conducted a series of tests on "fuel-rod failure" reported in early 1971 that previous analyses of how well the cladding would hold up might have been overly optimistic. The tests had produced two disturbing findings. One was the amount of swelling that might occur in the fuel cladding after a loss of coolant. This was an important consideration because if the high temperatures in the vessel distorted the shape of the cladding, it could block the circulation of water and prevent the ECCS from adequately cooling the core. Industry experiments had indicated that the swelling of the rods would not exceed 60 percent, but Oak Ridge researchers found that the extent of the swelling might exceed 100 percent.

The other major question that aroused concern was the temperature at which the fuel cladding became brittle and susceptible to shattering. If the cladding fractured and the core collapsed, the water from the ECCS would be unable to circulate through the molten mass that the loss of "core geometry" created. It was apparent that this could happen at temperatures below the melting point of zirconium because of metal-water reactions, but the level at which the fuel cladding would become brittle was less clear. Nuclear vendors claimed that as long as the temperatures in the core did not exceed 2700 degrees Fahrenheit, the cladding would not fail. But experiments at Oak Ridge yielded more pessimistic results. They indicated that, depending on the length of time that abnormally high temperatures existed in the core, the cladding could become brittle at levels considerably lower than 2700 degrees. Tests showed, for example, that significant embrittlement took place at temperatures of 2500 degrees for two minutes and 1950 degrees for thirty-five minutes. As long as core temperatures were lowered promptly, embrittlement of the cladding did not appear to be a major problem. But the findings of the Oak Ridge scientists added new uncertainties about the adequacy of existing assumptions and experimental information relating to the consequences of a loss-of-coolant accident.[44]

While the Hanauer task force was continuing its investigation, the AEC was attempting to prevent news about the semiscale and fuel cladding experiments from leaking outside the agency. Rumors about the semiscale tests had circulated among scientists since mid-February 1971, when a researcher attending a meeting on reactor safety had informed his colleagues that the budget for the LOFT support program was facing new reductions and appealed for funds from industry. Nevertheless, the AEC went to extraordinary lengths to keep the evaluation of the recent tests an internal matter, even to the point of not reporting information on them to the Joint Committee. The effort was futile. After two committee members heard about the potential problems with emergency cooling on a visit to the Idaho test site, Seaborg hastily sent a letter to Chairman Pastore. He told Pastore that "the results of recent preliminary safety research experiments have indicated that the predicted margins in ECCS performance may not be as large as those predicted previously." He disclosed that a "senior task force" was evaluating the issue and acknowledged that as a result, the AEC anticipated "that there will be delays in the licensing of some plants." A short time later, the AEC provided further information, in executive session, to the Joint Committee.[45]

The AEC realized that Seaborg's letter to Pastore was likely to stir press inquiries and that the effort to prevent information about the experiments on the effectiveness of ECCS from getting out to the public was doomed. After the Joint Committee released Seaborg's letter, a number of intervenors and other critics demanded that the AEC supply complete details about the experiments. Industry representatives also called for a full report from the agency. "How can we tell the public and our customers what is going on," complained one official, "if the AEC, which first identified the engineering problem, has said nothing?" On 26 May 1971 the *Washington Post* blew the story open with a front-page account of the semiscale tests. It suggested that the experimental failures could cause lengthy construction and licensing delays and "trigger a nationwide power shortage." The following day the AEC issued a press release, emphasizing that a loss-of-coolant accident was highly unlikely, that ECCS was only one of many safety systems in a power reactor, and that the semiscale tests did not accurately simulate the design or operation of an actual plant. It revealed that the regulatory staff's task force was considering whether improvements were needed in emergency cooling systems and predicted that it would complete its review within a few weeks. In the meantime, the AEC cautioned that some licensing delays might be necessary.[46]

The press release did not placate the AEC's critics, who assailed it for refusing to make information about the results of the semiscale tests public earlier. Some nuclear opponents called for a moratorium on licenses and the shutdown of the eleven operating nuclear plants until the ECCS question was resolved. Even friendly observers were troubled by the AEC's handling of the issue. Edward E. David, President Nixon's science adviser, told Seaborg: "This case does raise the question of the adequacy and effectiveness of the Commission's research program on safety. The safety program should provide us with answers to assure the public safety." Since the late 1960s, the AEC had tried to answer the question posed by the overseers of the federal budget: why is safety research needed if licenses are being issued for nuclear plants? The ECCS controversy stood that question on its head. It became, in the minds of a growing number of informed commentators: why is the AEC granting licenses if it has not fully resolved key safety issues?[47]

In the face of mounting criticism, the AEC attempted to clarify its position on ECCS and take action that would diminish the embarrassment the issue was causing. For years, despite the urging of the ACRS, the agency had declined to place a high priority on water-reactor safety research. The Commission had allowed the program to drift while it focused on matters it viewed as more pressing. But the ECCS controversy suddenly enhanced the importance of safety research in the eyes of the Commission. In an unprecedented step, Seaborg called the Office of Management and Budget (as the Bureau of the Budget had been renamed) and appealed for a restoration of funds for water-reactor safety research that it had cut from the AEC's budget. He cited the need to conduct research "on a crash basis because we have run into some problems." The budget office reluctantly offered an additional $2 million. Assistant general manager Kavanagh told the Joint Committee, however, that he thought the AEC required an extra $30 or $40 million.[48]

Meanwhile, the Hanauer task force that Price had formed was pursuing its study of the methods used to predict the performance of emergency cooling systems. It quickly became obvious that the task force could not complete a report within a month, as Price had requested. With the assistance of the Idaho Nuclear Corporation, the task force closely examined the use of computer programs and calculational techniques in designing and evaluating emergency cooling. The results of the study were enlightening but not always clear in their implications. Most members of the task force were satisfied that the work generally confirmed the conservatism of ECCS designs, but some of their colleagues contended that the computer programs were ambiguous, unproven, and unreliable on key safety issues. They argued that more empirical research was essential to establish the applicability of existing assumptions and computer programs. The complexity of the ECCS issue and the differing views within the task force prevented the early submission of a final technical report.[49]

The Commission did not want to wait for the results of new experiments or the completion of the Hanauer task force's investigation before it took a position on the ECCS question. Once the uncertainties about ECCS became a matter of public debate, it sought to answer the agency's critics and to reassure the public about reactor safety. The Commission and staff held a series of meetings in early June 1971 to consider their options on the issue. Based on the preliminary findings of the Hanauer task force, the Commission decided to issue interim criteria on emergency cooling systems, effective immediately, that plants would have to meet. The general requirements for all reactors stipulated that in the event of a loss-of-coolant accident, the temperature of the fuel cladding could not exceed 2300 degrees, the cladding had to be cooled before it became brittle, and no more than 1 percent of the cladding could react chemically with steam or water. Drawing on the results of the semiscale tests, the criteria specified that manufacturers should assume that coolant from the ECCS would bypass the core during the blowdown phase of the accident. This would mean that one-fourth to one-third of the total ECCS coolant would be lost (the remainder would enter the vessel after the blowdown ended and pressure fell).

In addition, the AEC's statement outlined ways in which reactor manufacturers could analyze their specific designs to make certain they complied with the criteria. The agency instructed owners of reactors licensed to operate before 1968 to analyze the capabilities of their rudimentary emergency cooling systems and make improvements in them before 1 July 1974. It directed the operators of the newer and larger plants to submit an analysis of the performance of their emergency cooling systems within three months. The regulatory staff hoped that the interim criteria would not only provide guidance to reactor manufacturers but also would impress the industry with the importance of addressing and resolving the ECCS issue.[50]

The AEC did not prescribe the methods for meeting the new requirements, but, in effect, it mandated that manufacturers and utilities place an upper limit on the amount of heat generated by the reactor. In some cases, this would force utilities to reduce the peak operating temperatures (and hence, the power) of plants in operation or expected to be soon, from an estimated 7 or 8 percent at two Turkey Point reactors in Florida to 2 percent at Indian Point II and the H. B. Robinson unit in South Carolina. The prospect of "derating" power levels, *Nucleonics Week* reported, "sent a shock wave through the industry," which was "totally opposed to it." The Commission was reluctant to impose derating on even a few plants, but viewed it as a temporary measure. In other cases, the ECCS criteria could subject utilities to extra costs by requiring more frequent refueling.

Despite complaints from Shaw that the criteria were overly conservative, representatives of reactor vendors did not view them as unreasonable. The ACRS expressed general approval of them, but urged the AEC to make clear that they were "an interim solution only" and that more research was needed on computer programs for and designs of emergency cooling systems. The AEC announced the ECCS criteria on 19

June 1971; the same day, Price told a press conference that although the AEC thought it was impossible "to guarantee absolute safety," he was "confident that these criteria will assure that the emergency core cooling systems will perform adequately to protect the temperature of the core from getting out of hand."[51]

The AEC was anxious to issue the interim criteria without waiting for the final report of the Hanauer task force or the results of further tests for a number of reasons. One was that it hoped to minimize construction and licensing delays in the more than fifty plant applications under review. This was a perpetual source of concern, intensified by the relentless pressure the AEC received from the Joint Committee and the industry to judge plant proposals and issue licenses promptly. But it took on increased urgency in early 1971 because a number of utilities in different sections of the country were facing shortages of electrical generating capacity and were worried about the availability of sufficient power. The twenty-nine members of the Atomic Safety and Licensing Board panel, from which the three-member boards for individual licensing cases were drawn, were meeting in late June 1971. Both the Commission and the regulatory staff wanted to be able to provide panel members with guidance and reassurance on ECCS performance because it was certain to be a prominent issue in licensing proceedings. A major consideration of the AEC in publishing the interim ECCS criteria was the realization, in Seaborg's words, "that we have an immediate problem to solve with respect to the attitudes of licensing boards."[52]

The AEC had additional reasons for acting immediately to deal with the ECCS issue. The Joint Committee had scheduled hearings on the licensing process to open on 22 June 1971, and the AEC was determined to arrive at a position on ECCS before then. This would enable the commissioners and staff to answer questions about the ECCS problem and also serve as a counterbalance to the criticisms that nuclear opponents were sure to air. "We couldn't avoid the issue if we wanted to," commented one committee staff member.[53]

The AEC's push to resolve, or at least to defuse, the ECCS controversy, also reflected its involvement in President Nixon's energy program. In the spring of 1971, Nixon was preparing a major message to Congress and the nation on his plans to develop sufficient energy supplies while at the same time protecting the environment. He requested the opinions of Seaborg, who emphasized the energy and environmental assets of nuclear energy and highlighted the potential advantages of breeder reactors. When Seaborg presented his views at a cabinet meeting on 13 April 1971, the president seemed greatly impressed with the promise of breeders. Nixon also asked questions about reactor safety; he disclosed that he had never worried about nuclear hazards until Southern California Edison published advertisements with pictures showing how close the western White House in San Clemente was located to the San Onofre nuclear plant. "Then," joked the president, "I began to worry." Seaborg assured him that the probability of an accident was "so small it's difficult to estimate."[54]

Nixon strongly endorsed the development of a breeder reactor, partly because of the advantages it offered and partly as a means to win the backing of Chet Holifield for a major government reorganization that the president was planning. Holifield, in addition to his position on the Joint Committee on Atomic Energy, was the chairman of the House Government Operations Committee. The president hoped that his support for the breeder would make Holifield more receptive to the reorganization proposal. In his energy message of 4 June 1971 the president cited the breeder reactor as "our best hope today for meeting the Nation's growing demand for economical clean energy." The ECCS controversy posed a serious threat to Nixon's program, however, by raising questions about reactor safety in general and by undermining the credibility of the AEC. White House officials expressed concern about the impact of the ECCS debate on the president's energy plan after CBS News contrasted Nixon's support of the breeder with complaints from AEC critics about the semiscale experiments. If the breeder reactor was going to receive the funding and continue to win the presidential patronage that the AEC so eagerly sought, it seemed essential to address questions about ECCS promptly.[55]

The AEC had few reservations about pushing ahead with its interim ECCS criteria because it did not view the technical problems as terribly serious. Seaborg's statement to Nixon on nuclear safety was one of many indications that he was not greatly troubled with the results of the semiscale or fuel cladding tests, at least as far as their technical aspects were concerned. He voiced similar opinions publicly. "The commission's regulatory staff is taking a very cautious position on this question," he declared. "This does not mean that there is a serious problem." Milton Shaw discounted the ECCS controversy as an issue blown out of proportion by "some people who have taken a little bit of data and made a big thing out of it."[56]

As Seaborg suggested, the regulatory staff was less inclined to dismiss the implications of the ECCS question. But it too was convinced that the technical issues would be resolved within a fairly short time. The staff was acutely aware of the uncertainties of the computer programs and the limitations of the empirical data on which ECCS designs were based. But it was confident that conservative application of what was known and compliance with the interim criteria would ensure the safety of operating plants until more complete information on ECCS operation was developed. The design of emergency cooling systems had improved greatly over the previous five years in terms of the amount and delivery rate of coolant they could make available, their ability to withstand damage to the core, and the redundancy of their components. Although a mock-up ECCS had failed to perform according to expectations in some experiments, the outcome of others had conformed with predictions. Tests on the performance of core spray systems, heat transfer from the core to the coolant, and other aspects of fluid and thermal dynamics, for example, had supported earlier calculations and assumptions (though by the fall of 1971, some regulatory staff experts were raising questions about the validity of the tests on heat transfer).

The regulatory staff regarded the results of the failed semiscale tests as serious and it believed that further research was needed to clear up the uncertainties they highlighted. But it did not view them as indications that existing designs were fundamentally flawed. It emphasized that its regulatory requirements reflected conservative engineering judgment, such as assuming that an accident would be caused by a double-ended pipe break, that no coolant would reach the core during the blowdown, and that all the fuel rods would become equally hot. An unidentified member of the regulatory staff told a *Science* reporter that until the AEC completed its water-reactor safety research program, "we are going to have to use rather more conservative bases for design judgments on plants, and we are going to have to make decisions with a certain lesser degree of cheerfulness, or confidence, than if we had the results of this research." The staff remained committed to resolving the technical uncertainties about ECCS performance before the plants in the licensing pipeline that had received construction permits were ready to apply for operating licenses. In the meantime it was satisfied that the interim criteria would provide acceptable margins of safety for operating plants and the four units under consideration for operating licenses.[57]

The AEC did not, however, issue its interim criteria primarily for technical reasons; if that had been its foremost concern it could have waited until more test data were available. It published the criteria so quickly primarily for political reasons, and in that regard it failed dismally to achieve its objectives. The criteria did not tone down the ECCS controversy or quiet the AEC's critics. In July 1971, the Union of Concerned Scientists (UCS), an organization established in 1969 to challenge the misuse of technology in the Vietnam war and the nuclear arms race, published a report on the ECCS issue that took sharp exception to the AEC's position. The UCS had recently turned its attention to environmental issues and had intervened in the licensing board hearings considering an operating license for the Pilgrim plant near Plymouth, Massachusetts. The organization's ECCS study concluded that emergency cooling systems were "likely to fail" in the event of a loss-of-coolant accident, which could cause "a peace-time catastrophe whose scale . . . might well exceed anything the nation has ever known."

The UCS report focused on the failure of the semiscale tests and stressed the existing uncertainties about ECCS performance. It rebuked the AEC for its "manifest failure to adhere to the vital and important procedure of establishing the safety of nuclear power plants *before* initiating its full-scale program for nuclear power plant construction." The UCS called for a moratorium on new operating licenses until the ECCS issue could be resolved and for an assessment by "a qualified, independent group" of the hazards posed by operating plants. Its evaluation provided no new information on the ECCS question, but it presented the problem in terms that the public, or at least the news media, could easily grasp. The UCS report received wide publicity, including stories on network television news programs. Three months later the UCS published a critique of the AEC interim criteria. It alleged that a maximum core temperature of 2300 degrees was excessive, that the computer codes used to predict ECCS performance were "highly inadequate," and that the criteria were "operationally vague and meaningless" because they assumed that no appreciable loss of core geometry would occur in a loss-of-coolant accident. In short, the UCS denied the AEC's claim that the interim criteria reflected conservative engineering judgment.[58]

Without endorsing the alarmist language of the UCS reports, some scientists in the AEC and national laboratories privately expressed some of the same reservations about the interim criteria. Questions about the reliability of computer programs continued to stir dissension within the Hanauer task force. William B. Cottrell, director of the nuclear safety program at Oak Ridge, advised the AEC of his staff's doubts regarding the soundness of the criteria. The Oak Ridge researchers worried that the 2300 degree temperature limit would not prevent embrittlement of the fuel cladding under all conditions and wondered how the criteria proposed to avoid a loss of core geometry. Cottrell recognized that "the wide gaps in our knowledge" precluded the development of definitive ECCS criteria, but he offered assistance in making necessary improvements in the interim requirements.

The adequacy of the interim criteria aroused misgivings not only among some scientific authorities but also among members of the Atomic Safety and Licensing Board panel. One of the primary objectives of issuing the criteria was to reassure licensing boards about the reliability of ECCS. In December 1971, the board considering an operating license for Indian Point II signaled clearly that the criteria had failed to accomplish this goal. It announced that in view of the uncertainties about ECCS and the interim criteria, it lacked sufficient information to grant a license for the Indian Point II plant. This action opened the way for intervenors in other proceedings to challenge the adequacy of ECCS regulations and introduced a new phase of the ECCS controversy.[59]

The realization by 1966 that, in the event of a core-melting accident, containment might not be sufficient to prevent the dispersion of large amounts of radioactivity into the environment had changed the emphasis of the AEC's reactor safety program from mitigation to prevention of accidents. The final line of defense in the worst-case accident shifted from dependence on containment structures to hold in the inventory of fission products to reliance on emergency core cooling systems to keep the core from melting. As the AEC placed the burden of ensuring reactor safety on the effectiveness of emergency cooling, it sought answers to the many outstanding questions about the performance of ECCS under accident conditions. It refocused its research on water-reactor safety to sponsor a variety of tests relating to the operation of ECCS. By 1971, those efforts had helped expand knowledge of the functioning of ECCS and improve the design of systems installed in nuclear plants.

But many questions remained unanswered in 1971. Despite the importance of ECCS in ensuring reactor safety, a series of delays postponed the completion of key research projects, particularly the LOFT program. The AEC did not approach water-reactor safety research with a sense of urgency and reduced spending on it in the face of federal budget stringencies in the late 1960s and early 1970s. It did not place great emphasis on resolving questions about ECCS performance until problems arose. After ECCS experiments produced unexpected and unwelcome results that raised questions about the validity of its approach to reactor safety, the agency panicked. It attempted to keep information about the semiscale and fuel-cladding experiments from reaching the public, and when that failed, hastily issued criteria that it hoped would contain the growing controversy. For reasons that the commissioners regarded as matters of priority, the AEC published the criteria even before its own task force could conclude its evaluation of the ECCS issue.

The AEC's response to the debate over the uncertainties of ECCS played into the hands of its critics. By withholding information about the unfavorable results of important experiments, it undermined its own credibility and fed suspicions about the adequacy of its regulatory program. The interim criteria that the AEC published, rather than resolving the ECCS issue, stirred further controversy. The debate over ECCS focused both on the results of the semiscale and fuel-cladding tests and the soundness of the interim criteria. Since many of the questions about them could not be answered with available data, the AEC relied heavily on public faith in its regulatory integrity and commitment to safety to win support for and confidence in its position. But that faith had been eroded by its handling of the ECCS issue.

By highlighting the tests that failed, the AEC's critics placed the agency on the defensive at the same time that its credibility on reactor safety was declining. Consequently, the debate over ECCS slighted many important points that the AEC, with limited success, tried to get across. It obscured the deficiencies of the semiscale tests as models for actual reactors, the worst-case assumptions that the AEC used in designing experiments and judging reactor applications, the small likelihood of a loss-of-coolant accident, the favorable outcome of other experiments on ECCS performance, and improvements in emergency cooling systems that had been made since 1966. It also blurred the fact that outstanding ECCS issues appeared to be, and eventually proved to be, resolvable, given time, money, and effort.[60] If necessary, the AEC always had the option of requiring that plants be operated at less than maximum power to reduce the chances that a major accident would cause a loss of containment.

Nevertheless, the results of the failed ECCS experiments were serious technical matters that merited prompt attention. The AEC elected to concentrate on the political rather than the technical aspects of the problem. By placing undue emphasis on trying to reassure the public about reactor safety, it took a worrisome technical question and turned it into a public relations disaster. Instead of frankly acknowledging the significance of the problem and outlining plans to resolve it, the AEC, before fully evaluating the technical uncertainties, attempted to deal with the implications of the ECCS quandary for winning public confidence in reactor safety. It gave credence to the charges of its critics that it was so determined to promote nuclear power and develop the breeder reactor that it was inattentive to safety. By moving so quickly on the issues raised by the semiscale and fuel-cladding tests, it gained little. But it paid a heavy cost by impairing its own credibility and fueling doubts about the safety of nuclear power.

CHAPTER IX

The First Line of Defense—And Other Safety Measures

Much of the AEC's regulatory agenda and safety research program focused on accidents that, although possible, were improbable. The regulatory staff recognized the importance of addressing those kinds of safety issues and taking steps to resolve them. But it also emphasized that emergency cooling systems, containment, and other equipment designed to protect the public from the effects of a severe accident were only a part of a nuclear plant's multiple safeguards. If a power reactor were properly designed, constructed, maintained, and operated, the chances of a serious accident and the likelihood of ever needing emergency core cooling systems would be reduced, in theory at least, to the point of being negligible. The AEC applied the concept of "defense-in-depth," which required that nuclear plants include a series of independent, redundant, and diverse safety systems.

The key to safe and reliable plant operation, the AEC believed, was a strong first line of defense. This depended on strict adherence to conservatism in design and quality assurance in construction. As an internal committee that the AEC established to study its licensing program declared in 1969: "The greatest emphasis should be placed on the first line of defense . . . so that [a plant] will perform during normal and abnormal conditions in a reliable and predictable manner. This assurance of quality is obtained only if safety requirements are clearly and adequately defined, plant designs meet those requirements without excessive complexity, construction is in accord with design, and operation and maintenance assure continuing conformance with safety criteria."[1] Promoting safety through careful design and rigorous quality control, though always a significant consideration, became even more critical after the size of plants increased in the mid-1960s. As a result, the AEC prepared regulatory guidelines that sought to clarify design objectives and define quality assurance requirements.

In the first decade of commercial nuclear power, the regulatory staff and the ACRS used a case-by-case approach to evaluate construction permit applications. Since the technology was still in its early stages and undergoing continual change, an effort to establish general guidelines that applicants would be expected to meet seemed pointless and perhaps counterproductive. By early 1965, however, the need for formal design criteria to apply in reviewing plant proposals seemed more pressing because of a number of considerations that converged at about the same time: a request of the Atomic Safety and Licensing Board in the Oyster Creek construction permit proceedings for more detailed technical information, a renewed AEC effort to simplify and accelerate the licensing process, and a recommendation by the first Regulatory Review (Mitchell) Panel that the agency provide fuller guidance to applicants on what it expected from them in their plant proposals.

The licensing board for the Oyster Creek construction permit application issued a ruling in December 1964 that called attention to the lack of general criteria to outline the amount of information and the level of precision the AEC required of prospective plant owners. The board granted a construction permit to the Jersey Central Power and Light Company, but it also attached conditions that, in the view of *Nuclear Industry*, promised to become "the regulatory cause celebre of the year." The board determined that the utility had not submitted sufficient data to show that the plant could be safely operated at the design power level of 1600 thermal megawatts (515 electrical megawatts), and it directed Jersey Central to provide additional supporting evidence within six months. It allowed site preparation to proceed but withheld final approval of the application until it received and reviewed the supplementary information.[2]

The board's decision elicited strong protests from the regulatory staff, the utility, and the designer and builder of the plant, General Electric. The staff had never required complete information about the design of a proposed plant at the construction permit stage, and it, along with the ACRS, had found Jersey Central's application to be satisfactory. The ruling raised a troubling question about the role of the licensing boards in the regulatory process: were they authorized to conduct an independent technical review of applications or was their jurisdiction limited to certifying that the staff had observed correct procedures and to adjudicating opposing positions in a contested hearing? This issue had been discussed but never clearly resolved when the AEC had established the licensing boards in 1962. Harold Price told the Commission that the Oyster Creek board had requested design details that had "not heretofore been required at the construction permit stage and some [that had] not been required at any stage." Representatives of General Electric complained that the decision "changed the existing ground rules for the scope of applications for provisional construction permits."[3]

The regulatory staff appealed the board's decision to the Commission. With support from Jersey Central and General Electric, it petitioned the Commission to overrule the board by finding that the construction permit application included sufficient information. The commissioners declined; in May 1965 they sustained the licensing board's ruling and denied that it had overstepped its authority. The Commission agreed with the regulatory staff that applicants were not required to supply full details of the design of their proposed plant to obtain a construction permit, but it also affirmed that licensing boards should exercise "considerable discretion." It suggested that the jurisdictional problem would be eased by the preparation of "more detailed guidance" that specified the information that the AEC expected in a construction permit application. Spurred in part by the disagreement over the Oyster Creek application, the regulatory staff had begun to draft criteria to provide such guidance.[4]

The AEC's effort to define design criteria also seemed more urgent by early 1965 because of redoubled interest in streamlining the licensing process. The regulatory staff anticipated that its licensing workload was likely to increase, and in light of the Oyster Creek experience, wanted to draw up general guidelines that it could use to judge applications. It received strong support and encouragement from Commissioner Ramey, who regarded the development of design criteria as a necessary step to accelerate licensing procedures and to make metropolitan siting more acceptable. Price maintained that criteria would be especially useful to reactor manufacturers and would mollify their complaints that they did not have a clear idea of what information the AEC wanted in construction permit proposals.[5]

Those views were endorsed by the Mitchell panel report of July

1965. The regulatory staff started work on design criteria in part because it had advance notice of the panel's conclusions. In recommending ways to improve the AEC's regulatory process, the Mitchell panel argued that there was an "immediate need for criteria in the various phases of the licensing procedure," particularly in the construction permit stage. It suggested that the development of guidelines would inform the industry of the AEC's expectations, ease the AEC's task of evaluating applications, provide "a framework for testimony at . . . public hearings," help in "limiting harassment by intervenors which is not based on relevant grounds," and increase public confidence in the licensing system. As a result, "the licensing process could be simplified, shortened, and made more exact and predictable."[6]

With those goals in mind, the regulatory staff drafted a series of design criteria for the consideration of the Commission and the ACRS. By May 1965 it had written a total of thirty-three general guidelines. Applicants for construction permits would have to present their plans and enough supporting technical data to convince the staff that their design would meet each of the criteria. The criteria spelled out broad requirements and left the means of achieving them to the discretion of the reactor manufacturers. The first criterion, for example, directed that reactor structures and equipment must be able to withstand, "without impairment of their capability to function, the most severe earthquakes, flooding conditions, winds, ice, and other natural phenomena anticipated at the proposed site during the lifetime of the proposed plant."

The list of criteria also instructed applicants to submit designs that would guard against, among other things, metal-water reactions, structural damage from internal missiles released by an accident, power excursions from excessive reactivity, radiation exposure by workers and the general public above the AEC's regulatory limits, and a loss of electrical power that would incapacitate a plant's safety equipment. Since the criteria were drafted before a core meltdown that breached containment became the principal regulatory concern, they placed more emphasis on ensuring containment integrity than on emergency cooling. They did, however, call for at least two independent cooling systems if the design of the proposed plant indicated that "engineered safeguards [would be] needed to prevent containment vessel rupture."[7]

The regulatory staff solicited the views of the ACRS on the draft criteria. Some committee members suggested that the criteria were premature or incomplete; others viewed them as little more than platitudes that expressed worthy objectives without helping much to achieve them.

Nevertheless, the ACRS thought that the draft guidelines were "a good beginning" and worked with the staff to sharpen and clarify their wording. Although committee members believed that the criteria would be of limited use to vendors and utilities with nuclear experience, most acknowledged that the design guidelines would provide some assistance to applicants without much background in nuclear technology and would offer a more clearly defined framework for evaluating applications.[8]

In November 1965, after several discussions with the ACRS and considerable redrafting, consolidating, and clarifying, the regulatory staff presented a list of twenty-seven design criteria to the Commission for approval to publish for public comment. It introduced the guidelines by stating that they did not necessarily apply to all reactor designs, especially unusual or advanced ones, and that the staff would necessarily continue to exercise engineering judgment in evaluating how well a plant proposal met the criteria. Price told the Commission that his staff had gone through a "long and tedious process" of securing the agreement of the ACRS on the criteria. He urged that they be published promptly to obtain the views of interested members of the public and the nuclear industry "while the criteria were in the formative stages of development." With Commission concurrence, the AEC issued the draft criteria on 22 November 1965. The accompanying press release pointed out that the draft guidelines represented a preliminary step and that "further efforts . . . will be necessary to fully develop these criteria." But it also suggested that they were "sufficiently advanced" to request public comments and to provide "interim guidance" to reactor vendors and utilities.[9]

One of Seaborg's staff assistants predicted that the nuclear industry would register strong objections to the proposed criteria on the grounds that they were "too restrictive and vague." But the response of industry groups to the draft criteria turned out to be quite favorable. The Atomic Industrial Forum recommended some changes in wording and organization. Otherwise, there were few complaints. In July 1966 Clifford Beck reported to the Commission that the "criteria had been well received by industry" and that they "were already being used by [the regulatory] staff on an informal basis in processing applications for construction permits."[10]

Despite industry's generally positive reaction to the proposed criteria, by the summer of 1966 the regulatory staff was working on major revisions. The primary reason was the increased concern over pressure vessel integrity and loss of containment from a core meltdown that had arisen since the earlier draft. After further discussions with the ACRS, the staff prepared a new version of the criteria. This one was longer; the number of individual items expanded to seventy, divided into nine broad categories. Many of the new criteria applied to the need to protect against pressure vessel failure or a loss-of-coolant accident. They placed much more emphasis than the criteria published in 1965 on the design of emergency core cooling systems (ECCS).[11]

Like the original version, the revised criteria listed broad requirements rather than detailed specifications. They told applicants what to do but not how to do it. One ECCS item, for example, called for two separate systems, "each with a capability of accomplishing abundant emergency cooling." The criteria demonstrated the effects of multiple authorship. The different perspectives and sometimes conflicting views of individual members of the ACRS, the regulatory staff, and other AEC officials in some cases produced criteria with compromise wording or shaded meaning. An early draft of the revised guidelines included an item that the ACRS strongly promoted. It directed applicants for construction permits to design plants so that the containment structure would prevent a large release of radioactivity even if the ECCS failed after a loss of coolant. The final version of the criterion, reflecting the lack of any existing designs for coping with a meltdown without an effective ECCS, was phrased more vaguely and open to greater interpretation. By contrast, the regulatory staff made other changes in the criteria because they were too ambiguous. It removed a criterion that required rapid insertion of control rods under "abnormal conditions," for example, because its implications were unclear. The staff discovered that some reactors could not meet the requirement if it were interpreted to mean that they needed equipment to drive rods, rather then dropping them, into the core.[12]

Despite the difficulties of writing generally applicable guidelines that satisfied different authors and users, the regulatory staff eventually produced design criteria that won the acceptance, if not the enthusiasm, of those involved in their preparation. ACRS member Stephen H. Hanauer described them as "pious platitudes" that would result in "no quantum jump in safety." But he added: "On the other hand, I see no great harm in promulgating these criteria." In July 1967 the AEC published for public comment the latest version of the design criteria. This time it received a less favorable response from the nuclear industry. The most common complaint was that the criteria too frequently used imprecise terms, such as "appropriate," "considerable," and "acceptable damage limits," that were subject to confusion and misinterpretation.

The regulatory staff explained that it deliberately avoided precise and restrictive terms that could limit flexibility in engineering judgment, but this did not placate the critics. The Atomic Industrial Forum and Westinghouse each suggested changes in sixty-eight of the seventy criteria; Commonwealth Edison was close behind with sixty-six. The responses of industry groups focused on terminology and interpretation rather than on the substance of the criteria, though some expressed concern that the new version would impose additional (and in the estimation of the commenters, unnecessary) requirements. The reaction of industry to the revised criteria, which was distinctly more antagonistic than its response to the initial draft two years earlier, reflected the new uncertainties over safety design and the complexities in evaluating reactor applications that had emerged in that period. Since 1965, questions about the reliability of pressure vessels, emergency cooling, and containment had made the licensing process more complicated and more problematic, and industry groups sought to make certain that the design criteria did not further that trend.[13]

In light of the comments it received, the regulatory staff, after consulting again with the ACRS and industry representatives, revised the draft criteria. It did not make extensive changes, but it clarified terminology and meaning, combined some items, broadened others, and added a few new requirements. Two issues generated considerable discussion. One concerned the supply of electricity from an off-site source if a plant lost the on-site power that was used to run safety systems. The staff thought that a criterion calling for one off-site transmission line was adequate, but the ACRS argued that two separate lines would be preferable. After lengthy review, the staff adopted the position of the ACRS on the need for two lines, though it also accepted industry's suggestion that two different sets of towers and rights-of-way for the lines were not essential. A second question that stirred debate was whether the criteria should direct applicants to protect against industrial sabotage. Industry officials objected to such a requirement on the grounds "that no one really knows what is necessary for an adequate design against sabotage," and the staff agreed to remove it from the list.[14]

In February 1971, the AEC issued a version of general design criteria that it intended to add to its regulations as Appendix A, Part 50, Title 10 of the *Code of Federal Regulations*. This time, there were fifty-five criteria divided into six broad categories. Even after six years of work, many discussions with the ACRS and industry representatives, and numerous drafts, the regulatory staff did not view the published criteria as final or complete. It emphasized that "certain important safety considerations . . . have not as yet been sufficiently developed and uniformly applied in the licensing process to warrant their inclusion in the criteria at this time." Those items included protection against failures that would disable even redundant systems and safeguards against industrial sabotage. The staff cautioned applicants that the omission of those and other safety matters in the criteria did not mean that they could be ignored in plant designs.[15]

The general design criteria that the AEC developed between 1965 and 1971 itemized the broad objectives that applicants should address in their plant proposals. They did not, however, spell out the nature and extent of the technical data that the AEC expected applicants to supply. Over a period of several years, the agency also prepared a list of the technical specifications, commonly referred to as "tech specs," that it needed to judge an application. Like the design criteria, tech specs were intended to inform plant vendors and owners about what data the AEC expected and to guide them in submitting acceptable proposals. The agency looked for general technical information and projections in a construction permit application and detailed tech specs in an operating license application.

The effort to write tech specs predated the preparation of general design criteria. In the earliest reactor applications, technical information about the design was included as a part of the hazards summary report, which was the primary safety analysis that the applicant submitted to the AEC. The regulatory staff evaluated the hazards report as a whole. It soon became apparent that this caused difficulties for both the agency and the reactor owner, because any change, even a minor one, in the design or operating conditions of the reactor required an extensive review and modification of the entire hazards summary report. In 1962, therefore, the AEC attempted to ease the burden of making changes in technical features and operating procedures by specifying those items that were vital to plant safety and could be altered only with the AEC's consent. This did not solve the problem, however, because the list of technical specifications still restricted flexibility by failing to distinguish clearly between what was essential for safety and what was not.[16]

In January 1964 Harold Price appointed a special panel, chaired by assistant director of regulation for nuclear safety Marvin M. Mann, to examine the question of tech specs. After working for eighteen months as time allowed, the committee prepared draft guidelines for the submission of technical data to the AEC. They not only defined more precisely what technical information was vital for evaluating the safety of the reactor but also provided for a range of permissible operating limits. As long as a plant stayed within those limits, the licensee could modify equipment or operating conditions without seeking the AEC's approval. If the limits were exceeded the agency would shut down the reactor until the problem was corrected.

The Mann task force drew up a sixty-four-page, single-spaced guide to what technical data should appear in the applicant's "safety analysis report" (as the "hazards summary report" had been renamed). The applicant was instructed, for example, to specify the "design bases" of the plant. They included data on "nuclear limits" such as fuel burnup, reactivity, stability, and power distribution, "reactivity control limits" such as shutdown margins, rod speeds, and emergency shutdown provisions, "thermal and hydraulic limits" such as fuel and clad temperatures, flow velocities, and hydraulic stability, and "mechanical limits" such as maximum stresses, fatigue limits, material selection, and shock loading. After receiving the comments of industry groups, which were generally favorable and called for relatively few changes, the AEC adopted a final version of the tech specs in November 1968. They became the basis for applicants to provide and the regulatory staff to judge the technical parameters of a proposed reactor.[17]

By explicating and codifying the safety information that the AEC required from applicants, the tech specs, like the general design criteria, sought to accelerate and rationalize the licensing process while at the same time improving safety. They were two of the major, but not the only, efforts of the AEC to define engineering standards for nuclear power plants. The regulatory staff worked with professional societies to establish equipment, fabrication, and inspection codes. It consulted with the American Society of Mechanical Engineers (ASME) in preparing guidelines on the reliability and testability of nuclear pressure vessels and in revising the ASME code. In addition, it applied standards developed by ASME, the United States of America Standards Institute, and the Institute of Electrical and Electronics Engineers for pipes, valves, pumps, reactor protection systems, and in-service inspection of safety equipment.[18]

After the effectiveness of ECCS emerged as a major concern, the staff drafted guidelines for emergency cooling systems that elaborated on the requirements cited in the general design criteria. On questions for which data or experience was insufficient to write standards, the AEC, beginning in November 1970, began to issue "safety guides." They lacked the legal force of regulations, but they served notice on applicants that consideration of potential problems had to be included in plant proposals. Within a short time, the regulatory staff prepared safety guides on a wide variety of issues, ranging from thermal shock to protection against industrial sabotage to assumptions used for estimating the radiological consequences of a loss-of-coolant accident. All of those standards, codes, guides, and criteria supplemented existing safety requirements on siting, radiation protection, and other matters.[19]

The AEC's design requirements and guidelines differed from one another in their legal status and level of detail, but they all served the same general purpose. They were intended to ensure that applicants for construction permits knew what the AEC wanted from them and that their proposals met exacting standards. The regulatory staff had no illusions that plants that conformed with all design guidelines were guaranteed to be safe or that the fulfillment of the requirements listed on paper replaced the need for engineering judgment in evaluating applications. But it believed that its guidelines and specifications significantly reduced the chances of an accident. The AEC hoped that the standards it adopted and the guides it issued would provide the basis for judging applications in a more uniform and expeditious manner. It also regarded them as a vital step toward achieving an even more fundamental objective—solidifying the first line of defense against the occurrence of an accident that would threaten public safety.

The preparation of plant designs in accordance with the criteria written by the AEC and the standards developed by professional organizations was only the first step in guarding against accidents. The safety of a plant depended not only upon a suitable design but also upon strict application of the standards and specifications called for in the plans. Assuring the necessary quality of reactor components and construction procedures was a major element of the "defense-in-depth" approach to nuclear safety, and like so many other issues, it took on greater importance when plant size increased. James T. Ramey, the most outspoken member of the Commission on the issue of quality assurance, told a meeting of the American Nuclear Society in November 1966 that the growing number of reactor orders and the larger size of plants made improved quality control measures essential. "We have recognized," he declared, speaking of the AEC, "that the rapid expansion of the nuclear power industry has imposed new requirements for a more effective quality assurance program to permit us to discharge our responsibilities for public safety."

Milton Shaw, director of the Division of Reactor Development and Technology, made similar arguments in a speech to the same audience. His views reflected his experiences in working with Admiral Rickover, who was even more vocal and uncompromising than usual on the subject of quality assurance. Shaw emphasized that quality control was not only a "safety imperative" for the nuclear industry but an "important economic consideration" as well. He pointed out that equipment or construction failures were extremely costly and could be avoided only by careful observance of specifications, standards, and procedures. "We cannot afford to jeopardize a technology effort because we are unable to procure . . . a good heat exchanger, a good valve, or do a proper welding job," Shaw warned. "None of us can afford to . . . keep paying the price in time and dollars and technical progress over and over again."[20]

Ramey and Shaw were not addressing abstract concerns or hypothetical problems. Both cited examples of serious lapses in quality control at commercial plants and AEC test reactors and complained about vendors who had sold components that failed to meet quality requirements. In a few cases, suppliers had provided reactor vendors with pipes, valves, heat exchangers, and other equipment that did not conform with specifications. To make matters worse, quality assurance procedures had not identified the substandard parts until after they were installed. Replacing the defective equipment was an expensive operation; it also caused substantial indirect costs by delaying the completion of the affected plants. The worst problems had occurred in the AEC's Advanced Test Reactor, a facility for conducting experiments on the behavior of nuclear fuel that was under construction at the Idaho test site. The quality of valves and heat exchangers originally furnished for the reactor was so poor that Shaw told the Joint Committee in 1967 that the situation was "deplorable." He added that the inspectors employed by the vendor who had checked the quality of the components "must have been blind." Shaw's experiences with the Advanced Test Reactor were a major consideration in his insistence on strict quality standards for the LOFT project.[21]

Ramey and Shaw told industry representatives that action to improve quality assurance was vital to the success of commercial nuclear power. As Shaw put it: "We have no choice but to insist on positive actions to place adequate emphasis on . . . meaningful engineering and quality assurance practices."[22] Within a short time, those concerns were under-lined by serious problems with quality control at the Oyster Creek plant. The situation at Oyster Creek underscored both the regulatory pressures and economic penalties that quality assurance deficiencies could impose on licensees. It also prompted the AEC to draft, for the first time, quality assurance regulations that would apply to all nuclear plants. By contributing significantly to the growing attentiveness to quality control and decisively to the effort to regulate it, Oyster Creek continued to set precedents for the industry, a tradition that had begun with its status as the first turnkey plant and carried on with its role in the development of general design criteria.

In December 1966 construction at the Oyster Creek reactor was about 55 percent complete and Jersey Central was hoping to operate it by April of 1968. Progress was delayed, however, by the AEC's insistence that improvements be made in the emergency cooling systems originally designed for the plant. The utility became increasingly frustrated by slippages in its schedule and alarmed about the possibility of insufficient generating capacity. It had been forced to curtail supplies to some industrial customers on occasion the previous summer because of inadequate capacity. In October 1967 company president William H. McElwain told Price that Jersey Central faced an "acute power supply problem." He appealed for expedited review of its application for an operating license so that power from the Oyster Creek plant would be available to meet demands for electricity by the summer of 1968. Price explained that the regulatory staff could not finish its evaluation of the application until Jersey Central complied with its request for fuller technical information on ECCS and other safety matters.[23]

Within a short time, new information about problems at Oyster Creek vindicated the regulatory staff's refusal to make an early decision on an operating license. On 29 September 1967, a few days before McElwain sent his letter to Price, workers at Oyster Creek detected a leak during tests on primary system components in the pressure vessel. By 25 October, inspections by the AEC's Division of Compliance had confirmed that the leak was a consequence of deficiencies in quality assurance. The Division of Compliance, a part of the regulatory staff, was responsible for reviewing and assessing a licensee's quality control programs during both construction and operation. During construction, its inspectors periodically witnessed fabrication of key components, surveyed records to make certain that correct procedures were followed and proper materials used, and observed tests of the quality of systems, components, and materials. In early 1967, the division employed a total of eighty-four professionals, most of whom reviewed medical, industrial, and other uses of radioactive substances under AEC licenses. A total of twenty-six professionals in the division were reactor specialists. They generally conducted spot checks on licensees' performance, but carried out more thorough inspection procedures during a plant's final construction and early operating stages.[24]

The problem at Oyster Creek was first detected during hydrostatic tests of the pressure vessel and connected piping. After further tests, AEC inspectors determined that the source of the leak was a crack in a stub tube weld. They also found smaller cracks in 108 of the total of 137 stub tubes that appeared to result from defective welds. The stub tubes, seven-and-a-half inches in diameter and ranging from a few inches to forty inches in height, were attached to the pressure vessel and held the control rod drive housings. The cause and severity of the defects were not immediately apparent, but it was clear that repairs would be needed and that delays in construction were likely. Evaluating the nature of the problem and fixing it were made more difficult by the limited access to the area in which the cracks occurred. Unless the core support structure was repositioned, a delicate operation that General Electric hoped to avoid, workers could repair the cracks only by crawling through a pipe twenty-six inches in diameter and climbing a rope ladder thirty feet high.[25]

After further examination, the problem appeared more serious than initial indications. It turned out that the cracks were not in the welds but in 123 of the stainless steel stub tubes. In addition, welds joining all of the 137 stub tubes to the control rod housings were found to be defective. The cause of the cracks remained uncertain, though it seemed to be exposure to some kind of a corrosive agent; the faulty welds were attributable to sloppy workmanship. The flaws in the stub tubes and welds, in themselves, were not safety matters of major consequence. They required attention, to be sure, but they were minor defects that, even undetected, were unlikely to have had severe consequences. Careful inspection of the pressure vessel walls, where flaws would have been much more alarming, revealed no cracks.[26]

Nevertheless, the lapse in quality assurance that the cracks and bad welds disclosed raised issues of fundamental importance for the AEC, Jersey Central, and General Electric. It underscored the need for rigorous quality control that AEC officials had emphasized while at the same time casting doubts on the quality assurance procedures that builders followed in nuclear plant construction. Assistant director of regulation Richard L. Doan complained that "the very large number of welding flaws . . . could readily have been detected early in the welding operation." He added that "if there was any process control, supervision or inspection at all during the field welding around the stub tube areas in the bottom head of the Oyster Creek pressure vessel, it was completely ineffective."

The discovery of the problems at Oyster Creek occurred shortly before the publication of the Ergen report on emergency cooling, which cited the importance of quality assurance for reactor safety, and at a time when the regulatory staff and the ACRS were focusing on the safety implications of larger power reactors. Oyster Creek was the first large boiling-water reactor to receive a construction permit, and it became a test case for the AEC's efforts to enforce quality assurance requirements. The effectiveness of quality control was essential to ensure a reliable first line of defense against accidents, and therefore, to support the foundations of the regulatory program. For those reasons, the Division of Compliance conducted inspections at Oyster Creek that were more extensive than usual to make certain that the deficiencies were corrected.[27]

The problems at Oyster Creek had far-reaching ramifications for Jersey Central and for General Electric. The utility's hopes for receiving an operating license in time to meet peak loads in the summer of 1968 were dashed. It planned to avert a critical shortage of generating capacity by buying power from two recently completed conventional plants in Pennsylvania. The price of doing so was estimated to be twice as much as the costs of producing power at Oyster Creek, which reduced earnings and triggered shareholder complaints.[28]

General Electric also suffered substantial penalties. It had contracted with Jersey Central to build the Oyster Creek plant at a fixed price, and it had never expected to make a profit. But the extra costs for labor and materials to correct the quality control problems helped drive expenses much higher than predicted. To make matters worse, similar quality control defects soon showed up at two other General Electric plants under construction, the Tarapur station in India and the Nine Mile Point reactor near Oswego, New York. The implications for the future of the nuclear industry were disturbing. General Electric had decided to build turnkey plants at a loss to stimulate orders and spur the growth of the nuclear power industry, but the problems with quality assurance endangered that objective. *Nucleonics Week* reported in November

1967 that industry executives were worried that the "latest Oyster Creek trouble, in combination with schedule delays of various other nuclear plants, might have a cooling-off effect on nuclear orders."[29]

After the defects in the stub tubes and welds were discovered, General Electric announced that it would complete repairs within a few weeks. It replaced all of the faulty welds. It ground out the cracks in the stub tubes, the precise cause of which remained uncertain, and performed stress analyses on them. It also added metal overlays to the exposed surfaces of the stub tubes to make them more resistant to corrosion. The process took longer than anticipated, partly because of the difficulty of the work and partly because of a labor strike. General Electric and Jersey Central still hoped that timely completion of the repairs would persuade the AEC to issue an operating license promptly. The regulatory staff, however, informed the utility in March 1968 that it would not even review the application until it received detailed information about the nature and safety implications of the defective components. Peter Morris, director of the Division of Reactor Licensing, reminded Jersey Central that his staff had been waiting for months for such a report and warned that the repair methods already being performed might be found unacceptable.[30]

Before the AEC and Jersey Central could resolve their differences, the utility discovered more quality assurance deficiencies. They included stress corrosion, poor welds, missing welds, and misalignment of components on steam separator assemblies and the core shroud support ring (the core shroud housed the core). Repair of those problems would require disassembly of the affected parts. Harold Price reported to the Joint Committee that the defects apparently occurred because of "inadequate quality control" by the subcontractors who provided the components. He also suggested that the failure to find the flaws sooner reflected poorly on quality assurance procedures at Oyster Creek and perhaps at other reactors as well. While General Electric proceeded with repairs, including the addition of a redundant shroud support ring, the AEC conducted several in-depth inspections. Finally, in December 1968, fourteen months after the quality assurance defects were initially identified, both the ACRS and the regulatory staff announced that they were satisfied with the repair methods carried out on the reactor. After further testing and inspection, the AEC issued a license authorizing low-power operation on 9 April 1969.[31]

But questions about quality control continued to plague the Oyster Creek project. In February 1969, a short time before the AEC issued the

low-power operating license, the Division of Compliance received allegations that substandard pipes and valves had been installed in the plant.
The source of the information was a steel pipe manufacturer in New
Jersey that had not supplied materials to Oyster Creek but claimed to
know of violations of quality control requirements there. Company
officials refused to provide specific details when AEC inspectors contacted them, but reiterated charges that the plant contained some pipes
that failed to meet specifications and some previously used or recondi-
tioned valves that had been falsely certified as new. The reported defects
did not create a risk for low-power testing of the reactor, and the AEC
allowed the utility to begin operation at a maximum power level of five
thermal megawatts while it investigated the latest quality assurance
snag. By late April 1969 the Division of Compliance had concluded that
many of the allegations about substandard materials were true; some
pipes did not comply with specifications and at least some of the valves
were "of questionable origin and history."[32]

The new findings came as a major blow to General Electric and Jersey
Central. They claimed that the components in question had passed mus-
ter in the preoperational testing and that additional testing or immediate
replacement of parts would be unnecessary and unreasonable. A utility
official, asked about plans to build a second Oyster Creek unit, responded: "We're not sure we won't consider other means of getting
capacity on the line which don't involve relying upon [the AEC]." Despite
vocal protests from General Electric and Jersey Central, the Commission
voted unanimously to demand, as a prerequisite for a full-power operating license, new tests of the pipes and valves of dubious quality by radio-
graphic or ultrasonic means. Once the Commission made its ruling, the
regulatory staff and General Electric undertook an extraordinary effort
marked by unprecedented cooperation to complete the work as quickly
as possible. The agency made certain that appropriate staff members
were available as needed for consultation and prompt decisions while
General Electric ran ultrasonic tests on piping and replaced all of the
suspect valves. In this way, it managed to complete the work in just three
weeks. The AEC issued an operating license on 1 August 1969 and Oyster
Creek began commercial operation in December.[33]

The protracted series of problems at Oyster Creek drew increasing
attention to the importance of quality assurance. In speeches and conferences, AEC and industry officials emphasized that quality control was
vital not only for the safety but also for the economic performance of
nuclear plants. Ramey returned to the subject in a speech to the Ameri-

can Power Conference in April 1968. He told his audience that "considerations of plant performance capability and reliability, as well as safety, make mandatory the application of sound quality assurance programs to assure the economic success of these plants."[34]

Concerns about quality control and references to the deficiencies at Oyster Creek received prominent mention in the proceedings of licensing boards considering applications for other plants. During hearings for a construction permit for the proposed Pilgrim reactor in Plymouth, Massachusetts in June 1968, for example, the board wanted to know what lessons had been learned from Oyster Creek and how they would be applied at Pilgrim. The board for the Zion plants postponed a decision on a construction permit in October 1968 until Commonwealth Edison provided more complete information on its quality control procedures. The board noted that despite the utility's experience with nuclear power, it was "not convinced that Commonwealth is sufficiently impressed with the necessity for well-defined procedures for inspection, testing, reporting, and auditing in a rigorous quality assurance program." *Science* magazine editor Philip H. Abelson told his readers in July 1968 that Oyster Creek offered a prime example of "a dramatic confrontation between rosy optimism and harsh reality." He suggested that nuclear power would overcome its problems and fulfill its promise, but he added: "How distant that day will be will depend mainly on how long it takes industry and labor to achieve new and higher standards of design excellence and quality control."[35]

Meanwhile, the AEC was preparing quality assurance criteria intended to provide guidance to the industry and to avoid repeating the problems that had occurred at Oyster Creek. The regulatory staff had recognized the need to draw up guidelines on quality control but made little progress on them until confronted with the series of lapses at Oyster Creek. In May 1968, the same month that defects in the reactor's steam separator assemblies and core shroud support ring had been discovered, the staff launched a "crash program" to draft quality assurance criteria. While the draft was under review within the AEC, the Division of Reactor Licensing required applicants to detail their quality assurance procedures and the Division of Compliance used the preliminary criteria on an interim basis for its inspections. Clifford Beck told a meeting of the Atomic Safety and Licensing Board Panel that even without formal approval of the criteria, the regulatory staff was "vigorously emphasizing to applicants their responsibility to set up quality assurance systems and to audit them on a regular basis."[36]

Existing regulations directed applicants for construction permits and operating licenses to provide a description of their quality assurance programs. The regulatory staff's objective in preparing more detailed criteria was to inform applicants about what information it expected and what procedures it required. The quality assurance guidelines performed the same purpose as the general design criteria and served as a supplement to them. In March 1969 the regulatory staff presented proposed criteria to the Commission for its consideration. The draft stipulated that adequate quality control was essential for a wide variety of safety-related activities, including "designing, purchasing, fabricating, handling, shipping, storing, cleaning, erecting, installing, inspecting, testing, operating, maintaining, repairing, refueling, and modifying."[37]

The proposed criteria made clear that the applicant exercised primary responsibility for quality assurance. The AEC spelled out the basic requirements and carried out spot inspections to check on their implementation, but it lacked the means or the motivation to conduct day-to-day on-site reviews. It assumed that both safety and financial considerations gave the applicant ample incentive to pay careful attention to quality control. The draft criteria directed that applicants carry out measures to ensure, among other things, the soundness of plant designs, the creation and maintenance of records to provide evidence of construction procedures, the control of purchased materials, equipment, and services, the performance of necessary inspection, tests, and audits, the calibration of instruments and testing devices, and the proper handling, storage, and shipping of materials and equipment.

Like the general design criteria, the quality assurance guidelines told applicants what to do but not how to do it. For example, the item on controlling the quality of purchased goods and services, an issue that had caused major problems at Oyster Creek, read in part: "Measures shall be established to assure that all purchased material, equipment, and services, whether purchased directly or through contractors and subcontractors, conform to the procurement documents. These measures shall include provisions, as appropriate, for source evaluation and selection, objective evidence of quality furnished by the contractor or subcontractor, inspection at the contractor or subcontractor source, and examination of products upon delivery."[38]

After Commission approval, the AEC issued the quality assurance criteria for public comment in April 1969. The response of industry groups was generally favorable but not uncritical. They endorsed the objective of the criteria while also suggesting changes in wording and emphasis. The comments were neither uniform nor consistent; some companies, for example, complained that the criteria were too broad to be meaningful while others found them too narrow and restrictive. Despite industry's acknowledgment of the importance of quality assurance, it was becoming increasingly concerned about burgeoning regulatory requirements. The agency was receiving reports in early 1969 that a growing number of utility executives, especially younger managers who were more likely to be the "impatient type," were convinced that the AEC was "such a bottleneck that they might as well stay away from nuclear power."[39]

After receiving and considering the comments on the quality assurance criteria, the regulatory staff made some minor revisions to sharpen wording and clarify meaning. It informed the commissioners that the proposed criteria, which applicants and the Division of Compliance were using even before formal adoption, had proven effective "in preventing and in minimizing the impact of unsatisfactory conditions." By requiring thorough examination of pipes and valves delivered to construction sites, for example, they had led in some cases to the identification of substandard components. The Commission approved the revised quality assurance criteria with little discussion and they were added to the AEC's regulations as Appendix B of Part 50 in July 1970.[40]

The AEC did not view the criteria as enough in themselves to solve the problem of quality assurance. The regulatory staff believed that an effective program to meet the provisions of the criteria would avoid the kind of deficiencies that it had encountered at Oyster Creek and other plants, but it recognized that issuance of the criteria would not guarantee that they would be followed. It sought to impress upon the nuclear industry the vital importance of rigorous quality assurance. While cautiously optimistic that industry would get the message, AEC officials viewed quality control as an ongoing problem that would require sustained attention. "We're receiving strong management support and involvement," Shaw remarked in July 1969. "But we have a long way to go."[41]

At the same time that the regulatory staff was working on its quality assurance and general design criteria, it was seeking to resolve another safety issue that involved substantial financial considerations for the industry. In a rapidly changing industry, an inevitable source of uncertainty and disagreement was the imposition of safety requirements that forced modifications in equipment that had already been installed in plants under construction or in service. This was known as "backfitting."

After the AEC determined that improved safety features, particularly upgraded emergency cooling capabilities, were needed in newer and larger plants, backfitting became a major concern to industry. A group of General Electric officials, for example, complained to the commissioners and Price in April 1968 that their costs on "retrofit" requirements had totaled $40 million on six plants, including $10 million on ECCS and other equipment at Oyster Creek. They were worried that they would be burdened with even more backfitting expenses.[42]

Backfitting was a difficult issue for the AEC. On the one hand, it was concerned about the costs of replacing equipment and the effects they might have on the ability of nuclear power to compete with coal. On the other hand, it wanted modifications made if information became available indicating that a new system or component could significantly enhance safety. On some matters the need for backfitting seemed clear. Despite industry's reservations, the regulatory staff and the ACRS were convinced by the fall of 1966 that improved emergency cooling systems were essential on larger plants, including those that already had received construction permits. The need for other kinds of equipment was often more ambiguous and raised questions about whether the gain in safety margins was commensurate with the expense of backfitting.

The regulatory staff made its judgments on requiring backfitting on a case-by-case basis, considering both the safety advantages and the difficulties the changes would create. By late 1968, however, it had decided to make a general statement of its policy on backfitting by amending Part 50 of the regulations. Price proposed to the Commission that backfitting be required only when it would provide "appreciable, additional protection." The burden of showing that a modification would meet this standard would fall on the regulatory staff. If an applicant or licensee wished to make changes in a plant's safety equipment, it would have to show that there would be no "significant, adverse effect." Although the amendment would only codify the policy that the regulatory staff had been following, Price thought it would offer some assurance to the industry that unreasonable demands would not be imposed. If an applicant or licensee objected to the staff's decision on a backfitting matter, it could appeal to a licensing board, which would make the final judgment.[43]

The Commission approved the backfitting amendment and after a public comment period, adopted it virtually unchanged in March 1970. The most significant revision was to require that the regulatory staff find that backfitting would deliver "substantial" rather than "appreciable" additional protection. Industry was generally satisfied with the amendment as a general statement of the AEC's policy. *Nucleonics Week* reported, however, that it might "not be the blessing it was first thought." Industry officials were concerned that as an alternative to backfitting, the regulatory staff would insist on settling unresolved safety questions at the construction permit stage. In fact, the staff was asking for more complete and precise information about safety systems in construction permit applications. This was an important objective of the general design criteria; conformance with them could reduce the need for backfitting. But the backfitting amendment was primarily intended to alleviate the anxieties of rather than increase the burdens on the industry. The ACRS was uneasy about the backfitting amendment for reasons quite different than industry. It feared that the rule would inhibit the staff from requiring backfits even when significant safety issues were involved. The amendment obviously did not resolve the uncertainty about the circumstances in which the AEC would find backfitting to be essential. The regulatory staff would continue to exercise its judgment about the need for backfitting, and, as usual, its position stood somewhere between the industry and the ACRS.[44]

While the AEC was developing its criteria on design and quality assurance and its policy on backfitting, it was also considering guidelines on another issue of growing importance—emergency planning. It hoped, of course, that an accident at a nuclear plant that threatened public health would never occur. But it acknowledged that the chances of a major accident could not be completely discounted, and therefore, that preparing emergency procedures was necessary. In the early days of nuclear power development, the AEC included in its regulations a broad requirement that applicants for operating licenses outline a plan for dealing with radiological emergencies. It offered training to state and local police, fire, and health departments and circulated a list of recommendations to local officials on how to handle accidents involving radioactive materials. The AEC also joined with other federal agencies to draft a program, the Interagency Radiological Assistance Plan, that spelled out the role that different departments would play in responding to a serious nuclear plant accident. The AEC assumed the primary responsibility for carrying out the plan in cooperation with thirteen other agencies. The requirements and arrangements for emergencies were vague, sketchy, and in keeping with the prevailing belief that even a severe accident would probably not release radioactivity into the environment, low in priority.[45]

After the size of nuclear plants increased and the assumption that containment would prevent the release of radioactivity in an accident became problematical, emergency planning took on greater importance. The ACRS first called attention to possible weaknesses in existing procedures. At a subcommittee meeting in December 1966, members noted that many applicants and licensees would rely heavily on local authorities to carry out an evacuation, if it should become necessary. But it was not clear in all cases that state and local governments had the knowledge, resources, or equipment to handle an evacuation during a radiological emergency effectively. Furthermore, there were no guidelines for judging when an evacuation of regions around a plant would be advisable. The ACRS decided that it should alert the AEC "to a problem area where little effort is being exerted." In March 1967, ACRS chairman Nunzio J. Palladino told the commissioners that in light of increasing plant capacity, the trend toward locating reactors closer to population centers, and the growing number of multireactor sites, the AEC should reexamine the question of emergency planning. The Joint Committee on Atomic Energy expressed similar views a short time later.[46]

Pressed by the ACRS and the Joint Committee, the AEC undertook a study of emergency plans and procedures. In late 1967, the regulatory staff asked the owners of test reactors and power reactors in operation or under construction to submit copies of their emergency plans. It also contacted some state and local agencies that would participate in carrying out the plans in the event of a severe accident. The most glaring deficiency that the survey revealed was that licensees assumed that state and local authorities had the capability to perform effectively in a radiological emergency. They made no effort to evaluate the validity of their assumption.

Drawing on the information it received, the regulatory staff drafted guidelines that called for more detailed emergency plans from applicants. Although the staff agreed with the ACRS and the Joint Committee that improved emergency planning was needed, it did not view the issue as a matter of urgency. It was swamped with reactor applications, the preparation or revision of several regulations, and growing controversy over the environmental impact of nuclear power. Those questions took precedence over emergency planning, especially since the regulatory staff remained opposed, at least for the foreseeable future, to metropolitan siting. Evacuation of the population living near a nuclear plant would be a particularly difficult problem in an urban setting.[47]

In April 1970 the regulatory staff presented its proposals for changing emergency planning requirements to the Commission. As a prerequisite for an operating license, it recommended adding a new Appendix E to Part 50 of the regulations that listed the items that applicants' emergency plans should contain. It stipulated that they should include, for example, specific details about who would exercise authority and perform assigned duties during an emergency, information about the means that would be used to determine the magnitude of the release of radioactivity, provisions for working with local, state, and federal agencies to notify the public and carry out an evacuation if necessary, and procedures for training employees and conducting drills to test emergency plans.

As an addendum to the draft regulation, the staff drew up guidelines to assist applicants in preparing their emergency plans. The guidelines instructed applicants to provide detailed information in ten different categories, from organizational responsibilities to medical arrangements. They required plant owners, for example, to make certain that appropriate state and local officials were informed of their duties in an emergency so that effective coordination would be achieved. They sought to ensure that licensees verified the capability of those agencies to perform necessary functions by calling for advance arrangements, in the event of an evacuation, for transportation, food, shelter, and sanitary facilities as well as traffic control, fire protection, medical support, and decontamination. The guidelines also told applicants to postulate accidents of varying severity and to prepare appropriate responses for different situations.[48]

Like other regulatory guidelines, the proposed appendix on emergency planning left to the discretion of the applicant the best means to carry out its provisions, subject to AEC approval. It cited the information that the agency wanted in applications while allowing prospective owners to tailor their plans to their facility and location. Like the general design and quality assurance criteria, it was intended not only to enhance public safety but also to ease the licensing process for both applicants and the regulatory staff. After Commission approval, the AEC released the emergency planning amendments for public comment in May 1970. It received only seven letters in response, most of which applauded the effort to clarify and strengthen previous requirements. After making a few minor revisions, the AEC adopted the new emergency planning regulation in December 1970.[49]

Another problem that took on increased importance and received greater regulatory attention after the mid-1960s was safeguarding nuclear materials. The objective was to prevent fissionable materials from falling into the wrong hands. Those "special nuclear materials," defined as plutonium, the isotope uranium–233, or the element uranium enriched in uranium–233 or uranium–235, were used, in different forms and levels of enrichment, to fuel both nuclear reactors and weapons. If not properly controlled, they could present a health hazard to workers and the general public, and, more ominously, could conceivably be surreptitiously obtained in sufficient quantities to build an atomic bomb. This was not a question of reactor safety, and unlike most of the profusion of new regulations that the AEC issued in the late 1960s and early 1970s, was not directly related to growing concern about the consequences of an accident in a large reactor. The safeguards issue aroused intense interest and prompted high-level scrutiny after the discovery of large discrepancies in inventories of special nuclear materials at an AEC-licensed plant exposed the need to strengthen existing requirements.

The Atomic Energy Act of 1954, while opening nuclear technology to private industry, stipulated that the government would retain title to special nuclear materials. The AEC leased special nuclear materials to licensees for their use in nuclear fuel or in medical and industrial applications. The question of safeguarding the leased materials generated considerable discussion as the AEC prepared its first rules to regulate the nuclear industry. The regulatory staff eventually decided that rather than imposing detailed safeguards and accountability procedures, it could depend on the intrinsic monetary value of special nuclear material to ensure that licensees would guard it carefully. "It is more valuable than gold," Harold Price told the commissioners at a meeting in 1955. "Banks know how to protect gold. We think [licensees] know how to protect this material."

Licensees would be assessed a heavy fee for the loss or damage of the materials they leased from the AEC. They were required to keep records of their inventory of special nuclear materials, to inform the AEC of any losses, and to submit semiannual reports on materials received, transferred, and possessed. AEC contractors, who were not subject to financial penalties, had to meet the same record-keeping requirements and provide guards, surveillance, secure areas, and other access restrictions that were not imposed on licensees. Along with the measures that it directed licensees and contractors to carry out, the AEC felt confident that severe criminal penalties for attempting to steal special nuclear material would offer adequate protection against unlawful activities.[50]

The safeguards requirements that the AEC prepared in the mid-

1950s remained in effect without drawing much notice for a decade. By 1965, however, the protection of special nuclear materials had stirred some renewed interest because of two developments. One was the passage of the amendment to the 1954 Atomic Energy Act allowing private ownership of special nuclear materials. Although the implications of changing the law for enforcing safeguards were not a major consideration during congressional deliberations, they had to be addressed once the amendment was enacted in 1964. The second development that called attention to the safeguards issue was the growing concern over the proliferation of nuclear weapons. By the mid-1960s high-level officials in the United States and other nuclear powers were becoming increasingly anxious about the possibility of more nations obtaining atomic weapons, particularly after China joined the nuclear club by testing a surprisingly sophisticated bomb in October 1964. This underscored the importance of adequate safeguards to prevent special nuclear materials from being diverted to foreign countries.[51]

Neither the private ownership of special nuclear materials nor the proliferation question generated a sense of urgency about domestic safeguards. The issue assumed much greater immediacy only after the AEC discovered that a large amount of enriched uranium was unaccounted for in a uranium processing and fuel fabrication plant. The plant, located in Apollo, Pennsylvania and owned by the Nuclear Materials and Equipment Corporation (NUMEC), operated under an AEC license and also processed fuel as an AEC contractor. During two inspections in 1965, the agency found that the NUMEC facility had experienced unusually high losses of uranium over a period of eight years. Part of the missing material disappeared through "known loss mechanisms," such as adhering to pipes and filters, blowing out of vents, sticking to shoes of workers, or being buried as scrap. Even after figuring the amount of material lost in this way, a discrepancy of nearly 100 kilograms (more than 200 pounds) remained.

The AEC believed that poor accounting practices and careless operating procedures on the part of NUMEC probably explained the discrepancy, and it assessed the company a penalty of over a million dollars. But it could not dismiss the possibility that the missing material, which was enough to make six atomic bombs, had been diverted to a foreign nation. There were suspicions within government and industry circles that Israel had acquired the uranium from the plant to use in developing an atomic bomb, though they could not be proven. The AEC conducted extraordinarily meticulous inspections without finding any indications of diversion. Its position was that while it "had no evidence that diversion had occurred, neither could [it] say unequivocally that the material had not been diverted."[52]

Whatever the reasons for the discrepancies at the NUMEC plant, it was clear that a reexamination of the AEC's safeguards procedures and requirements was essential. As assistant general manager Howard Brown told the Commission in February 1966, the NUMEC situation "convincingly demonstrated that fulfillment of a financial responsibility requirement might not really satisfy the AEC's interest in special nuclear materials unaccounted for." Depending on financial accountability alone to ensure that special nuclear materials would be protected no longer seemed sufficient, but determining a suitable course of action was a daunting task. The Commission initially took several steps to deal with the problem. It extended the controls that applied to leased special nuclear materials to those that were privately owned. At the same time it added requirements for improved bookkeeping and inventory procedures. They were intended to show that the AEC would place greater emphasis on preventing losses rather than simply assessing monetary penalties for them. The Commission decided not to impose physical security measures on licensees, such as the employment of guards and other access controls, without further study, but it instructed its inspectors to pay close attention to the need for tighter access restrictions while carrying out their regular surveys.[53]

In addition, the Commission decided to establish an independent panel of outside experts to review safeguards policies and procedures and to submit findings and recommendations. It did so not only because of its own questions about the soundness of its program but also because of reservations voiced by the Joint Committee. John T. Conway, the committee's executive director, first mentioned the NUMEC losses publicly, without naming the company, in a 1966 speech appealing for improvements in managing nuclear materials. Referring to the original AEC regulations on safeguards, he declared: "Looking back at that 1956 decision, applicable to privately owned plants, one can justifiably question its validity." To investigate the concerns of both the AEC and the Joint Committee, the review panel, officially known as the Advisory Panel on Safeguarding Special Nuclear Material, began its work in July 1966. Its chairman was Ralph F. Lumb, director of the Western New York Nuclear Research Center in Buffalo. Like Lumb, the other six panel members, who came from the nuclear industry, legal and accounting firms, and an AEC contractor, were recognized authorities on the subject of safeguards.[54]

The Lumb committee submitted its final report to the AEC in March 1967. It cited the rapid growth in the number of nuclear power plants ordered during the previous two years as a compelling reason for urging that "an effective world-wide international safeguards system be established quickly." It pointed out that, according to projections, by 1980 nuclear units throughout the world would produce plutonium as a by-product at a rate of 100 kilograms a day. Therefore, both international and domestic safeguards programs were essential. The panel acknowledged that safeguards alone could not thwart proliferation of nuclear weapons and that "no fool-proof system [could] be devised to prevent diversion." Nevertheless, it affirmed the importance of doing everything possible to discourage diversion. The committee concluded that the AEC had performed creditably in protecting special nuclear materials. However, without mentioning or even alluding to NUMEC, it complained that safeguards procedures had "not always received adequate attention at senior management levels within the AEC and within contractor and licensee organizations."

The Lumb panel offered a series of recommendations to improve existing safeguards programs and procedures. Some of them applied specifically to international efforts to detect and prevent diversion, such as intensifying efforts "to establish an effective universal safeguards system under the International Atomic Energy Agency" and the creation of an "International School of Safeguards." Several of the panel's recommendations focused on measures that the AEC should introduce to provide better domestic safeguards. They included a call for more emphasis on physical security in facilities that handled special nuclear materials, such as fences, locks, vaults, guards, and other means to limit access. Even without a formal requirement, many plants provided physical protection comparable to what the AEC imposed on contractors, but the Lumb committee advised that minimum standards be developed for all licensees. It also urged that the AEC prepare criteria, on an expedited basis, to guide licensees on the security and record-keeping procedures that were necessary for an adequate safeguards program. Advancing an idea that had been floated by the Joint Committee, it argued that stationing resident inspectors, who would be permanently assigned to a particular facility, could "make a contribution to safeguarding special nuclear materials."

The Lumb panel suggested a number of other actions that would enhance both international and domestic safeguards programs. It pressed for organizational changes within the AEC so that a single high-level office would formulate policy, coordinate the safeguards activities of different divisions, and serve as a liaison with other agencies. It called for stiffer criminal penalties for those involved in diversion or other illegal acts and for offers of rewards for persons who helped apprehend them. Finally, it recommended that the AEC expand research to develop new techniques and procedures for protecting special nuclear materials. It hoped that additional research efforts would find new ways to improve the effectiveness of safeguards programs.[55]

The AEC acted promptly to carry out or explore the feasibility of the Lumb committee's recommendations. Even before the panel submitted its final report, the agency made organizational changes intended to focus attention on and facilitate coordination of safeguards programs. It created an Office of Safeguards and Materials Management, directly responsible to the Commission, that would develop and evaluate policies, procedures, and standards for the AEC's international and domestic safeguards activities. In addition, the AEC established a new organization within the regulatory staff, the Division of Nuclear Materials Safeguards. It administered the safeguards program for AEC licensees, and reported to the director of regulation. At the same time it worked with the Office of Safeguards and Materials Management in the development of safeguards policies.[56]

While the Office of Safeguards and Materials Management and other AEC divisions focused on the international aspects of safeguards, the regulatory staff acted to implement the Lumb panel's suggestions on the domestic program. As a priority task, the Division of Nuclear Materials Safeguards drafted a guide that outlined the controls and procedures the AEC wanted licensees to provide. Like other AEC criteria, it cited in broad terms the information that the agency expected without specifying how the program should be carried out. Its purpose was to assist licensees in understanding what constituted adequate safeguards and in complying with AEC requirements. The guidelines did not cover the subject of physical protection of special nuclear material, but the AEC issued a separate regulation that directed licensees holding more than a specified amount of special nuclear material to restrict access to areas containing it and equip those areas with locks, alarms, safes, and other barriers.[57]

Beginning in July 1967, the AEC also established, on a one-year trial basis, a resident inspector program at three fuel fabrication facilities and the West Valley, New York fuel reprocessing plant. Of all AEC licensees, those units held and worked with the greatest quantities of special nuclear materials. The resident inspector program offered the advantages of on-site experts to observe safeguards procedures and verify measurements of inventories, but it also was costly at a time when the AEC faced severe staff shortages in its regulatory functions. Eventually, the agency kept the resident inspectors only at West Valley; it believed that an increased number of inspections and the new safeguards regulations would provide sufficient protection at other facilities. To further strengthen safeguards, the AEC recommended, and Congress agreed, that the criminal penalties for unlawful diversion of special nuclear materials be toughened. Finally, the AEC rapidly increased the amount it was spending on safeguards research to, in Seaborg's words, "help bridge the gaps between existing knowledge and the technology required for adequate control in the future." Between 1966 and 1967 it quadrupled its expenditures, from two hundred and thirty thousand dollars to about one million dollars.[58]

The new requirements that the AEC developed in response to the Lumb panel report substantially strengthened the safeguards program. They were not, however, a solution to the problem in themselves; their effectiveness depended on how well they were carried out. As an editorial in *Nuclear News* in late 1969 commented: "Nuclear materials safeguards are a serious matter—involving every man, woman, and child on this planet. . . . There is no denying that safeguards is a tremendously complex matter. But there is no choice; we must have a practical, reliable system soon." In June 1971, Charles Thorton, director of the AEC's Division of Nuclear Materials Safeguards, told a group of professionals in the field that the nuclear industry was not doing enough to protect against misuse of special nuclear materials. Charging the industry with "ineptitude" and a "lackadaisical" attitude, he complained that the cost of special nuclear material had fallen so low that plant managers and owners were satisfied with losing one to two percent of it rather than pay the costs of finding it. And this, he emphasized, could open the way for diversion or for the acquisition of "highly toxic materials" by "technically trained aberrant individuals."[59]

Between 1965 and 1971, in response to the growing number of reactor applications, the expanding size of proposed plants, and the discovery of serious problems at Oyster Creek, NUMEC, and elsewhere, the AEC issued a series of new regulations and guidelines. While

it recognized that those measures did not assure that the issues they
addressed would be fully resolved or replace the need for engineering
judgment, it viewed them as necessary and useful means to promote
safety. At the same time, the AEC sought to ease the licensing process by
informing applicants and licensees of what it expected of them in their
designs, quality assurance programs, emergency planning, and safeguards procedures.

The nuclear industry approved of the AEC's efforts in principle,
though it sometimes vocally protested the application of them. The AEC
was mindful of the costs its requirements imposed on the industry and
concerned about the impact they might have on the competitive position
of nuclear power. But it yielded little in disagreements over design and
quality assurance because it regarded them as vital both to safety and
the long-term economic welfare of the nuclear industry. It insisted that
maintaining a solid first line of defense against a major nuclear accident
was essential to protect the health of the public and the future of the
industry. Although experience with power reactors was still too limited
to prepare definitive regulations on many subjects, the AEC hoped that
the guidelines it issued would advance both objectives.

CHAPTER X

Regulating Mines and Mills

Health, Environmental, and Bureaucratic Hazards

The central focus of the AEC's regulatory program during the 1960s was reactor safety; siting and engineering questions relating to the prevention and mitigation of accidents occupied the time, attention, and expertise of the staff. But by the middle of the decade, other problems relating to the public health and environmental impact of the normal operation of nuclear power plants began to take on greater importance. As public interest in and concern over potentially harmful effects from routine, accident-free use of nuclear power increased, the AEC's performance on health and environmental issues became the target of growing criticism. In the long run, the attacks on the AEC—some justified, some exaggerated, and some groundless—fueled the emerging debate over nuclear power and greatly influenced public attitudes toward the technology. Among the earliest of several issues that aroused major controversies over the health and environmental consequences of the use of nuclear power was the role of the AEC and other government agencies in regulating the hazards of the "front end" of the nuclear fuel cycle—the mining and milling of uranium ore.

The radiation hazards of uranium mining had been a source of concern, but little effective regulatory action, during the 1950s. In the early post-World War II period, all of the uranium that the AEC used in its weapons programs came from abroad, a condition that caused considerable uneasiness as cold war tensions grew. To stimulate exploration for domestic ore, the agency announced in 1948 that it would pay a guaranteed minimum price for high-grade uranium-bearing ore and a bonus for discoveries of rich ore deposits. At first, the chances of finding substantial uranium ore deposits in the United States appeared slight, but in 1952 an independent prospector and geologist named Charles Steen struck a rich vein near Moab, Utah. Steen made a fortune on his discovery and inspired a feverish search for new deposits. The uranium rush eliminated the AEC's fears of insufficient supplies; it produced so much ore that the agency decided in 1957 to limit its purchases, causing consternation in the industry and ending the boom. Most of the uranium ore was found on the Colorado Plateau, an area that included sections of Colorado, Utah, New Mexico, and Arizona; smaller deposits occurred in other states as well. Some of the ore was close enough to the surface to be tapped by open-pit mining, but most of it could only be reached in underground mines.

Underground uranium mining posed serious risks from exposure to radioactivity. Radioactive decay of uranium yields, among other elements, radium–226, which, in turn, produces radon–222. Radon is a radioactive gas that is undetectable to the human senses and can be a health hazard if inhaled. An even greater danger comes from four "daughter products" that radon gives off—polonium–218, lead–214, bismuth–214, and polonium–214. The radon daughters readily cling to particles in the air and can easily be taken into the respiratory tract by breathing mine air. Once inside the body, the two polonium isotopes are especially hazardous because they lodge in lung tissue and emit alpha particles that can cause cancer. Uranium miners who worked in enclosed underground areas with limited air circulation were particularly susceptible if they were exposed to high concentrations of radon over an extended period of time.

By the early 1950s, health authorities had clearly recognized the dangers that high levels of radon and its daughters presented. But that in itself was not enough to ensure the adoption and enforcement of regulatory measures to protect miners by reducing radon concentrations. The primary responsibility for mine safety rested with state governments, which acted slowly and reluctantly to impose requirements on the uranium industry. A number of federal agencies also took an interest in mine conditions, but none had clear regulatory jurisdiction to enforce safety standards. The AEC, which was, by law, the sole purchaser of uranium ore, was not granted authority to regulate privately owned mines. Congress gave the agency jurisdiction over "source material" only after its removal "from its place of deposit in nature." In the Atomic Energy Act of 1946 it allowed title to unmined uranium ore to stay in private hands to encourage prospecting for new resources. Ore located on federal land was an exception; it remained government property.

Other federal agencies that were involved in the activities of mining industries or concerned with mine working conditions also lacked clear authority to enforce safety measures in privately owned uranium mines. The Bureau of Mines, a part of the U. S. Department of the Interior, was responsible for inspecting mines on Indian reservations and federal lands and was available for advice and technical assistance to federal and state agencies, the mining industry, and labor organizations. The U. S. Department of Labor administered the Walsh-Healy Public Contracts Act, which stipulated that federal contractors could not permit "working conditions which are unsanitary or hazardous or dangerous to the health and safety of employees." The applicability of Walsh-Healy to privately owned uranium mines was questionable, however, because even though the AEC was the sole purchaser of uranium ore, the mines did not operate under federal contracts. The U. S. Public Health Service, a part of the Department of Health, Education, and Welfare (HEW), offered assistance and expertise in studying, identifying, evaluating, and correcting health hazards but exercised no regulatory authority.

In 1949, Colorado, which took the lead among uranium-mining states on safety issues, asked the U. S. Public Health Service to conduct a study of the health effects of uranium mining. There was no firm standard for allowable concentrations of radon in mines at the time, but the Public Health Service had concluded that 100 micromicrocuries of radon per liter of air was an acceptable level. A curie is a unit that measures the rate of emission of subatomic particles from radioactive nuclei. The number of curies does not in itself tell how hazardous a radioactive substance is to human health because it does not indicate the nature of the radioactive emission. One curie of a certain substance, therefore, can be more or less dangerous than one curie of another. A micromicrocurie, later referred to as a picocurie, is one millionth of one millionth of a curie. In light of the limited scientific data, the Public Health Service's standard of 100 micromicrocuries of radon per liter of air was rather arbitrary and subject to change. It reflected the best available information and the judgment of agency experts about what was feasible and what seemed likely to provide adequate protection to miners.

The Public Health Service's surveys of mine conditions revealed radon concentrations far above the standard. In one study, it found that in

48 mines the median level of radon per liter of air was 3100 micromicrocuries and in 18 mines the median level of radon daughters was 4000 micromicrocuries. Another investigation showed that 77 percent of 157 mines had concentrations of radon greater than 100 micromicrocuries per liter of air. The Public Health Service recommended that the high concentrations of radon and its daughters be reduced by forced ventilation of uncontaminated air in the mines.

The findings of the Public Health Service generated concern among health officials in uranium-mining states, and in February 1955, delegates from seven states, mining companies, the AEC, the Public Health Service, and the U. S. Bureau of Mines met in Salt Lake City to discuss the problem. Industry representatives expressed doubt that forced ventilation could diminish radon to the levels sought by the Public Health Service and warned that strict application of the 100 micromicrocurie standard would cause many mines to close. Nevertheless, by the end of the meeting, the conferees agreed on a permissible "working level" in the mines of 100 micromicrocuries of radon and, in equilibrium with that level of radon, 300 micromicrocuries of radon daughter products. They viewed those values as target figures rather than inviolable requirements, but the working level became the basis for judging conditions in uranium mines.

Despite the concern about the health effects of radon that delegates to the Salt Lake City meeting expressed, conditions in the mines showed little or no improvement over the following five years. The Public Health Service's surveys showed that in many mines radon concentrations remained far above the working level and in some cases had gotten worse between 1955 and 1959. The states failed to take effective regulatory action. G. A. Franz, the deputy commissioner of mines in Colorado, admitted in May 1960 that up to 98 percent of the state's uranium mines would have to suspend work if forced to abide by the working-level standard, and added: "Our business is to keep the mines open." Even more ominous than the radon levels in the mines were early indications that miners were dying in disproportionately high numbers from lung cancer. The Public Health Service was studying the pathological effects of uranium mining, and although its sampling was too small to draw definite conclusions, the signs were deeply troubling.

Based on the preliminary findings of the Public Health Service, HEW secretary Arthur S. Flemming, with the concurrence of the AEC, the Department of Labor, and the Department of the Interior, called a meeting with the governors of uranium-mining states. The conference, held in December 1960, offered a forum that highlighted the health problems of uranium mining. A Public Health Service official reported that among a group of 907 miners who had worked underground for more than three years, five of forty-four mortalities had been caused by lung cancer. This was almost five times as high as the mortality rate for lung cancer among the general male population in Arizona, New Mexico, Utah, and Colorado. The governors who attended the meeting, though initially dubious about the severity of the dangers, became convinced that the situation was serious and required prompt corrective measures. They rejected the suggestion of Colorado governor Steve McNichols that the AEC should regulate the mines and resolved to improve state programs to deal with the problem.

After the 1960 conference, the uranium-mining states expanded their efforts to control radon levels, inspect the mines regularly, and close "high-hazard areas." Even as the states improved their regulatory performance, however, the Public Health Service published new statistics that further confirmed the susceptibility of uranium miners who worked underground for long periods to lung cancer. In August 1963, its findings showed that of 768 miners who had worked underground for 5 years or more through the end of 1962, 11 had died from lung cancer. This was "10 times the number expected on the basis of death rates for all white males of comparable age living in the states included in the study." Those figures suggested that an unusually high percentage of underground uranium miners would continue to die of lung cancer caused by exposure to excessive levels of radon.[1]

In 1963, then, the protection of uranium miners from radioactivity remained an unresolved problem. The mortality trends seemed clear but the allocation of responsibility for correcting the regulatory failures of the previous decade was still uncertain. The uranium-mining states enforced the working-level standard more rigorously, but limited resources hampered their programs. Furthermore, workers' compensation coverage and death benefits for dependents varied widely from state to state, raising questions, on the one hand, about the adequacy of protection for miners and their families and, on the other hand, about the sufficiency of state funds to pay claims. The role of federal agencies in regulating the mines remained ill-defined. Scientific data on the level of radon that presented significant hazards and the most effective means to protect miners was weak and imprecise. The working level that had been adopted in 1955 had little grounding in scientific evidence and was more an informal accord on what seemed reasonable than a well-defined level of safety. Although there was general agreement among interested state and federal agencies that action was needed to improve mine conditions, there was much less agreement about what should be done and who should take responsibility. Differing perspectives on how to approach the problem eventually led to public controversies and acrimonious interagency disputes.[2]

The Public Health Service's newest findings on lung cancer mortality rates among uranium miners, announced in 1963 and published in professional journals in 1964, intensified concern and prodded action on the federal level. The initial responses came from the Select Subcommittee on Labor of the U. S. House of Representatives' Committee on Education and Labor and from the Federal Radiation Council. For several years, the Select Subcommittee on Labor had been sponsoring legislation to improve safety in mines of all types. Representative James G. O'Hara introduced legislation in 1965 directing the secretary of the interior to develop safety standards in both metal and nonmetallic mines. During hearings on the bill, O'Hara and his colleagues made it clear that they were disturbed by the Public Health Service's statistics on mortality among uranium miners. Congress approved a revised version of the measure, the Metal and Nonmetallic Mine Safety Act, in 1966. For the first time, the U. S. Bureau of Mines assumed responsibility for setting safety standards in privately owned uranium mines and, if necessary, closing those that failed to comply.[3]

At the same time the Federal Radiation Council (FRC) was attempting to establish suitable limits for radon levels in uranium mines. President Eisenhower had created the FRC as a part of the White House staff in 1959 to advise the president on radiation safety. He acted largely because of a public outcry over radioactive fallout from nuclear bomb testing in the atmosphere and declining public confidence in the evaluation of fallout hazards by federal agencies, especially the AEC. The statutory members of the FRC were the secretaries of defense, commerce, labor, agriculture, and HEW, and the chairman of the AEC. The secretary of Health, Education, and Welfare was designated as chairman of the FRC and a small permanent staff offered logistical and technical support to its members. Paul C. Tompkins, a veteran of the Manhattan Project and formerly a radiation protection specialist with the U. S. Naval Radiological Defense Laboratory, the Public Health Service, and the AEC, served as the council's executive director. The purpose of the FRC was to provide "general standards and guidance" that would help protect public health from radiation hazards, but differing views among its members and its lack of binding authority over federal agencies severely limited its effectiveness on controversial issues.[4]

The FRC had begun a study of radiation risks in uranium mining in early 1961, a short time after the meeting between federal agencies and uranium-mining state governors. Progress on it had been stalled, however, by more pressing projects that the council undertook as a result of renewed controversy over fallout from atmospheric weapons testing. The reports of the Public Health Service on mortality rates among miners and congressional interest in improving mine safety prompted the FRC to return to its study of radon hazards. Its objective was to develop a standard for permissible exposure by uranium miners. In order to take advantage of the experience and expertise of the Bureau of Mines, the council invited the secretary of the interior to participate in its deliberations.[5]

It was apparent that the FRC's study would only be the first step in providing adequate protection to uranium miners. Deciding on a standard of exposure that would be satisfactory to the different agencies represented on the council as well as to the mining industry and labor unions who spoke for the miners was a daunting task, especially in light of the dearth of scientific data. Once the FRC settled on a standard, the question of enforcement would remain. The resources available to state and federal agencies to inspect the mines and ensure compliance with exposure standards were a continuing concern. Even with adequate resources, enforcement presented a formidable challenge. One reason was that controlling radon concentrations was often difficult. Ventilation was useful but not always effective in reducing radon to the desired levels. In some cases, it could not deliver enough uncontaminated air to all areas of a mine to assure that radon concentrations would not exceed the working level. Moreover, radon levels in one place could vary greatly from season to season, month to month, or even day to day. This meant that a mine or section of a mine that met the standards at one time might not do so at another time. Partly for those reasons, the radon levels in many mines, despite the redoubled efforts of state regulators and mine operators to control them, remained above the working level. Although conditions in the mines had greatly improved since 1961, an FRC committee estimated in 1966 that only about half of the underground uranium mines maintained an average concentration of one working level or less; the other half measured in the average range of two to three working levels (or 200–300 micromicrocuries).[6]

The attitudes and actions of miners often compounded the difficulties of protecting them from the effects of radon. Under the best conditions mining was a risky occupation, and the miners were inclined to dismiss hazards from a gas they could not see, taste, or smell. They were paid by the amount of ore they dug and resisted any requirements or equipment that slowed their pace of work and reduced their paychecks. For that reason, they generally refused to wear cumbersome respirators that would significantly decrease their inhalation of radon daughters. In cold weather, many miners objected to the introduction of fresh air for ventilation; the immediate dangers of circulating frigid currents of air seemed more serious than the long-term risks of radon exposure. Finally, there was strong, though not conclusive, evidence that smoking cigarettes greatly increased the perils of working in underground uranium mines. A large percentage of the miners who had died from lung cancer had been smokers, leading health experts to speculate that smoking and radon exposure might have a synergistic effect that magnified the risks of developing lung cancer. Given the difficulties of enforcement, deciding on a standard for permissible radon levels was only the first step in ensuring adequate protection to uranium miners. But it was a vital first step that aroused sharp differences of opinion.[7]

By December 1966, the FRC staff, after consultation with an interagency working group and other experts inside and outside the government, had completed a draft paper on uranium mine radiation hazards. It circulated the draft for the consideration of FRC members. Once they agreed on a position, they would send their recommendations to the president for approval. The draft proposed that the FRC adopt a standard of one working level as the permissible limit in uranium mines. It argued that the one working-level standard represented "a reasonable risk" and, based on available evidence, seemed "compatible with other hazards incident to underground mining." The staff paper acknowledged that a lower level might be desirable, but cautioned that it was probably not technologically possible and that no evidence existed to show the extent to which a more restrictive standard would reduce risk. The standard of one working level would not provide an absolute guarantee of protection from radiation hazards, but the FRC staff thought it struck "a reasonable balance between biological risk and impact on uranium mining."[8]

While the FRC staff circulated its proposal to member agencies for comments, it encouraged them to reach a decision in a timely manner. For several reasons, prompt action by the FRC seemed important. One was that, according to the Metal and Nonmetallic Mine Safety Act that Congress had passed in September 1966, the Bureau of Mines was required to issue safety standards for mines, including uranium mines. Paul Tompkins wanted the FRC to agree on its position "in time to be of assistance to the Department of the Interior." Another reason was that a new uranium boom seemed imminent. The industry had suffered a major downturn after the AEC cut back on its purchases, but the unexpectedly rapid expansion of nuclear power promised escalating demand for uranium. Some observers predicted that the coming uranium rush would even exceed the frenetic pace of prospecting and mining during the 1950s. The likelihood that the uranium industry would revive and that a growing number of miners would be working underground made the effort to set and enforce safety standards more urgent.[9]

The primary reason that a prompt FRC decision seemed necessary was that, for the first time, uranium mine safety became a headline issue beyond the Colorado Plateau. The *Washington Post* took the lead in publicizing the problem; it ran a series of prominent articles and hard-hitting editorials that focused on the failure of federal agencies, especially the AEC, to regulate the mines effectively. J. V. Reistrup, a young *Post* reporter, researched and wrote several news articles, and Leo Goodman, a veteran union official and long-time critic of the AEC, provided additional information to the newspaper's editorial writers. The first of Reistrup's stories told the sad tale of a veteran of underground uranium mining who was dying of lung cancer. The next day Reistrup reported that the FRC intended to adopt a "relatively strict" standard of one working level for permissible radon concentrations in uranium mines.[10]

The articles in the *Post* won high-level attention in the agencies interested in uranium mine safety. HEW secretary (and FRC chairman) John W. Gardner instructed the FRC staff to make certain that the recommended permissible levels were as sound and defensible as possible. The Department of Labor reexamined its role in enforcing mine safety standards under the Walsh-Healy Act. When Reistrup asked Secretary of Labor W. Willard Wirtz whether the act gave the department authority over the mines, Wirtz replied that he did not know and suggested that the reporter contact the solicitor of the Department of Labor. Charles Donahue, the solicitor, told Reistrup that the department did have power over the mines under its Walsh-Healy jurisdiction. As early as 1960 the department had concluded that Walsh-Healy applied to at least some uranium mines. In 1965, it affirmed and extended that opinion; it told the Select Subcommittee on Labor that the provisions of Walsh-Healy covered uranium mines that delivered ore to mills with which the AEC had a contract to purchase processed uranium. After

Donahue informed Reistrup of the department's legal opinion, the news-man passed it along to Wirtz. A short time later the *Post* published an article that revealed Wirtz's uncertainty about the applicability of Walsh-Healy to uranium mines and credited an unnamed reporter with alerting the secretary to his authority.[11]

Reistrup's articles contributed substantially to an emerging debate over uranium mine safety by spurring the Department of Labor to undertake a careful review of the problem. Assistant Secretary of Labor Esther Peterson, a native of Utah who had visited uranium mines to observe conditions first-hand and grown increasingly frustrated with the slowness of the FRC's procedures, told Wirtz that "none of the involved agencies has clean hands." She regretted that the "Department of Labor clearly was negligent in failing to exercise its jurisdiction." Citing the department's lack of technical expertise to inspect and evalu-ate mine conditions, she urged Wirtz to approach other federal agencies, starting with the Bureau of Mines, about working together "to remedy a bad situation." Wirtz took her advice, and a short time after he pro-posed to Secretary of the Interior Stewart L. Udall that they enter a cooperative arrangement in which the Bureau of Mines would supply inspectors to help enforce Walsh-Healy, the two agencies reached an agreement. It was apparent, however, that enforcement of the act could not be carried out until a standard for permissible levels of radon in the mines was issued.[12]

While the FRC was waiting for agency approval of its one working-level recommendation, labor organizations and some public health au-thorities began supporting a much stricter permissible level. Labor unions urged that in order to provide miners with an extra margin of safety, a 0.3 working-level (WL) standard should be adopted. They argued that this level was feasible if the industry was willing to spend "several million dollars," an amount that "the larger mines are well able to afford." The Public Health Service, to the surprise of other agencies, also advocated a standard of 0.3 WL. Its position was based on new calculations and projections that suggested that adopting a 0.3 WL limit would reduce the incidence of lung cancer among uranium miners to that of the general male population.[13]

As controversy over the standard for uranium mines was brewing, Reistrup stirred it with an article that appeared on the front page of the *Washington Post* on 14 April 1967. It cited a "highly restricted" esti-mate that 1150 uranium miners would die from lung cancer by 1985, the "great majority" from exposure to radon (the figure came from a

draft report that the FRC was circulating). The story was critical of
federal agencies, particularly the AEC, for not taking effective regulatory action, and noted that the secretary of labor had not even been
aware of his authority to set safety standards. The article infuriated and
embarrassed Wirtz, but it also reinforced his determination to impose
strict limits on radon levels. A few days later, the *Post* followed up on
Reistrup's article with an editorial that blasted the AEC. Titled "AEC
Death Mines," it declared: "A death warrant for perhaps 1150 uranium
miners has in effect been signed by the Atomic Energy Commission, the
sole purchaser of uranium in this country." It maintained that the standards being considered by the FRC "are widely regarded as inadequate,
even in the unlikely event that they were to be enforced." The *Post*
blamed not only the AEC, but also the Departments of Labor, Interior,
and HEW, and the Joint Committee on Atomic Energy for using the
"flimsiest pretexts of jurisdiction, cost, and security to avoid their responsibilities to the miners."[14]

The AEC was disturbed by the *Post's* attack; Seaborg commented
that it was an "unjustifiably critical editorial." The *Post* ignored the
constructive contributions that the AEC had made to mine safety during
the 1950s and early 1960s. It had given financial assistance to the
studies of mine conditions conducted by the Public Health Service, demonstrated that ventilation was an effective way to reduce radon levels by
imposing stringent requirements on mines located on federal lands that
were clearly within its jurisdiction, and encouraged the states to carry
out the regulatory authority that they claimed. But the fact remained
that the mines had been poorly regulated, and the AEC sought to do its
part to improve conditions and set suitable standards. The Commission
decided to approach the other agencies involved in mine safety to try to
agree on a joint program.[15]

The chances for a cooperative arrangement were thwarted, however,
by growing interagency dissension over mine standards. The FRC staff
continued to favor a basic standard of 1 WL; it argued that a 0.3 WL
limit was probably not achievable with existing technology. The Departments of Labor and Interior, however, lined up with the Public Health
Service in support of a 0.3 WL standard. High-level meetings among
agency officials failed to resolve their differences. Meanwhile, the Joint
Committee on Atomic Energy announced that it would hold hearings
on the radiation exposure of uranium miners. This gave additional
incentive to reach an interagency accord to present when the hearings
convened. The FRC met on 4 May 1967, five days before the Joint

Committee hearings were scheduled to open. The agencies on the council failed to compromise their differences. The Departments of Labor, Interior, and HEW voted in favor of a 0.3 WL standard while the Departments of Defense and Agriculture and the AEC voted in favor of 1 WL. The deadlocked council postponed further consideration of the issue.[16]

The following day, Wirtz announced that under the Department of Labor's Walsh-Healy authority he was imposing a radon standard in uranium mines of 0.3 WL, expressed as 3.6 "working-level months" (WLM) per year. A working-level month was the amount of radiation to which a miner was exposed in a month of full-time work. A standard of 1 WL, for example, would give an annual exposure of 12 WLM. Wirtz added that at his discretion, he might allow a grace period of eighteen months in which 12 WLM would be acceptable. His action, which would become effective after thirty days, set off a public interagency dispute over mine standards. Seaborg called him immediately to protest that "this was done too hastily and without consultation."[17]

The Joint Committee provided the forum to air the differing opinions on mine standards. The hearings opened on 9 May 1967 with committee members responding angrily to an editorial in that morning's *Washington Post*. The *Post* leveled another blast at the AEC and other agencies for their lack of action to protect uranium miners and accused the Joint Committee of being "marvelously unconcerned." Holifield complained that "this type of editorial is not responsible nor is it factual." He stacked volumes "a foot high" of hearings the committee had held on radiation hazards over the years to accentuate his point. His colleagues also bitterly attacked the *Post's* statements. Having begun on a sour note, the hearings continued in a similar vein. Secretary Wirtz was subjected to sharp and skeptical questioning. He explained that he issued his standard because he thought that federal agencies, including his own, had delayed taking action to protect miners long enough. Wirtz argued that the "basic issue here is very clear." While he and others supporting a 3.6 WLM standard were concerned only with the health and safety of miners, the "argument for the higher standard includes considerations of feasibility, considerations of enterprise, considerations of time and profit which I rule out, whether rightly or wrongly."[18]

Joint Committee members quizzed Wirtz closely on the scientific basis for the standard he supported. He credited the Public Health Service with supplying data showing that a standard of 3.6 WLM would reduce the incidence of lung cancer among miners to that of the general male population. The head of the Public Health Service, acting surgeon-general Leo J. Gehrig, however, did little to help Wirtz's case. He failed to provide a clear explanation of the Public Health Service's conclusions and, a short time later, backed away from them completely. He informed the Joint Committee that a group of epidemiologists and statisticians in his agency had determined that there were "serious limitations in the construction of the mathematical model which preclude its being used at this time to predict the outcome of past and future experience."[19]

The AEC, in its testimony before the Joint Committee, cited its objections to the standard issued by the Department of Labor. In the absence of conclusive scientific data that could distinguish the effects of exposure to 1 WL from 0.3 WL, the disagreement centered on the implications of the standard for the mining industry and, ultimately, for supplies of uranium. The fundamental difference between Labor and the AEC was that Wirtz did not need to worry about the economic consequences of the standard he imposed, at least as far as the functions of his department were concerned. The AEC, on the other hand, necessarily considered the long-range impact of Labor's requirements on the mining industry and the growth of nuclear power. It was committed to promoting mine safety and improving conditions, but it opposed a standard that might force the closing of a large number of mines. Even when the uranium-mining states had enforced a standard of 3 or more WL after 1961, they had closed many mines, either temporarily or permanently. Colorado shut down 137 mines temporarily and 18 permanently between 1961 and 1966, and New Mexico closed 11 large mines in 1966 alone.

Although the AEC supported Labor's standard as a long-term objective, it pointed out that there were no reliable instruments for accurately measuring exposure as low as 0.3 WL and that ventilation was still an uncertain means of controlling radon levels throughout all sections of uranium mines. Ramey told the Joint Committee that even the 1 WL standard would be difficult to meet in many mines. He suggested that federal agencies should act jointly and promptly to further reduce radon levels and to provide adequate medical care and workers' compensation for those miners who developed lung cancer from earlier exposure to high levels of radon. But he insisted that, at least for the present, Labor's standard was unattainable.[20]

The other witnesses in the hearings divided along the same lines as the Department of Labor and the AEC. Labor union representatives strongly supported Wirtz's standard. State mining officials and mine

industry spokesmen complained that a 3.6 WLM limit was impossible
to meet with existing technology and would force many mines to close.
Colorado's deputy commissioner of mines, G. A. Franz, estimated that
Wirtz's standard would "certainly close 95 percent of the underground
mines in the State." He added that this would "eliminate the means of
making a living for most of the miners." The mining industry thought
that even the 12 WLM limit was too restrictive and that a 36 WLM (3
WL) standard would be more realistic. Two months earlier it had protested against a 12 WLM limit on the grounds that it would cause many
mine closures.[21]

The Joint Committee hearings did not resolve the interagency differences but they did force Wirtz to reconsider the standard he had issued.
He was in an awkward position; he had argued eloquently that the 3.6
WLM limit was essential to save the lives of many miners, only to be
undercut by the Public Health Service's retreat from the data on which
he had relied. But after weighing the testimony in the Joint Committee
hearings, he revised his stance only slightly. Wirtz reaffirmed his commitment to the 3.6 WLM standard, calling it "the right standard from
the health standpoint." He acknowledged the difficulties of complying
with it, and in the major departure from his previous policy, offered a
blanket (rather than discretionary) allowance of an eighteen-month
grace period. He insisted that mine operators abide by the 12 WLM
limit and make progress toward the 3.6 WLM standard during the
interim period. Wirtz's statement was greeted cooly by the Joint Committee. Congressmen Holifield and Melvin Price responded that the
recent hearings had revealed "that there is virtually no factual basis for
a standard as low as [3.6 WLM]."[22]

Although Wirtz continued to favor the 3.6 WLM standard, his allies
on the Federal Radiation Council changed their positions to support the
12 WLM limit. The FRC, which was still considering the issue after its
deadlocked meeting on 4 May 1967, sought to reach unanimous agreement on a permissible level. After extensive discussion, it settled on a
compromise. Wirtz went along with the standard of 12 WLM that the
other members backed, but he won their endorsement to review the
standard again after one year (instead of the five years that had been
planned). The council voted unanimously for this arrangement. The
FRC's action was consistent with Wirtz's recent announcement that he
would wait eighteen months before enforcing a stricter standard. President Johnson approved the council's recommendations on 27 July
1967. The FRC's decision was a milestone in the effort to provide

adequate protection to uranium miners. For the first time the federal agencies involved in the problem agreed on a standard for permissible limits of radon in the mines and formally adopted the working level that had been accepted as a target figure twelve years earlier. But the compromise between the agencies represented on the FRC was a temporary accommodation that did not settle their differences or end the controversy over mine safety.[23]

The FRC's decision triggered sharp attacks from its critics. The *Washington Post* editorialized that the "FRC has demonstrated, beyond all but a bombmaker's doubt, its disinterest in the public health." Reistrup coauthored an article in the *Bulletin of the Atomic Scientists* charging that the FRC "voted to give the mine owners, not the miners, the benefit of the doubt," an action that would require miners to "continue their deadly love affair with the daughters of radon." W. A. (Tony) Boyle, president of the United Mineworkers of America, declared that the deaths of uranium miners from lung cancer "underscores once again the brutal cost that this nation has paid to develop the atomic industry." The miners themselves were more ambivalent about the impact of mine regulation. While concerned about the risks of lung cancer, they were also worried that stringent standards could close mines and cost them their jobs. An article in the *Wall Street Journal* posed the question that the miners faced: "Would you gamble your life for a good salary, low taxes, and cheap housing?" On the Colorado Plateau, it found, "the answer is yes." One miner commented that "if I'm going to get cancer, I'm going to get it," and another suggested that "those fellows who died with cancer probably had weak lungs and shouldn't have been in the mines in the first place."[24]

Even before the FRC agreed on the 1 WL standard, federal agencies had stepped up their efforts to promote mine safety. Following up on the suggestions that Ramey outlined in the Joint Committee hearings, the AEC urged the secretaries of Labor, Interior, and HEW to join with it in a cooperative program. They were receptive, and in a meeting attended by Gardner, Wirtz, and other high-level officials, Seaborg stressed that effective action to protect uranium miners depended on interdepartmental coordination. The areas that appeared to be most important were: continuing to improve conditions in the mines, conducting research on instruments to measure radon levels more accurately and on the medical implications of radon hazards, and providing adequate workers' compensation for miners. There was no disagreement on the need for the interested agencies to combine their resources and expertise, but the Department of Labor was suspicious that the AEC intended to undermine the enforcement of its standard. Esther Peterson told Wirtz that the AEC had given her "a clear impression of wanting to 'move in' and control all that happens from here on out."[25]

The first step in the federal effort to improve mine conditions was to inspect the mines and take corrective measures where necessary. Although the uranium-mining states had improved their regulatory performance since 1961, their activities were still limited. Colorado, which had the largest number of mines and employed five inspectors, had the most aggressive program; using a 3 WL limit, it closed 8 mines and removed miners from areas of 35 more in 1967. The other states had, at best, only one person surveying their uranium mines. Under an arrangement between the Departments of Labor and Interior, the Bureau of Mines, which employed sixteen qualified inspectors, also evaluated uranium mine conditions to enforce the Walsh-Healy Act. In 171 mines that it inspected in 1967 it found that over 60 percent had radon concentrations of 1 WL or less, and the number of miners exposed to more than 3 WL had decreased by nearly two-thirds. An AEC survey produced equally promising results. The Department of Labor contacted the owners of mines subject to Walsh-Healy that exceeded the 1 WL standard and directed them to reduce radon levels. It reported that "not a single mine has failed to indicate some plan for corrective action." The surveys of uranium mines did not mean that the problem of radon had been solved but they did indicate encouraging progress.[26]

In the other areas requiring attention by the agencies involved in the mine problem, there was less progress. An interdepartmental committee formed to work on mine safety agreed that research was needed to provide additional information on medical and epidemiological issues and on controlling radon levels. The Public Health Service sought better data on a variety of questions, such as the relationship between the levels of radon to which a miner was exposed and the dose he actually received, the effects of smoking on mortality trends for miners, and the development of tests to enable early diagnosis of lung cancer. It hoped that new data would offer a clearer basis for determining radon standards in the mines. The Bureau of Mines planned a series of research projects on the effectiveness of ventilation in controlling radon levels, methods of designing mines and sealing off unoccupied areas to improve protection, and filtering mine air to remove radioactive contaminants. The AEC sponsored projects to design better instruments for

9. Scientists testing face mask filters in a uranium mine in New Mexico. (National Archives 434-SF-63-10)

detecting and measuring radon in mines and improved dosimeters for showing the exposure of individual miners.[27]

There was no dispute among the interested agencies about the desirability of those research programs, but there were questions about how they would be funded. The AEC, the Public Health Service, and the Bureau of Mines requested that Congress provide money to raise the total amount spent on research by all three agencies from $1,720,000 to $3,627,000 in fiscal year 1968. The AEC offered to transfer $500,000 from its budget to assist the Bureau of Mines in its research. This proposal was rejected by Michael J. Kirwan, chairman of the Subcommittee on Public Works of the House Committee on Appropriations, who suggested that if the Department of the Interior needed the funds it should ask for them through normal channels. Funds for research on uranium mine safety were further curbed by federal budget stringencies. Despite the appeals of the AEC and the protests of the Joint Committee, the money allocated to the Public Health Service and the Bureau of Mines by their departments was reduced sharply. The FRC reported in January 1968 that because of "fiscal restraints," important research projects had been "seriously curtailed and indefinitely postponed."[28]

Federal efforts on the third problem of concern to the interdepartmental committee—workers' compensation for uranium miners—were even less fruitful. The provision of workers' compensation and death benefits was normally an obligation assumed by individual states, but coverage for uranium miners ranged from uncertain to inadequate. Workers' compensation for all persons exposed to radiation in their jobs had been a subject of concern for years, but it remained an unresolved issue. The greatest hardship in claiming compensation was proving that injury or death was the result of occupational exposure to radiation rather than other factors. This was especially difficult for diseases that were "nonspecific" in origin, such as cancer or leukemia. Furthermore, many states imposed time limitations for making claims, which could preclude benefits for radiation-induced illnesses with a long latent period.

The states on the Colorado Plateau offered limited coverage to radiation workers, including uranium miners. The Department of Labor's Esther Peterson told Wirtz that "compensation laws in the uranium-producing States are woefully inadequate in terms of level of benefits." Only one state, Colorado, paid compensation to families of miners whose deaths seemed clearly linked to exposure to radiation. Utah rejected a claim by the widow of a miner in August 1967 and no applications for benefits had been filed in other states. In light of the prospects for an increasing number of claims, Colorado was deeply concerned about the solvency of its compensation insurance fund. It appealed for assistance from the federal government.[29]

Proposals for federal subsidies to the states received sympathetic consideration from the agencies involved in the uranium mining problem. Ramey believed that the federal government had "a moral if not a legal responsibility" to the miners, and the interdepartmental committee suggested that it owed them "an unusual social and financial responsibility." After consultation with the AEC and the Departments of HEW and Interior, the Department of Labor drafted legislation to supplement state compensation programs. It was one of several different bills introduced in Congress in 1968 offering federal funds to ensure that uranium miners afflicted with lung cancer received adequate compensation. The insurance industry led the opposition to the proposals. Its objections centered on questions about administration of federal compensation, the ill effects of federal assistance on persuading states to improve their coverage, the advisability of singling out one group for benefits, and the fear that congressional approval of legislation would set a precedent for federal intrusion in a function traditionally carried out by the states. The Select Committee on Labor took no action on the bills to increase compensation for uranium miners. The failure of federal legislation meant that the only available alternative for interested federal agencies was to encourage the uranium-mining states to extend their coverage. Federal officials were not optimistic about the prospects for influencing the states to improve their compensation programs.[30]

Meanwhile, the FRC, in preparation for the new review of mine standards on which its members had agreed in July 1967, was seeking new data and soliciting fresh appraisals of the problem from outside experts. The council had decided to reexamine the subject after an interval of one year, and it became increasingly apparent that the controversy would resume. None of the member agencies had modified its position, and officials in the Bureau of the Budget worried that the battle would be waged "with perhaps even greater intensity than last year." They reported that at least two members of the Joint Committee feared that the issue "might explode" during the election campaign of 1968.[31]

The FRC's efforts to inform the debate and find grounds for agreement met only limited success. Tompkins asked the National Academy of Sciences, a prestigious independent body of scientific authorities that was chartered by Congress, to evaluate the existing data on radon hazards. The National Academy appointed a special committee to consider the evidence, and, not surprisingly, it reached no unequivocal or incontestable conclusions. It found that miners who were exposed to a total of 100 to 400 WLM during their lifetimes (this would be equivalent to exposure of approximately 0.3 to 1 WL over a period of thirty years) showed a slightly higher than normal risk of lung cancer. The National Academy committee attributed this statistical increase at least partly to radiation in the mines but also submitted that other factors might "account for all or part of the increase." It emphasized the dangers of smoking, noting that the data suggested "that cigarette smokers among the uranium miners are particularly susceptible to lung cancer." The report, though a useful summary of the issues, made no distinction between the risks of 0.3 WL (3.6 WLM) and 1 WL (12 WLM), and therefore, did little to resolve outstanding interagency differences.[32]

The FRC also commissioned a study of another issue that caused divisions among its members—the economic impact of tightening standards. It hired a contractor, the Resource Management Corporation, to investigate the question. The company's report largely supported the arguments of the AEC and other advocates of a 12 WLM standard. It concluded that a 3.6 WLM standard did not "appear technically feasible at this time," but it suggested that incremental improvements over time could approach that level "without impairing the economic health of the industry." Industry representatives were not convinced. An official of Union Carbide, which produced about 10 percent of the uranium that came from underground mines and employed about 19 percent of the miners, declared: "We do not believe it will ever be possible to continually control all mine areas below the 1 W-L [limit]." Mine operators continued to make good progress in lowering radon concentrations; by May 1968, 85 to 90 percent of 120 mines surveyed were meeting the FRC's standard. But owners insisted that reducing the limit further would force mines to close.[33]

By the early fall of 1968 the FRC was making plans for a meeting at which the existing standard would either be reaffirmed or changed. In deference to the Joint Committee, the meeting was delayed until after the fall elections. The committee staff worried that if the FRC failed to reach a unanimous decision, the controversy could become an issue in the reelection campaigns of some members. Once the election was over, prompt action by the FRC was needed. The grace period that Wirtz had allowed uranium mine operators when he announced his radon standards under Walsh-Healy would end on 1 January 1969. Moreover, Udall wanted to issue Interior Department regulations under the Metal and Nonmetallic Mine Safety Act by early January.[34]

While waiting for a meeting to be scheduled, the FRC staff and member agencies sought to achieve another compromise agreement. With the exception of HEW, which had decided to support Labor's position, the division of views had not shifted since July 1967. Labor and HEW called for an immediate 3.6 WLM standard and the other agencies favored retaining the 12 WLM limit until an unspecified future date. Rather than face an impasse, FRC members hoped to reach an accommodation. The Department of Labor recognized that its leverage was diminishing. In 1970 all government contracts to purchase uranium would expire, which would deprive the department of enforcement power under the Walsh-Healy Act. In order to exert what influence it could, it wanted to persuade the Department of the Interior to adopt strict standards under the Metal and Nonmetallic Mine Safety Act. But this required compromising its position because the staff at Interior backed the 12 WLM limit. The AEC was willing to compromise to avoid the immediate imposition by Labor of a 3.6 WLM standard, which it contended would "cause unwarranted hardship on both the uranium miner and the industry without commensurate benefit to the health of the miner."[35]

After much negotiation, the FRC failed to find a compromise that was acceptable to all members. At a meeting on 29 November 1968, the Labor Department proposed a slightly revised standard of 4 WLM, effective on 1 January 1969. It was defeated by a vote of five to two, with only HEW supporting Labor. By another vote of five to two, the recommendation of the FRC staff carried. It provided for retaining the 12 WLM limit for the time being and set goals of reaching 8 WLM by 1 January 1971 and 4 WLM by 1 July 1974. The reduced levels would not automatically go into effect; the FRC would reconsider epidemiological data, exposure records, costs, and feasibility before making a final deter-mination. The FRC's decision included another new feature. Stressing the possible synergistic relationship between radon exposure and smok-ing, it urged that a "concerted effort be made by all concerned to discourage cigarette smoking among underground uranium miners."[36]

The FRC's action did not end the controversy. Wilbur J. Cohen, HEW secretary and FRC chairman, still sought to reach a unanimous accord that was closer to his own position. Without consulting the AEC, he offered a new agreement to the other members of the FRC. It would make the 4 WLM standard effective on 1 January 1971. Cohen also took his proposal to the president, who sent it along to the Bureau of the Budget. The bureau, in turn, contacted FRC members to lobby for a unanimous agreement. The AEC reluctantly acquiesced because of its understanding that all the other member agencies had accepted Co-hen's proposal. To its dismay, it soon learned that the Department of

Labor had refused to go along with Cohen because the 4 WLM limit would still not be "iron-clad." Wirtz issued the department's regulation on 28 December 1968, calling for an immediate limit of 4 WLM but, until 1 January 1971, allowing for variances up to 12 WLM on a case-by-case basis. President Johnson signed the FRC's new recommendation on 11 January 1969 and the Bureau of Mines published its standards a few days later. The three measures were similar in the levels they permitted—12 WLM until 1 January 1971 and 4 WLM thereafter. The Labor Department's standards were more binding and less negotiable than those of the FRC and Interior but would lose much of their impact when the AEC's uranium purchase contracts expired.[37]

The outcome of the latest effort to agree on mine standards left the agencies most involved in the issue frustrated and disgruntled. The Labor Department was isolated in its position and losing its influence. The AEC had been outflanked and misled into accepting a standard it did not support, at least for the short term. The FRC staff had been by-passed by Cohen in his negotiations and made to look weak, laggard, and superfluous. Joint Committee members assailed both the Labor Department and the FRC, the one for its impertinence and the other for its impotence. Holifield accused Wirtz of acting out of "personal feelings and emotion," and Congressman Wayne Aspinall complained that "the FRC could not—or would not—live up to its obligations."[38]

When the Nixon administration entered office in January 1969, the question of mine standards remained unsettled. The AEC hoped that the departure of Wirtz and Cohen would open the way for a reconsideration of the position they had advanced. There were early indications that the new administration would be less inclined to enforce the regulations that Wirtz had issued; a Labor Department official announced that his agency was "not going to close any mines." Neither the new secretary of labor, George P. Shultz, nor the new secretary of HEW, Robert H. Finch, was ready to take prompt action on uranium mines, however, and during the two years before the 4 WLM standard was scheduled to take effect, familiar patterns on the mine safety issue repeated themselves. Finch established a new interagency committee to study the problem and report to the FRC. The Joint Committee held hearings that attempted, with limited success, to clear up the confusion caused by the differences in the regulations of various agencies. Committee members aired their reservations about the advisability of the 4 WLM standard and criticized Wirtz and Cohen for pushing it.[39]

Spokesmen for the uranium mining industry continued to insist that a 4 WLM standard was impossible to achieve with existing technology. They argued that the progress they had made in meeting the 12 WLM limit, which nearly 90 percent of the mines were doing, did not mean that with greater effort the mine operators could reduce levels to 4 WLM. For over half of the mines, a mining company official told the Joint Committee, this "presents difficulties not yet solved." The results of the Bureau of Mines' inspections supported the industry's claims. Although mine conditions showed dramatic improvement after 1967, the pace of progress had slackened considerably by the end of 1969. For its part, the AEC continued to worry that strict enforcement of 4 WLM would force mine closures and undermine the growth of nuclear power. "Present indications are that existing underground uranium mines are reaching a point of diminishing returns in the use of ventilation to control radon," Rafford L. Faulkner, director of the Division of Raw Materials, told the Commission in February 1969. This raised the possibility that "nuclear power growth may be seriously affected by the economic results of overly stringent regulation." The Commission decided to work for a change in the FRC standard unless new evidence showed it to be "clearly necessary to protect the miners' health."[40]

The FRC, once again, tried to resolve outstanding issues before deciding whether to recommend that the 4 WLM limit go into effect on 1 January 1971. As that date approached, however, it had to postpone the deadline for six months while it waited for the results of new studies on epidemiological trends and the economic impact of tightening standards. As with earlier reviews of those subjects, the new submissions were useful but not definitive. The epidemiological report prepared by the Public Health Service echoed the conclusions of the National Academy of Sciences in 1968. It found that miners exposed to a total of 120 to 360 WLM in their lifetimes faced an increased risk of cancer. The results of the new study on the economic impact of a 4 WLM limit were more surprising. Prepared by the Arthur D. Little company, it contended that the economic consequences of the stricter standard would be limited. It maintained that operators could meet a 4 WLM standard without undue difficulty, though it acknowledged that they would need about two years to do so. The report further suggested that enforcing the standard would have little adverse effect on the cost of or demand for uranium.[41]

By the time that the new reports were completed, the FRC had been abolished. Its functions were transferred to the newly created Environmental Protection Agency (EPA). EPA administrator William D. Ruckelshaus weighed the views of different agencies in determining whether to affirm the previously established standard (4 WLM as of 1 July 1971) or to make a different recommendation to the president. There was no consensus among the agencies; HEW wanted to retain the immediately effective 4 WLM while the other former members of the FRC called for 8 WLM as of 1 July 1972. Three departments, Labor, Interior, and Agriculture, supported making the 4 WLM limit effective 1 July 1973, but the AEC refused to commit itself to a firm date for the implementation of that standard. It was still concerned about the impact of 4 WLM on the economic well-being of the industry and thought the Little report was overly optimistic. Ruckelshaus eventually settled on a compromise. He found the evidence of the Public Health Service and the analysis of the Little report convincing and, therefore, elected to stick with the 4 WLM standard. But in recognition of the difficulties of immediate compliance, he proposed that the Bureau of Mines, which would enforce the standard, postpone taking action against violators for an unspecified time after the regulation took effect on 1 July 1971.[42]

Ruckelshaus's decision elicited predictable protests from the mining industry, the AEC, and the Joint Committee. Of greater potential consequence were reservations expressed by the Bureau of Mines, which had opposed immediate imposition of the 4 WLM standard. It announced that it lacked authority to allow exceptions or variances to the standard, and therefore, that it had to enforce it immediately. This position sent "a shock wave" through the industry. Eventually, the bureau modified its regulations to allow mine operators six months, with the possibility of further extension, to comply with the 4 WLM standard.[43]

The actions of Ruckelshaus and the Bureau of Mines finally ended the lengthy, controversial, and tortuous quest to establish a permissible standard for radon exposure in uranium mines. After years of bureaucratic dissension, the standard that went into effect was very close to what Willard Wirtz and his allies had advocated since 1967, though the deadline for compliance remained indefinite. Wirtz deserved much credit for taking steps that greatly improved conditions in the mines. Once he became aware of the problem of excessive radon concentrations, largely as a result of stories in the *Washington Post*, he acted promptly to force mine operators to reduce radon levels. At his initiative, the federal government for the first time enforced standards in the mines. As a result, radon levels were reduced sharply; the threat of federal regulation seemed to make a greater impact on mine owners than the efforts of the states alone. Although mine conditions had improved after the states began to enforce standards, progress was more rapid and more impressive after the federal government stepped in.

The adoption of a standard close to what Wirtz favored was a tribute to his department's persistence in the face of strong opposition. It also obscured the fact that Wirtz's 3.6 or 4 WLM limit was an arbitrary requirement that had little basis in scientific evidence. No one disputed the desirability of achieving the standard established by the Labor Department, but the feasibility of doing so in the time frame it imposed was less clear. Wirtz dismissed the objections of the industry, the AEC, and others by asserting that they valued the lives of miners less than he did. One Department of Labor press release began with a lead-in that defined the issue in stark terms: "Wirtz Puts Human Life Above Economic Values."[44] This approach won Wirtz plaudits from the *Washington Post* and labor unions, but it did little to solve the practical problems of reducing radon levels.

Other agencies and the mining industry faced the more taxing challenge of improving health conditions without closing a large number of mines. The standard that the AEC and several other agencies supported as an immediate requirement, though less stringent than the limit that Wirtz favored, was a major improvement in reducing allowable radon levels in the mines, and one that the industry strenuously protested. The standard they advanced was an attempt to strike a balance between protecting the health of miners while also protecting the jobs of miners. This was an inherently difficult task that was further burdened by the suggestions that those who opposed the Labor Department's position placed economic considerations above the welfare of the miners. The issue was not that simple, and the controversy between agencies over uranium mine standards would have been less acrimonious in the absence of such posturing.

Nevertheless, the existence of bureaucratic strife was an indication of the increased level of concern and commitment to regulatory action on the part of the federal government. Mine conditions became much safer as a result of federal involvement. Ironically, however, the federal government failed to do much to help those miners who had worked during the 1950s and earlier, when radon levels were extremely high and regulatory responses were inadequate. One measure that would have offered a modicum of assistance to those who would suffer from lung cancer caused by exposure to radon was financial support for state workers' compensation programs. But proposals to subsidize state payments to miners died in Congress.

   Although the radioactive hazards of uranium mines presented the
most acute health problem arising from the front end of the nuclear fuel
cycle, dangers caused by the processing of ore in uranium mills also
generated concern and controversy. The mills crushed and ground uranium ore, chemically extracted the uranium, and produced a partially
refined concentrate called yellow cake. The ore residues that were left
after the separation of uranium were finely ground, sand-like materials
called tailings, which were stored in huge piles at mill sites. The tailings
still contained traces of uranium and thorium that gave off small
amounts of radium and radon. To prevent the slightly radioactive tailings and the radon gas they emitted from escaping off-site, the piles
were controlled by covering or wetting them.
   The AEC licensed uranium mills and exercised regulatory authority
over them. It required mill operators to comply with its regulations on
permissible exposure to radiation by workers and by the general public.
A short time after the agency issued its first radiation-protection regulations in 1957, it conducted inspections of several mills and found that in
many cases they had "not exerted sufficient effort to comply." The AEC
directed the mills to correct their deficiencies, and by May of 1960 was
persuaded that they were making satisfactory progress toward improving health and safety conditions.[45]
   The AEC regulated mills as long as they were operating, but it dis-
claimed responsibility for tailings piles once a mill closed. This emerged
as a problem after the uranium boom ended and about half of the
twenty-five licensed mills were shut. The AEC's statutory authority
extended only to source materials that could be used to make special
nuclear materials; the agency defined source materials as ores that contained by weight more than one-twentieth of 1 percent (0.05 percent) of
uranium, thorium, or a combination of the two. That definition generally did not apply to mill tailings. The AEC regulated tailings piles as
long as a mill remained in business, since they were "an integral part of
the milling operations." Once a mill closed, however, the AEC considered its authority over materials with a concentration of less than 0.05
percent uranium and/or thorium to be terminated. The radium that the
tailings contained fell outside the AEC's regulatory jurisdiction because
the 1954 Atomic Energy Act did not apply to naturally occurring radioactive materials, including radium. Therefore, the AEC's legal staff concluded that the agency lacked any authority to enforce its regulatory
requirements in abandoned mills.[46]
   Despite the low levels of radioactivity in mill tailings and its denial of

authority over them in closed mills, the AEC, along with the U. S. Public
Health Service, sought more conclusive information than what was
available about the potential hazards that tailings posed. Public Health
Service officials were disturbed by reports that tailings were being used
in construction materials, land fills, sand traps on golf courses, and even
children's sand boxes. A greater source of concern, at least initially, was
that tailings piles would no longer be controlled in abandoned mills,
allowing radioactive particles to be carried by winds into surrounding
areas or to contaminate adjacent bodies of water. Moreover, the piles
were unsightly. This was an especially important consideration for sites
located near cities and towns; one mill, for example, sat on the outskirts
of Salt Lake City and another on the edge of Durango, Colorado.[47]

The AEC undertook a study of the possible dangers of airborne
radioactive tailings carried off piles by the wind, and the Public Health
Service conducted a survey of the effects of tailings on rivers and
streams near mills. After weighing the information it collected and the
data that the Public Health Service provided, the AEC determined that
the tailings at abandoned mills did "not constitute an unreasonable risk
to health and safety of the public," and that "controls over such tailings
for radiological safety purposes are unnecessary." The AEC based its
position on data indicating that if tailings were swept off piles or
leached into waterways, their concentrations of radioactivity would
remain far below permissible limits. Even if a pile collapsed and slid into
an adjacent stream or river, it found that radiation exposure standards
would not be exceeded. In addition, the legal staff reaffirmed its opinion
that the AEC lacked regulatory jurisdiction over tailings at closed mills
that contained less than 0.05 percent uranium and thorium. In short,
after conducting its investigation, the AEC concluded that tailings did
not pose a radiological hazard and that control of their use or disposition rested with state governments.[48]

The Public Health Service, using the same data, reached different
conclusions in a report it drafted. It agreed with the AEC that radiation
from tailings presented "no significant immediate hazard." But it expressed much greater concern about long-term effects of the tailings and
called for "caution and a conservative approach to the problem of
disposition and control of [them]." It urged that action be taken
promptly to prevent "the erosion and spread" of tailings and that arrangements be made to ensure permanent maintenance of tailings piles.
The Public Health Service not only was more troubled than the AEC
about erosion but also about the possibility that a restraining dike could

break or a pile could collapse, dumping tens of thousands of tons of tailings into an adjacent waterway. The AEC thought that the Public Health Service overstated the radiological hazards of tailings and the need for control of the piles, but it sought to avoid a public confrontation over the issue. It approached Public Health Service officials about meeting to discuss and perhaps resolve their differences and it pledged to press mill owners informally to take further steps to stabilize tailings piles. Before representatives of the two agencies could discuss their viewpoints privately, however, the Public Health Service aired its position at a conference on water pollution in Grand Junction, Colorado in December 1965. The local press headlined the problem of radioactive contamination of the Colorado River that the Public Health Service had emphasized.[49]

Within a short time, the issue received national attention. In March 1966 the *New Republic* ran an unsigned article that accused the AEC of trying to prevent public release of the Public Health Service's report because of fears that it would create "public hysteria." The article also depicted mill tailings as a serious radiological hazard. The AEC was dismayed by the *New Republic* piece, partly because the charge of attempting to suppress the Public Health Service's report was groundless and partly because the description of the tailings problem was distorted and exaggerated. The AEC was even more disgruntled when the article was used as a basis for other stories on the subject. The Federation of American Scientists published an excerpt in its newsletter, and CBS News ran stories on both radio and television. Correspondent Terry Drinkwater told television viewers that if the tailings were not covered or removed, "another generation may well look at these radioactive man-made mountains as monuments to the carelessness of . . . the early years of the Nuclear Age."[50]

The publicity about mill tailings and the release of the final version of the Public Health Service's report stirred the interest of Senator Edmund S. Muskie of Maine, chairman of the Subcommittee on Air and Water Pollution of the Senate Committee on Public Works and a prominent advocate of rigorous antipollution laws. Muskie called a hearing on water pollution from mill tailings, and quizzed representatives from the Federal Water Pollution Control Administration (a part of HEW), the Public Health Service, and the AEC. The HEW representatives reiterated their contention that tailings were not an immediate threat but could become one over time. The AEC restated its opinion that they did not present a long-term hazard and that it lacked regulatory authority over them, views that Muskie challenged but did not shake. The Muskie hearings generated additional news stories about the problem of mill tailings.[51]

The flurry of publicity in the spring of 1966 and the differing positions of the AEC and the Public Health Service aroused concern among state health authorities. Several of them expressed reservations about the AEC's assessment of the radiological hazards of mill tailings and urged the agency to take action to make certain that the piles were properly controlled. The AEC responded that although it recognized that tailings piles were unsightly and disagreeable nuisances, it lacked authority to require that they be stabilized. It pledged to continue to encourage mill owners to do so voluntarily and it advised state governments to enact regulations requiring the stabilization of tailings piles. Colorado became the first state to take action by adopting regulations that directed mill owners to maintain control of the piles and most other states with uranium mills followed suit. In December 1966, the Federal Water Pollution Control Administration, the Public Health Service, and the AEC agreed on a joint statement that compromised their differences. It acknowledged that inactive tailings piles "should be stabilized and contained to prevent water and wind erosion" while placing responsibility for doing so on mill owners and the states.[52]

Just when it appeared that the questions of what should be done about mill tailings and who should do it were largely resolved, a new concern arose. Health authorities had considered the primary hazard of the tailings to be radium, but the Public Health Service in late 1966 began to express uneasiness about the dangers of radon emitted from the tailings. It worried that even if the piles were stabilized, radon released into the air around them would cause a health risk. This occurred at about the same time that radon hazards in uranium mines emerged as a topic of growing attention and debate. At the prodding of Congressman Aspinall, the AEC and the Public Health Service undertook a joint study of radon that tailings piles emitted. Over a period of a year they took samples of air around the piles at three mills to check radon levels under different conditions. They concluded in a March 1969 report that "no significant radiation exposure to the public" resulted from the radon released by mill tailings. The findings of the joint study on radon were reassuring, and, in general, control of the piles had greatly improved by the late 1960s. Still, some piles resisted efforts to stabilize them, the question of dike breaks or collapsing piles remained unresolved, and the potential health and environmental hazards of tailings piles continued to be a matter of some concern to public health officials.[53]

One way to reduce the size of the piles was to find useful applications for the tailings. As concern about the problems created by the piles diminished, however, anxiety about the consequences of distributing tailings for constructive purposes increased. This issue emerged in 1966 when the Colorado Department of Health, checking the performance of new instruments developed to monitor radon gas in uranium mines, detected the presence of radon in newly constructed buildings in Grand Junction, Colorado, a city of about 20,000 on the Colorado Plateau. Several newspapers ran stories reporting that cinder blocks containing mill tailings were used in building a high school and a sporting goods store. One article cited the dangers of radon escaping from the blocks and quoted an unnamed U. S. Public Health Service official as saying that "living in a building made of these cinder blocks would be like living in a poorly ventilated uranium mine." The AEC found that tailings had in fact been used as land fill in the construction of the store and the school, but not in cinder blocks. In both cases, the tailings had been covered with other materials. The AEC also learned from the Public Health Service that a reporter had misinterpreted the statements of its representatives.

The AEC was concerned about the reports but did not view the problem they cited as a matter under its jurisdiction. It had informed the states in 1961 that it would exercise no authority over the use of mill tailings, adding that "the radium content of these tailings may be such as to warrant control by appropriate state authorities." It had on occasion reaffirmed its position in response to requests from state officials about the advisability of using mill tailings in road beds and basement fills for buildings. The AEC regarded such applications of tailings as generally acceptable and, if done properly, unlikely to pose an appreciable risk to public health. But it told state health authorities that they must determine whether they would allow the use of tailings in specific cases. In response to the situation in Grand Junction, the Colorado Department of Health in 1966 prohibited any further use of mill tailings in construction.[54]

The following year, the Colorado Department of Health undertook a survey of radon levels in buildings in Grand Junction where mill tailings had been used for foundation fill. By 1969 it had found about sixty homes where indoor radon concentrations in spots exceeded 0.01 WL. A reading of 0.01 WL was regarded as acceptable because it was close to the natural background level of radon that was present in Grand Junction. Basements in a few of the affected homes showed radon levels that were above the standard the FRC had recently set for uranium mines. A Colorado health official announced that "there is no reason for panic and no immediate danger," and promised that the state would conduct further studies to learn more about the magnitude of the problem.[55]

The news about Grand Junction was featured in newspaper, magazine, and TV accounts, many of which portrayed the hazards as more alarming than Colorado health officials had indicated. There was no evidence to confirm the particularly unsettling reports that tailings had been used in children's sandboxes, or as *New York Times* columnist Tom Wicker put it, that the AEC "gave away the substance" for such a purpose. But information about the use of mill tailings in construction was accurate, and it attracted more press attention than it might otherwise have done because of its relationship to other well-publicized stories. It drew comparisons with the controversy over radon hazards in and standards for uranium mines. It also occurred at about the same time as a major fire at the Rocky Flats nuclear weapons production site near Denver and the detonation of a nuclear blast used to tap natural gas resources in the Rocky Mountains. Furthermore, by late 1969 a new debate had arisen over the adequacy of the AEC's radiation standards for protecting public health from nuclear hazards. The AEC was the object of sharp attacks on all of those issues; one Colorado School of Mines professor charged that the AEC was running "the risk of exterminating the human race." In that atmosphere, the agency was further assailed for not preventing the use of mill tailings for construction purposes. The most prominent critic was H. Peter Metzger, a biochemist and chairman of the Colorado Committee for Environmental Information. He accused the AEC of "dereliction of duty" and called its assessment of radiation hazards "a calculated risk—the AEC does the calculation and the public gets stuck with the risk."[56]

Much of the criticism of the AEC focused on its denial of authority to regulate the use of mill tailings. Colorado health officials informed the state Board of Health that they did not realize until 1966 that the AEC assumed no responsibility for evaluating the safety of mill tailings in construction projects. Paul W. Jacoe, director of the health department's Division of Air, Occupational, and Radiation Hygiene, declared: "We should have gotten into it sooner. We thought the AEC had jurisdiction but they decided they didn't." In response to queries, the AEC told reporters that Harold Price had sent a form letter in March 1961 to notify states with uranium mills that it exercised no authority over the transfer of tailings and that regulation of their use was a state responsibility.[57]

Metzger called the AEC's explanation "an outright lie." He acknowledged that "if this letter had been sent, it would go a long way to vindicate the A.E.C. for their many years of inattention and inactivity." He contacted the state health departments to whom the AEC had sent the letter, and all responded that they could not locate it in their files. Metzger charged that "somebody at A.E.C. is making an attempt to rewrite the history books so as to evade ultimate responsibility." The regulatory staff undertook a rather frantic search for the letter. Staff members located a copy in the proper place in the AEC's files, but could not be positive that it was ever sent out. State officials confirmed that they did not have the letter in their files, though one remembered having seen it.

The AEC found that the *Atomic Energy Clearing House*, a compilation of legal public documents published by a private company, included the Price letter in an issue of 20 March 1961. This supported the AEC's argument that it had made its position on regulating mill tailings known in 1961. The AEC also pointed out that subsequently, in response to requests for its views on the use of tailings, it had sent letters to individual states reiterating its stand. The reasons that the states did not have copies of the March 1961 letter remained unclear. The AEC speculated that it could have been lost, misplaced, or disregarded by state health departments, especially since it was a form letter. In any event, the missing letter was an embarrassment to the AEC. It fueled allegations that the AEC had failed in its responsibilities and then made a clumsy attempt to cover its tracks.[58]

Meanwhile, the AEC, the U. S. Public Health Service, and the state of Colorado were trying to find ways to deal with the problem of tailings in Grand Junction. The state health department continued its survey of radon levels in buildings in the city, progressing at a pace of about 170 locations per year. By August 1970 it had checked 534 buildings, finding that 65 exceeded 0.05 WL and 30 exceeded 0.1 WL. The Public Health Service had recently recommended that "remedial action" be taken where readings were higher than 0.05 WL, that it might be needed at levels between 0.01 and 0.05 WL, and that it was not required at levels below 0.01 WL. But the questions of what remedial action to take and who would pay for it remained to be addressed. The surest solution was to remove the tailings from under and around buildings regardless of radon levels, but the costs of such an undertaking were estimated to be very high. They would far exceed the resources of the state and federal agencies involved in correcting the problem.[59]

The AEC did not view the health risks of the tailings as cause for serious concern and continued to insist that it lacked jurisdiction over the use of them in construction. It recognized, however, that its position would not placate its critics or end the controversy over radon in Grand Junction. As one staff member noted, this was especially true in light of "the current problems that AEC has in every field involving radiation and the public." By 1971, the agency had concluded that even in the absence of legal liability it had a moral responsibility to assist property owners "because the tailings were generated . . . to fulfill a Government defense need." The AEC provided funds for a feasibility study of how best to remove the tailings and technical assistance to the survey of radon levels. It also urged the Office of Management and Budget to allocate federal funds for a clean-up program in Grand Junction.[60]

As the proper course of action in Grand Junction continued to be studied and debated in interagency meetings, congressional hearings, and the press, two main questions caused controversy. One was whether virtually all of the tailings in the city should be removed or only those around buildings with radon levels in the range at which the Public Health Service had recommended remedial action. The other question was who would pay for any corrective measures that might be taken. Roy L. Cleere, executive director of the Colorado Department of Health, stated bluntly that the problem at Grand Junction was the "full and complete responsibility of the AEC" and argued that it "should pay for the cost of the removal of the tailings." James R. Schlesinger, who replaced Seaborg as AEC chairman in August 1971, offered a different opinion. While he affirmed that the tailings presented "no immediate danger" he acknowledged that in some locations radiation levels were "higher than we would prefer." This, he said, was a "moral responsibility" of the AEC and the federal government. But he also asserted that there had been "no shortage of mistakes on the part of all parties," including the state, the mill owners, and the contractors who used the tailings. Eventually, Congress enacted legislation that reflected the AEC's view. A law passed in June 1972 authorized the AEC to enter an agreement with Colorado to pay 75 percent of the costs of remedial action, the need for which would be based on the guidelines of the Public Health Service.[61]

In the cases of both abandoned mill tailings and uranium mines, the AEC claimed no jurisdiction to regulate conditions or impose safety requirements. In both instances it adhered to its legal stance but eventually acknowledged that corrective action was required. It suffered blistering attacks for its positions on both mine safety and control of tailings, from an accusation of creating "death mines" to an allegation of "dereliction of duty" in the distribution of tailings. Those charges were exaggerated and one-sided; the AEC's performance was hardly above reproach but that of other federal and state agencies was similarly flawed. Still, the critics of the AEC made some telling points that helped move the agency to action. The AEC, after determining that the regulation of the mines and mill tailings was not its responsibility, had left the task to state governments that lacked the money, resources, and/or the interest to carry it out. Only after the problems worsened and press assaults intensified did the AEC look beyond its statutory restraints for solutions.

As the agency primarily responsible for the development and regulation of atomic energy applications, the focus of criticism fell on the AEC. The AEC's disavowal of regulatory authority over uranium mines and mill tailings, however sound legally, made it appear indifferent to their hazards. Most commentators drew no distinctions between the AEC's limited jurisdiction in those matters and its generally prevalent role in atomic energy programs. The charges that the AEC refused to carry out its responsibility for regulating the health hazards of uranium mining and milling contributed to an increasingly common view that the AEC declined to deal fully and forthrightly with the health and environmental problems created by the use of nuclear power. This image of the AEC became further entrenched during debates over the environmental effects of nuclear power generation that emerged after the mid-1960s.

CHAPTER XI

The AEC and
the Environment

The Thermal Pollution Controversy

The decade of the 1960s witnessed rapidly intensifying public concern over the ravages of industrial pollution and steadily increasing political activity to protect America's environment. Events ranging from the publication of Rachel Carson's exposé of the dangers of pesticides in her book *Silent Spring* in 1962 to the widespread observance of Earth Day in 1970 demonstrated that the condition of the nation's environment had taken on growing urgency as a public policy issue. A series of controversies over the effects of substances such as DDT, mercury, and phosphates, ecological disasters such as a huge oil spill off the coast of California and the death of Lake Erie from pollution, and easily visible evidence of foul air and dirty water fueled public alarm about the degenerating quality of the environment. They also prompted legislative measures to deal with the threat. Congress enacted or tightened a number of laws designed to preserve areas of exceptional natural beauty and to combat air and water pollution during the 1960s, culminating in the National Environmental Policy Act, which gave federal agencies a broad mandate to protect against environmental abuses.[1]

At the same time that the environmental crisis commanded increasing attention, questions about the availability of electrical power triggered deepening concern. Since the early twentieth century, the use of electricity in the United States had expanded by an average of 7 percent per year, but in some years during the 1950s and 1960s it grew even faster. During 1955, for example, sales of electricity increased by an astonishing 17 percent, and during the late 1960s by about 9 percent annually. Utility and government planners anticipated that the pace of growth would continue at 7 percent or higher in the near future. A report prepared by the White House Office of Science and Technology in 1968 predicted that the nation would need about 250 "mammoth-sized" new power plants by 1990. Long-term requirements were an ever-present worry, but short-term problems were even more pressing. Although there was no lack of electrical generating capacity on a national scale, some utilities and local areas faced potentially serious shortages during times of peak load. Power blackouts and brownouts in which utilities turned down their voltage and asked consumers to switch off appliances became increasingly commonplace in the latter half of the 1960s, graphically illustrating the discomfort and inconvenience caused by inadequate electrical supplies.[2]

The commitment of utilities to provide electricity to their customers was inseparably linked to environmental issues because generating stations were major polluters. Fossil fuel plants provided over 85 percent of the nation's electricity in the 1960s, and in the process spewed millions of tons of noxious chemicals into the atmosphere annually. Coal, by far the most commonly used fuel for producing electricity, also placed a far greater burden on the environment than other fossil fuels. The sulfur dioxide that coal plants released formed sulfuric acid mists that posed a serious threat to the welfare of humans, vegetation, and property, and nitrogen oxides they emitted were a major ingredient in smog. Although carbon dioxide was not in itself a harmful pollutant, its growing presence in the atmosphere from the combustion of coal raised the possibility of harmful climatic changes over a long period of time. Oil-fired plants also contributed to atmospheric degradation, though their effects were substantially less damaging than those of coal facilities. The environmental impact of natural gas, the other fossil fuel used to generate electricity, was so limited that, at least in comparison to coal and oil, it was regarded as practically benign.[3]

The air pollution problems caused by burning fossil fuels defied easy solutions. Expansion in the use of natural gas was environmentally advantageous, but it was limited by increasingly tight supplies. After 1967, gas was consumed faster than new reserves were opened. Oil with a low sulfur content was an attractive alternative, but its availability to meet growing demand was, at best, uncertain. Low-sulfur coal was much less harmful to the environment than coal with a high sulfur content, but most low-sulfur coal reserves were in the western part of the United States, and transportation to load centers in the East was expensive. Further, the efficiency of plants designed to burn high-sulfur coal suffered significantly if they used low-sulfur coal instead. Efforts to design equipment that would remove sulfur from coal plant effluents were progressing, but the costs and effectiveness of desulfurizing devices remained questionable.[4]

The growing public and political concern with environmental quality and the continually increasing demand for electricity placed utilities in a quandary. It was, as a report of the Conservation Foundation phrased it, "a most vexing dilemma: How do we protect the environment from further destruction and, at the same time, have the electricity we want at the flick of a switch?" An article in *Fortune* magazine depicted the same problem in even starker terms: "Americans do not seem willing to let the utilities continue devouring . . . ever increasing quantities of water, air, and land. And yet clearly they also are not willing to contemplate doing without all the electricity they want. These two wishes are incompatible. That is the dilemma faced by the utilities."[5]

After the mid-1960s, utilities increasingly viewed nuclear power as the answer to that dilemma. While conforming with their plans to meet demand by achieving "economies of scale," it promised the means to produce sufficient electricity without fouling the air. Environmental concerns were a major spur to the growth of the "great bandwagon market," and industry voices emphasized the environmental benefits of nuclear generation. In a rare editorial, titled "Let the Public Choose the Air It Breathes," *Nucleonics Week* concluded in 1965 that in comparison with coal, "the one issue on which nuclear power can make an invincible case is the air pollution issue." In a memorandum to senior staff members, public information officials of the nuclear vendor Atomics International itemized environmental assets of nuclear power. Describing it as "safe, clean, quiet, and odorless," they observed that a nuclear plant did "not release harmful amounts of pollutants to the atmosphere . . . [or] to water." Other nuclear vendors also stressed the cleanliness of atomic power; it was an important selling point in their effort to expand their markets.[6]

As the buyers of generating facilities, many utilities found the case for the environmental advantages of nuclear power compelling. Sherman R. Knapp, chairman of the board of Connecticut Light and Power, told an American Nuclear Society meeting in February 1965: "Atomic power is bound to be increasingly attractive to communities as concern over air pollution intensifies." Other utility executives echoed the same sentiments, and took actions that proved the accuracy of Knapp's prediction. Northern States Power of Minneapolis, for example, decided in 1967 to build a 550 megawatt nuclear unit because of environmental considerations, even though the estimated costs were higher than for a comparable fossil fuel plant. Richard D. Furber, vice president of the utility, explained that Northern States had just suffered through a lengthy controversy over the construction of a coal plant and added: "Many times during this three-year controversy the opposition indicated they would lay off if we would convert this plant to a nuclear plant."[7]

This was not an isolated case in which environmentalists declared their preference for nuclear power, though they clearly were less enthusiastic about the technology than were industry representatives. While acknowledging the advantages of nuclear power in combating air pollution, some environmentalists cautioned that radioactive effluents could also pose a serious problem. Malcolm L. Peterson, a spokesman for the Greater St. Louis Committee for Nuclear Information, declared in 1965: "Because nuclear power plants do not pollute the air with smoke, not produce any of the ingredients of photochemical smog, they are regarded as 'clean,' but it should not be forgotten that radioactivity, though invisible, is also a contaminant." Another prominent environmental organization, the Sierra Club, was ambivalent in its position on nuclear power; as a policy matter it neither endorsed nor opposed the construction of nuclear plants. The attitudes of environmental groups was perhaps best summarized in the equivocal assessment of Thomas E. Dustin, president of the Izaak Walton League of America, in 1967: "I think most conservationists may welcome the oncoming of nuclear plants, though we are sure they have their own parameters of difficulty."[8]

The attitudes of the general public about the environmental effects of nuclear power were seldom evaluated. One poll published in early 1966 suggested that many members of the public lacked strong views or informed opinions about the subject. The survey, conducted with residents of Buchanan, New York (site of the Indian Point nuclear plant), Philadelphia, and Atlanta, asked, among other questions, "How 'clean' are nuclear plants in operation?" In each location, from 40 to 50 percent of the respondents had no answer. Those who did respond, however, overwhelmingly expressed a favorable outlook on the cleanliness of nuclear power.[9]

Officials of the Atomic Energy Commission actively promoted the idea that nuclear power provided the answer to both the environmental crisis and the energy crisis. Seaborg told the National Conference on Air

Pollution in 1966 that in light of expanding demand for electricity and deteriorating air quality, "we can be grateful that, historically speaking, nuclear energy arrived on the scene when it did." Although he acknowledged that nuclear power had some adverse impact on the environment, he insisted that its effects were much less harmful than those of fossil fuels. In comparison with coal, he once declared, "there can be no doubt that nuclear power comes out looking like Mr. Clean." Other AEC officials expressed the same sentiments. Ramey, for example, emphasized the environmental virtues of nuclear power on numerous occasions. "It is needed," he told a meeting of the International Atomic Energy Agency in 1970, "and, in environmental as well as economic and resource terms, it is the best hope for the world's power needs."[10]

Other than radiation protection, the AEC did not view environmental issues as a central part of its responsibilities, at least until the early 1970s, when congressional and public pressure forced it to broaden the scope of its activities. Although it expressed concern about environmental matters in general, it insisted that its statutory mandate for regulating its licensees did not extend beyond radiation hazards. The agency conducted numerous research projects around its own installations to seek information about the impact of nuclear weapons tests, underground explosions, and reactor wastes on the natural environment and animal life. In nearly every case, the projects focused on the effects of radiation; the major exception was a series of studies done on heated water in the Columbia River from the reactors on the Hanford reservation, which had been carried out since 1946.[11]

The AEC cooperated on an informal basis with other government agencies in assessing the environmental aspects of reactor licensing and operation, particularly the U. S. Public Health Service, a part of the Department of Health, Education, and Welfare, and the Fish and Wildlife Service, a part of the Department of the Interior. Under a 1961 interagency agreement, the AEC provided the Public Health Service with copies of applications for power reactor construction permits so it could evaluate the possible effects of radiation releases on the environment of areas surrounding the proposed plants. The Public Health Service advised the AEC of its opinion, but the principal use of its report was to offer information and guidance to state health officials. In addition, the Public Health Service conducted a number of studies of operating plants to measure the kinds and amounts of radioactive isotopes that they discharged into the atmosphere and to estimate their impact on the environment.[12]

The AEC shared similar arrangements with the Fish and Wildlife Service. Beginning in 1961, it furnished copies of reactor applications to the Fish and Wildlife Service, which reviewed each of them to evaluate the potential radiological effects of the proposed plant on the animal and marine environment. It sent copies of its report to the fish and game department of the state in which the reactor would be built, and to the AEC, which considered Fish and Wildlife's findings as a part of the licensing process. Those procedures were incorporated in a memorandum of understanding that the AEC and the Department of the Interior signed in March 1964, providing for consultation between the two agencies on reactor applications.[13]

The views of the Fish and Wildlife Service and the AEC did not clash over the radiological impact of proposed nuclear facilities, but sharp differences arose between them over other issues. The Fish and Wildlife Service began to suggest in the mid-1960s that the AEC should take nonradiological environmental effects into account in licensing cases, especially the consequences of discharging large quantities of heated water for aquatic life. The AEC responded that it lacked authority to set requirements for any nonradiological impact that a nuclear plant might have on the environment. What began as a dispute between the AEC and the Fish and Wildlife Service soon flared into a major public debate over "thermal pollution." The controversy not only embroiled the AEC in a conflict with Interior but also antagonized some prominent members of Congress, generated unfavorable publicity, and raised questions about the extent to which nuclear power was environmentally superior to fossil fuels. As a result of the thermal pollution issue, nuclear power, rather than being seen as the answer to environmental degradation from electrical production, appeared to a growing number of observers to be a part of the problem.[14]

Thermal pollution resulted from cooling the steam that drove the turbines to produce electricity in a fossil fuel or nuclear plant. The steam was condensed by the circulation of large amounts of water, and in the process the cooling water was heated, usually by ten to twenty degrees Fahrenheit, before being returned to the body of water from which it came. This problem was not unique to nuclear power plants; fossil fuel plants also discharged waste heat from their condensers. It was more acute in nuclear plants, however, for two reasons. Fossil fuel plants, unlike nuclear ones, dispelled some of their heat into the atmosphere through smokestacks. More importantly, fossil plants used steam heat more efficiently than nuclear ones, meaning that nuclear plants generated 40 to 50 percent more waste heat than did comparably sized fossil plants. The cooling water that nuclear power stations released was not radioactive; it circulated in a separate loop from the water used to cool the reactor core.[15]

The problem of thermal pollution was not new in the mid-1960s, but it created more anxiety at that time because of the growing number of power plants being constructed, the greater size of those plants, and the increasing inclination of utilities to order nuclear units. Those trends combined to amplify concern about the effects of waste heat on the environment. Although the precise impact of thermal pollution was uncertain, there appeared to be ample cause to be disturbed about its implications. Some scientists suggested that waste heat deposited in lakes and rivers from steam-power plants posed a grave threat both to fish and to other forms of aquatic life.[16]

The effects of thermal discharges on fish were worrisome because many species were highly sensitive to changes in temperature. A rise in water temperature could alter their reproductive cycles, respiratory rates, metabolism, and other vital functions. A drastic or a sudden shift in temperature could be lethal. Between 1962 and 1967, the Federal Water Pollution Control Administration found at least ten cases in which fish were killed by waste heat from fossil fuel power stations. The most serious incident occurred in the Sandusky River in Ohio, where over 300,000 fish died in January 1967; the others were much less severe. It was more common for fish to be killed indirectly by heat discharges. In the Hudson River around the Indian Point nuclear power station, for example, tens of thousands of striped bass died in 1963 after being attracted during cold weather to the warm currents coming from the plant. As nearly as experts could determine, the fish got caught in the water intake system of the plant and died from exhaustion or from contact with pumps or other equipment. Large fish kills attracted a great deal of attention, but the more subtle threats to the marine environment were at least as troubling. As one writer argued: "In the long run temperature levels that adversely affect the animals' metabolism, feeding, growth, reproduction and other vital functions may be as harmful to the fish population as outright heat death."[17]

The concern about thermal pollution extended not only to its hazards for fish but also to other potential consequences. It could disrupt the ecological balance by killing certain kinds of plant life while causing other kinds to flourish. Water warmed by thermal discharges, for example, contained relatively greater quantities of blue-green algae than of other species, and an excess of blue-green algae made water look, taste, and smell unpleasant. Rising temperatures also reduced the capacity of water to retain dissolved oxygen, which was needed to chemically convert waste matter into innocuous forms. As the amount of oxygen in the water diminished, the amount of undesirable wastes and pollutants increased.[18]

The nature and severity of the environmental damage attributable to waste heat depended on variables that differed widely from place to place, including the size and efficiency of the power plant, the type and adaptability of the fish and plant life in the affected body of water, the rate and volume of water flow, and the natural thermal characteristics of the water. While many questions about thermal pollution remained unanswered, the prospect that scores of new power plants, over half of them nuclear, would be built within two decades generated substantial alarm about its long-term effects. An article on the subject in *Scientist and Citizen*, the publication of the Committee for Environmental Information (the successor to the Committee for Nuclear Information) declared in 1968: "We cannot continue to expand our production of electric power with present generating methods without causing a major ecological crisis." Television newsman Edwin Newman informed his viewers of an even drearier prognosis. "The gloomiest forecast we know of about the future of our water resources is that by the end of the decade our rivers may have reached the boiling point," he reported in 1970. "Three decades more and they may evaporate." Newman added: "This vision of an ultimate cataclysm is based on the assumption that we will continue to discharge heat into our rivers at the rate at which we're doing it now." Most warnings about thermal pollution were far less apocalyptic than the one that Newman cited, but anxiety about the dangers of waste heat from power plants was widespread among both experts and laymen.[19]

Some observers, however, found less cause for concern. Although they acknowledged that thermal pollution was a problem, they also argued that its threat to the environment had been exaggerated. Scientists who took this point of view noted that laboratory experiments demonstrating serious effects of waste heat sometimes conflicted with actual field experience. They also showed that, contrary to the impression that newspaper and magazine articles often gave, only a small percentage of fish kills were caused by heated water from power plants. In addition, some scientists argued that heated water from generating stations could be beneficial. While certain kinds of fish were adversely affected, others thrived in warmer water. The Pacific Gas and Electric Company, citing studies by the California Department of Fish and Game, asserted in 1970: "Fishermen rarely criticize utility companies for the warmer temperature of water near power plants. That's where the fishing is likely to be best." Heated water offered other potential advantages. Glenn Seaborg, for example, suggested that waste heat could be put to work irrigating fields to extend the growing season and reduce frost damage. In this regard, he and others maintained that the proper term for waste heat was not "thermal pollution" but "thermal enrichment."[20]

Even those who were most sanguine about the implications of waste heat recognized that its harmful effects could not be ignored. The disagreement of opinion arose over the severity of the problem, not the existence of one. In order to find out more about the consequences of thermal pollution and ways to control it, several government agencies and a number of utilities sponsored research programs. But most utilities could not wait for research to produce conclusive results about waste heat; they needed to build plants immediately to meet anticipated demand for electricity. Public concern about thermal pollution and newly established state water quality standards made it imperative for many of them to act promptly to curb thermal discharges.[21]

Technical solutions were available to deal with the problem of waste heat, but they required extra expenses in the construction and operation of steam-electric plants. Gradually and often reluctantly, a growing number of utilities decided to pay the costs of mitigating the effects of thermal pollution. To do so, they built systems to replace their traditional, and preferred, practice of "once-through cooling," in which water was drawn into the plant, used to cool steam in the condenser, and then directly returned to its source. Utilities generally elected to use alternatives to once-through cooling because the volume and flow of the water available for cooling was insufficient, environmental groups raised vocal protests, and/or limits set by state agencies required them to reduce the temperature of waste heat discharges. The federal Water Quality Act of 1965 encouraged states to establish water quality standards for interstate streams and coastal waterways, and they moved promptly to control water temperatures. The increasing concern about environmental quality, the imposition of state standards, the growing number and size of power stations, and the paucity of good sites for plants accelerated the trend away from once-through cooling, though utilities still employed it where they could.[22]

Utilities could choose from several options to reduce the effects of waste heat. The cheapest and easiest approach was to limit the environmental impact of heated water without building a separate system. This could be done, for example, by pumping more water through the condenser, which raised its temperature less, or by providing a long channel to discharge the heated water into different sections of the source body of water. In many cases, however, a more elaborate system was essential. The available alternatives offered the means to resolve or greatly alleviate the problem of waste heat, but also exacted significant costs. One method was to dig a cooling pond, where contact with air would cool the heated water on the surface. The primary disadvantage of a cooling pond was that it required a sizable area of land. A large plant would need a pond of several hundred acres (the rule of thumb was two acres for every megawatt), and except in rural regions, the cost of that much land was prohibitive.[23]

Utilities generally found it more economical to build cooling towers. Several different designs were available, but the most commonly used were natural draft or mechanical draft towers. Either type of tower dumped waste heat into the atmosphere as warm vapor or warmed air. A natural draft tower could rise as high as a thirty-story building. It worked like a chimney, drawing air warmed by contact with heated water upward and out the top of the tower. This process cooled the water, some of which evaporated and the rest of which either was recirculated in the condenser or returned to its source. The principal drawback to a natural draft tower was its cost, estimated in 1967 to be four thousand to ten thousand dollars per megawatt. Mechanical draft towers used fans to circulate air and cool the water from the condenser. They were less expensive to build than natural draft towers because they did not need to be nearly as high, but they were more expensive to operate. In addition to their costs, cooling towers posed other problems. They reduced the generating capacity of the plant by a small, but not negligible, amount. The water that cooling towers added to the atmosphere raised concern that they would cause localized fog and icing conditions, though there was little evidence that this was a common occurrence. Finally, natural draft towers were aesthetically objectionable to those who disliked the way they dominated the skyline for miles around.[24]

The problem of cooling waste heat discharges was not peculiar to nuclear plants, but it was particularly troublesome in them. A utility that considered building a nuclear unit in the late 1960s inevitably

10. Vermont Yankee Nuclear Power Station on the Connecticut River at Vernon, Vermont. At right center are two mechanical draft cooling towers. (National Archives 434-SF-39-41)

11. The Trojan Nuclear Power Plant, located on the Columbia River near Prescott, Oregon, under construction. The plant's natural draft cooling tower (foreground) rises above the dome-shaped reactor building (center). (National Archives 434-SF-39-25; courtesy of Portland General Electric Company)

confronted the issue of thermal pollution. In 1967, only a handful of power companies planned to use cooling towers, but by early 1970, over half of the eighty-five plants on order or under construction were designed with cooling systems. Most of those without cooling apparatus were located on oceans, bays, or the Great Lakes, where the threat of waste heat seemed less acute. Although the trend was clear, it did not emerge without major controversies over the effects of thermal pollution and the role of the AEC in regulating them.[25]

Control over thermal pollution was, in the phrase of a writer for the trade journal *Nuclear Industry*, "a jurisdictional 'no man's land.' " The Department of the Interior, including both the Federal Water Pollution Control Administration and the Fish and Wildlife Service, took particular interest in the problem, but its statutory power extended only to advising other federal agencies and state governments on the protection of aquatic life. Enforcement of water standards remained a function of the states, but their regulations were not always adequate or uniform. Some members of Congress and officials of the Fish and Wildlife Service suggested that the AEC should assume greater responsibility over thermal discharges from nuclear plants, but it denied that it had the statutory authority to do so.[26]

The AEC's refusal to regulate thermal effects stirred private expressions of concern, and later, unusually blunt protests from the Fish and Wildlife Service. The differing views of the two agencies emerged clearly, and publicly, in a disagreement over an application for a construction permit for the Millstone Nuclear Power Station in Waterford, Connecticut. In November 1965, the AEC, as a part of its customary procedures, sent a copy of the Millstone application to the Fish and Wildlife Service for comment. It, in turn, forwarded the document to one of its subdivisions, the Radiobiological Laboratory of the Bureau of Commercial Fisheries. Theodore R. Rice, director of the laboratory, prepared an evaluation of the possible effects of the proposed plant on fish in the vicinity. He concluded that the reactor could be operated without radiological injury to fish. Rice appended a section cautioning that thermal discharges from the plant might have adverse consequences, but he accepted the AEC's view that its jurisdiction was "limited to matters pertaining to radiological safety."[27]

To that point, the comments of the Fish and Wildlife Service had followed well-established patterns. Clarence F. Pautzke, head of the Service, made a major departure from routine procedures, however, when he sent Rice's report to the AEC in March 1966. Pautzke announced that even though his agency had in the past submitted comments similar to Rice's, it had changed its position because of growing federal concern for environmental quality. "We wish to make clear that Dr. Rice's statements . . . concerning the jurisdiction and responsibility of the Atomic Energy Commission in regard to thermal pollution," he declared, "[do] not represent the policy of the Fish and Wildlife Service." Pautzke asserted that the AEC's regulatory authority covered thermal pollution and suggested that it ask the Department of Justice to review the question. If Justice supported the AEC, he thought that "legislation to provide this necessary authority should be sought by the Commission."[28]

Pautzke's letter caught the AEC by surprise. Harold Price complained that the Fish and Wildlife Service had not only sent it to several state agencies but also had "openly and publicly challenge[d] the position of the Commission with respect to authority over thermal effects." The Joint Committee was equally startled by the implied effrontery and concerned about the possible effect of Pautzke's arguments. It had recently heard similar criticism from Representative John D. Dingell, chairman of the House Subcommittee on Fisheries and Wildlife Conservation, who suggested that the AEC was evading the provisions and the intentions of the Fish and Wildlife Coordination Act. "The effect of this has been," he charged, "that they have proceeded without due care for either the enhancement or the preservation of fish and wildlife values." In response to a request from Chet Holifield and for its own information, the AEC reviewed its legal stance on regulating against thermal pollution and applying the Fish and Wildlife Coordination Act to its activities.[29]

Howard K. Shapar, who as assistant general counsel for licensing and regulation was the AEC staff's authority on the legal aspects of regulatory issues, reaffirmed the agency's position in a lengthy analysis. He argued that the Atomic Energy Act of 1954 and its subsequent amendments restricted the AEC's regulatory power to hazards peculiar to nuclear facilities, and that therefore, its statutory mandate extended only to radiological health and safety.

Shapar further contended that the Fish and Wildlife Coordination Act, which Congress had passed in 1934 and strengthened in 1958, did not apply to AEC licensees. The act required federal agencies to consult with the Fish and Wildlife Service "with a view to the conservation of wildlife resources" if they undertook or licensed activities in which water would be "impounded, diverted, . . . controlled or modified."

Shapar submitted that nuclear plants simply circulated and returned water to its source "essentially unchanged." They did not impound, divert, control, or modify it in the way that dredging, irrigation, or flood control projects did. Shapar acknowledged that a nuclear facility would raise the temperature of the water it used, but he did not view that as sufficient grounds to require AEC compliance with the Fish and Wildlife Coordination Act. Moreover, the act did not expand the regulatory authority of the AEC or any other federal agency. Consequently, even if the AEC were to agree that the law was binding, it would apply "only with respect to the radiological effects of licensed activities."[30]

Shapar's brief demonstrated that the AEC could make a strong legal case for not regulating thermal pollution. But the problem remained, and the AEC offered no alternative approaches for dealing with waste heat. When several members of Congress introduced legislation to resolve the issue by explicitly subjecting the AEC to the provisions of the Fish and Wildlife Coordination Act, it objected. One reason was that the bills did not grant the AEC any new regulatory authority, so that, in its view, its jurisdiction would still be limited to radiological hazards. A more important consideration was that the agency feared that the proposals, if enacted, would discriminate against nuclear power. Since fossil fuel plants were not licensed by federal agencies, they would not be required to meet the same conditions as nuclear plants to control thermal discharges.[31]

In hearings held on 13 May 1966, Representative Dingell grilled AEC officials about their views on thermal pollution. He opened the hearings by lamenting the "grossly inadequate protection now being afforded fish and wildlife resources," and the AEC's explanation of its position did not mitigate his anxiety. Harold Price told him that the AEC was "very much in sympathy" with programs intended to protect fish and wildlife, but stressed that it opposed measures that would affect nuclear but not fossil fuel plants in doing so. When Dingell asked whether the agency assumed any responsibility for or took any interest in nonradiological environmental problems, Price replied that its authority was restricted to radiation hazards but that it was "very much interested in" preserving fish and wildlife resources. Dingell wondered if the AEC had proposed any legislative solutions to vest "in your agency power to correct the hazard that is clearly apparent?" Price said no, that he believed the problem was "not peculiar to atomic energy plants, and it ought to be attacked more broadly." Dingell inquired about what the AEC would do if a proposed plant would obviously heat a river enough to be "enormously destructive." Price responded that "we would be very unhappy," but that the AEC "could not, under the law, deny the license on that ground." Although Dingell was unfailingly polite to Price and other AEC representatives, he did not conceal his annoyance that their expressions of concern about thermal pollution did not convey a willingness to suggest anything they might do about it.[32]

The AEC was aware of the problem but uncertain of how to handle it. Agency officials agreed that thermal pollution required regulatory action, but they opposed any solution that would place nuclear power at a competitive disadvantage with fossil fuel plants. None of the several legislative measures proposed between 1966 and 1969 resolved that dilemma. The AEC did not want to exercise authority over thermal effects of nuclear plants unless fossil facilities had to meet the same conditions. It also objected to granting the secretary of the interior regulatory jurisdiction over thermal discharges from atomic power stations. As an alternative, it continued to consult with the Fish and Wildlife Service, which, for its part, stopped insisting that the AEC already had the necessary authority to regulate waste heat from nuclear plants. It asked that the AEC urge applicants to take action to control thermal discharges and to cooperate with interested state agencies. The AEC passed on the views of the Fish and Wildlife Service, and through it, the Federal Water Pollution Control Administration, to nuclear plant applicants as a normal part of the licensing process. But the recommendations were strictly advisory; compliance with them was not mandatory for receiving a construction permit.[33]

Meanwhile, public and congressional concern about thermal pollution continued to grow. The focal point of the enlarging controversy was the proposed Vermont Yankee Generating Station. In November 1966, the Vermont Yankee Nuclear Power Corporation, a consortium of ten utilities, applied for a construction permit for a 514 electrical megawatt plant on the Connecticut River at Vernon, Vermont. The situation in Vermont with regard to energy needs and environmental concerns reflected the national outlook in particularly sharp relief. The state had so little generating capacity of its own that it imported about 80 percent of its power, and its out-of-state suppliers were unable to provide for Vermont's rapidly increasing demand. At the same time, residents and state officials were committed to protecting Vermont's environmental resources from the threats posed by industrial development and population growth. The Vermont Yankee plant was intended to serve both energy and environmental requirements, but it soon aroused a sharp debate over the issue of thermal pollution.[34]

Officials in Vermont and adjacent states were gravely concerned about the threat of thermal pollution. James B. Oakes, attorney general of Vermont, insisted that the plant would need cooling towers to prevent ecological damage from waste heat. New Hampshire, across the Connecticut River from the proposed plant, and Massachusetts, five miles south of the site, expressed equally deep apprehensions about the environmental impact of Vermont Yankee. Elliot L. Richardson, attorney general of Massachusetts, complained: "Vermont will receive a million-dollar injection into its economy. Massachusetts will receive hot water." All three states protested the AEC's refusal to regulate thermal effects as a part of its licensing process.[35]

The Vermont Yankee Nuclear Power Corporation initially rebuffed suggestions that it add cooling towers to its plant by maintaining that they were unnecessary and too costly. Within a short time, however, the utility relented in the face of determined opposition from Vermont, New Hampshire, and Massachusetts. The company's concession was not enough to end the controversy. It made plans to use "open cycle" towers, in which the water from the condenser would circulate through the cooling system and then be returned to the river. This would enable the plant to meet Vermont's water standards by raising the temperature of a "mixing zone" in the river by a maximum of four degrees. But this was not sufficient to conform with the water standards of New Hampshire and Massachusetts, which required that even at the point of discharge the plant could not heat the river water at all. This could be done only by building a "closed cycle" system, in which the condensate water returned to the condenser after running through the cooling towers. The drawbacks of the closed cycle system were not only that it would be more expensive to build but also that it would reduce plant efficiency substantially. The issue was still unresolved in December 1967 when the AEC granted Vermont Yankee a construction permit, once again disclaiming responsibility for regulating thermal pollution.[36]

Edmund S. Muskie, chairman of the Subcommittee on Air and Water Pollution of the Senate Committee on Public Works and a leading advocate of measures to improve environmental protection, observed the Vermont Yankee proceedings with growing impatience at the AEC's position. On 20 September 1967, he wrote to Seaborg, questioning the legal basis for the AEC's refusal to consider thermal effects in its licensing actions. Muskie asserted that an executive order of July 1966, implementing sections of the Federal Water Pollution Control Act of 1965, had instructed all agency heads to combat water pollution from federal government activities. He wondered how the AEC could justify its denial of authority. Harold Price replied to Muskie, pointing out that the executive order did not expand the AEC's regulatory jurisdiction. He contended that it applied only to installations operated by federal agencies and not to licensees of the AEC.[37]

Muskie was visibly irritated by Price's letter; one of his staff members commented that the senator thought that the AEC was "thumbing its nose at the intent of Congress." He fired off another letter to Seaborg, reasserting his contention that the executive order and the Federal Water Pollution Control Act required the AEC to regulate thermal pollution. He also noted that his concern over the issue had been further piqued by the application of the Maine Yankee Atomic Power Company to build a plant in his home state. On 4 November, ten days after Muskie's letter, Seaborg responded. He reiterated the AEC's standard arguments on why it believed that its authority did not extend to thermal discharges, but he promised that the agency would seek the opinion of the Justice Department about the legal soundness of its position. In the meantime, Muskie had announced that he would hold hearings to investigate the AEC's practice of granting licenses "without giving due consideration to the effect of waste heat."[38]

In hearings he conducted in Montpelier, Vermont on 14 February 1968, Muskie heard representatives of Vermont, New Hampshire, and Massachusetts denounce the AEC for its refusal to exercise jurisdiction over thermal pollution. The governor of Vermont, Philip H. Hoff, after declaring that his state was "blessed with a matchless environment," went on to attack the AEC's position. "We were dismayed during the Vermont Yankee hearings when the AEC decided that thermal pollution was none of its concern," he said. "When it ignored the issue of thermal pollution . . . I think it declared itself to be a promotional agency—in effect, a publicly financed lobby." Officials of the other two states expressed similar opinions in language that was only slightly less blunt. The consensus clearly favored regulatory action by the AEC or some other federal agency. Muskie agreed, observing at one point that the AEC was "about as arbitrary in rejection of responsibility [as] I can recall in [my] experience with federal agencies."[39]

Despite the vocal objections to its denial of authority, the AEC received support for its legal stance from two important sources. The first came from the Justice Department, which the AEC had asked, in response to Muskie's queries, to review the question of whether or not it had statutory jurisdiction over thermal discharges. In April 1968 the Justice Department reported that it concurred with the AEC's view. After examining the provisions of the Atomic Energy Act, the Federal Water Pollution Control Act, and the executive order implementing sections of that act, Justice Department attorneys concluded that the AEC did not have authority to regulate against thermal pollution.[40]

The AEC's legal claims also received support from the U. S. Court of Appeals for the First Circuit in Boston, which sustained the agency's position but viewed its policy implications with an obvious lack of enthusiasm. After the Atomic Safety and Licensing Board granted a construction permit for the Vermont Yankee plant, the state of New Hampshire filed an appeal for a rehearing, which was turned down by the Commission. New Hampshire then took its case to court, arguing, in terms similar to those of Senator Muskie, that the AEC had the statutory obligation to consider thermal pollution in its decision to issue the permit to Vermont Yankee. The court of appeals denied that assertion in a ruling of 13 January 1969. It agreed with the AEC and the Department of Justice that existing legislation did not assign authority to regulate the thermal effects of licensed plants. But the court also declared: "We confront a serious gap between the dangers of modern technology and the protections afforded by law as the Commission interprets it. We have the utmost sympathy with the appellant and with the sister states of Massachusetts and Vermont." The court expressed its regret that Congress had not resolved the issue by "requiring timely and comprehensive consideration of non-radiological pollution effects." New Hampshire appealed the decision to the U. S. Supreme Court, which allowed the lower court ruling to stand by refusing to hear the case.[41]

Although the AEC won its battle in court, it was left in an uncomfortable position. It had clear judicial support for its argument that it lacked jurisdiction over thermal pollution, but it was under attack from critics who accused it of indifference to the environment. The once widely held assumption that nuclear power would provide both electricity and environmental protection was being questioned because of the emerging debate over thermal pollution. From the AEC's perspective, the best way out of this predicament was to support legislation that would clarify the roles of federal agencies in regulating waste heat discharges. But the agency favored legislation only if it did not discriminate against nuclear power or give the Interior Department final authority to decide thermal issues for nuclear plants. None of the several bills that were introduced during 1968, some granting the AEC and some the Interior Department responsibility over thermal pollution, won enough backing in Congress for passage, and the impasse continued.[42]

While the issue remained unresolved, criticism of the AEC became increasingly more pointed and more frequent. For a time, the attacks were sporadic and localized, largely limited to several members of Congress, a handful of environmentalists, and critics in the specific areas of a few proposed nuclear plants. But the problem of thermal pollution and the AEC's position on it captured expanding national attention after the publication of an article in the high-circulation *Sports Illustrated* in January 1969. The article was written by Robert H. Boyle, a senior editor for the magazine, devout fisherman, conservationist, and author of a book on the natural history and resources of the Hudson River. One year earlier, Boyle had coauthored another article in *Sports Illustrated* that included a passage charging the AEC and the nuclear industry with a lack of concern for thermal pollution. It supported that allegation by quoting Harold Price's remarks in his testimony before Congressman Dingell's subcommittee in 1966. Price at that time had attempted to explain the legal bases for the AEC's denial of jurisdiction over waste heat, but the article, by tailoring his comments, made the AEC look totally insensitive to the problem. Price's protests received little sympathy from the magazine's outdoors editor, Arthur Brawley, who asserted that if the AEC was really concerned about thermal pollution, it should ask Congress for the authority to prevent its occurrence.[43]

Boyle's 1969 article, titled "The Nukes Are in Hot Water," was even more disparaging of the AEC than the previous report. "What literally may become the 'hottest' conservation fight in the history of the U. S. has begun," it opened. "The opponents are the Atomic Energy Commission and utilities versus aroused fishermen, sailors, swimmers, homeowners, and a growing number of scientists." Boyle went on to describe the threat of thermal pollution to aquatic life and to water quality. He assailed the AEC for refusing to take responsibility for the problem, attributing its inaction to a fear of the "financial investment that power companies would have to make . . . to stop nuclear plants from frying fish or cooking waterways wholesale." He jibed at Seaborg, suggesting that even though the AEC chairman had won a Nobel prize for finding plutonium, he had "yet to discover hot water." Boyle predicted that since "more than 100 nuclear plants are on the drawing boards, . . .

almost every major lake and river and stretches of Atlantic, Gulf and Pacific coasts are likely to become battlegrounds." The article was in many ways distorted and unfair; it misrepresented the AEC's position to the point of caricature. Boyle obviously had no intention of writing a balanced scholarly treatise, and his tone of indignation and incredulity was an effective way to advance his own point of view.[44]

Although the precise impact of Boyle's article was impossible to define, it clearly broadened and called attention to the thermal pollution controversy more than any previous discussion had done. Debate over the issue was well under way before the article appeared, but after its publication, and to an appreciable degree because of its publication, thermal pollution became the subject of elevated interest and heightened concern on a national scale. One indication was the reaction and commentary that the article stirred. For three consecutive weeks *Sports Illustrated* ran letters to the editor that both commended and criticized the article. Representative Tim Lee Carter of Kentucky inserted it into the *Congressional Record*, hoping that it would "begin a rational discussion of what might be a tremendous problem in the future." Chet Holifield, worried that some of his colleagues "may have taken Mr. Boyle's utterances at face value," countered the assertions presented in "that esteemed technical journal, *Sports Illustrated*." He defended the AEC from the charge that it did not care about the effects of waste heat and maintained that nuclear power was essential for achieving the twin goals of producing sufficient electrical power and preserving the environment.[45]

Some industry spokesmen reaffirmed the same view, but the article dismayed many of their colleagues. According to one knowledgeable observer, the story "shocked many in the industry," especially since "many utilities had in fact chosen nuclear power largely *because* of environmental advantages." The AEC's director of public information concluded in March 1969 that although "public acceptance of nuclear power remains at a high level," the "biggest problem today is the question of thermal effects." He added that "until some positive action is taken to place responsibility for thermal effects this question will continue to give us trouble." One major source of concern was that even if the environmental impact of a single nuclear plant was relatively inoffensive, the consequences of placing several plants that discharged waste heat into the same body of water might be ruinous. This was a point that Boyle highlighted, and the projections for the rapid growth of nuclear power fed those fears.[46]

The spreading alarm about thermal pollution was evident in protests

against a number of proposed nuclear plants. Although the specifics
varied widely from place to place, the general patterns of the debate
followed similar lines. Like the controversy over the Vermont Yankee
reactor, they usually started when state officials or conservationists
raised questions about the thermal effects of a plant, became matters of
dispute when a utility refused to build cooling towers or take other
action to mitigate waste heat discharges, and ended only after considerable acrimony and/or concessions by the power company. In several
cases, what began as local issues received widespread attention as a part
of growing national concern over environmental quality in general and
thermal pollution in particular.

One of the first examples of this pattern was a dispute over a reactor
that the New York State Electric and Gas Corporation proposed to
build on Cayuga Lake, the second largest of the celebrated Finger Lakes.
The site for the projected plant, named the Bell Power Station, was
sixteen miles north of Ithaca, home of Cornell University. When the
utility first announced its plans in June 1967, it received little adverse
reaction. A few months later, a group of Cornell faculty members, most
of them aquatic and fishery biologists, issued a pamphlet suggesting that
thermal discharges from the plant could cause severe and irreversible
ecological damage to the lake. They urged the utility to construct cooling towers or a cooling pond to reduce the harmful effects. New York
State Electric and Gas declined, however, citing the high cost of towers,
which it estimated at $21.3 million, as well as the possibility of in-
creased fogging and icing and undesirable aesthetic consequences. The
critics replied that the company's cost projections seemed too high
(those at Vermont Yankee were about $6.5 million), and that even in
the worst case, the average rate increase to consumers would only be
five or six dollars a year. The disagreement gradually hardened into a
contentious controversy. It ended in April 1969, when the utility decided to postpone work on the plant indefinitely to study the economic
and environmental impact of cooling systems.[47]

Thermal discharges also emerged as a major issue in the construction
of nuclear plants on Florida's Biscayne Bay. When the Florida Power
and Light Company announced plans to build two nuclear facilities at
Turkey Point, twenty-five miles south of Miami, to go along with two
oil-burning plants already operating there, state officials got worried.
They expressed their concern to the AEC, which, as always, denied any
authority over thermal pollution. It did provide funding to the University of Miami to study the effects of power plant effluents, both nuclear

and fossil, on marine organisms. The Federal Water Pollution Control Administration assumed a more active role, reporting that waste heat from the oil plants had definitely caused environmental damage and recommending that action be taken to limit the upper levels of water temperatures in the bay. The utility agreed that thermal pollution was a problem at the site, and proposed to solve it by digging a canal that would send the heated water into nearby Card Sound. This was unacceptable to both the water pollution control administration and the state government, who argued that the canal would not lower temperatures sufficiently and that it would pollute the sound. After a series of meetings, hearings, court rulings, and negotiations between the utility and environmental groups over a period of two years, Florida Power and Light agreed to build a 4000 acre closed-cycle cooling canal system at a cost it had initially deemed to be excessive.[48]

The thermal pollution question generated even more controversy over plans to place several nuclear plants on Lake Michigan. The primary source of concern was the effect that waste heat from the ten plants operating, under construction, or on order would have on the ecology of the lake. There were two catalysts instrumental in stirring this debate. One was the Metropolitan Sanitary District of Greater Chicago, a Cook County agency that sued to stop construction of Commonwealth Edison's Zion units. The other was a proposal of Murray Stein, the chief enforcement officer of the Federal Water Quality Administration (as the Federal Water Pollution Control Administration had been renamed). At a meeting in May 1970 with representatives of the four states that bordered on Lake Michigan, Stein urged that they adopt standards that would prohibit a rise in water temperature of more than one degree at the point of discharge. Both actions underscored the divisions over thermal issues. Commonwealth Edison responded to the Metropolitan Sanitary District by pointing out that its plans for releasing heated water into the lake had been approved by federal and state agencies, and it successfully resisted efforts by the county, environmentalists, and even some of its own stockholders to force it to add cooling towers to the Zion plants. Utilities greeted Stein's recommendation by describing it as "arbitrary," "ill-advised," and "ridiculous." The water quality administration soon retreated from Stein's proposal, but it continued to press for strict water quality regulations.[49]

Of all the nuclear units sited on Lake Michigan, the one that aroused the strongest opposition was the Palisades plant, thirty-five miles west of Kalamazoo, Michigan. A group of intervenors appealed to the

12. Researchers from Argonne National Laboratory take measurements of the thermal discharge plume from the Big Rock Nuclear Power Plant on Lake Michigan. (National Archives 434-SF-11-94)

Atomic Safety and Licensing Board in June 1970 to deny the application of Consumers Power Company for a low-power operating license. They charged that the plant provided insufficient protection against thermal pollution and radiation. The attorney for the intervenors, Myron M. Cherry, argued that the AEC was obligated to regulate waste heat discharges, regardless of the Vermont Yankee decision, and that its radiation regulations were outmoded and inadequate. With construction of the plant complete, Consumers was anxious to secure its operating license, but it resisted making concessions to the intervenors. Finally, in March 1971, after numerous delays, several hearings, and sharp exchanges between the attorneys for both sides, the utility decided that it preferred a settlement with its opponents to the prospect of further costly delays. It agreed to build cooling towers and to virtually eliminate the discharge of liquid radioactive wastes into the lake. In return, the intervenors dropped their action against the plant and opened the way for its full-power operation.[50]

There were exceptions to the general pattern of the debates over thermal pollution. Commonwealth Edison prevailed over the objections to building the Zion units without cooling systems; other utilities elected to forestall criticism by including towers or ponds in their plans for new plants. Whatever the variations in the tone, process, or outcome of the controversy in individual cases, the issue had assumed vital importance as a focus of public concern. The AEC continued to take a fundamentally passive position by disclaiming any statutory responsibility for thermal pollution, but it also responded to growing distress over the problem in more positive ways. Agency officials believed that the thermal pollution question had obscured the environmental advantages of nuclear power, and they decided to be more aggressive in pointing out those benefits and in rebutting critics. To achieve its goals, the AEC undertook a number of initiatives, which included appearances at public meetings, speeches by Seaborg and other commissioners, and pamphlets and films on nuclear power and the environment. In their determination to highlight the assets of nuclear technology compared to fossil fuels, agency officials sometimes accentuated cosmetic over substantive issues. Seaborg, for example, repeatedly insisted that "thermal pollution" was an inaccurate term and should be replaced by "thermal effects," a distinction that not only understated the potential blight of waste heat but also offered support for the suspicions that the AEC was indifferent to the problem.[51]

While the AEC was redoubling its efforts to explain the environmental benefits of nuclear power, it was sponsoring research to enhance scientific understanding of the ecological consequences of waste heat. Much of the debate over the issue was based on limited knowledge; even the experts were uncertain of the magnitude of the threat posed by thermal discharges. The AEC supported ecological studies not only at Biscayne Bay but also at plant sites at rivers in Illinois, Massachusetts, and Maryland to assess the impact of waste heat. In keeping with the growth in concern about the problem, the AEC doubled its expenditures on research between 1969 and 1970. Investigations funded by the AEC, the Department of the Interior, and several utilities promised to provide a clearer picture of the effects of thermal pollution, but the wide variations in the environments surrounding different plants made unequivocal or universal conclusions an impossibility, at least in the short term.[52]

Even in the absence of definitive scientific findings, however, the controversy over thermal pollution largely died out by the early 1970s. The defusion of the issue occurred for a number of reasons. One was that the results of the first meticulous studies of thermal effects, though far from conclusive, were encouraging. An investigation of the Connecticut River in the vicinity of the Connecticut Yankee nuclear plant, which opened in 1967 and did not have cooling towers, demonstrated "no significant deleterious effect on the biology of the river," according to an article published in 1970. Scientists who traced the consequences of thermal discharges from the AEC's plutonium reactors at the Hanford reservation on the Columbia River made a similar assessment, finding "no demonstrable effect" on the salmon or trout in the river. Neither study claimed to evaluate the long-term effects of waste heat or the implications of placing many plants on a single body of water. Their findings, therefore, played only a limited role in alleviating concern over thermal pollution.[53]

A more important influence was that after years of fruitless efforts, Congress passed legislation that assigned federal agencies a clearly defined role in regulating water quality. Since much of the thermal pollution controversy had centered on the AEC's denial of statutory authority, congressional action removed one of the leading sources of dispute. In January 1969 Senator Muskie introduced a bill requiring applicants for AEC construction permits or other federal licenses to present certification from appropriate state or interstate agencies that the plant could meet the water quality standards of their jurisdiction. Members of the House proposed similar legislation. Although the measures did not extend the direct authority of the AEC, they required that it formally consider thermal effects as a part of its licensing process. The legislation applied to any activity that could lower water quality; it did not aim specifically at nuclear power or thermal pollution.[54]

On 3 March 1969, Commissioner Ramey announced that the AEC supported the Muskie bill, a position he reaffirmed three days later when testifying on similar House proposals. He explained that although the AEC had objected to earlier measures on the grounds that they discriminated against nuclear power, it believed that Muskie's bill addressed the problem satisfactorily if not completely. The agency had discovered that fossil fuel plants that were at least partially constructed on navigable waters required a permit from the U. S. Army Corps of Engineers, and therefore, came under the provisions of the bill. In 1967, two-thirds of the large fossil plants licensed would have been included in this category.

There were other reasons that the AEC found Muskie's proposed legislation attractive. Compared to previous bills, it diminished the role of the Department of the Interior, which allayed the AEC's concern that Interior would exert undue influence in its licensing actions. Most importantly, the AEC was anxious to see a law governing thermal effects passed, because even though the new proposals were not drastically different than earlier ones, the political atmosphere was. As *Nuclear News* pointed out: "For the AEC, the sooner adequate and appropriate legislative control can be established over thermal effects the better. . . . The rash of adverse public opinion stirred up recently by the national news media (and by the Muskie hearings themselves) has made early and appropriate control mandatory." The AEC's endorsement of the Muskie bill was not enough in itself to ensure its enactment, but it also won the backing of key members of the Joint Committee on Atomic Energy and others who had opposed previous measures. In March 1970, after clarifying amendments and after more than five months of discussion in conference committee, Congress passed the final version of the bill as the Water Quality Improvement Act. A broader measure, the National Environmental Policy Act, signed into law on 1 January 1970, provided further assurance that federal agencies would treat the problem of thermal pollution.[55]

The most important reason that thermal pollution ceased to be a major focus of environmental concern was that utilities increasingly took action to curb its consequences. Most nuclear plants being built on or planned for inland waterways by 1971 included cooling systems. Although power companies initially resisted the calls for cooling equip-

ment, they soon found that the costs of responding to litigation, endur-
ing postponements in construction or operation of new plants, or suffering loss of public esteem were less tolerable than those of building
towers or ponds. Even though a cooling system added substantially to
the expense of a plant, it was still usually a small percentage of the total
cost of the facility. Utilities increasingly saw it as a part of the price they
had to pay to fulfill their primary objective, which was to meet the
growing demand for electricity. Once they reached that conclusion and
began to act on it, the issue of thermal pollution lost much of its potency
and immediacy.

Even after the thermal pollution controversy largely faded from view,
its legacy lingered on. The most important effect of the debate from the
perspective of the AEC and the nuclear industry was that the image of
nuclear power as the antidote for the environmental hazards of electrical production was irreversibly tarnished. Thermal pollution was the
first issue to raise widespread skepticism about the environmental benefits of nuclear power, and it laid the foundations for subsequent controversies over the dangers of the technology. It played a vital role in
transforming the ambivalence that environmentalists had demonstrated
toward nuclear power into strong and vocal opposition. By the end of
the 1960s environmental groups spearheaded protests against plans for
many nuclear plants, and thermal pollution was a key element in their
arguments.

In a similar manner, the issue wakened doubts among the general
public about the environmental impact of nuclear power. Before thermal pollution was featured in a plethora of news stories, public attitudes
seemed to be at worst uninformed and ill-defined and at best highly
favorable. As the public became increasingly concerned about environmental problems, however, it increasingly viewed nuclear power as one
more threat to environmental integrity. Although no opinion polls on
the subject were published in the late 1960s or early 1970s, a survey
conducted in 1975, several years after thermal pollution had ceased to
be a headline topic, indicated that 47 percent of the public thought that
"the discharge of warm water into lakes and rivers" from nuclear plants
was a "major problem"; another 28 percent thought it was a "minor
problem."[56]

The AEC was convinced that nuclear power offered the means to
provide both ample electricity and environmental protection, and it was
slow to respond to those who questioned this view. The agency came

under increasing attack for failing to weigh the effects of thermal pollution in its licensing procedures, and its protestations that it lacked authority sounded like insensitivity to the environment to its growing legion of critics. In fact, the AEC was concerned with environmental quality, but it was even more concerned with ensuring that atomic power was available to meet the nation's escalating energy requirements. It feared that the outcry over thermal pollution and the demands for cooling systems would undercut that goal. As a matter of priorities, the AEC gave greater attention to the need for power than to the problem of thermal pollution, partly because it believed that the possibility of a shortage of power was a more acute danger and partly because it had long been inclined to emphasize the development over the regulation of nuclear electricity.

Despite its commitment to environmental quality, the AEC did not act aggressively to combat thermal pollution. It did not oppose regulating against the effects of waste heat, but it insisted that the same standards must apply to fossil plants. Its argument that imposing regulations only on nuclear plants would imperil the technology's growing use was more of an intuitive assumption than a result of studied analysis. When quizzed about the impact of adding cooling towers on the relative economic advantages of fossil and nuclear plants, Ramey acknowledged in 1968: "I don't know that this would be a significant difference in [their] competitiveness."[57] The AEC made a strong legal case that it lacked the statutory jurisdiction to force licensees to observe water quality standards, but a growing number of observers wondered why the AEC was so passive in its approach to the thermal pollution problem. The answer was that the agency feared that taking forceful action would discourage the growth of nuclear power. Ironically, in its view, this would lead to greater use of fossil fuels and harm the environment by causing more air pollution. In the thinking of the AEC, it was providing an important benefit to the environment by licensing new plants.

The AEC's reasoning was not clear or convincing to those whose priorities were different. Critics portrayed it as apathetic to environmental needs and therefore, loath to force its licensees to comply with water standards. The complaints were on solid ground in pointing out that the AEC's primary interest was not environmental protection, though they were often oversimplified and sometimes overwrought. Still, they sounded persuasive to many people in a time of growing concern over environmental quality and growing outrage against those who abused it. As a result of the thermal pollution controversy, the AEC and the nuclear industry frequently found themselves included among the ranks of the enemies of the environment. In a period of a few years, the image of nuclear power changed from being the solution to the dilemma of producing electricity without ravaging the environment to being a significant menace to the environment, a perception that endured long after the debate over thermal pollution ended.

CHAPTER XII

Radiation Standards

*Debate Within and Challenges
from Without*

The most serious environmental and health hazard that nuclear power presented, and the one on which the AEC focused its regulatory efforts, was radiation. The primary purpose of the agency's siting policies and safety requirements was to guard against radiological contamination that could threaten public health. Unlike the regulation of uranium mines or thermal pollution, the AEC's responsibility to protect the public and plant workers from excessive exposure to radiation from nuclear power was unambiguous. What was less clear, even after years of controversy and scientific scrutiny, was the level of radiation that posed a significant danger. The AEC's radiation protection standards were designed to provide an ample margin of safety that would prevent injury to workers or harm to the public. But the adequacy of the AEC's regulations remained a source of concern both inside and outside the agency, and by the latter part of the 1960s had become the subject of a major national controversy.

Scientists learned about the risks of exposure to radiation within a short time after the discovery of x-rays and natural radioactivity in the 1890s. They concluded that heavy doses could not only cause obvious effects, such as loss of hair, skin irritations, and severe burns, but also much more insidious consequences, such as sterility and cancer. In response, radiologists, physicians, and equipment manufacturers formed organizations and issued guidelines intended to offer safeguards against harmful amounts of radiation from x-rays or the highly radioactive element radium. Those efforts culminated in the establishment of the International X-Ray and Radium Protection Committee in 1928 and an American counterpart, the Advisory Committee on X-Ray and Radium Protection, the following year. In 1934, both groups took an unprecedented step by recommending a quantitative "tolerance dose" of radiation. Although the recommendations of the two organizations were based on incomplete data and flawed research, they were an important milestone in the theory and practice of radiation protection.

The success of the Manhattan Project in splitting the atom during World War II made radiation safety a vastly more complex task. One reason was that nuclear fission created many radioactive elements and isotopes that did not exist in nature. Instead of dealing only with x-rays and radium, health physicists, as professionals in the field of radiation protection called themselves, had to consider the potential hazards of new radioactive substances about which even less was known. Furthermore, the number of people exposed to radiation from the development of military and civilian uses of atomic energy was certain to grow dramatically, which meant that scientists had to think in terms of large groups instead of the relatively few individuals who had worked with x-rays and radium before the war.

Under the radically altered circumstances, both the American and the international radiation-protection committees lowered their suggested exposure limits. The American body, renamed the National Committee on Radiation Protection (NCRP), reduced its recommendations on allowable levels by a factor of two for those who worked with radiation. It also adopted new terminology, replacing "tolerance dose" with "maximum permissible dose." It defined the permissible dose as that which, "in light of present knowledge, is not expected to cause appreciable bodily injury to a person at any time during his lifetime." The NCRP acknowledged, however, that in some cases individuals could suffer harm from radiation even in amounts below the permissible limit. The international group, renamed the International Commission on Radiological Protection (ICRP), took action similar to that of the NCRP.

The exposure limits that the NCRP and the ICRP recommended applied to radiation workers and not to the general population. After the mid-1950s, however, the question of how low levels of radiation could affect the public-at-large became a subject of great concern and heated debate. It arose because atmospheric testing of hydrogen bombs by the United States, the Soviet Union, and Great Britain produced radioactive fallout that spread to populated areas far from the sites of the explosions. Scientists disagreed sharply about how serious a risk fallout posed for the population. The controversy greatly increased public consciousness of the potential hazards of low levels of radiation. It also prompted both the NCRP and the ICRP to further reduce their suggested exposure limits for radiation workers and to recommend for the first time an allowable dose for the public. Scientists never fully resolved the dispute over the consequences of fallout, though it largely faded from public view after the Limited Test Ban Treaty of 1963 prohibited atmospheric testing. The fallout debate left a legacy of ongoing scientific inquiry and latent public anxiety about the health effects of low-level radiation.[1]

Although the hazards of low-level radiation remained arguable, by the early 1960s scientists had greatly expanded their knowledge of the ways in which radioactivity jeopardized human health. They had determined that harmful consequences arose from the ionizing effect of radiation on human cell structure. Radiation has high levels of energy, whether in the form of x-rays from machines or in the form of alpha particles, beta particles, or gamma rays, which are emitted as the atomic nuclei of radioactive elements undergo spontaneous disintegration. The products of this radioactive decay differ from one another in mass, electrical charge, and power of penetration. When radiation passes through matter it deposits energy in that matter, which can alter the structure of atoms by stripping electrons from them. If this occurs, the total negative electrical charge of the electrons no longer balances the total positive charge of the protons in the atom's nucleus and the atom is left with an electrical charge. Such charged atomic fragments are called ions. Those changes in the composition of a cell's atoms can lead to mutations and ultimately to serious biological injury. The damage caused by ionizing radiation clearly depends on the dose received, but scientists also identified other factors that could affect the severity of the injury, including the sensitivity of different body organs and the form of radiation absorbed.

Gamma rays from natural radioactive decay and x-rays from man-made machines—both energetic forms of light—can penetrate far inside the body from external sources. The more massive beta particles and the much heavier alpha particles, by contrast, transfer their energy before penetrating deeply from outside. But if an element that emits alpha or beta particles is breathed or swallowed and lodges in internal organs, it poses a serious biological risk. Health physicists, therefore, had to consider the potential hazards of both external and internal radiation. Using a number of different devices, it is relatively simple to measure external radiation; the exposure to the body is the dose that shows on the instruments. It is much more difficult to determine the exposure from "internal emitters" that enter the body. Since this exposure could not be precisely measured, radiation-protection organizations sought to control the risks it presented by limiting the concentrations of radioactive isotopes in air and water as well as the internal "body burden" of radioactivity. Scientists were particularly concerned about the dangers of internal emitters because once they get into the body they are often difficult to eliminate, they irradiate the organs in which they settle continuously, and they remain in close proximity to affected body tissue.[2]

Radiation experts took into account not only the effects of various forms of radiation but also the different sensitivities of body organs. They were concerned with both genetic consequences, which would affect future generations, and somatic injury, which would affect only the individual exposed. As early as the 1920s, experiments performed with fruit flies by geneticist Hermann J. Muller indicated that reproductive cells are highly susceptible to damage from even small amounts of radiation and that mutant genes could be inherited from a parent with no obvious radiation-induced ailment. Muller's research persuaded scientists that, at least for genetic effects, there was no threshold below which exposure to radiation was biologically innocuous. The genetic implications of radiation were particularly troubling because if the use of atomic energy were not carefully controlled, over time a significant portion of the population could undergo harmful mutations that would enter the genetic pool and afflict untold numbers of unborn children.

The maximum permissible doses recommended by the NCRP and the ICRP were designed not only to protect the general population from genetic trauma but also individuals from somatic injury. The most serious long-term somatic effects of radiation that scientists had identified included leukemia, bone, lung, and other forms of cancer, cataracts, and possibly premature aging. Radiation protection guidelines sought to curb the chances of those diseases occurring by setting limits for "whole-body" exposure to external sources. This was measured by the exposure of the "most critical" tissue in the blood-forming organs, head and trunk, and gonads. Higher limits applied for less sensitive areas of the body. Internal emitters pose especially severe hazards once they enter the body because certain isotopes tend to concentrate in specific organs. For example, strontium–90, a common ingredient in radioactive fallout from nuclear bombs, lodges in bone tissue, and iodine–131, another by-product of nuclear fission, collects in the thyroid. The goal of the NCRP and ICRP, therefore, was to set permissible concentrations in air and water that would minimize the risk to anyone exposed.[3]

Scientists developed several units to measure levels and amounts of radiation. Radioactive elements continuously and spontaneously "decay" at a rate characteristic for each element. This is expressed as the time required for a radioactive substance to lose 50 percent of its radioactivity through decay. The half-lives of elements range from a few seconds to billions of years. To indicate the amount of radiation delivered to human tissue, a unit called a "rad" is employed. A rad measures the "absorbed dose" of radiation as defined by the amount of ionization it causes under prescribed conditions. Another unit, the "rem," indicates the comparative effectiveness of different kinds of radiation in producing biological injury. It applies to chronic low-level exposures and is a factor by which the dose in rads is multiplied to take into account the "relative biological effectiveness" of different kinds of radiation.[4]

Despite the knowledge they had acquired about radiation over the years, scientists remained uncertain about many aspects of the effects and risks of low-level exposure. In the absence of a universally accepted value for "low-level" radiation, they applied the term during the 1950s and 1960s to doses of less than 50 rads because they found no observable short-term effects below that level. They agreed that acute doses of more than 50 rads were progressively more injurious and that acute exposure to between 600 and 1000 rads would be lethal to nearly everyone receiving it. The biological impact of low-level radiation was more difficult to assess. Information on which to base scientific judgments came from a limited number of sources: the survivors of the atomic bombings of the Japanese cities of Hiroshima and Nagasaki; Marshall Islanders accidentally exposed to fallout from bomb testing; patients irradiated with heavy doses of x-rays to treat cancer or other diseases; workers exposed to acute radiation in accidents on the job; and research on animals. The evidence from those sources was often ambiguous or unreliable. Experiments with animals, while useful, could not be easily applied to humans, and extrapolating from the effects of radiation on those who had received heavy doses was of dubious value in understanding low-level hazards.

A number of important questions about the effects of low-level radiation evaded sure answers and stirred debate among scientists. One was whether a threshold existed for somatic radiation injury. If so, it indicated that there was a level at which exposure to radiation was safe, at least for somatic effects. If not, it implied that exposure to radiation in any amount increased the chances that a person would develop leukemia, cancer, or other radiation-induced illness, proportional to the dose received. Experts were also uncertain about the impact of the length of time over which a dose was absorbed (the "dose rate") on human cell structure. Research on the subject suggested but did not unambiguously show that the same amount of radiation caused less harm if delivered over a long period of time. Scientists tried to determine the ability of both somatic and germ cells to repair damage caused by radiation, but again, their findings were inconclusive. Those and other unanswered questions underscored the limitations of scientific knowledge about the risks of radiation exposure.[5]

In preparing their recommendations for maximum permissible doses, both the NCRP and the ICRP took a conservative approach. They assumed that there was no threshold for somatic effects and that no amount of radiation was certifiably safe. They based their recommendations on the belief that even though it was possible to suffer injury from exposure below the allowable levels, in the words of the NCRP, "the *probability* of the occurrence of such injuries must be so low that the risk would be readily acceptable to the average individual." The fundamental limit on which scientific authorities agreed by the late 1950s was whole-body exposure to external radiation of an average of 5 rem per year for radiation workers. Using this value as the starting point, they also offered guidance for other kinds of exposure. For internal emitters, they calculated the concentrations in air and water that seemed likely to meet the 5 rem limit if inhaled or ingested. For the general public (as opposed to radiation workers), they rather arbitrarily reduced the permissible limit for individuals to one-tenth of the occupational levels for external and internal exposure. In addition, they recommended an average limit of 5 rem over thirty years (0.17 rem per year) to large population groups in an effort to curtail genetic consequences. This was done so that even if some individuals received more than the allowable dose, the genetic peril to the larger population would be mitigated. The permissible doses that radiation experts suggested did not include "background radiation," which is a small (estimated at about 0.15 rem annually in the early 1960s) but unavoidable part of environmental exposure that arises from cosmic rays, radioactive elements in rocks and soil, and other natural sources.[6]

The fundamental objective of the NCRP and the ICRP in establishing permissible doses was to keep the probability of radiation injury to a minimum without inhibiting the constructive use of radiation sources. Neither group viewed its exposure limits as final or definitive. Both stressed that the numerical values they proposed should not be regarded as absolute or inviolable; a person who received more than the allowable dose would not necessarily suffer any harm while a person exposed to less than the limit could incur radiation injury. But the NCRP and the ICRP believed that the permissible doses were conservative enough to make the chances of serious consequences statistically slight. They sought to balance the benefits and the risks of machines that produced radiation by recommending exposure levels that seemed generally safe without being impractical.

The exposure limits applied not only to nuclear power but also to the many industrial uses of radioactive isotopes, such as measuring the thickness of sheet metal, rubber, and plastics and studying the wear qualities of gears, dyes, and tires. The allowable doses also provided guidance to protect medical technicians from overexposure, though they were not applicable to patients who required heavy doses to combat disease. "The goal of the health physicist is to keep exposure levels as low as practical and still obtain the benefits from the use of ionizing radiation," wrote Karl Z. Morgan, director of the Health Physics Division of Oak Ridge National Laboratory and a member of both the NCRP and the ICRP, in 1963. "His guiding principle is that any unnecessary man-made exposure is too much exposure . . . but he believes that constant vigilance and intelligent planning can reduce radiation damage so that the evident advantages of the proper use of nuclear energy and other sources of ionizing radiation may be obtained."[7]

By the early 1960s, several organizations and government agencies were actively involved in radiation protection. The NCRP and the ICRP remained enormously influential. Both included leading experts in the field of radiation protection and their authority derived from the respect that their members commanded. They were relatively small, informal, and unaffiliated organizations that were determined to maintain their independence from government agencies. Their recommendations carried no statutory authority, but they were adopted by government bodies throughout the world.

The AEC incorporated the NCRP's recommendations for maximum permissible doses into its radiation protection regulations. They were intended to provide safeguards against excessive exposure from installations licensed by the AEC to both radiation workers and the public.

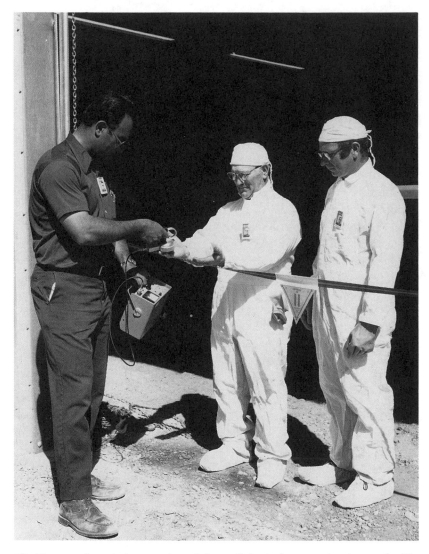

13. Two workers being monitored for radiological contamination at the National Reactor Testing Station in Idaho. (National Archives 434-SF-12-72)

They employed the 5 rem-per-year average as the basic value for workers and one-tenth of that level for individuals outside the boundaries of the plant. Like the NCRP, the AEC believed that the exposure limits provided ample protection for most individuals, though it did not guarantee "that a particular individual may not be harmed by exposure to radiation below the limits established in the regulation." While it could not offer absolute assurances about the safety of its regulations, its experience with its own facilities at least provided encouraging evidence that the limits could be met by its licensees. Statistics compiled for over 150,000 employees of AEC contractors during 1959 and 1960 showed that 94.5 percent received doses of less than 1 rem per year and 99.9 percent less than 5 rem per year.

The NCRP's recommendations were also applied by other federal agencies with responsibilities for radiation protection. In addition to their importance to the AEC, they were particularly critical to the Federal Radiation Council (FRC), which was established as a part of the White House staff in 1959. The function of the FRC was to offer expert advice on radiation safety to the president and to supply "general standards and guidance" to federal agencies. The practical effect of the creation of the FRC on federal agencies was limited, in part because it had no legally binding authority over them and in part because it also relied on the levels proposed by the NCRP. There were no fundamental discrepancies between the exposure limits adopted by the FRC and those used by the AEC. Other agencies involved with radiation protection, though in less central ways than the FRC and AEC, also drew on the numerical recommendations and general guidance offered by the NCRP.[8]

In an effort to learn more about the effects of radiation on human cells, the AEC sponsored a variety of research projects. During the latter half of the 1960s it spent over $90 million per year on biomedical research, and in 1969, for example, it supported about 90 percent of the total national outlay for investigations of the health and environmental hazards of radiation. The major projects it funded included experiments with beagles on the long-term somatic effects of low-level radiation, studies of ecological responses to radiation by analyzing rodent and reptile populations near the Nevada bomb testing site, work with monkeys on the cancer-inducing properties of radiation, and investigations of the genetic consequences of radiation in mammals through elaborate experiments with hundreds of thousands of mice. Those and other research projects provided the basis for much of what scientists knew about the hazards of radiation, though they left many questions unresolved and subject to further study.[9]

The effort to increase understanding of radiation effects spawned some projects whose scientific value or ethical foundations were more ambiguous than the major programs the AEC sponsored. This was especially true of research on human subjects. During the 1960s and early 1970s, the agency provided funding for several experiments in which a small number of volunteers swallowed, inhaled, or received injections of radioactive isotopes. The purpose was to trace the human body's absorption, retention, and excretion of radioactive substances. Another series of experiments sought information on how radiation affected the functioning and fertility of human testes. Over a period of eight years, sixty-seven inmates at the Oregon State Prison and sixty-four at the Washington State Prison voluntarily submitted to irradiation of the testes in doses ranging from 8 to 600 rads. The AEC spent a total of about $1.6 million on the two projects. The experiments with human subjects during this period were restricted to volunteers, small in scale, and openly reported in scientific literature. The scientific results they produced were modest. As the investigators were acutely aware, the ethical questions they raised, especially in the prison studies of testicular functions, were sensitive and potentially controversial. The doses given to some of the prisoners were extremely high, and although all of them agreed to vasectomies to prevent genetic effects, the chances that they would suffer somatic injury troubled some observers. The University of Washington withdrew support from the project in its state in 1969 because it concluded that the benefits to science did not justify the risks to the participating inmates. The experiments in Oregon ended after the principal investigator suffered a debilitating illness and the state discontinued research projects in its penitentiaries.[10]

During the middle and late 1960s, after the Test Ban Treaty of 1963, the effects of radiation were matters of concern mostly to scientists and health professionals. Once the fallout issue faded from view, radiation was no longer a source of national controversy or widespread public anxiety. It did, however, occasionally reappear as a part of public policy and health debates in local areas. Fear of radiation was a key element, for example, in the protests against the proposed Ravenswood and Bodega Bay nuclear plants. It also emerged as a prominent issue in sections of Utah and Nevada situated in the path of fallout from the nearby atmospheric testing of atomic bombs. For a time in the mid-1960s, the AEC and the U.S. Public Health Service engaged in a sharp dispute over radioactive contamination of the Columbia River by the reactors located on the AEC's Hanford reservation. It finally ended after the U. S. Geological Survey, at the request of the AEC, conducted a study and found that levels in the river had "always been well within the established standards."[11]

The radiation issue that attracted the most attention between 1963 and 1969 was entirely removed from the jurisdiction of the AEC, but more than anything else in this period it fueled public anxieties about the risks of exposure. In May 1967 the General Electric Company announced a recall of about 100,000 color television sets that had inadequate shielding against the emission of x-rays. The vacuum tube in most television receivers was a potential source of x-rays, and the more powerful tubes in color sets presented more serious hazards than black-and-white models. Color televisions came into wide use in the United States for the first time in the mid-1960s and by early 1967 about fifteen million had been sold.

General Electric denied that the defective sets created a significant risk to the public, but the problem caused a public uproar. One result was that other manufacturers' televisions were checked and in some cases found to have similar flaws. Another was that Congress passed the Radiation Control for Health and Safety Act of 1968, which instructed the Department of Health, Education, and Welfare to set limits for radiation from televisions. In late 1971, the department announced that radiation from new sets did not pose a major health hazard. This highly publicized episode was a reminder to the public of the dangers of radiation, though it did not cause enough alarm to slow the rapidly expanding sales of color televisions.[12]

While radiation was making occasional appearances as a public issue, the AEC quietly considered changes in its radiation protection regulations, which were published as Part 20 of Title 10 of the *Code of Federal Regulations*. The AEC's standards for licensees, originally implemented in 1957 and revised in 1959, embodied the numerical limits recommended by the NCRP. The permissible dose for a worker in a plant licensed by the AEC was 5 rem per year, which could be extended to a maximum of 12 rem in a single year if the person's lifetime total beyond the age of eighteen did not exceed an average of 5 rem annually. The allowable individual dose for those who did not work at the plant but could be exposed to radiation from it was one-tenth of the basic level, 0.5 rem (or 500 millirem) per year.[13]

In establishing the limit for routine releases of radiation outside a nuclear facility, the AEC measured the exposure of the public to external radiation by assuming that a person stood outdoors at the boundary of the plant for twenty-four hours a day, 365 days a year. To restrict public exposure to internal emitters, the agency used similar assumptions. The amount of radiation that a plant added to cooling water, for example, had to be low enough before being diluted by the body of water into which it was discharged to allow a person to drink it for a lifetime without exceeding the permissible limits for individuals. To calculate the release rate for iodine–131, the AEC assumed that a person received his or her entire supply of fresh milk, the most likely source of exposure, from cows continuously grazing near the area around the plant where radiation levels were highest. The conditions and assumptions that the AEC imposed on licensees in computing allowable releases of radiation from a plant were very conservative, but operating experience during the late 1960s showed that most licensees met them easily. In 1968, for example, only one of eleven operating power reactors exceeded 20 percent of the permissible levels of radioactivity in liquid effluents, and seven measured less than 3 percent. In the case of gaseous effluents, one plant released 57 percent of the limit, but all the others measured less than 3.4 percent and nine of those less than 1 percent.[14]

Despite the commendable performance of operating reactors in limiting their releases, the regulatory staff of the AEC sought to tighten the standards for public exposure to radiation from nuclear plants. It did so partly to comply more fully with the recommendations of the Federal Radiation Council. The AEC's regulations had been prepared before the creation of the FRC, and although the basic levels used by the two agencies conformed, there were some discrepancies. One was that the FRC's radiation protection guidelines recommended that the average dose for population groups be restricted to 0.17 rem (170 millirem) per year. AEC regulations included a similar requirement, but it was less specific and mentioned only exposure to internal emitters. Another was that the FRC urged that federal agencies attempt to keep radiation doses as far below its guidelines "as practicable." The AEC standards did not contain such a provision, and regulatory officials thought that adding it might discourage licensees from assuming that approaching the permissible limits as a matter of course rather than trying to keep levels to a minimum was acceptable.

Emphasizing the need to minimize radiation releases seemed particularly vital in light of the expanding use of radiation sources, which was the primary reason that the regulatory staff considered amending the AEC's regulations. The public, potentially at least, received ever-increasing exposure from color televisions and other consumer goods, more frequent air travel, the possible application of nuclear explosions for peaceful purposes, and the growing number of nuclear power plants. The staff wanted to make certain that releases from nuclear plants did not, when combined with radiation from other sources, push public exposure to undesirable and perhaps hazardous levels.[15]

In July 1968, Forrest Western, director of the Division of Radiation Protection Standards and a leading advocate of revising the regulations, drafted a proposal for changes that he and other members of the regulatory staff had discussed for some time. Western held a Ph.D. in physics and had worked as a physicist in industry and at Oak Ridge National Laboratory before joining the AEC in 1951; he was well-known among professionals in the field of radiation protection for his work with the NCRP, ICRP, and FRC. He urged that the FRC limit of 170 millirem for "a suitable sample of the most highly exposed population group" be added to the AEC's regulations. This would broaden, to include external radiation, an existing stipulation restricting exposure of population groups to internal radiation to one-third of that for individual members of the public. Western also suggested that the AEC "put the nuclear industry on notice" that it was contemplating revisions in the regulations that would weigh exposure to radiation from all possible sources and that would specify that doses from power plants "be kept as low as practicable." Citing the potential sources of public exposure to radiation, he commented: "Such activities can continue to develop without undue exposure of the public to radiation only if technological advances in activities producing sources of radiation continue to be accompanied by the development of appropriate measures to control exposures of the public to radiation."[16]

Western proposed only modest changes in the existing regulations. Applying the 170 millirem average population limit to external and internal radiation would be more a clarification of the AEC's standards than a major revision, and introducing a requirement that exposures be held as low as practicable would be a codification of unstated AEC assumptions rather than a new departure. Other divisions of the regulatory staff generally agreed with Western's recommendations, though some argued that the existing standards already served the same purposes. At least one influential official outside the regulatory staff expressed skepticism about the changes. Milton Shaw, director of the Division of Reactor Development and Technology and a potent voice within the AEC, did not openly oppose Western's proposals, but he urged that any action be delayed until a "thorough assessment of the need" for tightening the regulations could be conducted. He intimated that the need was not apparent.[17]

While the regulatory staff was circulating its proposals internally, some observers outside the agency were raising more fundamental issues. A number of critics suggested that the AEC's regulations were insufficiently rigorous and that they should be substantially revised. The growing controversy centered around the Monticello Nuclear Generating Plant, under construction on the Mississippi River about thirty-five miles northwest of Minneapolis. Northern States Power Company, the utility building the 545 electrical megawatt plant, received a construction permit from the AEC in June 1967. A short time later, it applied for a permit from the recently established Minnesota Pollution Control Agency (MPCA) to discharge radioactive wastes into the Mississippi River. The company affirmed that it would comply with AEC regulations in releasing effluents from the plant.[18]

The utility's application elicited a number of protests that soon grew into a major public debate. At a public hearing in February 1968, a spokesman for Minneapolis Mayor Arthur Naftalin urged the MPCA to refuse a permit because radiation in river water would pose "a serious threat to the health of the people of the Twin Cities." Two University of Minnesota scientists offered similar views, charging that AEC standards offered inadequate protection and calling for further study of the possible effects of the proposed plant. One of them, Dean E. Abrahamson, a professor of anatomy in the university's medical school, assumed a leading role in raising questions about the safety of nuclear power in general and the Monticello facility in particular. Abrahamson, who held both a doctorate in science and a medical degree, argued that much remained to be learned about nuclear reactors and radiation hazards, and he was not satisfied with the assurances provided by Northern States and the AEC. "I am not against nuclear power reactors," he once declared. "I simply don't have enough answers to be for or against them. But I insist that these answers be supplied and weighed before we go on."[19]

Northern States strongly denied the allegations that radiation from the Monticello unit would threaten public health, and a spokesman for General Electric, which was building the plant, told the MPCA that the levels of radiation that the plant would add to the Mississippi were lower than found naturally in comparable quantities of beer, ocean water, or salad oil. The conflicting arguments placed the pollution agency in an awkward position. It had no way to resolve the issue and its legal authority even to consider radiation dangers was dubious; the utility and the AEC contended that the states had no jurisdiction over radiological matters involving nuclear power facilities. Meanwhile, the

MPCA came under attack in the local press; the *St. Paul Pioneer Press* complained that it did not seem to "know what is going on" and urged it to "call on every scientific resource available for help in making its decision." The MPCA decided to defer a ruling on Northern States' permit application until it received a report on radiation hazards from an expert consultant. After a long search it hired Ernest C. Tsivoglou, a former U. S. Public Health Service official who had challenged the AEC's position in the recent interagency disagreement over the safety of radiation levels in the Columbia River. Tsivoglou had left the Public Health Service to become a professor of sanitary engineering at the Georgia Institute of Technology.[20]

Tsivoglou submitted his report to the MPCA in January 1969. In some ways his analysis was reassuring. He conceded that "responsible national and international agencies have adopted a quite cautious policy with regard to radioactive pollution control." He found a tendency in Minnesota and elsewhere to "lose perspective" about radiation and to exaggerate its dangers compared to other pollutants. Tsivoglou suggested that the nuclear industry contributed many social benefits along with "some small potential risk." Nevertheless, he emphasized that it was essential to keep the risk to a minimum. He recommended a number of ways in which he thought this could be done without placing unreasonable demands on the nuclear industry. As a general proposition applicable to all reactors, he urged the MPCA to eliminate the distinction between individual dose and population dose that the Federal Radiation Council made and to require nuclear plants to meet the stricter standard (170 millirem per year). Tsivoglou also made a series of recommendations that applied specifically to Monticello. He advised the MPCA to lower the permissible concentrations of radiation in plant effluents from the levels approved by the AEC. The effect of his proposals would be to reduce the allowable doses to an estimated 2 or 3 percent of the AEC's regulatory limits.[21]

Tsivoglou's report commanded a great deal of attention and intensified the controversy in Minnesota. Critics of plans for the Monticello plant who favored a requirement of zero release of radiation found his recommendations disappointing. Northern States, by contrast, called Tsivoglou's proposals "unnecessarily strict." It pointed out that the plant would normally stay within 1 percent of the AEC's limits, but added that during periods of low river flow, the concentration of radiation would increase. The company worried that at those times the plant could exceed the levels that Tsivoglou recommended, forcing it to shut down. One company official commented: "We think it is improper to set up an absolute legal limit so different from the Atomic Energy Commission standards that we could conceivably be forced to cease operations."[22]

Tsivoglou's report not only rekindled debate in Minnesota but also raised issues that extended far beyond the Monticello case. One was the challenge to the radiation standards of the AEC; as an article in *Science* noted, it "cast doubt on the adequacy of existing AEC regulations to cope with radioactive effluent from the expected proliferation of new reactors." This was a scientific problem of continuing importance, but it was soon overshadowed by a legal question: did the state of Minnesota have the authority to impose standards stricter than those of the AEC?[23]

During the 1950s the role of the states in nuclear regulation had emerged as a major controversy. The AEC claimed exclusive jurisdiction while the states insisted that they should play a part in regulating the health and safety aspects of nuclear power. The AEC resisted the states' appeals principally because it feared that they would pass a perplexing maze of different requirements that would seriously impede the growth of the nuclear industry. In 1959, Congress enacted an amendment to the 1954 Atomic Energy Act that granted the states an opportunity to participate in a limited way in nuclear regulation but left control over power reactors solely in the hands of the AEC. The amendment clarified federal and state jurisdiction and, in the opinion of the AEC and Northern States, precluded any attempt by a state to employ regulations different than those of the AEC. Even the attorney for the MPCA expressed doubt that it could depart from federal radiation standards. But some legal experts suggested that even though the law unambiguously prohibited states from adopting limits that were less restrictive than those of the AEC, it did not clearly prevent a state from requiring levels that were more restrictive.[24]

On 12 May 1969 the MPCA granted Northern States a waste discharge permit on the condition that it comply with the effluent limits that Tsivoglou had recommended. Despite its rejection of the legal argument that it lacked jurisdiction and the stringent requirements it imposed on Northern States, the pollution agency's action was greeted by angry cries from an audience of about two hundred. The one board member who voted against the permit received a standing ovation when he urged that a decision be postponed until "all doubts are erased" about the health effects of radiation. The utility declined to comment on the ruling until it could determine whether it would accept the MPCA's requirements or seek legal recourse.[25]

The AEC, unlike Northern States, did not hesitate to make its position clear. It viewed the MPCA's ruling as an unwarranted intrusion on its exclusive authority to set radiation standards for nuclear power plants. Minnesota governor Harold LeVander, who had told Seaborg a year earlier that he had "complete confidence" in the AEC, now asked the support of the agency for the MPCA's decision. Seaborg replied in a letter that was worded much more sharply than his usual matter-of-fact tone. He rejected LeVander's request, not only citing legal arguments but also questioning Minnesota's ability to monitor its own requirements. He further suggested that the state's regulations "could be unduly burdensome without making a meaningful contribution to the public health and safety." The Joint Committee on Atomic Energy was equally incensed by Minnesota's action. Holifield had told the MPCA before its decision that it lacked legal authority to set radiation limits. He insisted that the federal government properly exercised sole jurisdiction because the AEC commanded greater expertise than individual states and because of the need for uniformity in regulations. After the MPCA's ruling, Holifield urged the AEC to consider contesting it in federal court.[26]

The AEC weighed the possibility of taking legal action; the likeliest tactic was to join Northern States if the utility elected to challenge the MPCA's decision in court. But the AEC refrained from making a commitment before consulting the White House and the Justice Department. It received very little support for the idea from the White House Office of Science and Technology. S. David Freeman, director of the Energy Policy Staff, told Lee A. DuBridge, head of the Office of Science and Technology and presidential science adviser, that he found the AEC's case unsound. Freeman thought that as a policy matter it was unwise to deny the right of the states to place tighter restrictions on nuclear plants. "It is my feeling that, while the AEC may be correct as a technical legal matter," he wrote, "they are taking on a fight that in the end they cannot win and perhaps should not." DuBridge agreed, and informed Seaborg that he believed that "if a State wishes to impose more severe regulations than the Federal Government, the State should have the right to do so." The AEC protested that a failure to oppose Minnesota's requirements would encourage other states to take similar action, with the result that "the nuclear power program will be stopped, or greatly curtailed." The Office of Science and Technology remained unconvinced, a position that was consistent with the Nixon administration's effort to increase the responsibilities of the states relative to the federal government. For its part, the Justice Department deferred taking a formal stand on the question until Northern States decided whether or not it would bring suit.[27]

The issue came to a head in August 1969, when Northern States announced that it was taking its case to court. The utility was, in the words of one report, an "unwilling gladiator." It feared that a court battle would further damage its public image, which was already blemished by the attacks of Monticello opponents. In an effort to explain its position to its customers, the company bought newspaper advertising in which it called the MPCA's regulations "arbitrarily restrictive" and argued that it could not "serve two masters" in carrying out safety requirements.[28]

The AEC strongly endorsed the utility's decision and promptly requested that the Justice Department intervene on its behalf in support of Northern States. Justice demurred, partly because of its own reservations about the policy implications of the AEC's position but largely because the White House did not want the federal government to join a suit against the right of a state to set its own standards. To make matters worse from the AEC's perspective, *Electrical World* magazine reported that Assistant Attorney General William D. Ruckelshaus had suggested that Justice might decide to intervene on the side of Minnesota. This story triggered angry queries from Congressmen Holifield and Hosmer, both of whom affirmed their belief in the legal and policy merits of the AEC's position. Ruckelshaus responded that his views had been misrepresented and told Holifield that he thought that the chances that Northern States would win its case were "about eighty per cent certain." But the Justice Department, caught between the administration's philosophical commitment to states' rights and its legal interpretation of the strength of the argument for exclusive federal jurisdiction, took no action to intervene.[29]

Although the AEC was unable to join the suit in support of Northern States, it took an active interest in the case. Indeed, it was more adamant in pressing for an affirmation of its exclusive authority to regulate radiation emissions than was the utility. When Northern States informed the AEC that it was trying to arrange a compromise settlement with the MPCA, the agency protested. Ramey, in a letter approved by the entire Commission, told Northern States Chairman Earl Ewald that he believed that a ruling from the court and a clear resolution of the debate was essential. A compromise that left the legal question unresolved, he maintained, "could certainly make AEC's position in relation to other

States, and the position of the nuclear industry much more difficult."
Since the AEC could not present its own case in court, it had to depend
on the utility to do it. Northern States, however, was even more concerned with getting the Monticello plant on line. In the fall of 1969 the
unit was nearing completion, and the company was becoming increasingly anxious to reach an accord with the MPCA. But negotiations
proved fruitless. Finally, in February 1970, the company agreed to comply with the radiation limits set by the MPCA, but it also proceeded
with its litigation denying the authority of the state to impose those
standards.[30]

The dispute over the Monticello reactor drew national attention and
sparked a great deal of criticism of the AEC. The National Governors'
Conference of 1969 unanimously adopted a resolution introduced by
Minnesota's LeVander asserting the right of states to set radiation standards. LeVander told his colleagues that before the AEC tried to prevent
state action on the issue, it "should attempt to repair its own house as a
first measure of business." The National Association of Attorneys General passed a similar measure in support of Minnesota. Many newspapers and magazines published stories about the controversy. The
New York Times, for one, endorsed Minnesota's position in an editorial, commenting that the "standards set by Federal agencies, the A.E.C.
included, too often have proved inadequate." Several members of Congress took the same view; Congressman Jonathan Bingham of New
York sponsored a bill that would authorize the states to set radiation
limits stricter than those of the AEC.[31]

At a widely reported conference on "nuclear power and the public,"
held at the University of Minnesota in October 1969, Washington University biologist and environmentalist leader Barry Commoner stated
bluntly: "I would hope . . . that the AEC will itself take immediate steps
to relinquish its control over standards of radiation contamination." At
the same meeting, George Washington University law professor Harold
P. Green, declared: "Although the atomic energy establishment is prone
to dismiss those who are concerned about the health and safety implications of nuclear power plants as ignorant of the facts, overly fearful, or
in cahoots with the coal interests, the fact of the matter is that there is a
legitimate basis for apprehension. . . . There is no assurance that [radio-
active effluents] will not result in harm." Conveying the same view with
a lighter touch, a satirical singing group in Minnesota called the Hill-
Dillies recorded a number titled "Atomic Power, Monticello-Style" that
sold briskly. Despite the group's breezy tone, its message was somber. It

warned of the "insidious poison" of radioactivity and compared the construction of the Monticello plant with the atomic bombing of Japan.[32]

Meanwhile, the suit filed by Northern States against the MPCA was slowly approaching a decision by the United States District Court for the District of Minnesota. The utility contended that the federal government had preempted regulatory authority over radiation standards, while the state, supported by briefs from seven other states, argued that it retained the power under the tenth amendment to the Constitution to protect public health by imposing stricter requirements. In a decision handed down on 22 December 1970, Judge Edward J. Devitt ruled in favor of Northern States. Making clear that he was not weighing the merits of the differing federal and state standards, he determined that on the basis of the relevant statutes, particularly the 1959 amendment to the Atomic Energy Act, "the State of Minnesota is without authority to regulate the release of radioactive discharges from plaintiff's Monticello Nuclear Power Plant." The state took its case to the U. S. Court of Appeals for the Eighth Circuit, which upheld the District Court in a 2–1 decision. Minnesota then appealed to the U. S. Supreme Court, which refused to review the case and summarily affirmed the lower court's ruling.[33]

The federal court rulings settled the legal arguments and jurisdictional claims advanced in the Monticello case, but the technical issue that had started the debate—the adequacy of the AEC's regulations—remained a subject of dispute. In addition to divergent opinions on radiation limits, the controversy highlighted a number of related questions that AEC critics were raising. One was how AEC regulations applied when two or more reactors were located at the same site. The agency responded to this concern by explaining that the same limits applied to a site no matter how many reactors were placed on it. The total effluents from multiple plants had to stay within the same permissible levels as a single plant. A related question was how the AEC dealt with many reactors situated on the same body of water but at different sites. In this case, the regulations stipulated that the AEC reserved the right to tighten the standards if required. Harold Price told the Joint Committee in October 1969 that the agency could enforce the provision "if we faced the problem of 25 reactors up and down a river," but that it had not been necessary to that time.[34]

Some observers, including staff members of the Fish and Wildlife Service, expressed concern that some fish and animals might be especially sensitive to radiation, and that even if the AEC regulations provided ample protection for humans they might not do so for wildlife. The AEC's answer was that it assumed that if the amount of radiation found in fish and wildlife was low enough that they could be used as a source of food without exceeding permissible human intake levels, it was also low enough not to cause them harm. This raised the issue of "reconcentration," in which some radioactive isotopes became concentrated in certain living organisms as they moved through the food chain. The effect of reconcentration had received a great deal of publicity as a result of an uproar over the use of the pesticide DDT, which fed anxiety that levels of radiation could become increasingly hazardous in fish and animals before being consumed by humans. The AEC pointed out that only a few isotopes posed such a threat and that its regulations took this into account. The allowable releases of iodine–131, for example, which concentrated in cow's milk, were reduced by a factor of 700. The AEC recognized that several other isotopes concentrated in fish but did not make special provisions for them because they were not major ingredients in discharges from nuclear power reactors.[35]

While the AEC was answering questions and offering assurances about the adequacy of its radiation standards, the regulatory staff was continuing its efforts to tighten them. Forrest Western of the Division of Radiation Protection Standards had drafted proposals on this matter in July 1968, but before the AEC took any action, the Monticello controversy infused the question with greater urgency. Apparently, it also prompted the regulatory staff to present more drastic revisions. In March 1969, a few weeks after Tsivoglou submitted his report to the MPCA urging the adoption of radiation limits far below the AEC's, Western, with the assistance of his deputy Lester R. Rogers, prepared a new proposal to amend the AEC's regulations. It called for a reduction of concentrations of radioactivity in effluent water from power reactors to 1 percent of existing permissible levels. A week later, another draft added a provision cutting the allowable exposure of an individual to radiation in gaseous effluents to 170 millirem a year, which decreased the limits by two-thirds and conformed with Tsivoglou's recommendation that the distinction between individual and population limits be eliminated. Together, the proposals of the regulatory staff reduced the permissible releases from power reactors even further than Tsivoglou was urging for the Monticello plant. A later draft lowered the allowable concentrations of gaseous effluents outside the plant boundaries still further, to 10 percent of existing levels.[36]

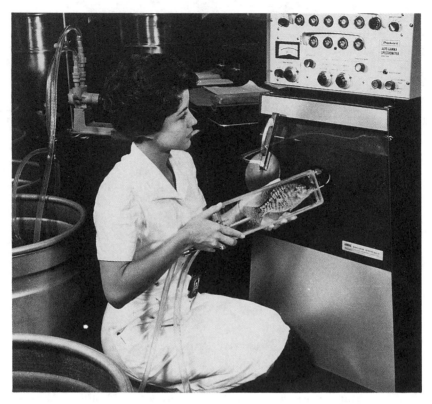

14. An ecologist at Oak Ridge National Laboratory uses a radiation-detecting instrument to measure radioactivity in the body of a live fish. (National Archives 434-SF-51-57)

The reasons that Western and other regulatory officials cited for recommending the revisions in the regulations were the same as those given in the initial proposals in 1968: to follow the Federal Radiation Council's advice to reduce radiation levels as much as practicable and to make certain that the population did not receive an undue portion of its exposure from nuclear plants. Operating experience with reactors indicated that they could meet the revised standards without great difficulty or expense, and making the changes would avoid the "undesirable position" of approving releases of radiation far larger than were necessary. The timing and the substance of the regulatory staff's new proposals suggested that an unstated motivation was the growing controversy with Minnesota. Tightening the regulations could answer the increasingly frequent questions about the adequacy of the AEC's standards and perhaps silence the critics by demonstrating the agency's commitment to safeguarding public health.[37]

The outcome of the regulatory staff's proposals depended on the receptiveness of the commissioners, and their views were, for a time, uncertain. They listened without much comment to a staff briefing on revising the regulations, and at their first meeting on the subject they failed to come to any conclusions. One member of the Commission, however, made his opinions unmistakably clear; Theos J. (Tommy) Thompson argued forcefully and repeatedly against adopting major changes in the radiation standards. Thompson was the newest member of the Commission, having joined it in June 1969. But he had worked in the field of nuclear energy for nearly two decades and was a well-known and well-respected member of the nuclear community. He had received his bachelor's and master's degrees from the University of Nebraska, where he also played quarterback on the football team and made an appearance against Stanford in the 1941 Rose Bowl. Thompson earned a Ph.D. in nuclear physics from the University of California at Berkeley, where he knew Glenn Seaborg, and worked for a time on reactor design at Los Alamos. He joined the faculty of the Massachusetts Institute of Technology in 1955 and became a professor of nuclear engineering three years later. He served on the Advisory Committee on Reactor Safeguards from 1959 to 1965.[38]

Thompson's appointment to the AEC was greeted warmly by the nuclear industry, but he did not hesitate to criticize the performance of the industry if he believed that it failed to meet high standards. "It is important that each of the corporations who are responsible for major nuclear reactor design and for the construction and operation of those facilities recognize that it, as a whole, has a corporate engineering social responsibility," he once declared. He also described himself as "an ardent conservationist," an outlook that grew from his love of camping and fishing. While a member of the Commission, he was also a member of the Appalachian Mountain Club. He was convinced that nuclear power was an asset to the environment. As an engineer, Thompson exhibited a deep first-hand knowledge of reactor systems and safety. As a former professor, he was curious and inquisitive, and, in the thinking of some acquaintances, "a trifle pedantic." As a former quarterback, he showed a take-charge attitude that contrasted with Seaborg's passive leadership style. As a commissioner, he was unusually outspoken, frequently argumentative, and sometimes impatient with what he viewed as overly zealous regulation. One of those instances was his reaction to the proposals of the regulatory staff to tighten radiation standards.[39]

In September 1969, Harold Price outlined two approaches that the AEC could take in revising the regulations. One would be to adopt the staff's latest draft by requiring a sharp reduction of permissible exposures. This would limit water effluents to 1 percent and gaseous effluents to 10 percent of existing levels. An alternative would be to make minor changes similar to the original staff proposals. This would include specifying that the average exposure of population groups to any form of radiation could not exceed one-third of the limit for individuals (170 millirem) and stating as a matter of principle that releases of radiation should be kept "as low as practicable." Thompson led the way in registering strong opposition to the first approach. He acknowledged that the AEC standards were under increasing attack and that if the AEC did not lower its limits, other states were likely to follow the example of Minnesota. He also agreed that some changes in the regulations might be necessary to make certain that several reactors located near one another did not exceed the permissible levels and to ensure that the growing number of reactors in operation across the country did not appreciably raise the average exposure of the entire population. But Thompson argued that deliberations on such measures could be carried out in a "careful and judicious manner." He denied that any "clear and present danger" existed that required prompt action by the AEC.[40]

Not only did Thompson believe that reducing the permissible limits was unnecessary, he suggested that in several ways it would be foolish. He pointed out that the standards reflected the results of many years of research and discussion by the world's foremost experts on radiation protection, and that no ill-effects had been demonstrated at the levels of exposure in the existing regulations. He maintained that the regulatory staff's proposals would threaten electrical supplies by removing the margin of error that allowed a plant to keep operating even if it experienced a problem that temporarily increased the radioactive content of its releases. Thompson questioned the need for reducing the limits to a level too low to be measured, making it "difficult, if not impossible, to carry out . . . effective enforcement."

Thompson warned that although the staff recommendations applied only to power reactors, they would inevitably be adopted, at least informally, for other radiation sources as well. This would cause problems and additional costs in uranium mines, power plants, research reactors, and reprocessing facilities. Moreover, it would create serious difficulties for other AEC programs. Thompson was particularly concerned about the consequences for the AEC's plans for peaceful nuclear explosions, called the Plowshare program. Plowshare projects that were contemplated included large-scale excavations for canals and harbors and setting off nuclear blasts to free reserves of natural gas that could not be tapped by conventional means. The AEC attached great importance to Plowshare, and Thompson feared that reducing permissible radiation levels would curtail it.[41]

Thompson's arguments received powerful support from staff offices in the AEC other than Regulation. They agreed that even though the revised regulations would apply only to nuclear power plants, "there probably would be strong pressure to apply any new limits to all nuclear operations." The Division of Production, which was responsible for running the reactors that made plutonium and tritium for weapons at Hanford and Savannah River, complained that the changes would result in major expenses for the AEC "without any corresponding increase in safety." The Division of Peaceful Nuclear Explosions feared that tighter radiation restrictions would threaten the existence of the Plowshare program. Rickover's Division of Naval Reactors "emphatically" opposed changes in the AEC's standards. Other operational units within the AEC were equally adamant. They insisted that there was "no valid health and safety reason for reducing the limits at the present time" and suggested that amending the existing limits would be, "in effect, an assertion that these limits are hazardous." They urged that if the regulatory staff thought it necessary to revise its standards, it do so on an *ad hoc* basis that would apply to individual cases rather than to all reactors. In short, the regulatory staff's proposals met strong and vocal opposition not only from Commissioner Thompson but also from the most influential staff offices in the AEC.[42]

Thompson's colleagues on the Commission endorsed his view that the AEC's radiation limits should not be sharply reduced. At the same time they sought a way to blunt the attacks of critics of the standards. At a meeting on 23 October 1969 the commissioners unanimously decided "in principle" that the best approach was to accept the second alternative that Harold Price had outlined the previous month—stipulating that licensees must keep the levels of radiation released from plants "as low as practicable." At that point, the recommendations of the regulatory staff for major revisions in the regulations seemed to be dead. The commissioners saw no compelling benefits in drastically decreasing allowable releases and, as itemized by Thompson and several staff divisions, many disadvantages in doing so. They wanted to make certain they agreed on their position before they appeared at hearings on the environmental effects of producing electricity that the Joint Committee on Atomic Energy was convening the following week.[43]

The Joint Committee had been planning the hearings for several months. On 3 June 1969, a short time after the Minnesota Pollution Control Agency had required that Northern States Power comply with the radiation standards proposed by Ernest Tsivoglou, Seaborg and Ramey met with Holifield to discuss scheduling hearings on radiation safety. Seaborg thought that hearings could put the problem, "which is now the subject of so much emotional attack in the press," in proper perspective. Holifield agreed, and the committee proceeded with preparations for the hearings. They had, as one committee staff member noted, "both announced and un-announced objectives." The announced goal was to provide a forum for an informed review of the environmental impact of generating electricity with fossil as well as nuclear fuel. The unannounced purposes included demonstrating "that the responsible Federal agencies are regulating the civilian nuclear power industry in a conservative manner." The committee viewed the hearings as an opportunity to respond to critics of the AEC and to show that the agency was scrupulously fulfilling its regulatory responsibilities.[44]

The hearings, beginning on 28 October 1969, attracted more attention than any held by the Joint Committee since its widely publicized sessions on fallout during the 1950s and drew overflow audiences for each of their first three days. The first phase of the hearings featured officials from federal agencies, though the Joint Committee made it clear that it would invite representatives of private organizations and the public to appear in the second phase. Spokesmen for the AEC got a warm reception from the committee. Seaborg led off by calling opponents of nuclear power development "irrational" and adding: "In the years ahead, today's outcries about the environment will be nothing compared to cries of angry citizens who find that power failures . . . have plunged them into prolonged blackouts." The other members of the Commission testified in detail about the AEC's regulatory procedures and its radiation standards. They made no mention of their recent decision to amend the regulations, but they emphasized that the permissible limits provided a large margin of safety for the public. Congressman Hosmer pursued that point, asking Thompson why the AEC could not tighten the standards if the "cushion" was so large. Despite the attempts of Thompson, Ramey, and Milton Shaw, Hosmer did not receive a satisfactory response and finally gave up trying.

As the committee intended, the first phase of the hearings provided a great deal of information on the environmental impact of electrical production. The extent to which they carried out the committee's goal of enhancing the credibility of the government's, and especially the AEC's, efforts to protect public health was less clear. Members were unfailingly friendly and at times deferential to AEC officials, as Hosmer's reluctance to press hard for an answer on reducing permissible limits illustrated. One committee aide complained that the hearings had turned out to be "a whitewash for AEC."[45]

As the AEC was explaining the background of and the rationale for its radiation standards before the Joint Committee, it continued to consider amending them. The Commission decision to make no substantive changes other than formally requiring that licensees keep releases "as low as practicable" did not end the debate within the AEC. The Environmental Subcommittee of the Advisory Committee on Reactor Safeguards questioned the value of the proposed revision. It suggested that it be redrafted to clarify what "as low as practicable" meant and that a thorough study of the regulations be undertaken immediately. Joseph M. Hendrie, acting chairman of the ACRS, added that he thought the existing limits were "too high to be justified." Those comments helped persuade Seaborg that doing more to tighten the regulations was necessary, though his colleagues on the Commission remained opposed to the idea.[46]

With the comments of the ACRS in mind, the regulatory staff drafted a new proposal for revising the regulations. It tried to define more clearly how it would implement the "as low as practicable" principle. One way would be to establish design criteria specifying that plants should contain filters, traps, hold-up systems, and water treatment equipment that would reduce the radioactive content of effluents to a minimum. The staff also reinserted numerical limits that licensees would be expected to meet, though they were cited as objectives rather than absolute requirements. The exposures for an individual were the same as previous proposals for water effluents (1 percent of existing levels), and even tougher than previous proposals for gaseous effluents (3 percent of existing levels).

The regulatory staff's draft emphasized that the regulations must allow for flexibility, because "as low as practicable" did not necessarily mean "as low as possible." They would not, for example, require that a plant separate tritium, a radioactive isotope of hydrogen, from ordinary hydrogen in effluent water because it would take heroic efforts without clear gains in protecting public health. In addition, reactors might on occasion discharge levels of radiation above the design objectives. This would be acceptable as long as the higher-than-normal emissions were a transient condition and remained well within the maximum limits (which were fifty to one hundred times greater than the proposed design objectives). It was apparent that the enforcement of "as low as practicable" would depend on the judgment of the staff. Finally, the paper reiterated that the basic purpose of revising the regulations was to help ensure that the likelihood of growing public exposure from nuclear plants and other sources of radiation did not become a public health problem. This was the only major feature that the latest draft had in common with the original regulatory staff proposals of sixteen months earlier.[47]

The recommendations of the regulatory staff won limited support from the commissioners. Amid growing criticism of the AEC's radiation standards, Seaborg became convinced that more stringent requirements were needed to earn public confidence. Ramey seemed to agree, but Thompson vehemently differed and Commissioners Clarence E. Larson and Wilfrid E. Johnson leaned the same way. Larson was a former president of the Nuclear Division of the Union Carbide Company, and before joining the company, had served as director of Oak Ridge National Laboratory. A Ph.D. in biochemistry, he became a commissioner in September 1969. Johnson was a former general manager of the Hanford Works for General Electric; he had started as a design engineer in the household refrigerator department of the company and risen through the ranks to his position at Hanford. He was appointed as an AEC commissioner in 1966, a short time after he retired from General Electric.

Seaborg hoped that the Commission could reach a compromise on the issue of revising radiation standards. The difficulty was that the dissenting commissioners opposed any numerical limits, even as design objectives, because they feared that those values would be viewed as inviolable canons. "The general public will not look upon them as expressing what it is possible for the industry to do," admonished Johnson, "but rather as levels which cannot be exceeded without undue risks to public health and safety."[48]

In an attempt to find grounds for an agreement among the commissioners, the regulatory staff prepared yet another draft. The major change was that in the place of citing fixed limits as design objectives it offered a range of values that plants would be expected to meet under normal operating conditions. The levels were still a small percentage of the maximum permissible limits for an individual member of the general public—ranging from 1 to 2 percent for water effluents and 3 to 6 percent for gaseous effluents. But they underscored the point that the new numbers were goals and not inflexible limits. The regulatory staff's latest draft won the support of Larson and Johnson. On 20 February 1970, with only Thompson dissenting, the Commission decided to accept the recommendations of the regulatory staff and tighten its radiation standards.

It was a remarkable turn-around. Just four months earlier the same commissioners had voted against reducing the numerical limits in the regulations. The reversal came as a result of the protests of the Advisory Committee on Reactor Safeguards, the arguments of the regulatory staff and Seaborg (once he changed his mind), and above all, the perception that the AEC must respond affirmatively to the crescendo of criticism of its radiation-protection regulations. An extreme but not isolated example of the attacks was a flyer that named nuclear power plants as "the greatest threat to the environment of metropolitan New York, New England, and the Northeast" because the "reckless promotion of the AEC and the electric utilities . . . threatens to increase cancer, leukemia, and defective births." The AEC took its action despite the complaints of industry representatives, who argued that revising the standards would not only be unjustified and unnecessary but would impose costly maintenance procedures on nuclear plant owners.[49]

After the decision to revise the regulations, Seaborg wrote in his diary that the AEC would inform the Joint Committee, "and it is expected that we will meet some opposition there." This proved to be a monumental understatement. Commissioners Ramey, Larson, and Johnson and Harold Price met with Holifield and Hosmer the day after the Commission vote, and came away with the impression that the two congressmen would go along with the revisions. But when the committee staff saw the changes the AEC planned to make, it strongly objected, largely on the same grounds that the Commission had been reluctant to approve the reduced limits. One staff member complained that the design objectives would become frozen in place as requirements: "The history of AEC regulations will show that . . . new regulations which are imposed for future use can never be made less conservative once they are put out for public comment." Committee aides convinced Holifield and Hosmer to

15. The commissioners of the AEC in 1970. Left to right: Theos J. Thompson, Wilfrid E. Johnson, Glenn T. Seaborg, Clarence E. Larson, and James T. Ramey. (National Archives 434-SF-19-11)

oppose the inclusion of any numerical values in the revised standards, whether they were objectives or firm requirements.

On 26 February, Seaborg and Ramey explained the AEC's reasons for proposing the amendments, but both congressmen were unconvinced. Holifield responded that the AEC "would be letting the Joint Committee down after all the effort they had expended in defending the AEC and its standards." He ended the meeting with an emotional statement that if the AEC made the changes, it "would so undercut his effectiveness that he would no longer be [the AEC's] supporter in Congress on any other matter that required his help." At that point, the regulatory staff's effort to tighten its radiation standards by cutting the permissible limits, even as flexible objectives, was defeated. It had overcome the vocal opposition of other AEC divisions and the reluctance of the commissioners only to sink under the weight of the hostility of the Joint Committee.[50]

Unable to win the endorsement of the Joint Committee for any numerical tightening of its radiation standards, the AEC elected to add an "as low as practicable" provision to the regulations and to require applicants for licenses to describe precisely how they would accomplish that end. Those changes were not inconsequential, but they were far less ambitious than the Commission had approved before the intervention of the Joint Committee. Indeed, by removing any reference to numerical objectives for radiation releases, they were, at least in a formal sense, less rigorous than the modest proposals Forrest Western had originally made in July 1968.

Ramey announced the revisions at a crowded press conference on 27 March 1970. He pointed out that an individual living near a nuclear plant could expect to receive an exposure of only a few millirem per year, compared to more than 100 millirem the same person received from natural background radiation. He denied that the AEC had acted in response to its critics, an assertion that met with obvious skepticism from reporters. Both Ramey and Harold Price insisted, in response to questions, that the changes were more than "just cosmetic." Price argued that the regulations would now ensure that licensees would install the equipment needed to keep releases as low as practicable. He also maintained that another addition to the regulations, a requirement that licensees submit a semiannual report on their effluent releases, would allow the AEC to monitor plants more precisely and spot problems more easily. Despite the AEC's protestations, it was apparent that its opponents on the issue of radiation standards were unimpressed. One dismissed the revisions as "changing a comma in the rules," and Minnesota governor LeVander commented that "this appears an admission that the AEC regulations have not been realistic."[51]

In contrast to the critics of the AEC's action, the nuclear industry did not voice objections, perhaps because plants were easily meeting emission standards and the new requirements would not change the numerical limits. Within a short time after the AEC revised its regulation, Westinghouse announced that it would offer a plant design that featured an "essentially zero release" of radioactivity during normal operation. Utilities showed little interest. A survey conducted in late 1970 disclosed that only about one-quarter of the power companies building nuclear units were considering "minimum release" designs. They were allocating much more money to thermal pollution and aesthetic considerations than to equipment that would reduce radiation emissions further than the existing designs.[52]

The changes in the regulations that the AEC intended to make, pending final action after a public comment period, were limited in scope and impact. Still, despite the many revisions they underwent after Western's initial proposals, they fulfilled the objectives he had outlined. They provided a measure of additional protection to prevent the public from receiving an undue portion of its total radiation exposure from nuclear plants. They also brought the AEC's regulations closer to conformity with the Federal Radiation Council's guidelines. In the atmosphere in which Western had drafted his recommendations, those goals seemed sufficient and appropriate. The effluents from reactors in operation were well within the AEC's limits, and on its own initiative the regulatory staff took steps to keep them from rising to higher levels as more plants went on line.

The atmosphere soon changed, however, largely because of the controversy over the Monticello plant. As a result of the challenge to its standards, the regulatory staff drafted new proposals that were considerably more rigorous. But they encountered stern opposition from other AEC divisions and from the commissioners. Only after the level and the intensity of the attacks increased did the commissioners recognize that the issue had become more a matter of public perception than of scientific merit, and only then did they decide to amend the regulations. The passionate arguments of Theos Thompson and others that there was no scientific basis or technical justification for tightening the AEC's numerical standards were largely beside the point. As public anxiety about the effects of radiation and the adequacy of the AEC's standards increased, the agency needed not only to provide ample protection but also to demonstrate its commitment to doing so to the public. This was underscored in the dispute with Minnesota, where the AEC won its jurisdictional battle but was markedly unsuccessful in inspiring confidence in its regulatory performance.

During the Monticello debate and the growing controversy over radiation hazards, it was clear that the assumptions of earlier periods were no longer persuasive to many observers. Throughout the 1950s and most of the 1960s, the AEC and other organizations responsible for radiation protection had acknowledged that even low exposure involved some risk, albeit a very small one. But they also submitted that the benefits of the many uses of radiation more than compensated for the risks. By the late 1960s, however, an increasing number of people wondered whether the benefits of radiation were worth the risks. As public concern about environmental and health issues intensified and faith in the performance and good-will of the federal government diminished, confidence in the adequacy of federal radiation standards visibly declined.

Radiation experts had long affirmed the safety of the existing permissible limits by pointing out that there was no evidence that anyone had ever been harmed by exposure at the established levels. But some critics turned that argument on its head by suggesting that there was no evidence that receiving a permissible dose of radiation was innocuous. As Ernest Tsivoglou told the Joint Committee in January 1970: "In essence, we do not appear to have positive evidence that any specific level of radiation exposure is completely harmless, though there may be presumptive evidence that currently recommended limits should be inherently safe."[53]

In this changing atmosphere the AEC's standards and performance became a natural target. Not only were the number of nuclear plants increasing rapidly but the thermal pollution issue had stigmatized the AEC as unsympathetic to environmental concerns. If the AEC had acted promptly to revise its regulations, it might have arrested the rising level of criticism. But it moved slowly and grudgingly. After it finally did resolve to make major changes, the Joint Committee overruled the Commission's decision. The irony was that the committee, the increasingly uncritical defender of the AEC, did the agency a grave disservice. The proposed amendments that the AEC planned to make might not have won a warmer reception from critics. But it was apparent that the revised regulations the AEC did announce, even if they improved plant performance in important ways, did little or nothing to enhance its credibility. The same was true for the Joint Committee's environmental hearings, which were designed to highlight the AEC's commitment to public health and safety. Thus, major efforts by both the AEC and the Joint Committee to win public confidence were largely fruitless. By the time the AEC announced its revisions, the radiation controversy had become more bitter and more polarized over new charges that the AEC was failing its regulatory responsibilities and that nuclear power was a growing peril that threatened to take a heavy toll in human health and life.

CHAPTER XIII

Fallout over Radiation

As the controversy over the adequacy of the AEC's radiation standards developed, debate over a related question intensified the growing concern about radiation exposure. The dispute over radiation standards initially treated the potential consequences of exposure in general terms. In the absence of definitive scientific evidence on the effects of low-level radioactivity, those who discussed the issue made no effort to specify what the health implications of lowering, or not lowering, permissible limits might be. Beginning in 1969, however, several scientists suggested that the hazards were far more severe than the AEC and other government agencies had assumed in setting standards for radiation protection. They argued that exposures to low levels of radioactivity had caused tens of thousands of deaths and that the growth of the nuclear industry would lead to many more unless corrective measures were taken. Those assertions triggered a bitter and acrimonious exchange of views. The debate produced more angry allegations than scientific evidence, but it received a great deal of attention, raised new anxieties about the effects of low-level radiation, and eventually helped prod the AEC to further tighten its regulatory standards.

The controversy over the effects of low-level radiation resurfaced in the late 1960s, like a major public debate on the issue a decade earlier, as a result of concern over radioactive fallout. It revived because of the claims of Ernest J. Sternglass, a professor of radiation physics at the University of Pittsburgh School of Medicine, that fallout from the atmospheric testing of nuclear weapons in the 1950s and early 1960s had caused the deaths of staggering numbers of unborn and infant children in the United States. He had begun to worry about the health effects of radiation in 1947, when his infant son became critically ill. The baby died at age two from a genetically-induced disease, and Sternglass was troubled by the nagging thought that his father's heavy use of x-rays in his medical practice might have caused the genetic defect that led to the tragedy. After receiving a doctorate in engineering physics from Cornell University in 1952, Sternglass worked for fifteen years for Westinghouse Research Laboratories before moving to the University of Pittsburgh. The fallout debate of the 1950s heightened his concern about the health implications of low-level radiation and spurred him to investigate its effects.[1]

Sternglass's interest quickened after he read articles suggesting that low-level exposure had caused a significant increase in cancer mortality among children. In articles published in 1956 and 1958, Alice Stewart of Oxford University and colleagues maintained that children of mothers who had received diagnostic x-rays during pregnancy had a mortality rate from leukemia and other forms of cancer that was roughly twice as high as normal. To test Stewart's findings, which evoked criticism on methodological grounds, Brian MacMahon of Harvard University undertook a massive survey of hospital and physicians' records of x-rays administered to pregnant women. In 1962, he corroborated Stewart's thesis that x-rays received by children *in utero* greatly increased the risk of cancer. He disagreed with Stewart, however, on the degree of the additional risk; he estimated the incidence to be 40 percent rather than two times higher than normal. Another article published a short time after MacMahon's also caught Sternglass's attention. Ralph E. Lapp, a well-known physicist, writer, and expert on fallout, suggested that unusual weather conditions had deposited radioactive debris from a 1953 bomb test in Nevada in heavy concentrations around the cities of Troy and Albany, New York, and that the health effects in that region should be carefully studied.[2]

Drawing on the evidence presented by Stewart, MacMahon, and Lapp, Sternglass proposed a new view of the consequences of exposure to radioactive fallout. He submitted that the incidence of childhood cancer reported by Stewart and MacMahon indicated that comparable levels of fallout would produce similar effects. He estimated that the recent atmospheric testing conducted by the United States and the Soviet Union would cause an increase in mortality from cancer, especially leukemia, among children of between 2.5 and 10 percent. Sternglass further suggested that risks would be higher in areas that had received heavy local fallout, such as the Troy-Albany region.[3]

Those findings, published in the prestigious journal *Science*, aroused considerable criticism. The New York State Department of Health declared that its data did not show an elevated incidence of leukemia among children born in 1953 in the Troy-Albany area. An official of the U. S. Public Health Service thought that Sternglass's argument was, at best, overstated, and, at worst, "baseless." The AEC complained that Sternglass ignored evidence that conflicted with his claims and that his use of data in calculating the doses that fetuses might have received was flawed. Brian MacMahon told the Joint Committee on Atomic Energy: "The position that Dr. Sternglass took may or may not turn out to be the correct one. However, his substantiation of the position was, in my view, very weak." Sternglass's critics, despite their reservations about his evidence and objections to his conclusions, agreed with him that more research was needed.[4]

Sternglass continued his own efforts to resolve unanswered questions about the effects of low-level radiation. He caused a new stir with a paper he delivered before the Health Physics Society in Denver in June 1968. Building on his *Science* article, he contended that heavy fallout in the Troy-Albany area in 1953 had doubled the number of leukemia cases among children. The trend had been obscured, he argued, because the fallout not only affected children *in utero* but also the reproductive cells of the parents of children born many years later. Thus, without knowing of the genetic damage from fallout, parents passed on defects that made their offspring more likely to develop leukemia. Sternglass's presentation, like his earlier article, was greeted with skepticism. The New York State Department of Health again denied an increase in the rate of childhood leukemia in the Troy-Albany vicinity. The AEC suggested that Sternglass had "misconstrued or misinterpreted the data to the point that his entire proposition becomes invalid." It dismissed his thesis that tiny amounts of radiation, far smaller than those cited by Stewart and MacMahon in their studies of x-rays, had doubled the incidence of leukemia. Expert readers who evaluated the paper for *Science* found it unpersuasive and the journal rejected it for publication.[5]

Sternglass, convinced that he had made a discovery of major importance, pursued the subject in an attempt to answer his critics. While examining a volume of vital statistics he came across what he later described as "incredible findings." He noticed that the rate of fetal mortality in the United States, which had steadily declined after 1941, leveled off after 1952. The rate did not increase but it stopped falling as rapidly as it had during the 1940s. Because data for fetal mortality was of questionable reliability, Sternglass checked the statistics for the mortality rates of infants less than a year old. He found the same pattern; the decline in the frequency of infant mortality had leveled off in 1950. Sternglass contended that there was a close correlation between nuclear testing and the slowing in the reduction of fetal and infant death rates, and he concluded that the logical causative agent was radioactive fallout. In October 1968 he announced that between 1951 and 1966 fallout was responsible for the deaths of 375,000 children in the United States who would have lived if the decline in infant mortality had continued at the rate that prevailed between 1935 and 1950.[6]

Sternglass elaborated his thesis in an article that the *Bulletin of the Atomic Scientists* published in April 1969. He suggested that the 375,000 "excess infant deaths" demonstrated not only the special sensitivity of fetuses and young children to radiation but also the genetic damage caused by fallout. In this article and in two papers he delivered a short time later at professional meetings, he cited the findings of Swedish scientist K. G. Luning, who had conducted experiments with mice on the effects of strontium–90 on reproductive cells, to support his case. Sternglass argued that "the human ova, sperm and fetus may be considerably more sensitive to internal radiation from certain radioisotopes than had been expected." He maintained that the result of the genetic injury to parents was that children were more susceptible to fatal illness.[7]

Sternglass's findings were alarming and sensational enough to command attention under any circumstances. But they probably received even more notice than they otherwise would have because of their implications for a public debate over building an antiballistic missile system (ABM). It was intended to defend the United States from attack by blasting hostile missiles out of the air. Since the early 1960s supporters and critics of the ABM had battled fiercely over its cost, technical feasibility, and repercussions for national security and the nuclear arms race. The issue became more divisive and contentious as Congress deliberated over funding the ABM in 1968 and 1969. Sternglass, who had argued in his April *Bulletin* article that his findings made the cessation of all nuclear testing imperative, a short time later issued dire warnings about the ABM. In a June 1969 letter to the editor of the *New York Times*, he announced that the effect of the ABM, by escalating the arms race, "would be to seal the biological doom of mankind."[8]

An editor for *Esquire* magazine noticed Sternglass's letter and asked him to convert it into an article. Less than three weeks later, the fastest the magazine had ever gotten a manuscript into print, *Esquire* promoted Sternglass's article in full-page advertisements in major newspapers and rushed it to members of Congress. The article, added to the magazine as an insert, was titled "The Death of All Children." Sternglass repeated his claims that fallout had caused an increased incidence of childhood leukemia, uncounted numbers of fetal deaths, and some 375,000 infant fatalities. The meaning of this, he emphasized, was that a full-scale nuclear war, which the ABM would encourage, "would most likely be sufficient to insure that few if any children anywhere in the world would grow to maturity."[9]

Sternglass's article had no apparent impact on the ABM vote, but along with his other statements, it made him a celebrity and his views on the effects of low-level radioactivity a focus of concern. A Toronto newspaper ran a story about Sternglass under the banner "It's Mega-death Time for the Kiddies," and the *New York Post* headlined "1 A-Attack Fatal to Man: Expert's View." Sternglass explained his thesis on NBC's "Huntley-Brinkley Report," the CBS "Morning News," the "To-day Show," a series of local television shows, and programs in Canada, Australia, and New Zealand. He was featured on the front page of the London *Observer* and wrote articles for a variety of popular and technical publications.[10]

Sternglass's views provoked sharp dissent. The least surprising sources of criticism were government agencies and private organizations whose activities or programs were directly involved in evaluating the risks of low-level radiation. The AEC found his data weak, his reasoning imprecise, and his conclusions unconvincing. Although at first it refrained from initiating public comments on Sternglass's suggestion that fallout had killed 375,000 infants, internal staff reviews took issue with his findings. John B. Storer, deputy director of the Division of Biology and Medicine, for example, thought that the most serious flaw in his *Bulletin of the Atomic Scientists* article was "the assumption that because two phenomena can be superficially associated there is necessarily a cause and effect relationship." Sternglass's statistics, he asserted, were no more meaningful than the fact that "the average hair length of males under 30 in the United States has increased at about the same rate as the use of communications satellites."[11]

After Sternglass's theories received headline treatment, the AEC, at the urging of the Joint Committee, moved more aggressively to counter his statements. It contacted several reporters to challenge Sternglass's arguments and arranged for Storer and Leonard A. Sagan, a former AEC staff member who was the associate director of the Department of Environmental Health at the Palo Alto, California Medical Clinic, to appear on the "Today Show" to dispute his position. The AEC and those who agreed with its views objected to Sternglass's conclusions on a number of grounds. In addition to Storer's criticism of his use of statistics, they rejected his claim that leukemia rates had risen as a result of fallout. If one accepted Sternglass's correlations for the Troy-Albany area, it seemed impossible to explain why leukemia had not risen in other places hit with fallout. The AEC also denied that fallout was the probable cause for the leveling off of the decline in fetal and infant mortality. It submitted that the decrease between 1935 and 1950 occurred because of major advances in medical care, such as sulpha drugs and antibiotics, and that there was no reason to view the decline in that period as the norm. It contended that Sternglass's computation of "excess infant deaths" based on the assumption that the same rate of decline should have continued indefinitely was invalid.[12]

Other organizations involved in radiation protection took a similar position. The Public Health Service undertook a formal examination of Sternglass's arguments after he urged the Department of Health, Education, and Welfare to develop chemical processes to remove strontium—90 from milk. It found them unpersuasive. One staff member asserted: "The simplistic approach that infant and fetal mortality rates in the United States will behave in a purely mathematical fashion over a period of 33 years, independent of epidemics, advancement in medicine, changes in socio-economic levels, war, depressions, birth rate patterns and other such controlling factors is scientifically untenable." An *ad hoc* committee of the New York chapter of the Health Physics Society, after stating that its members opposed the ABM, declared: "Dr. Sternglass has ignored, misinterpreted . . . and misused available data and principles of scientific research. . . . Certainly he has not proved his hypothesis."[13]

Some of the most damaging assessments of Sternglass's findings came not from government agencies and health physicists, whose antipathy was predictable, but from scientists who might have been expected to sympathize with his theories. Although Sternglass implied that those who disagreed with him were merely reciting an official line, in many cases this clearly was not true. Writing in *Environment* magazine, frequently a source of harsh criticism of the AEC, two members of the Scientific Division of the Committee for Environmental Information concluded that although Sternglass had raised important questions, "the balance of evidence, at this time, would appear not to support [his] thesis." Patricia J. Lindop and Joseph Rotblat, prominent British scientists whose views on nuclear issues generally conflicted with the AEC, worried that if Sternglass's claims were taken seriously and later discredited, it "could easily lead to a relaxation of the rules governing radiation protection." Asserting that his thesis had "no scientific justification," they pointed out that if one took Sternglass's assumption about infant mortality rates and extrapolated backwards, "we find that less than 100 years ago the infant mortality in the United States was 100 per cent!"[14]

Scientists whom Sternglass cited to support his arguments also expressed grave reservations about his conclusions. Alice Stewart commented: "This is not the first time that a reputable scientist has fallen into the trap of over-confident extrapolation, or asked his readers to believe in an implausible situation." K. G. Luning, the Swedish expert upon whom Sternglass relied for his claims about the genetic effects of strontium–90, dissented even more sharply. Declaring that he agreed with the criticism aired by the AEC, he added: "We have surely found small genetic damage after radiation of mice with strontium. But the effects are very small, and the doses given the mice were at least 1000 times stronger than the doses a human can obtain after a nuclear test."[15]

Despite the prevailing skepticism about Sternglass's theories, a number of prominent scientists, including some who rejected his specific arguments, thought that the questions he posed deserved serious attention. "The evidence is not sufficient to prove that Sternglass is right," observed Nobel laureate Freeman J. Dyson in June 1969. "The essential point is that Sternglass may be right." The *American Journal of Public Health* offered a similar opinion in an editorial: "Stripped of the emotional and political arguments utilized by Professor Sternglass to focus attention on his hypothesis, the facts appear to present a case for further investigation." The need for research on the risks of fallout and other low-level radiation was not disputed; the AEC and other government agencies agreed that more work on the subject was essential. The AEC, in fact, had been sponsoring a full-scale investigation of the health effects of fallout for a number of years, and after Sternglass raised the issue to new levels of concern, it hoped that its project would provide evidence to refute his findings. Instead, those conducting the research, while rejecting most of Sternglass's arguments, submitted that parts of his analysis might have some validity. This precipitated a new and even more acrimonious controversy over the consequences of releasing radioactive effluents from nuclear power plants.[16]

The investigation of radiation risks that set off a new phase of the debate was based at the Lawrence Radiation Laboratory in Livermore, California, a facility run by the University of California and funded by the AEC. Located about fifty miles east of San Francisco, the Livermore Laboratory was one of two AEC-supported research centers (the other was on the campus of the University of California at Berkeley) named for Ernest O. Lawrence, a pioneering nuclear physicist at Berkeley. Livermore had been established in 1952 primarily to accelerate work on the hydrogen bomb and other nuclear weapons. It also took the lead in conducting research on Project Plowshare, the use of nuclear explosions for peaceful purposes.[17]

In 1963, the laboratory assumed responsibility for a new program. The atmospheric nuclear testing conducted by the Soviet Union and the United States in 1961 and 1962 had revived public and congressional concern about fallout, and had underscored the many unanswered questions about the severity of the health risks it presented. The AEC acted to augment its existing research programs by establishing a new one at Livermore. Its primary purposes were to study the effects of fallout and other radiation releases on the natural and human environment and, if possible, to develop methods for "minimizing the impact of radiation and radioactivity release on man." The AEC realized that the mandate that it gave the laboratory was imprecise, but it hoped that the program would introduce novel approaches and new data. It was especially interested in analyses of short-lived radioactive elements (such as iodine–131) released by nuclear explosions and in the "early fallout" that followed a detonation within hours or days. Although the AEC's Division of Biology and Medicine retained the overall responsibility for fallout studies, the agency believed that Livermore could make important contributions to understanding the ways in which radioactivity jeopardized human health. It anticipated that within five years the program would cost more than $10 million annually and employ 300 to 400 people, particularly professionals in biomedical fields.[18]

To head the biomedical program, Livermore recruited John W. Gofman, who held both a doctorate in chemistry and a medical degree from the University of California. He had studied for his Ph.D. with Glenn Seaborg at Berkeley, where he had codiscovered the artificially produced radioactive element uranium–233. Seaborg described him as "one of his most brilliant students." After working for a time on the Manhattan Project, Gofman entered medical school. A short time after he completed his internship in internal medicine, he joined the faculty at Berkeley as a professor of medical physics. During the late 1940s and 1950s he conducted research on cholesterol, arteriosclerosis, and heart disease that won wide recognition. The AEC staff thought that he was "an excellent choice" to direct Livermore's new biomedical division.[19]

As the program developed, it undertook research on a wide variety of problems, principally involving the dispersion and distribution of radioactive isotopes from the use of nuclear energy in the environment. Livermore scientists also studied the biological impact of low-level human exposure to those isotopes, though this question was, in the words of Michael M. May, director of the laboratory, "one of the most difficult in biology." Gofman told a conference on the Plowshare program in 1964 that gathering sufficient data on the effects of low-level radiation would require a survey of one to twenty million people around the world for at least ten years. While acknowledging the Herculean difficulty of such an undertaking, he cautioned that that "radiation guides, or standards, . . . are painfully uncertain. . . . We have no direct valid information on the subject of injury to be anticipated in humans either for the generation receiving such radiation or for their descendants."[20]

The arguments advanced by Ernest Sternglass in the spring and summer of 1969, focusing on the general subject that Livermore had investigated for several years, aroused interest and skepticism among laboratory scientists. Shortly after reviewing Sternglass's theories on fetal and infant mortality, Arthur R. Tamplin, a group leader in the biomedical division, undertook an evaluation of and response to them. Tamplin had earned a doctorate in biophysics at Berkeley, where he studied under Gofman. After working for the Rand Corporation and the Thiokol Chemical Corporation, he joined the Livermore program in 1963. His major research projects included estimating the doses of iodine–131 that children had received from nuclear testing and the amounts of radioactivity released by Plowshare explosions.[21]

In April 1969 Tamplin delivered a paper assessing Sternglass's thesis at an informal seminar for Livermore employees. Like other critics, he charged Sternglass with misusing or manipulating data to suit his conclusions. Tamplin contended that the two major influences on fetal and infant mortality rates were socioeconomic conditions and the introduction of antibiotics. He pointed out that middle-class Americans had the lowest infant and fetal mortality rates in the world while the poor suffered from disproportionately high rates. The most effective way to reduce the frequency of mortality among unborn and young children, he suggested, was to improve living standards for the poor. But Tamplin did not entirely dismiss fallout as a contributing cause, albeit a tertiary one, to fetal and infant death rates in the United States. He argued that Sternglass had overstated the effect of fallout by a factor of at least one hundred. But this still meant that fallout was responsible for "a not negligible number of fetal and infant deaths." According to Tamplin's calculations, fallout could be held accountable for as many as 8000 fetal and 4000 infant deaths.[22]

In response to requests for copies of Tamplin's paper, Livermore Laboratory printed 750 copies for distribution. Tamplin also sent it to *Environment* magazine, whose editor, Sheldon Novick, promptly agreed to publish it in revised form. Novick told Tamplin that *Environment* had rejected the article that Sternglass had later published in the *Bulletin of the Atomic Scientists*, but he believed that the issues Sternglass raised still deserved serious attention. A short time later, in response to an announcement that Tamplin submitted on the availability of his seminar paper, the editor of the *Bulletin*, Richard S. Lewis, offered to publish an abbreviated version of his critique.[23]

Although many observers found Tamplin's assessment of Sternglass to be an effective and persuasive rebuttal, staff members of the AEC's Division of Biology and Medicine were disturbed by his estimates of the number of fetal and infant fatalities caused by fallout. On 13 August, division director John R. Totter expressed his reservations in a telephone conversation with Tamplin and Gofman, who was Tamplin's boss at Livermore. Totter contended that Tamplin's critique of Sternglass's theories was a valuable contribution, but that his calculations of mortality rates were unproven and disputable. He heartily approved of Tamplin's intention to publish his evaluation of Sternglass in *Environment*. But Totter suggested that Tamplin separate the section of his paper explaining his own estimates of fetal and infant mortality and submit it to a more technical journal, such as *Health Physics*, that used expert referees to review manuscripts for publication. Tamplin rejected that idea, arguing that *Environment* was "a far better journal than *Health Physics*."[24]

Totter and members of his staff were concerned that Tamplin's analysis of the effects of fallout, though far less shocking than that of Sternglass, was still alarming enough to heighten public concern about low-level radiation. Although Tamplin largely discredited Sternglass's specific findings, he supported his general thesis that fallout, and by implication, other sources of low-level radioactivity, were serious public health problems. For that reason, Totter objected to publication of Tamplin's estimates in quasi-popular journals such as *Environment* and the *Bulletin of the Atomic Scientists*, at least until more supporting evidence was available. He insisted that Tamplin made unwarranted assumptions about the causes of fetal and infant mortality and failed to account for experiments with mice and dogs that seemed to refute the magnitude of the mortality rates from fallout that he proposed. Tamplin replied that his estimates were upper limits that, even though necessarily imprecise, were within the realm of possibility. Totter remained unconvinced. "My contention is that . . . your estimates are not upper limits," he wrote to Tamplin. "They are simply incorrect."[25]

The disagreement between Totter and Tamplin produced sharp exchanges but no revisions in their views. Tamplin went ahead with his plans to publish his critique of Sternglass and his own estimates of fetal and infant mortality in *Environment* and the *Bulletin of the Atomic Scientists*. The AEC, despite reservations about the scientific merits of Tamplin's article and concern about its impact on the radiation debate, made no effort to prevent him from doing so. Ironically, *Environment* decided not to publish Tamplin's article after it found that the *Bulletin* had scheduled it "before we could possibl[y] be in print." The article appeared in the *Bulletin's* December 1969 issue. It focused on Sternglass's thesis and the effects of fallout, but it also suggested that radiation from other sources could cause "considerable human tragedy."[26]

By the time that the *Bulletin* article was published, Tamplin had turned his attention to the potential health effects of nuclear power. He made explicit what to that point had only been an implicit question raised by Sternglass's theory: if the low level of radioactivity in fallout could cause the damage that Sternglass, or Tamplin, believed, were not radioactive effluents from nuclear plants a major peril to public health? In the fall of 1969, Tamplin, joined by Gofman, answered that question in an alarming affirmative; they maintained that the growth of nuclear power could result in the deaths of thousands of Americans every year from cancer. Their analysis, coming at the same time that the controversy over the Monticello plant was attracting notice, added a new dimension to the debate over radiation.

At about the same time that he was skirmishing with Totter about his work on the consequences of fallout, Tamplin was asked to speak at two public meetings on nuclear power. The person who arranged the invitations was Dean Abrahamson, who was spearheading the effort to force the Monticello plant to comply with the radiation standards of the state of Minnesota. Both Tamplin and Gofman had met Abrahamson a few months earlier, and the Minnesota professor had impressed them with the importance of carefully examining the health hazards of nuclear plants and the adequacy of existing radiation protection regulations. At a conference at the University of Vermont in September 1969, Tamplin challenged the statement of AEC officials that the health effects of radioactivity from nuclear units were statistically too small to be detectable. He was more outspoken at a symposium on "nuclear power and the public" at the University of Minnesota the following month. In a paper that made Commissioner Thompson "quite upset," Tamplin disclosed that he viewed the growth of the nuclear industry "with a great deal of anxiety." He argued that the health of a person who received the maximum permissible concentrations of certain radioactive isotopes could be seriously imperiled, and therefore, "plants should be designed so as to approach absolute containment of the radioactivity."[27]

Tamplin's remarks in Minnesota previewed even more disquieting conclusions that he and Gofman presented less than three weeks later. In a paper delivered at a nuclear science symposium in San Francisco sponsored by the Institute of Electrical and Electronics Engineers, they contended that if the entire population of the United States received the permissible dose of 0.17 rads per year throughout their lifetimes, the result would be 17,000 additional cases of cancer annually. Basing their figure on extrapolations from the effects of high doses of radiation, they suggested that the risks of low-level radioactivity had been badly underestimated. Tamplin and Gofman admitted that "the population has not received anywhere near 0.17 Rads per year from atomic energy activities thus far," and they hoped that their calculations overstated the hazards of low-level exposure. But they insisted that in the absence of definitive knowledge about the consequences of exposure and the prospective growth of the nuclear industry, permissible levels should be made more conservative. While affirming their support for nuclear power, they urged that the Federal Radiation Council, and by implication, all federal agencies, lower allowable doses by "at least a factor of ten." An account of the paper appeared in the *San Francisco Chronicle*, which also quoted Gofman as saying that "to continue the present [exposure] guidelines is absolute folly."[28]

Senator Edmund S. Muskie provided Gofman and Tamplin with a more prominent platform a short time later by asking them to testify

16. John W. Gofman (Lawrence Livermore National Laboratory)

before his subcommittee on air and water pollution. The Livermore scientists repeated the arguments they made in San Francisco, arguing that they were presenting "hard evidence" that permissible doses of radioactivity would greatly increase the risk of leukemia and lung, thyroid, and breast cancer. Muskie thanked them warmly for a "very lucid

17. Arthur R. Tamplin (Lawrence Livermore National Laboratory)

and helpful statement" that "even a layman like myself understands."
Gofman and Tamplin won a less cordial reception from members and
staff of the Joint Committee on Atomic Energy, who asked to see them
while they were in Washington for the Muskie hearings. Holifield wondered why they were speaking in public forums rather than taking their
views first to other experts. Gofman replied that he and Tamplin had to
go public because the AEC refused to take their position seriously. One

witness reported that Holifield "gave Gofman a hard time," but that Gofman "wasn't all that cowed by it."[29]

Gofman did take pains, however, to explain his position to Glenn Seaborg, his former mentor at Berkeley. He submitted that the AEC chairman had been "dangerously misled" by his staff. He complained about Totter's "humiliating" effort to convince Tamplin to "whitewash" his estimates of fetal and infant mortality, about other AEC staff criticism of his and Tamplin's findings, and about the Joint Committee's "insult, veiled and unveiled intimidation, ridicule, sarcasm, and jokes." Gofman pointed out that he and Tamplin had refrained from attacking the AEC at the Institute of Electrical and Electronics Engineers symposium, though with "great forbearance and with much conciliation." He denied that they sought to undermine atomic energy. "Hell, Glenn, I know we need electric power," he wrote. "I am only asking for responsible action to achieve this." Gofman reiterated that he was convinced that the effects of low-level radioactivity would prove to be much more serious than the AEC acknowledged. "I believe that false efforts to provide you with 'pleasing' information is going to be a hell of a boomerang," he argued. "I suggest you look inward to the AEC staff for some objectivity so we can all work together to resolve this thorny problem in a healthy fashion."[30]

AEC staff members found Gofman's and Tamplin's analysis unpersuasive on a number of grounds. One was their use of extrapolations from high doses of radiation to estimate the hazards of low-level exposure. This assumed that small doses of radiation delivered over a long period of time caused as much damage as an acute dose. Although the evidence was not definitive, there were strong indications that low doses over an extended period were less harmful than heavy doses in a short span of time. The AEC staff also noted that other experts had considered the same data as Gofman and Tamplin and judged the risks to be much lower. The Livermore scientists' conclusions were not derived from new findings or original research. The difference lay not in the "hard evidence" that they claimed to have but in the interpretation of existing data. The AEC denied that Gofman and Tamplin had proven that their interpretation was more compelling than that of other radiation experts. It saw no way, for example, that the entire population of the country could be exposed to radiation in amounts close to the permissible limits, which applied at the boundaries of a plant. For those and other reasons, the staff found that the "opinions and scientifically questionable derivations of Gofman and Tamplin do not make a case for revision of radiation protection standards."[31]

Despite the objections that the AEC presented, Gofman's and Tamplin's recommendations were similar to the proposals for reducing the permissible limits that the AEC's regulatory staff was advancing. Indeed, the regulatory staff called for revisions that were more radical than the reduction, by a factor of ten, that Gofman and Tamplin urged. The staff members who criticized Gofman and Tamplin seemed unaware of the regulatory staff's position, and they addressed the technical merits rather than the policy implications of the Livermore scientists' views. The regulatory staff's proposals had met strong opposition within the AEC and it was far from clear at that juncture that they would receive the approval of the commissioners. Yet the irony remains that, at least for a time, Gofman and Tamplin and the AEC regulatory staff, though coming from different starting points, agreed that the regulations should be revised to provide an extra measure of protection for the public. Eventually, the Commission voted in favor of the regulatory staff's recommendations, but by then the questions raised by Gofman and Tamplin had launched a bitter, angry, and highly visible dispute.

At first, both sides in the growing rift between Gofman and Tamplin and the AEC kept their differences on a plane of professional disagreement; their debate was sharp but civil. But it soon deteriorated into an unbecoming quarrel in which Gofman and Tamplin, feeling that they had been mistreated, heaped *ad hominem* abuse on those who took issue with them. The breach began to widen irreversibly after Roger E. Batzel, associate director of Livermore, pressed Tamplin to make changes in a paper he was preparing for a meeting of the American Association for the Advancement of Science. Batzel's action ran counter to the wishes of the AEC commissioners. In a discussion of Gofman's and Tamplin's public statements, the Commission had considered restricting them and other employees of AEC laboratories to publishing in scientific journals or, if they chose to air their views to the general public, to clear them with the AEC. Seaborg and Clarence Larson argued vehemently against such a requirement. Eventually the commissioners "agreed that the best solution to the problem was to have individuals . . . criticized by their peers rather than restricted by the supporting agency."[32]

Batzel either was unaware of the commissioners' decision or elected to ignore it, which he could do because of the semiautonomous status of the Livermore Laboratory. He was concerned that Tamplin's analysis would appear to be an official position of the laboratory and/or the AEC. He was especially worried about a section in the paper dealing with reactor safety and calling for a moratorium on construction of new nuclear plants. Gofman and Tamplin later charged that Batzel threatened to withdraw funding for Tamplin's trip to the meeting in Boston unless he revised the paper, though at the time Tamplin told *Nucleonics Week* that laboratory managers had only asked him "to cool it or say things in a different way." He and Gofman also later claimed that the "Batzel-censored manuscript was returned to Tamplin with little left in it but the prepositions and conjunctions." That was an exaggeration; the paper that Tamplin delivered was a forceful reiteration of the points he had stressed in earlier appearances. It also repeated two fundamental errors that Tamplin had made in discussing federal radiation standards; he contended inaccurately that existing regulations assumed a threshold below which radiation exposure was safe and that they failed to account for concentration of the radioactivity of certain isotopes in the food chain. But Batzel's action, no matter how much he objected to Tamplin's arguments, was clumsy and counterproductive. His censorship of a well-regarded scientist was not only indefensible on its merits but also needlessly intensified an increasingly acrimonious dispute.[33]

Although the AEC had decided as a matter of policy not to place restrictions on Gofman and Tamplin, it was determined that their position should not go unchallenged. It countered Tamplin's presentation at the conference of the American Association for the Advancement of Science by distributing copies of staff comments on the Gofman-Tamplin position. The staff had prepared its evaluation at the request of both Muskie and the Joint Committee. It denied that the Livermore scientists had proven their case for an immediate tenfold reduction of permissible limits. A summary of the AEC's rebuttal went out on the Associated Press wire service and appeared in many newspapers. A short time later, John Totter declared in a feature article on Gofman and Tamplin in the *Los Angeles Times*: "I don't think they can raise any new questions because they themselves have not conducted any experiments . . . which have made any contribution to radiobiology. . . . They've been using other peoples' data almost exclusively."[34]

Since Gofman and Tamplin had chosen to report their views in public forums, publication of the AEC's response in itself was unexceptionable. But coming in the wake of Batzel's enforced manuscript revisions, it seemed to incense Gofman and Tamplin. They became noticeably more strident and self-righteous in their arguments, even to the point of intolerance for anyone who disagreed with them. Gofman complained to Commissioners Thompson and Larson that "the blatantly stupid 'refutations' of our work emanating from the DBM [Division of Biology and Medicine] Staff of AEC were doing incalculably more harm to the cause of peaceful atomic energy than our criticism of FRC standards." Without specifying who accepted a threshold theory as a basis for radiation protection standards, Gofman told *Newsweek*: "The statement that there's some number that's safe is an absolute, unmitigated lie." In response to University of Utah nuclear physicist Charles W. Mays, a sharp critic of the AEC's statements on fallout who strongly contested his and Tamplin's analyses, Gofman declared: "I am really inclined to tell you how idiotic you truly must be to write the brash, insulting letter you have written."[35]

When Lauriston S. Taylor, president of the National Council on Radiation Protection and Measurements (NCRP), advised Senator Muskie that Gofman and Tamplin had "presented no new data, new ideas, or new information" that "highly experienced" experts had not already considered, the two Livermore scientists denounced "this fraudulent, hypocritical, and incompetent document," and added: "Incompetence in the extreme is our only possible evaluation of Lauriston Taylor and his cohorts." And in another shot at the AEC, Gofman informed a special committee on environmental protection of the City Council of New York on 4 March 1970 that "the difference between us and the AEC is that we are not willing to play Russian roulette with human lives." He elaborated at a luncheon of the National Committee to Stop Pollution the following day, predicting that the AEC would soon relax its radiation standards and the result would be a 30 percent increase in cancer. "There is no morality," he said of agency officials, "and there is not a shred of honesty in any one of them." This statement came about a week after the AEC's behind-the-scenes effort to tighten its permissible limits had been squelched by the Joint Committee.[36]

Gofman and Tamplin received a great deal of popular and professional attention that was mixed in its reaction to their conclusions. Many scientists were skeptical. "An analysis of their work reveals that most of the assumptions they use in making predictions can be neither proved nor disproved," reported an article in *Science* in February 1970. "But the consensus of their peers is that at least some of their assumptions are wrong." A group of twenty-nine recognized radiation experts, most of them based at universities and national laboratories, sent a petition to Holifield, declaring that existing standards reflected the "great competence" and "concern for public health" of the organizations that formulated them. The petitioners complained that "a tiny minority of experts" was causing unwarranted public alarm and that an "adequate rebuttal" required a "lengthy and technical reply unsuitable for publication in the press." An editorial in *Physics Today* lamented that Gofman and Tamplin failed "to place their points of disagreement in proper perspective before the public" or to compare the risks of nuclear power with the hazards of air pollution from fossil fuel plants.[37]

Gofman's and Tamplin's arguments made a much greater impact in other circles. This was apparent, for example, in the second phase of the Joint Committee's hearings on the environmental effects of producing electrical power in January 1970. Several witnesses, including Minnesota governor Harold LeVander and Vermont attorney general James M. Jeffords cited the Livermore scientists to support their own objections to AEC policies. Other opponents of planned nuclear stations also drew on the Gofman-Tamplin analysis to substantiate their views. A Maryland state senator, concerned about the effects of the proposed twin Calvert Cliffs plants on the Chesapeake Bay, asserted that Gofman's and Tamplin's findings confirmed that "the State of Maryland is being lead down the primrose path." The Livermore scientists received sympathetic treatment in many newspaper articles and television interviews. *Nucleonics Week*, which was not inclined to accept Gofman's and Tamplin's views, acknowledged that despite reservations about them among scientists, "no one doubts that they have scored verbal points."[38]

Gofman's and Tamplin's challenge to the AEC and other bodies responsible for radiation protection was formidable, and to many observers, persuasive, for a number of reasons. For one thing, they both had excellent professional qualifications. Although the AEC and others occasionally pointed out that their position was not grounded in original research on radiation effects and that they were relative newcomers to the field, this hardly detracted from the strength of their credentials, especially in the eyes of nonscientists. Indeed, it sounded rather like cranky nitpicking on the part of their critics. Gofman's and Tamplin's credibility was further enhanced by their positions as AEC insiders. This appeared to give them special knowledge and insight; the fact that they were taking on their employers made their views more credible, particularly to those who harbored their own doubts about the adequacy of existing regulations. When their ideas were questioned by the AEC or censored by Livermore, it enabled Gofman and Tamplin to claim persecution. This charge had some foundation, and although they greatly embellished it, they sounded very convincing as martyrs.

Both Gofman and Tamplin were articulate, confident, and impressive in outlining their opinions in personal appearances. Gofman, who gradually supplanted Tamplin as the more outspoken and visible of the two dissenters, could be counted on for a command performance when he spoke before groups—cocksure and compelling in presenting his own views and witty and sarcastic in attacking his opponents. Even Holifield commented favorably on Gofman's demeanor before the Joint Committee: "The conviction with which you express yourself and the dynamic way in which you present your ideas are very impressive."[39]

The attention and regard that Gofman and Tamplin commanded was a product of more than just their personal and professional attributes. Their arguments were tailor-made headline material; their general assertions were easy to understand and certain to arouse public interest. They were also difficult to refute, partly because the precise effects of low-level radiation exposure were still an open scientific question and partly because there was no simple way to explain the basis for existing radiation standards. Gofman's and Tamplin's projections were based on mathematical calculations of presumed conditions and consequences, and although nobody disputed the accuracy of their computations, most radiation experts rejected their assumptions. But Gofman and Tamplin insisted that the only responsible course of action was to take the most conservative possible position on permissible limits and that to do anything less was gambling in serious ways with public health. The AEC, the NCRP, and other organizations responded that their standards were already extremely conservative and were providing more than adequate protection, but their explanations were usually too technical and always less dramatic than Gofman's and Tamplin's allegations.

The political atmosphere of the times worked to the advantage of Gofman and Tamplin and to the disadvantage of the AEC and other agencies. The Livermore scientists' rise to prominence came at about the same time that a number of events were aggravating public concern about the impact of technology on the environment and on public health. A huge oil spill off the coast of Santa Barbara, California, public hearings about the risks of the insecticide DDT, a ban on the artificial sweetener cyclamate, and passage of the National Environmental Policy Act all occurred during 1969. They focused attention on environmental hazards, and Gofman and Tamplin seemed to be highlighting yet another peril that had profound implications for public welfare. At the same time, the credibility of the statements and the performance of federal agencies in general and the AEC in particular was declining. On environmental questions, the AEC's assurances that it was carefully safeguarding public health were increasingly suspect, and the doubts had been intensified by the thermal pollution and Monticello controversies. In less turbulent times, Gofman's and Tamplin's harsh rhetoric might have undermined their own position, but in 1970 fierce attacks on government agencies and officials were something of a norm and hardly a cause for a loss of faith among those who endorsed the Livermore scientists' views. Gofman and Tamplin launched a frontal assault on the established assumption that the benefits of nuclear power were well worth the risks, and they won support from growing numbers of state and local officials, environmentalists, and other citizens who worried that the hazards of radiation were much greater than they had ever been told.

The impact of Gofman's and Tamplin's conclusions on the debate over radiation hazards became apparent early in 1970. On 23 January, Secretary of Health, Education, and Welfare Robert H. Finch, acting in his capacity as chairman of the Federal Radiation Council, told Senator Muskie that in response to the questions raised by Gofman and Tamplin, the FRC would undertake "a careful review and evaluation" of the risks of low-level radioactivity. Finch made clear that "we do not agree with all the premises, conditions and extrapolations used by Gofman and Tamplin," but he suggested that it was appropriate for the FRC to examine developments over the decade since its establishment. The FRC promptly made arrangements for the National Academy of Sciences to conduct a study that would reassess the scientific basis for radiation standards and estimate the risks they presented. The study was expected to take two years. Although one of Finch's aides insisted that the new survey was "not intended to cast doubts or darkness on present standards," Gofman and Tamplin seized on it as vindication for their arguments. "The A.E.C. says our work has no merit," declared Tamplin, "but Robert Finch has overruled it and ordered a sweeping review of their standards." *Nucleonics Week* commented: "The planned FRC study represents a triumph for Gofman and Tamplin."[40]

The impact of Gofman and Tamplin on the radiation debate was also evident when the AEC announced on 27 March that it was revising its radiation standards. The growing criticism of the AEC's regulations had prompted the Commission to approve lower numerical limits as design objectives, but the opposition of the Joint Committee had overturned that decision. As an alternative, the AEC proposed to add the "as low as practicable" provision to its regulations. In explaining the draft revisions to a crowded press conference, Ramey emphasized that the "existing standards provide a high degree of protection to the public." In light of the increasing use of nuclear power, however, the AEC wanted to ensure an extra measure of protection. Ramey asserted that attacks on the AEC's standards "are not justified and many of them are irresponsible." The AEC's proposed revisions were too limited in scope and too late in publication to achieve any perceptible increase in public confidence. Critics dismissed them and the controversy over radiation continued unabated.[41]

The rift between Gofman and Tamplin and the AEC became increasingly bitter and unbridgeable. The Livermore scientists began to assert that their estimates of the increase in the incidence of cancer from nuclear plant emissions had been too low; they revised the figure from 17,000 to 32,000 cases every year if the entire population received the permissible dose. They also suggested that "all of the major forms of cancer and leukemia are induced by radiation." Based on those determinations, they urged not that allowable limits be reduced by a factor of ten but that they be lowered to zero. Complaining that they had tried unsuccessfully to present their findings "in a polite, scientific way," they condemned the AEC in a series of appearances for complacency, inaction, and indifference. "The issue is simple," Tamplin declared at a meeting of the American Cancer Society. "The AEC wants to release more radioactivity into the environment than is safe."[42]

The AEC continued to contest the validity of Gofman's and Tamplin's claims. Commissioner Thompson faulted them in speeches and interviews for failing to compare the hazards of nuclear stations with those of fossil plants and for misrepresenting the AEC's radiation regulations. He estimated that the radiation released by nuclear plants in operation, under construction, or planned would cause, statistically, less than one extra case of cancer or leukemia a year, or one fifty-thousandth of what Gofman and Tamplin alleged. Victor Bond, associate director of Brookhaven National Laboratory, submitted that the Livermore scientists based their conclusions on a "fictitious dose" delivered to a "fictitious population," because it was inconceivable that the entire population could be exposed to the maximum permissible dose. When the Atomic Industrial Forum printed one of Bond's critiques, Gofman fired off an angry letter to the organization. "The AIF, the

AEC, and Dr. Bond all seem to believe that a stupid set of lies will enable them to ram ill-considered atomic programs down the throats of the American public," he declared. "The more you all lie, hide the facts, and deliberately and unashamedly distort every responsible criticism, the earlier will be the celebrated demise of your outrageous activities."[43]

This letter so troubled Livermore director Michael May that he told Gofman to either stop sending "such libelous letters" or leave the laboratory. He also wrote to Paul Turner, public affairs manager of the Atomic Industrial Forum, to express regret for the "personal attacks made by Dr. Gofman" and released a letter he had sent Seaborg two months earlier emphasizing that Gofman and Tamplin did not speak for Livermore Laboratory in publicizing their hypothesis. At about the same time, the AEC, convinced that Gofman and Tamplin were wrong and concerned about their impact on public views of nuclear power, considered what might be done to curb or halt their attacks. The staff prepared a paper outlining the advantages and disadvantages of taking "definitive action with respect to Gofman and Tamplin." The benefits of removing their status as "AEC scientists" and perhaps diminishing their appeal were balanced against the drawbacks of congressional and press criticism and loss of credibility in the scientific and academic communities. Seaborg reemphasized his opinion, as voiced to May, that it "would be very counterproductive to fire them, and the only real solution lies in the area of answering their accusations, no matter how intemperate, in the open field of public debate." Despite Seaborg's admonition, AEC and/or Livermore authorities apparently concluded that something must be done to undermine Gofman and Tamplin.[44]

On 5 July 1970 the *Washington Post* reported that the AEC was retaliating against Gofman's and Tamplin's dissent, most seriously by slashing their budgets and staffs. Gofman lost two of his twelve staff members, which he did not consider a reprisal because other offices had been cut as much or more. But eleven of the twelve people on Tamplin's staff were transferred, leaving him "with the lonely feeling that he was no longer wanted at Lawrence Radiation Laboratory." Tamplin told the *Post*: "I used to be a group leader with 12 people and a budget of more than $300,000 a year. But I guess you can't be a group leader if you don't have a group." Other newspapers quickly picked up the story, and consumer advocate Ralph Nader asked Muskie to investigate the situation that the *Post* described. "The available indications are that Gofman and Tamplin have been accused of heresy," he wrote, "by an agency so committed to the promotion of atomic energy that it has insisted that radiation risks be treated more as articles of faith to be intoned than propositions to be examined continually."[45]

Both the AEC and Livermore issued public statements denying that they were penalizing Gofman and Tamplin for their statements about radiation standards. Speaking for the laboratory, Roger Batzel disputed the claim that Gofman and Tamplin had been mistreated. He pointed out that the budget of the entire laboratory had been cut, and he contended that seven of Tamplin's staff members had been transferred by mutual agreement. Three others had been reassigned when it became "clear that support for [their] work was not warranted in the light of budget and scientific priorities." Batzel further argued that Livermore's insistence that Gofman stop making personal attacks or leave the laboratory was justified. "We do not believe that this kind of attack by Dr. Gofman on the character and motives of those who disagree with him," he asserted, "is in the interest of science, the Laboratory, or enlightened public debate." In its statement, the AEC also emphasized that all its national laboratories had undergone "substantial budget cuts" and that the reduction in Tamplin's staff reflected Livermore's "judgment of relative program priorities." It denied that Gofman's and Tamplin's freedom to express their views on radiation hazards had been curtailed, noting dryly: "There is no indication that Drs. Gofman and Tamplin have been inhibited in their public criticisms."[46]

At Holifield's request, the AEC also hastily prepared a lengthy response to Gofman's and Tamplin's charges that their opinions were being suppressed. It admitted that the transfer of Tamplin's staff to other projects might look like reprisal. But elaborating on the points in its earlier statement, the report insisted that the reassignments took place because of "budgetary reductions, allocation of resources to programs of higher priority, and a judgment of relative scientific productivity." Furthermore, it stressed that the shifts in staffing had little impact on Gofman's and Tamplin's work on radiation standards, which drew upon published literature rather than experimental research. The AEC's explanation was greeted with skepticism by agency critics. Gofman called it a "shallow, glib, obviously false effort by AEC Staff to whitewash one of the greatest scandals in American Science." Tamplin made a similar observation in more subdued terms: "I am proud to be on the side of a rational approach to public health and at odds with the AEC." Muskie complained to Seaborg that the AEC's account did "not appear to be an unbiased review of the allegation made by Drs. Gofman and Tamplin." He asked the

American Association for the Advancement of Science to investigate the charges and report its findings to his subcommittee.[47]

The AEC's protestations of innocence of Gofman's and and Tamplin's allegations were not entirely convincing. Although it is unclear whether the Commission voted formally or some AEC and Livermore officials decided informally to take action, it seems apparent that the reduction of Tamplin's staff was retaliatory in intent and effect. Although it was certainly true, as the AEC emphasized, that the budget of all the national laboratories and other programs at Livermore also suffered cuts, the impact on Tamplin was disproportionately great. The explanation that his program had a lower priority reflected the AEC's disgruntlement with Tamplin, and that had begun after he started to air his views on the hazards of nuclear power. If the AEC had a case that his work was suddenly unworthy of support it did not make it, or even try very hard to do so. The denials that the removal of Tamplin's staff was related to his attacks on the AEC strained the agency's credibility. They fed suspicions that it was not only trying to silence debate but also refusing to admit it.

Gofman's and Tamplin's outspoken dissent placed the AEC in an extremely awkward position. Agency officials feared that attempting to curtail the Livermore scientists' freedom of expression would backfire by greatly increasing the difficulty of recruiting and retaining the scientific talent the AEC needed. As Seaborg told Muskie: "The national nuclear energy program depends upon the technical and moral support of the entire scientific community." Yet the AEC strongly objected to the manner in which Gofman and Tamplin advanced their ideas. What they claimed were incontrovertible facts were actually arguable opinions, based on questionable assumptions and inaccurate assertions about existing radiation standards. And their angry attacks on their critics, apparently arising from what Gofman described as their "emotional outrage," indicated that they were more intolerant of opposing points of view than was the AEC. This violated what Seaborg called scientists' "special responsibility for careful, reasoned, and accurate accountings to the public of their findings." Faced with distasteful options no matter what it did, the AEC elected to try to discourage Gofman's and Tamplin's activities without being flagrantly heavy handed. Inevitably, perhaps, it failed to achieve its objectives. Instead, the reduction of Tamplin's staff offered support for his and Gofman's allegations and enhanced their own credibility at the expense of the AEC's.[48]

Gofman's and Tamplin's dispute with the AEC over staff reductions increased their already considerable visibility in the growing controversy over the environmental effects of nuclear power. They received prominent and sympathetic coverage on radio and television as well as in newspapers and magazines. They achieved a milestone rare for reputable scientists in reaching a popular audience when their claims were featured in a story in the *National Enquirer*. The Livermore scientists were less successful in winning the acceptance of other scientists for their views. The harshness of their attacks seemed to diminish their impact on their professional colleagues, perhaps more than with nonscientists. Philip M. Boffey, in an article in *Science* that treated Gofman's and Tamplin's position respectfully, nevertheless suggested that they damaged their case "by indulging in verbal overkill that alienates their peers and undermines their credibility." Although some authorities agreed that a reduction in permissible limits was advisable, they denied that the consequences of maintaining the existing standards would be nearly as severe as Gofman and Tamplin contended. Most radiation experts, however, saw no need, or at least no urgency, to revise the standards.[49]

The two scientific bodies most responsible for radiation protection standards, the International Commission on Radiological Protection and the National Council on Radiation Protection and Measurements, did not consider major changes in their recommendations to be required. Gofman and Tamplin, in contrast to their denunciations of the NCRP, warmly praised the ICRP, which was curious in light of the fact that both groups took positions on radiation hazards that were virtually the same. When questioned about the ICRP's assessment of Gofman's and Tamplin's hypothesis, the chairman of the organization was noncommittal. He commented only that the ICRP would continue to review the scientific bases for its recommendations, adding that it did not "make pronouncements except in the fullness of time."[50]

For its part, the NCRP announced in January 1971 that after a ten-year study it had determined that its recommendations for basic exposure levels—500 millirem a year for individual members of the public and 170 millirem for population groups—were sound. Lauriston Taylor denied that the NCRP's conclusions were a response to Gofman and Tamplin, noting that the council had reached its conclusions before the Livermore scientists had launched their attacks. He affirmed that the NCRP had then examined their arguments but found them unpersuasive. The most important revision in the NCRP's report, in response to studies of fetal sensitivity to radiation, was to lower permissible doses for pregnant women. The council reemphasized that radiation levels should be kept "as low as practicable." The director of the study, former Hanford manager Herbert M. Parker, told a press conference: "I think we all agree that the best exposure level is zero. It is idealistic, however, to expect zero release. . . . So the best realistic level is the lowest practicable level for each particular set of circumstances."[51]

While the debate over radiation standards attracted increasing attention, the AEC was reviewing public comments on changes it planned to make in its regulations. The revisions, announced in March 1970, proposed to add a requirement that licensees keep radioactive releases "as low as practicable." The regulatory staff received eighty responses to the publication of its draft, most of which supported tightening the regulations. A few environmental groups urged the AEC to reduce radioactive emissions to zero, but the most common complaint was that the meaning of "as low as practicable" was too vague. Twenty-five comments, including nineteen from nuclear utilities and vendors, advised the AEC to specify numerical limits as design objectives. The problems that the proposed wording could cause were itemized in the Westinghouse Corporation's statement: "Interpretation difficulties due to the present vague wording will lead to uncertainties for the systems designer; major disagreements between applicants and regulatory personnel, hearing boards and parties to hearings; increased intervention; lengthening of the licensing processes; and uncertainties in reporting requirements." The AEC had included quantitative design objectives in an earlier draft of its revisions, only to remove them at the insistence of the Joint Committee. After the public comments offered such a clear message, the regulatory staff again suggested that it develop "definitive criteria on design objectives." The Commission agreed; it made the proposed regulation effective 2 January 1971, with the provision that the regulatory staff would immediately take action to define "as low as practicable."[52]

The staff promptly arranged meetings with industry representatives and environmentalists to explain the revised rules and to solicit advice in setting numerical exposure limits as design objectives. It also consulted with Joshua Lederberg, a Nobel prize-winning geneticist at Stanford University, who had published a number of newspaper articles on the radiation controversy. Lederberg suggested that for genetic reasons the AEC restrict exposure of individual members of the public to less than 10 millirem per year. Seaborg and other AEC officials responded favorably to this proposal and sought ways to incorporate it into the new revisions.[53]

On 30 March 1971 the regulatory staff sent a series of recommendations to the Commission. It suggested that a licensee would meet design criteria if effluents from its plant were less than 5 percent of natural background radiation. This was about 1 percent of the regulatory limit of a maximum exposure of 500 millirem by an individual member of the public, a level that remained in effect. Although the AEC's new numerical guidelines, if implemented, would not be inflexible requirements, the staff made clear that it would expect plants to comply with design objectives under normal operating conditions and would take enforcement action against those that did not. As an alternative to the 5 percent of background exposure, the regulatory staff proposed that a licensee would meet design objectives if it ensured that an individual living at the boundary of a plant did not receive more than 5 millirem per year. And to provide additional protection for population groups, the staff introduced a new concept (though one that was widely used in Europe) for measuring exposure, the "man-rem." It submitted that a plant would conform with design objectives if the exposure of the population within a fifty-mile radius did not exceed 100 man-rem per year for each 1000 megawatts of nuclear capacity. The man-rem, rather than assuming a uniform dose for an entire population group (as the existing standard did), estimated the exposure to those who lived within different concentric areas from a plant. It was computed by multiplying the average dose received by members of a large group by the number of people in that group. If members of a population group of 100,000, for example, were exposed to 5 millirem apiece, the total would be 500 man-rem. This method of measurement not only provided additional assurance that population exposure would remain very low but also undercut Gofman's and Tamplin's calculations, which were based on the assumption that every person in the United States received the allowable population exposure limit of 170 millirem.[54]

Seaborg thought that the staff recommendations "would be a tremendous step" if the Commission accepted them. Ramey and Larson were hesitant, but the commissioners soon agreed on the proposals. The Joint Committee, on the other hand, was still strongly opposed. Holifield had an "extremely adverse emotional reaction" and threatened, as he had done a year earlier when the AEC informed him of its plan to set numerical guidelines, to withdraw his support from the agency. Other committee members also protested, though more mildly

than Holifield. Although major vendors supported, or at least toler-
ated, the AEC's proposals, other industry representatives expressed
keen disapproval. Several utility executives complained that the AEC
had "capitulated to the demands of Gofman and Tamplin." The
agency made some changes in its draft, revising, for example, the 100
man-rem goal to 400 man-rem. But in contrast to its previous submission to Joint Committee pressure, it kept the basic proposals for numerical design objectives intact. "The force of the argument is so
great," noted Seaborg, "that the Commission feels it simply must go
ahead and make this improvement."[55]

A number of considerations made the "force of the argument" seem
so great. One was the commitment of the AEC to provide a wide margin
of safety from the hazards of civilian nuclear plants. The objectives it
prepared were intended to reduce the possibility of injury to the public
to a minimum without forcing overly stringent requirements on the
nuclear industry. The limits the AEC proposed, despite some industry
objections, were technically achievable and were generally being met by
operating plants. Harold Price told a press conference that there were
two reasons for the AEC's new proposals: "We think it's right and we
think it's technologically and economically feasible." But the AEC was
also influenced and motivated by its critics. Price denied that the AEC
was acting in response to Gofman and Tamplin, but it was unlikely that
the agency would have disregarded the opposition of the Joint Committee and leading industry representatives if it had not felt pressed to
demonstrate the adequacy of its regulations. The design objectives it
published for public comment in June 1971 would effectively cut permissible limits for the public by a factor of one hundred, which was much
more conservative than Gofman's and Tamplin's original call for a
reduction by a factor of ten. The AEC's credibility had suffered as a
result of its dispute with Gofman and Tamplin and it hoped to recover
some lost ground by amending its radiation regulations. Finally, the
AEC wanted to tighten its standards because it feared encroachment on
its authority by another federal body, the recently established Environmental Protection Agency (EPA).[56]

In July 1970 President Nixon had announced that he was establishing the EPA to consolidate many federal pollution control programs.
Divisions of the Departments of Health, Education, and Welfare, Agriculture, Interior, and other agencies that had performed environmental
functions were transferred to the EPA. In the area of radiation standards, the new agency took over the responsibilities of the Federal

Radiation Council and the role of the AEC in determining exposure limits for the general population. The EPA assumed authority for radiation standards outside the boundaries of nuclear plants, but the precise division of duties between the EPA and the AEC remained ill-defined. The AEC continued to set occupational standards and to regulate radioactive effluents, and the ambiguity of where its responsibilities ended and the EPA's began soon led to dissension. The AEC had already clashed with the EPA over thermal pollution and uranium mine safety standards, and it was concerned that the new agency planned to announce a reduction in exposure limits before the AEC could publish the proposed revisions in its regulations.[57]

At a luncheon meeting with the head of the EPA, former assistant attorney general William D. Ruckelshaus, Seaborg argued that the AEC should announce its proposed revisions because it had been working on them before the EPA had even been established. He contended that if the AEC "deferred to EPA, it would have an adverse effect on AEC." Ruckelshaus disagreed on the grounds that the "EPA, as the standards setting agency, should do it first in order to establish credibility with the public." Eventually the staffs of the agencies worked out compromise wording that emphasized that the AEC was acting under its authority to regulate plant effluents. This was a satisfactory solution for both the AEC and the EPA, but it obscured the fact that one consideration of the AEC in amending its standards was the fear that if it did not take decisive action it would surrender stature, jurisdiction, and an opportunity to improve its public image to the EPA.[58]

The AEC's announcement of its proposed revisions in June 1971 elicited a generally, but not unanimously, favorable reaction. A number of prominent environmental leaders applauded the AEC's action, and even Tamplin conceded that it was "a step in the right direction." But critics of the agency remained skeptical of its motives for and commitment to the proposed regulations. Some were concerned that they only applied to light-water power reactors rather than to emissions from any kind of nuclear facility. More commonly, critics expressed doubt that the AEC would enforce the design objectives, especially since the agency had issued them as guidelines that did not replace the old standards. The changes failed to "alter the picture of concern over nuclear power at all," declared Gofman. "As long as the numbers are still on the books there is no doubt that they will go ahead with a melange of stupid and sensible nuclear power programs that will eventually add up to 170 mrem." Some scientists objected to the AEC's proposals from a diametrically opposite perspective. The Standards Committee of the Health Physics Society, for example, found them to be "unnecessary and unwise." In its view, "to formally incorporate these extremely restrictive design criteria into the AEC Rules will tend to create a psychology of disbelief in the adequacy of any other higher stated exposure criteria."[59]

The AEC's intention to make sharp reductions in its exposure guidelines neither impressed its harshest critics nor ended the controversy over radiation hazards. Although it planned to ensure that the radioactivity in plant effluents was kept at levels even lower than its most outspoken opponents had urged, it received less credit for its radiation regulations than it deserved. As AEC officials frequently pointed out, the agency had taken the initiative to tighten its standards before the public controversy had begun. The revisions it considered at that time had been modest, but the AEC had reacted positively to responsible criticism by taking steps to cut exposure levels, effectively if not officially, by significant margins. By the middle of 1971, those facts had largely been obscured by the bitterness and doubt that prevailed in assessing the AEC's performance in safeguarding the public from radioactivity. The AEC's credibility on radiation protection was undercut by a number of other issues and events that were occurring at about the same time that it was revising its regulations. The agency's reluctance to take an expansive view of its environmental responsibilities under the National Environmental Policy Act, its denial of authority to combat thermal pollution, growing opposition to nuclear plants, and a series of highly critical books and articles all laid fertile ground for Gofman's and Tamplin's charges by casting suspicions on the commitment of the agency to public health and safety. The doubts about radiation standards seemed likely to linger at least until the National Academy of Sciences and the Environmental Protection Agency completed their reviews of the subject.

Gofman's and Tamplin's contribution to the debate over radiation was more rhetorical than substantive. Their arguments and the attention they commanded helped prod reluctant commissioners to lower numerical design objectives for effluent releases in early 1970, and even more clearly, to override the Joint Committee's objections to the AEC's proposed amendments a year later. But it was the Livermore scientists' fervent blasts against the AEC that received the most notice and made the greatest impact on the public. While Gofman and Tamplin succeeded in helping prompt the AEC to tighten its regulations, they also needlessly alarmed the public with their implausible estimates of cancer potentially caused by nuclear power and their allegations that the AEC was indifferent to the hazards of radioactivity. They did not offer their views as constructive criticism but as immutable and incontestable truth. Gofman and Tamplin had some valid complaints about their treatment by the AEC, or more specifically, by the Livermore Laboratory, but this hardly seemed to justify the self-righteous, uncharitable, and often inaccurate allegations they leveled at the AEC and others who questioned their positions. Rather than enlightening the public or clarifying the complexities of radiation protection, they simplified and polarized an issue that even under the best of circumstances was difficult to understand and impossible to evaluate with certainty.

CHAPTER XIV

Environmental Law and Litigation

From NEPA to the Calvert Cliffs Decision

At the same time that the AEC was embroiled in contentious debates over thermal pollution, low-level radiation, and other environmental issues, it was establishing procedures to carry out the requirements of the National Environmental Policy Act (NEPA), which Congress passed in December 1969 and President Nixon signed into law with much fanfare on 1 January 1970. In its response to NEPA, the AEC attempted to strike a balance between the mandate for environmental protection embodied in the new law and its determination to avoid excessive delays and disruptions in the licensing process. As a result, it advanced a narrow interpretation of its responsibilities under NEPA, which aroused the protests of environmentalists and prompted legal action by them. In a suit over the AEC's environmental rules as they applied to the Calvert Cliffs nuclear power plants, under construction on the Chesapeake Bay in rural Maryland, environmentalists won a stunning and decisive victory. Ironically, the Calvert Cliffs ruling, by leading to a lengthy suspension of plant licensing, caused precisely what the AEC had tried to prevent when formulating its environmental regulations.

The National Environmental Policy Act was the culmination of months of legislative drafting, negotiation, and revision in Congress. The original bills that evolved into NEPA were introduced by Henry M. Jackson in the Senate and John D. Dingell in the House. Other members of Congress, especially Edmund S. Muskie, played a key role in shaping the final version of the act. Remarkably, the most comprehensive and far-reaching environmental measure that Congress had ever enacted became law with a minimum of hearings, lobbying, floor debate, or controversy. It vividly testified to the growing concern in Congress and the entire nation with the quality of the environment. It also reflected the recognition that the projects of federal agencies were contributing to environmental deterioration. In a rush of business before adjourning for the Christmas holidays in 1969, most members of Congress, anxious to pass legislation to protect the environment, did not pay close attention to the implications of the bill for which they were voting. Both houses approved NEPA overwhelmingly; the Senate did so without even a roll call vote.[1]

The basic purpose of NEPA was to require federal agencies to consider the environmental effects of their activities and to administer their policies and regulations in accordance with its provisions "to the fullest extent possible." The text of the law began by declaring "a national policy which will encourage productive and enjoyable harmony between man and his environment." NEPA instructed federal agencies to prepare a "detailed statement" for "major Federal actions significantly affecting the quality of the human environment." The law stipulated that the statements assess: the "environmental impact" of the proposed project; any adverse but unavoidable environmental effects of undertaking the proposed project; possible alternatives to the proposed project; the relationship between local, short-term advantages of the project and "the maintenance and enhancement of long-term productivity"; and finally, "any irreversible commitments of resources which would be involved in the proposed action." NEPA also created a statutory Council on Environmental Quality in the executive office of the president to evaluate the efforts of federal agencies to comply with the law.[2]

The act was a clear and concise statement of national policy, but it left unanswered a large number of questions about how its provisions would be implemented. It gave federal agencies broad discretion in deciding how to carry out their environmental mandate, and the boundaries of their new authority were vague and confusing. George F. Trowbridge, a prominent attorney in nuclear licensing cases, described it, in terms of legislative drafting, as "an atrocious piece of legislation" that was "poorly thought out and ambiguous at all the crucial points." He added that "the final product is an invitation to litigation for the next decade."[3]

Uncertainty about the meaning and the implications of NEPA extended to the Atomic Energy Commission. The AEC had monitored the progress of the environmental law through Congress with sentiments ranging from uneasiness to foreboding. The AEC's objections to the pending legislation were much the same as its concerns about proposals to give it authority over thermal pollution—agency officials feared that the measure would discriminate against nuclear power and delay licensing procedures. Commissioner Ramey discussed the AEC's position with Senator Jackson and members of his staff, and for a time Ramey thought that he had succeeded in exempting nuclear plant licensing proceedings from having to consider environmental impact statements. But other members of Congress, especially Muskie, were determined that the law would apply fully to the AEC's licensing process. Referring to the AEC and the U. S. Army Corps of Engineers by name in a speech to the Senate, Muskie complained that "these agencies have always emphasized their primary responsibility making environmental considerations secondary." In a compromise he reached with Jackson over jurisdictional and other issues, both senators agreed to language designed to subject the AEC to the legislation's requirements. "It is time," commented Jackson, "that [the] AEC be given a larger mandate against which to weigh the environmental impact of its planned and proposed activities."[4]

Despite the statements and the intentions of two of the leading proponents of the environmental bill, the AEC remained uncertain that the measure would apply to its licensing procedures. The general counsel of the agency, Joseph F. Hennessey, suggested that the language of the proposed legislation did not enlarge the AEC's regulatory authority beyond radiological effects or empower it to deny licenses on other environmental grounds. He told the staff director of the Senate Committee on Interior and Insular Affairs, which Jackson chaired, that the AEC opposed legislation that would force compliance by nuclear plants but not by fossil facilities. Such a law, he said, "might well distort utility judgments in selection of plant types and delay the early utilization of nuclear power." Hennessey urged that the bill be amended to postpone its applicability to licensing actions for two years, or to state categorically that it would not expand the regulatory jurisdiction of federal agencies. Congress ignored those recommendations, but even after final passage of the bill, the AEC continued to express the opinion that NEPA did not necessarily affect its licensing or rulemaking functions.[5]

The AEC's narrow interpretation of its responsibilities under NEPA was an initial and instinctive reaction that was hardly definitive. Like many other federal agencies, it undertook a careful examination of the measure's provisions to try to determine precisely what their impact would be on its programs. While the subject was under review, the AEC simply avoided making any statements about its authority on nonradiological environmental issues. In response to a congressional inquiry about thermal pollution, for example, Harold Price neither affirmed nor denied the AEC's jurisdiction, but blandly recited the obvious fact that NEPA had recently become law.[6]

The AEC's uncertainty about the regulatory implications of NEPA as well as its hope that the law's impact would be slight were apparent in an analysis that the general counsel's office presented to the commissioners in February 1970. In a summary of the issues that prefaced the paper, Howard K. Shapar, the assistant general counsel for licensing and regulation, concluded that NEPA required that the AEC perform at least two tasks: preparation of a detailed statement on the environmental impact of major regulatory actions, including granting construction permits or operating licenses and adopting new or revised rules; and review of the existing regulatory authority of the agency to determine whether it needed to be amended to enable full compliance with the new law. Other than those two items, Shapar found the mandate that NEPA provided the AEC to be ambiguous. It was uncertain, he maintained, whether or not the act expanded the regulatory jurisdiction of the AEC so that it would have to consider the environmental impact statements as a part of its licensing or rulemaking decisions. While arguing that "there is reasonable support on both sides of the question in the legislative history and in the statute itself," he believed that if taken to court, "it is probable that the question would be resolved in the affirmative."[7]

Even if the AEC anticipated such a court ruling by assuming that NEPA expanded its regulatory authority, Shapar pointed out, vital questions about the extent of its jurisdiction remained to be answered. The law stated that federal agencies should carry out its purposes "to the fullest extent possible," an ill-defined phrase that gave them broad discretion. Shapar contended that at least for the immediate future, an AEC decision not to deny or to place conditions on license applications on the basis of nonradiological environmental effects would be "reasonable and defensible." But he also argued that the AEC's legal position and its public image would be strengthened if it did more than simply file a statement on the environmental impact of the proposed facility. He suggested that the AEC require that licensees observe federal and state standards for environmental protection. This offered the advantage of taking action beyond the minimum to comply with NEPA but without imposing an "undue burden" on applicants. The chief disadvantage was that if licensees violated existing laws and regulations, the AEC could be expected to enforce the standards developed by other federal and state agencies.[8]

At a meeting on 20 February, the commissioners discussed the points that Shapar raised in his paper. Seaborg agreed with the recommendation that the AEC should direct licensees to abide by relevant federal and state environmental regulations. He argued that the agency should adopt this policy partly because "it is the right thing to do" and partly because "if we do not take such initiative, I feel it is inevitable that such action will be forced upon us." The other commissioners concurred. In early April, the AEC announced its plan to implement NEPA, at least on an interim basis. It proposed to add an appendix to Part 50 of Title 10 of the *Code of Federal Regulations*, which governed the licensing of power plants, fuel reprocessing facilities, and other nuclear installations. The amendment soon became known inside and outside the AEC by the shorthand term "Appendix D."[9]

Appendix D spelled out the procedures the AEC would follow in evaluating environmental issues. It would send copies of applications for construction permits or operating licenses for power reactors and reprocessing plants to other agencies with environmental expertise for comment. Drawing on their views as well as its own assessment of radiological effects, the regulatory staff would prepare a detailed environmental impact statement. The statement would become a part of the public record of the licensing review process. Appendix D also specified that the AEC would require that licensees observe federal and state environmental standards. It made clear, however, that the AEC did not intend to make independent judgments about the validity of those standards or the ways in which applicants for licenses proposed to meet them. As outlined in Appendix D, NEPA's impact on licensing procedures would be limited. "We are simply going to get the recommendations from these various agencies, which is pretty much what we've been doing," one AEC official explained, "and require—instead of just suggesting—that the applicants comply with whatever requirements those agencies impose."[10]

The AEC's first effort to clarify and codify its responsibilities under NEPA produced as many questions as it answered. The largest unresolved issue was how the AEC would weigh the environmental statements; it was unclear how the reports of other federal and state agencies would affect the deliberations of the Atomic Safety and Licensing

Boards. If the views of two agencies clashed in a licensing case, for example, how would the board handle the dispute? Would nonradiological environmental matters even be admissible, given the fact that the boards had traditionally refused to consider issues outside the AEC's statutory jurisdiction? Since the AEC insisted that NEPA had not substantially enlarged its authority and that it would not exercise independent judgment on nonradiological questions, Appendix D caused a great deal of confusion. "We're scratching our heads over what the hell they said," grumbled a Department of the Interior attorney. In a meeting with members of the regulatory staff, industry representatives expressed uncertainty not only about those points but about others as well. They sought guidance, for example, on whether or not license applicants were expected to prepare their own environmental impact statements and on the extent to which the AEC planned to enforce the federal and state standards it instructed licensees to observe.[11]

The prevailing confusion was compounded by passage of the Water Quality Improvement Act, signed into law on 3 April 1970. It required that applicants for federal licenses present certification from appropriate state or interstate agencies (or in the absence of adequate state regulations, the secretary of the interior) that the proposed facility could meet water quality standards. The measure allowed a grace period of up to three years for compliance by plants already under construction. Its enactment raised the question of how its provisions affected the broader environmental mandate that NEPA assigned.[12]

In May 1970 the AEC's regulatory staff revised the version of Appendix D it had published the previous month. It sought to clarify some of the points that had caused confusion, to declare its position on the implications of the water quality act, and to incorporate guidelines recently issued by the Council on Environmental Quality on the preparation of environmental statements. The updated Appendix D answered some procedural questions by directing applicants for construction permits to submit their own statement on the environmental consequences of the proposed plant to the AEC. The AEC, in turn, would send copies of the applicant's report to federal agencies designated by the Council on Environmental Quality and to state officials. It would also publish a notice of the availability of the report in the *Federal Register*. Federal agencies would have thirty days and state or local bodies sixty days to return their comments to the AEC. The regulatory staff would then draw up the detailed environmental statement required by NEPA. At the operating license stage, applicants would not need to submit another full environmental statement but discuss only the items that departed significantly from the original report. The revised Appendix D also declared that the Water Quality Improvement Act superseded NEPA on the issues it covered. With respect to water quality, including thermal pollution, applicants would be expected only to show that they had secured certification of compliance with applicable standards.[13]

The AEC's new statement, issued on 3 June 1970 for public comment as a step toward a final rule, addressed some of the sources of confusion about its environmental policy but left the major ones unresolved. It was notable, one observer remarked, "for what it didn't say rather than what it did say." It did not make clear the AEC's view of the extent of its jurisdiction on environmental matters. Although it cited the Council on Environmental Quality's guidelines on procedural issues, it pointedly ignored the Council's declaration that the only limitation on the authority of federal agencies to implement NEPA was an express prohibition in existing law. The role of the environmental impact statements in the licensing process remained undefined, as did the steps the AEC would take to enforce the standards of other governmental units, whether federal, state, or local. As *Nucleonics Week* commented: "The precise AEC stance on the environmental issue remains fuzzy."[14]

The AEC's vagueness about its position on implementing NEPA was partly intentional. It wanted to avoid adopting a definitive policy that could limit its options in the future. Until it decided on a course of action or was forced to extend its jurisdiction, it preferred to maintain as much flexibility as possible in its approach to environmental issues. But the agency's lack of clarity was more a result of its own uncertainty about how to carry out NEPA. The AEC, one unnamed observer remarked, "doesn't have the slightest idea yet on how they're going." Although it was more prompt than other federal agencies in drafting its response to NEPA, it was also more cautious and restrictive in defining its responsibilities.[15]

There were a number of reasons that the AEC did not take a bolder and more expansive view of its environmental mandate under NEPA. One was its conviction that the routine operation of nuclear plants was not a serious threat to the environment. AEC officials acted on their deep-seated belief that nuclear power, compared to burning fossil fuel to produce electricity, was beneficial to the environment. Further, regulation of the two major products of nuclear power that affected the environment, radiation releases and thermal discharges, were covered by other legislation. Therefore, NEPA, while expressing admirable objectives, seemed to be of rather peripheral importance to the AEC. In fact, aggressive implementation of the law could undercut other goals that the AEC deemed more pressing, especially evaluating the safety of reactors and issuing licenses for them. "So much emphasis has shifted to public and official concern about pollution and the environment, that the proportion of attention . . . to safety matters has greatly decreased," Clifford Beck told a congressional briefing on nuclear power in June 1970. "One major . . . reason the AEC was set up in the first place was to assure the safety of the public from the real hazard of major dimensions that could arise from accidents in nuclear reactors. This was—before pollution, is now, and will continue to be the objective on which a major portion of our regulatory effort is expended."[16]

Beck's statement reflected the AEC's concern that NEPA would divert limited human resources from tasks that were more central to the agency's mission. Even before the law was passed, the regulatory staff was "all but overwhelmed" by its workload in processing the unprecedented number of applications for construction permits and operating licenses it had received. Although the size of the regulatory division had grown by about 50 percent between 1965 and 1970, it could not keep pace with increasing demands. In April 1970, the AEC was reviewing twenty-seven applications for construction permits and twenty-two more for operating licenses. At the same time, it faced expanding obligations in its quality assurance programs. The Nixon administration, however, was trying to cut back on the size of the federal bureaucracy and was not sympathetic to appeals for the creation of new positions. In this situation, the prospect of having to devote considerable staff time to environmental reviews, monitoring, and liaison with other agencies was distressing to the AEC.[17]

The training and expertise of the regulatory staff focused on reactor science and engineering rather than the general range of environmental problems that NEPA encompassed. To the AEC, it made sense to rely on the technical judgment of federal and state agencies with knowledge and experience in fields other than nuclear safety. The AEC did establish a small Office of Environmental Affairs, reporting to the general manager, in June 1970. Its function was to coordinate and monitor agency activities, not to perform the wide variety of the AEC's expanded environmental responsibilities. In light of its personnel shortages and limited experience, the AEC wanted to keep its environmental tasks within confines as narrow as possible.[18]

The AEC's view of environmental problems and its staff limitations, both qualitative and quantitative, were important considerations in its response to NEPA, but they were not the controlling ones. The decisive factor was the agency's fear that the ultimate result of weighing environmental issues would be unwarranted delays in licensing plants. The extra workload on the regulatory staff and the additional time it would take to evaluate safety issues could extend the already increasing amount of time that applicants had to wait for an AEC review. By placing time limitations on the comments of other agencies and by restricting its own role, the AEC hoped to prevent major new disruptions in the licensing pipeline. As it was, the average time required for the review of an application for a construction permit had increased from about ten months in 1967 to about eighteen months in 1970. The AEC worried that NEPA would cause what Ramey called a "quantum leap" in the length of the process. The agency's draft rules on implementing NEPA were an effort to strike a balance between environmental concerns and energy needs. In the AEC's view, meeting the demand for electricity was a more important and immediate problem than carrying out all the conceivable ramifications of NEPA.[19]

The AEC's proposed rule elicited objections from critics whose priorities and perspectives were different. Several environmental groups contended that the regulation did not adequately account for the environmental impact of nuclear plants under construction. They petitioned the AEC to demand that utilities file an environmental report as soon as possible and to order that backfitting be carried out if it would provide additional environmental protection. Other observers faulted the agency for refusing to undertake an independent evaluation of the possible effects of plant construction and operation. The primary, though not the only, concern in this regard was the AEC's interpretation of its responsibilities under the Water Quality Improvement Act. Its argument that applicants for construction permits need only submit certification from an appropriate agency to fulfill the requirements of both the water quality act and NEPA aroused sharp condemnation.[20]

Congressman Joseph Karth of Minnesota, a prominent participant in the controversy over the Monticello plant, complained that "the AEC seems to be demonstrating a rather unique arrogance" that "raises the question whether the AEC considers itself above the law of the land." The New York Times protested in an editorial that the AEC "has the notion that in licensing nuclear plants it has no authority even to consider a threat of thermal pollution." It suggested that "the AEC should be made accountable for its decisions affecting the environment before scores of

nuclear plants further endanger the life of the nation's waters." Senator
Philip A. Hart of Michigan also objected to the AEC's reliance on the
water quality act because it included a three-year grace period for compliance. He was particularly worried about the impact of the Palisades plant
on Lake Michigan, and expressed his dismay with the prospect that
during that time the objectives of NEPA would be disregarded.[21]

The AEC attempted to alleviate the confusion over its environmental
policies and to address at least some of the criticism of them as it
prepared the final version of its regulation on implementing NEPA. A
paper from the regulatory staff and the general counsel's office informed the commissioners that public comments on the policy statement published in June were sharply divided. Utilities and reactor vendors argued that the AEC had gone too far to comply with NEPA and
that the result would be licensing delays. Environmentalists, in contrast,
suggested that the AEC's proposed position was an inadequate response
to the mandate provided by NEPA. It remained uncertain if the AEC's
position would survive a legal challenge, but general counsel Hennessey
suggested that in light of several recent decisions on the scope of the
law, "it is likely that a reviewing court would hold that our proposed
policy statement does not go far enough in carrying out the directive of
NEPA."[22]

The staff paper proposed four alternative courses of action that the
Commission could elect to follow in its final policy statement. The first
was to adopt the rule largely as originally published. The major change
would be to state explicitly that environmental impact statements
would not be considered at Atomic Safety and Licensing Board hearings
on applications for construction permits or operating licenses. The primary advantage of this option would be to limit the opportunities for
delay of the licensing process and to ease the burden on the staff. The
principal disadvantages would be the likelihood of losing a battle in
court over NEPA and aggravating the already unfavorable public image
that the proposed Appendix D had given the AEC. The second alternative was the same as the first except that it would require applicants and
the staff to discuss water quality in their environmental statements.
Even if the AEC took no action on water quality, including it in the
reports "might serve to alleviate or diminish the criticism of AEC that
otherwise may be expected."

The third alternative that the staff outlined would broaden the AEC's
consideration of nonradiological environmental issues. Any party in a
licensing proceeding would be permitted to raise questions about

whether or not there was "reasonable assurance" that an applicant would comply with federal and state standards. In this case, the board would weigh any adverse effect on the environment that seemed likely against the benefits of the proposed plant, "including the need for an adequate supply of electric power." Under this option, the AEC would not evaluate water quality issues but would rely entirely on state certification. The main advantage of this option, the staff argued, was that it would "reflect a higher degree of responsiveness to the requirements of NEPA," and in that way, reduce the chances of an adverse court decision. It could also "contribute favorably to AEC's public image by providing a more open forum for the airing and resolution of environmental matters." The disadvantages were the increased potential for retarding the licensing process and more work for the staff. The fourth alternative was the same as the third with the addition of considering water quality a part of the AEC's responsibilities under NEPA.[23]

The commissioners discussed the four alternatives at a meeting on 14 October. They agreed that the third alternative was the best course to follow, assuming that the Department of Justice supported the AEC's view of its authority (or lack of it) under the water quality act. They asked the staff to prepare yet another paper on the subject before they adopted it as the final version of Appendix D. Seaborg thought that the Commission had interpreted its responsibilities on environmental matters "rather broadly." For the first time it had unequivocally included nonradiological environmental issues within the AEC's jurisdiction. But the option that the commissioners favored still fell short of what AEC critics were urging, and it did not envision a decisive role for NEPA in the licensing process.[24]

In response to the Commission's request, the staff submitted a paper detailing the new procedures. It made clear that nonradiological environmental matters could be raised during licensing proceedings. The AEC would not submit its own assessment of the validity of an applicant's assurances that a proposed plant could meet federal or state standards, but if disputes arose, they would be resolved by the licensing board. Water quality issues would be discussed in the environmental impact statements of the applicant and the regulatory staff, though they would not be considered during licensing hearings. The AEC would reject appeals that work on plants under construction be suspended until an environmental statement had been approved and that plants under construction or in operation be required to backfit equipment for environmental purposes.[25]

The latest attempt to delineate the AEC's treatment of environmental issues generated considerable debate among the commissioners. Ramey was concerned about the impact of the proposed rules on plants being built. Theos Thompson was adamantly opposed to requirements that would further encumber the licensing process. He was particularly apprehensive about the effects of considering subjective matters such as noise or aesthetic qualities. He asked at one meeting what the AEC would do in the hypothetical case of an elderly lady who objected to the hum of electrical transformers or complained that a nuclear plant looked like "a huge phallic symbol." His examples were meant to be humorous but his point was not. The commissioners and senior staff decided to meet Thompson's objections by tightening the language in the draft to lessen the possibility of subjective issues arising in board hearings. They realized, however, that this remained a murky problem that would have to be dealt with on a case-by-case basis. To curb Ramey's uneasiness about the effect of the new Appendix D on plants under construction, they extended the grace period before environmental issues could be raised in licensing hearings. With those changes, Thompson and Ramey reluctantly agreed to vote for the proposed environmental policy.[26]

After months of confusion, criticism, and controversy, the AEC published its revised version of Appendix D on 4 December 1970. Agency officials hoped that it would meet the requirements of NEPA and win the endorsement of reviewing courts without adding major obstacles to the licensing process. The AEC's "main concern," the statement declared, "has been to find out and strike a reasonable balance" between the demands for electricity and environmental protection. Appendix D reiterated that the AEC would rely on federal and state agencies to evaluate the environmental impact of a proposed plant. It expressed the AEC's view that the two principal effects of nuclear power generation, radiation and thermal discharges, lay outside the jurisdiction of NEPA. It also specified that environmental issues under NEPA could only be raised in licensing proceedings for which a notice of hearing was published after 4 March 1971. This was done "in order to provide an orderly period of transition" and "to avoid unreasonable delays in the construction and operation of nuclear power plants." Within those constraints, the new Appendix D imposed several requirements on applicants for permits or licenses. They had to file a detailed environmental impact statement, which could be challenged in a licensing board hearing if a party to the proceeding thought it misrepresented the consequences of building the plant. The board would resolve any disputes between parties, and could grant, deny, or place conditions on a permit or license on the basis of environmental considerations. Owners of plants already under construction should file an impact statement "as soon as practicable" if they had not already submitted one.[27]

The AEC's policy statement received strong support from a number of interested observers; others were less enthusiastic. The Joint Committee and the Justice Department expressed approval of the purposes of the regulation and the AEC's interpretation of its legal authority. Timothy Atkeson, general counsel of the Council on Environmental Quality, said he was "delighted" with the new Appendix D and conveyed the Council's opinion that "the revised AEC statement was by far the best policy statement that any Agency had issued." Industry representatives, by contrast, voiced misgivings about the new rule because of its potential for lengthening the licensing process. Environmentalists were pleased that it included some of the provisions they had sought, but they also concluded that it did not go far enough in fulfilling the objectives of NEPA.[28]

The reservations of environmentalists about the AEC's policy were apparent in hearings that one of their leading spokesmen in Congress, John Dingell, held on the implementation of NEPA. Dingell told Ramey that he was disturbed that the AEC planned to rely on the judgment of other agencies on the environmental impact of proposed plants. "If this is the only thing that AEC is going to [do]," he declared, "then let me assure you that AEC is in great trouble." Ramey responded that it was a practical way for the AEC, lacking adequate staff and expertise, to carry out NEPA, but Dingell was unconvinced. The AEC fielded the same question in a meeting it held for representatives of twelve prominent conservation organizations in January 1971. The session was cordial and constructive, but left no doubt that environmentalists remained skeptical of the AEC's approach to NEPA.[29]

One leading environmental group, the Natural Resources Defense Council, spelled out other concerns in detailed comments on Appendix D. It faulted the AEC for several aspects of its plans to gather information about and admit public comments on the environmental effects of a proposed plant. The organization criticized, for example, the limited number of federal and state agencies from whom the AEC would seek advice, the failure to publish a list of the agencies the AEC would consult on a given licensing case, the assumption that an agency supported the issuing of a permit or license if the AEC did not receive comments from it, and the lack of a provision for public hearings on the environmental impact of plants under construction. The complaints of the Natural Resources Defense Council centered not on those procedural matters, however, but on the AEC's lack of attention in Appendix D to enforcement of NEPA requirements. This applied to the AEC's plan to accept the standards of federal and state agencies without attempting to evaluate their adequacy or completeness, and to the lack of a mechanism to ensure compliance by nuclear plants with environmental regulations. "The AEC should include . . . specific provisions which describe in detail the methods by which it will monitor and secure compliance with the permit and license conditions for environmental protection." The comments of the Natural Resources Defense Council, like those of environmentalists generally, demonstrated how little confidence they had in the AEC. Appendix D required a measure of faith that the AEC would carry out its responsibilities under NEPA, but most environmentalists lacked such faith. They suspected the AEC's intentions and disagreed with its priorities, and therefore, wanted Appendix D to spell out issues clearly to minimize the discretion it would allow the AEC.[30]

The differences between the AEC and environmentalists over NEPA inevitably led to court. The specific case on which the controversy focused involved the twin Calvert Cliff plants being built by the Baltimore Gas and Electric Company (BG&E) on the Chesapeake Bay. The site was about forty-five miles southeast of Washington, D. C. in a sparsely populated area near the hamlet of Lusby, Maryland. When complete, the 880 megawatt units would stand at a point where the bay was six miles wide and a maximum of 110 feet deep. By any standard, the tract was a scene of exceptional natural beauty.[31]

At first, BG&E's plans for the Calvert Cliffs plants attracted little interest or opposition. A prominent environmental group, the Chesapeake Bay Foundation, even offered to cooperate with the company to provide the public with objective information about the facilities. By early 1969, however, when BG&E was applying to the AEC for a construction permit, the foundation had joined with other environmentalists in raising questions about the possible impact of the Calvert Cliffs plants. They based their objections on what by that time were increasingly familiar considerations. They suggested that AEC regulations did not necessarily afford adequate protection against radiation hazards and that thermal pollution posed an undefinable but serious ecological threat to the bay. A group of seven Johns Hopkins University scientists spearheaded the opposition, urging that the AEC defer issuing a construction permit until more information about the effects of the plants on the bay was available.[32]

The AEC did not address the questions about thermal pollution because at that point it still considered them to be outside its statutory jurisdiction, and it denied that radiation from the plants presented a public health or environmental danger. In testimony before the licensing board for the plants and in subsequent statements, the Johns Hopkins scientists had stressed the hazards of tritium, a radioactive isotope of hydrogen. They asserted that tritium was particularly worrisome because as it moved through the food chain its concentrations in living organisms would increase. The ultimate result would be that people who ate seafood from the Chesapeake could ingest a much higher level of tritium than would seem possible from a simple calculation of the amounts that the power stations would release into the bay. After investigating this argument, the AEC found no evidence that tritium became concentrated in living organisms; it insisted that the levels of the isotope from the Calvert Cliffs units would remain only a few millionths of what radiation experts considered acceptable. The licensing board found the position of the intervenors unconvincing, and on 30 June 1969 it authorized the issuance of the construction permits for the plants.[33]

The licensing board's action only intensified the controversy over the plants' effects on the bay; many Marylanders agreed with Edward P. Radford, one of the Johns Hopkins scientists, that the Calvert Cliffs hearing had been "pure window-dressing." A few days after BG&E received its permit, the governor of Maryland, Marvin Mandel, announced the appointment of a seventeen-member commission to study the impact of nuclear power plants on the environment. Chaired by William W. Eaton, a prominent industrial consultant and nuclear physicist, the commission included academic scientists and administrators, state and federal officials, and businessmen. In its report, published in December 1969, it sharply criticized procedures that allowed the utility to begin work on the plants before receiving all the required state and federal authorizations. But the commission concluded that the operation of the plants would not "seriously impair the quality of the Chesapeake Bay environment." It argued that the consequences of thermal pollution would be "very small," and it agreed with the AEC that tritium and other radioactive isotopes presented no major health hazards.[34]

At the same time that the governor's task force was preparing its report, the Maryland Academy of Sciences was conducting a study on the same general topic. It reached a similar conclusion, submitting that "in all probability, the plant will not of itself represent a major environmental threat." It expressed uneasiness, however, about the future use of the bay as a site for power stations, and it urged a moratorium on the construction of any other plants until appropriate environmental studies could be carried out and evaluated. The academy's report, published in January 1970, also advised that the state restrict radiation emissions from the Calvert Cliffs facilities to one-tenth of the AEC's permissible levels. Those recommendations reflected the widespread concern that even if one nuclear plant on the bay was not a problem, a proliferation of them would be. The Maryland Department of Water Resources, in granting BG&E a water use permit, went even further than the Academy of Sciences suggested in seeking to reduce the impact of the plants on the bay. It set allowable radiation releases at 1 percent of the AEC's standards, though the state's legal authority to do so was dubious. It further stipulated that the utility would have to meet temperature limits for waste heat at the point of discharge without using a mixing zone.[35]

The reassuring expert assessments of the probable environmental impact of the Calvert Cliffs plants and the stringent regulatory requirements imposed by the state might have been enough to allay public concern and end the controversy. But the enactment of NEPA and the AEC's response to it kept Calvert Cliffs in the center of the debate over nuclear power and the environment. In June 1970, less than four weeks after the AEC issued its draft of Appendix D for public comment, three environmental organizations joined in an appeal to the AEC to force BG&E to comply immediately with NEPA. The petitioners included two nationally recognized and long-established groups, the National Wildlife Federation and the Sierra Club, and the recently founded Calvert Cliffs Coordinating Committee, a coalition of individuals and conservationist organizations "specifically concerned with the environmental effects of the proposed Calvert Cliffs plant." The environmental groups had initially planned to sue the AEC and other agencies over the thermal pollution that the plants would cause in the bay. They hired Anthony Z. Roisman, a thirty-two year old Harvard Law School graduate who along with two other young lawyers had set up a public interest legal firm the previous year. Roisman, who at that time had only limited experience in environmental law, pointed out that a major suit could be extremely costly and recommended, as a

18. University of Maryland researchers check the size, number, and sex of crabs captured in the Chesapeake Bay to gather data for assessing the environmental impact of the Calvert Cliffs nuclear plants. (National Archives 434-SF-11-83)

first step, submitting a rule-making petition to the AEC based on the requirements of NEPA.[36]

The petition requested that the AEC apply NEPA to plants under construction, require backfitting if necessary on those plants, and order BG&E to show cause why its construction permit should not be suspended until the the environmental impact of Calvert Cliffs could be fully investigated. Roisman argued that NEPA obligated the AEC to assess the environmental impact of plants under construction. As written, Appendix D required consideration of the environmental effects of those facilities only when their owners applied for an operating license. Roisman contended that to wait that long violated NEPA because it required that "the consideration of environmental factors is no longer to be postponed until irreparable harm results or economic pressures preclude a full evaluation of alternatives." Suggesting that the "one inescapable conclusion . . . is that virtually nothing is known about the environmental impact of the proposed Calvert Cliffs plant," he called on the AEC to direct BG&E to show cause why construction should not be suspended. Once the issue was decided, "as we are sure it will be," in favor of halting construction, the AEC could undertake a detailed study of environmental issues, hold public hearings, instruct the utility to prepare an environmental statement, and, if necessary, order modifications in or relocation of the plants. The AEC denied the petition on the grounds that it was in the process of drawing up its rules on implementing NEPA. Although it pledged to consider the points raised by the petitioners as a part of its rule-making proceedings, it deferred any action on their request until after the final regulations had been published.[37]

In late November 1970, the Calvert Cliffs Coordinating Committee, the National Wildlife Federation, and the Sierra Club requested that the U. S. Court of Appeals for the District of Columbia Circuit review the AEC's refusal of their appeal for environmental action on Calvert Cliffs. Two weeks later, a few days after the AEC issued its final version of Appendix D, the three organizations challenged the agency's rules in a suit filed in the same court. The two suits, which the court combined into a single case, not only reiterated the earlier arguments for immediate consideration of the environmental effects of the Calvert Cliffs plants but also disputed the AEC's entire approach to carrying out NEPA. The questions they raised, noted *Nuclear Industry*, were "fraught with far-reaching consequences for nuclear power plant licensing."[38]

The petitioners' brief, written by Roisman, elaborated on previous environmentalist complaints with the AEC's response to NEPA. It emphasized that the AEC had failed to fulfill the purposes of the act because it planned to rely on the standards of other agencies in evaluating environmental issues. Roisman jabbed the AEC for its willingness to accept the judgment of the states on environmental matters after fighting so hard against the efforts of Minnesota to set its own radiation standards. A vital question that the AEC left unanswered, he suggested, was whether state or local "standards and requirements by their very existence will totally foreclose any examination of adverse environmental effects which will occur even when the standards and requirements are met." The petitioners also objected to the AEC's refusal to require backfitting on plants under construction for equipment or design changes that would improve environmental protection and its provision that environmental issues under NEPA could only be raised in licensing cases where a notice of hearing was published after 4 March 1971. In short, they argued that Appendix D fell far short of full compliance with NEPA.[39]

In its brief, the AEC, as it had done in publishing the final version of Appendix D, stressed that it was seeking to take a "balanced approach" between environmental and energy needs. It argued that the nation faced a serious shortage of electrical power that necessarily entered into its policy deliberations. Since NEPA specified that environmental goals should be pursued in a way that was "consistent with other considerations of national policy," the AEC was following the dictates of the law. The AEC once again pointed out that the major environmental effects of nuclear plants were covered by statutes other than NEPA. It dismissed the allegations of the petitioners that it was doing nothing about environmental problems as unsubstantiated "hyperbole." The brief further contended that the AEC's plan to utilize the evaluations of other federal and state agencies on environmental issues was "wholly reasonable," especially since its own expertise focused heavily on radiological health and safety. Anyone who wished to challenge the standards or recommendations of those agencies had the opportunity to do so in the licensing hearings. On another matter raised by the petitioners, the AEC declared that backfitting for environmental purposes, if found necessary, was covered by regulations other than Appendix D. Finally, it insisted that deferral of formal consideration of environmental issues other than radiation and water quality was necessary to avoid power shortages.[40]

The AEC's brief made the strongest possible case for its plans to implement NEPA. The general counsel's office had pointed out from the beginning of the AEC's deliberations over NEPA that it might lose a court challenge, but senior agency officials hoped that they had broadened Appendix D sufficiently to win the case. They were particularly encouraged by the Council on Environmental Quality's warm reception of the final version of the rule. Staff lawyers, however, feared that the

AEC's arguments would not fare well in court because the brief emphasized policy considerations rather than legal precedents. The concerns arising from the frailty of the AEC's legal position were heightened by the composition of the panel of judges selected to decide the case. The judges, J. Skelly Wright, Edward A. Tamm, and Spottswood W. Robinson III, seemed likely to give the environmentalists a sympathetic hearing. "The luck of the draw was with us," Roisman commented later. "We got just a super panel."[41]

The AEC's presentation during the oral arguments before the panel, held on 16 April 1971, did not improve its chances of winning the case. Marcus A. Rowden, who as AEC solicitor made the oral argument, heavily emphasized the need for power and the importance of avoiding delays in licensing plants, at least partly to compensate for the weakness of the AEC's legal stance. Judge Wright became visibly annoyed when Rowden suggested that the AEC was specifically prohibited from imposing water standards stricter than those approved by the Federal Water Quality Administration, but failed to cite any statutory language that would support such a conclusion. Although the oral arguments were probably not decisive in the outcome of the suit, they clearly did not enhance the AEC's case. Within a short time, *Nucleonics Week* reported that lawyers following the case thought it likely that the AEC would lose. In that event, the impact of the decision would depend on whether the court accepted any of the agency's major arguments and "on the way in which the court couches its ruling."[42]

The court's decision, handed down on 23 July 1971, was a crushing defeat for the AEC. Not only did the ruling categorically reject the agency's arguments, but it did so in language that was extraordinarily harsh. Judge Wright, who wrote the opinion, opened by declaring that the court's duty was to make certain that "important legislative purposes, heralded in the halls of Congress, are not lost or misdirected in the vast hallways of the federal bureaucracy." Suggesting that the AEC's policies undermined the objectives of NEPA, he stated the court's position: "We conclude that the Commission's procedural rules do not comply with the congressional policy."

Wright explained why the court found each of the AEC's arguments unconvincing. In his most widely quoted phrase, he faulted the agency for requiring environmental statements but not necessarily considering them during the licensing process: "We believe that the Commission's crabbed interpretation of NEPA makes a mockery of the Act." The law, he wrote, required that the AEC take the initiative to conduct independent evaluations of the environmental effects of proposed plants. "Its responsibility is not simply to sit back, like an umpire, and resolve adversary contentions at the hearing stage." Wright's opinion applied the same reasoning to the AEC's reliance on the environmental standards of other agencies. He agreed that their views should be solicited, but he denied that NEPA authorized a "total abdication to those agencies." Wright rejected the AEC's contention that certification from appropriate federal or state officials carried out the intentions of NEPA on water quality issues, and he affirmed that nothing precluded the AEC from setting standards that were stricter than those of the certifying agencies. He stressed that NEPA obligated the AEC to weigh the benefits of a proposed plant against the environmental costs of building it. By "abdicating entirely to other agencies' certifications," the AEC "neglects the mandated balancing analysis."

Judge Wright sternly reproached the AEC for its refusal to consider nonradiological environmental issues in proceedings for which a notice of hearing was published before 4 March 1971. "Such a time lag," he said, "is shocking." While some delay in implementing NEPA regulations was justifiable, it was clear that the period of transition "must proceed at a pace faster than a funeral procession." Wright did not think that the AEC's emphasis on the need for power was a defensible basis for deferring NEPA requirements. "Whether or not the spectre of a national power crisis is as real as the Commission apparently believes, it must not be used to create a blackout of environmental consideration in the agency review process." In the specific case of the Calvert Cliffs plants, Wright also found the AEC's environmental actions deficient. He saw no reason to wait until the operating license stage to weigh environmental issues, and suggested that the AEC "consider very seriously the requirement of a temporary halt in construction . . . and the 'backfitting' of technological innovations." Although this would cause delay, it would carry out more promptly and more completely the purposes of NEPA. Thus, the court's decision and Wright's opinion rejected not only the AEC's legal arguments but questioned its efforts to balance environmental and energy needs as well.[43]

The court's ruling did not come as a surprise to those who had followed the case, but the tone of Wright's language and the totality of the AEC's defeat were unexpected. The decision was, reported *Nucleonics Week* in a comment representative of the general reaction, "a stunning body blow." Once the initial shock wore off, the primary question facing the AEC was how to respond. There were four different courses of action it could follow, none of which was very appealing. It could seek review of the case by the Supreme Court or ask for a rehearing *en banc* by the full Court of Appeals. But the chances of a reversal of the decision seemed remote, and in the period that an appeal was in process, construction of nuclear plants would be plagued with uncertainty and challenged in litigation. Another alternative was to seek a legislative remedy by amending NEPA, but the prospects for success did not appear hopeful.[44]

The remaining choice was to comply with the ruling, which the Joint Committee, the Council on Environmental Quality, and eventually the AEC viewed as the best way to proceed. The AEC delayed taking action, partly because it wanted to weigh its alternatives carefully but mostly because of the appointment of a new chairman and commissioner. Two days before the court decision, President Nixon had disclosed that James R. Schlesinger, assistant director of the Office of Management and Budget, would replace Seaborg, and that William O. Doub, chairman of the Maryland Public Service Commission, would fill a vacant seat on the Commission. Both Schlesinger and Doub were determined to make the AEC more responsive to environmental concerns and to improve its public image. As a part of those efforts, the AEC announced on 26 August 1971 that it would not appeal the Calvert Cliffs decision.[45]

Arriving at that conclusion was easier than carrying out its ramifications. The environmental requirements that *Calvert Cliffs* imposed on the AEC placed an enormous burden on the regulatory staff. It was obvious that rewriting Appendix D, reviewing license applications, revising environmental impact statements, holding additional hearings, and training new staff members would cause major delays in the regulatory process. Shortly after the court decision, Harold Price told the Commission that it would affect sixty reactor cases involving eighty-three units. He hoped that if he added sixty technically qualified people to his staff he could hold the delays to six months. Less optimistic predictions from the staff and from industry envisioned a *de facto* licensing moratorium of up to a year.[46]

The first priority was to revise Appendix D in accordance with the directives of the court. Within six weeks after the decision, the AEC published a new version of the rules. It delineated the procedures that applicants for new construction permits would be expected to follow to carry out NEPA. One new feature that the AEC required in an applicant's environmental impact statement was an analysis of the consequences of several categories of postulated nuclear accidents. The regulatory staff ranked accidents according to severity in nine classifications. They ranged from trivial occurrences (class one) to hypothetical accidents that exceeded the design bases for the plant (class nine). The AEC concluded that the consequences of a class one accident were too slight to affect the environment and that a class nine accident was so improbable that applicants did not have to discuss its potential impact. The AEC requested an evaluation of the probability and the environmental consequences of all the other classes of postulated accidents. The agency also instructed applicants to draw up a detailed comparison of the environmental costs and benefits of building proposed plants. It recognized that estimating the environmental effects of projected accidents and balancing costs and benefits of new facilities would be unavoidably imprecise and speculative, but it decided that such calculations were necessary to comply fully with NEPA.

In assessing an applicant's impact statement, the licensing staff would independently consider every possible environmental effect of the proposed plant and prepare its own cost-benefit analysis. It would give its opinion to the licensing board about whether the application should be issued, denied, or revised on environmental grounds. The new Appendix D required additional NEPA reviews for plants under construction and allowed for new hearings in those cases. It also instructed affected permit holders or licensees to show cause why construction or operation of their facility should not be suspended until completion of the NEPA evaluation.[47]

On 27 August 1971, the AEC held a meeting with officials of industry, government agencies, and environmental groups to explain and solicit comments on a draft of Appendix D. The majority of those attending represented the nuclear industry, and they were, according to one report, "as enthusiastic about the task ahead as patients awaiting major surgery." They were particularly concerned that the new rules would cause major delays in plants that were nearly ready for operation. John T. Conway, executive assistant to the chairman of the board of Consolidated Edison, complained that new environmental reviews and hearings would postpone the start-up of the almost-completed Indian Point II reactor. This, he said, would jeopardize the well-being of eight million New Yorkers who needed electricity from the plant, and, ironically, force the company to use older equipment that was much more damaging to the environment. He and others at the meeting urged the AEC to issue operating licenses to plants in the position of Indian Point II while the environmental reviews were proceeding. Anthony

Roisman responded to this suggestion by declaring that environmental organizations would strongly oppose any such action. The tone and the substance of the meeting made clear that while the Calvert Cliffs decision marked the beginning of a new phase in the regulatory process, it did not signal an end to the controversy over nuclear power and the environment.[48]

In retrospect, the AEC committed a major blunder by not taking a broader view of its environmental responsibilities at the time that NEPA became law. Other agencies did so; the U.S. Army Corps of Engineers, for example, whose developmental functions were even more unambiguous than those of the AEC, responded much more willingly and expansively to the law.[49] The AEC's narrow interpretation of its duties under NEPA not only aroused the opposition of environmentalists but also undermined its own objectives. It sought above all to prevent the law from impeding the licensing process, but the Calvert Cliffs decision led to delays that were more disruptive and disorganized than would have been likely if the agency had immediately complied fully with NEPA. In the end, the harvest of its reluctance to fulfill the requirements of NEPA was a great deal of ill-will, further suspicion that it was indifferent to environmental quality, an embarrassingly one-sided setback in federal court, and an involuntary licensing moratorium.

A shock as unsettling as the Calvert Cliffs decision was probably necessary to shake the AEC out of its complacency toward environmental issues. Contrary to popular perceptions, the AEC was not insensitive to environmental needs, but it was convinced that it was doing its part to protect the environment. In its view, NEPA was at best a nuisance and at worst a threat to other goals that it deemed more pressing. The AEC had compelling reasons for not proceeding more aggressively to carry out NEPA. Its staff shortages were an ever-present problem and power supplies were increasingly problematical in some areas. The principal effect of *Calvert Cliffs* on the AEC was to force it to seek ways to deal comprehensively with environmental questions at the same time that it was striving to issue licenses promptly. In a time of growing public concern about the environment, the AEC could no longer confine its regulatory reviews to radiation protection without eliciting sharp criticism. NEPA and *Calvert Cliffs* thrust the AEC, gradually and grudgingly, into an era of environmental awareness and anxiety in which full consideration of the impact of power plants on their natural surroundings was a strict imperative.

CHAPTER XV

The Public and Nuclear Power

The AEC and the nuclear industry devoted careful attention and considerable resources to winning public support for the use of nuclear power. When the technology was still new and few plants were operating, the prospects for securing widespread public approval were auspicious; the introduction of nuclear power occurred with little protest and with overwhelmingly favorable press notice. As the number of nuclear plants that were planned, under construction, or operating expanded during the 1960s, however, public attitudes toward the technology became more ambivalent. Although opinion polls continued to show strong backing for nuclear power, a growing legion of critics and a profusion of books, articles, speeches, and demonstrations voiced questions about the safety of the technology and the regulatory performance of the AEC. The AEC and industry representatives countered by intensifying their efforts to persuade the public of the advantages of nuclear power. But to their dismay and frustration, they found that their campaigns achieved limited results and that the arguments of their opponents were eroding the public support that they viewed as essential to the future of the technology.

In the early days of nuclear power development, public attitudes toward the technology were highly favorable, as the few opinion polls published on the subject showed. In a February 1956 survey, 69 percent of those questioned had "no fear" of having a nuclear plant located in their community, while only 20 percent expressed apprehension. In the spring of 1960 a national telephone survey asked for responses to the proposition that "atomic power should be used to produce electricity." Sixty-four percent of those questioned gave positive responses and only 6 percent were negative. At about the same time, Consolidated Edison commissioned a sampling of public opinion in the region of its first Indian Point reactor, then under construction. It found that 63 percent of the respondents thought that the plant was a "good idea" and only 12 percent "felt any reservations."

Press coverage of nuclear power, though not as naively exuberant as reports of the prospective applications of the peaceful atom that appeared immediately after Hiroshima, continued to emphasize the benefits it would offer. This was evident, for example, in Walt Disney's widely viewed 1956 film, "Our Friend the Atom." A lengthy and lavishly illustrated article in *National Geographic* two years later acknowledged that the use of nuclear energy was costly and potentially hazardous but conveyed little doubt that those problems would be overcome. It concluded that "abundant energy released from the hearts of atoms promises a vastly different and better tomorrow for all mankind." The journal *Nucleonics* later observed that the indications of strong support or, at worst, widespread indifference, fed a prevailing conviction within the nuclear industry "that public acceptance of nuclear power would not be a significant problem."[1]

In the late 1950s and early 1960s, however, as a result of the controversy over fallout from the atmospheric testing of nuclear weapons, the American public became increasingly troubled about the risks of exposure to radioactivity. The possibility that radiation from bomb tests could cause birth defects, cancer, and other diseases called attention to and raised misgivings about the use of all sources of radiation. Consequently, plans for the construction of nuclear power plants began to arouse greater skepticism in the areas in which they would be located. The opponents of the proposed Ravenswood and Bodega Bay plants, who were vocal, determined, visible, and ultimately successful, highlighted comparisons between the hazards of radioactive fallout and those of nuclear power. So did critics of other reactor projects in the early 1960s. The growing public awareness of radiation dangers that developed during the fallout debate fueled the first grass-roots protests against nuclear power and introduced a new ambivalence in public attitudes toward the technology.

Most critics did not object to nuclear power in general; rather they attacked plans to build reactors at specific sites. Their campaigns remained localized and focused on conditions of particular concern in their own areas, such as the density of the population around Ravenswood and the threat of an earthquake at Bodega Head or Malibu. The opposition to nuclear plants, even if it applied only to conditions at an individual site, caught nuclear proponents by surprise and shook their confidence that the public would accept the technology with equanimity. In June 1963, in the middle of the controversies over Ravenswood and Bodega, *Nucleonics Week* urged the nuclear industry and the government to deal with public concern about the technology "fully and forthrightly." It argued that although many of the claims of the critics had been "based on extreme misunderstanding and distortion of the facts," including misleading comparisons of nuclear plants with bombs and fallout, their objections had to be taken seriously and their questions answered thoroughly. The journal's editors expressed a widely held view by suggesting that the future of nuclear power depended on how well the industry and the government satisfied public concerns. "If the public does not accept nuclear power," they concluded, "there will be no nuclear power."[2]

The key to ensuring public support for nuclear power, in the minds of officials in both industry and the AEC, was to educate the public and improve its understanding of the technology. In that way, they reasoned, it would not be susceptible to inaccurate information advanced by poorly informed or dishonest critics. The Atomic Industrial Forum established a Public Understanding Committee in April 1963, and it focused its efforts on drafting a booklet on nuclear safety that was aimed at "reasonable, intelligent, interested laymen." It also worked, in cooperation with the AEC, on preparing a film titled "Atomic Power Today: Safety and Service." While nuclear advocates supported those efforts, some insisted that more should be done. Westinghouse vice-president John W. Simpson warned in November 1963 that a "continued barrage of ill-informed, emotional or highly motivated attacks can exert such pressure on legislators, administrators and utility companies that they will be reluctant to support atomic power developments. . . . Our whole industry is being attacked; the whole industry should answer."[3]

William E. Shoup, in a speech he delivered as president of the American Nuclear Society in 1965, submitted that a major part of the problem facing the nuclear community arose from its terminology. By overusing "exotic vocabulary" and technical jargon in addressing the public, he argued, "too often we either scare them or bore them to death." He maintained that when the public heard phrases like "maximum credible accident," "radioactive vapor container," "going critical," or "scram," its "primitive fears" of a "mysterious, invisible danger" were aroused. Shoup urged that the nuclear industry stop discussing its safety procedures before the general public because the mere mention of safety concerns unsettled those unfamiliar with the technology. Instead, he recommended that it adopt a symbol to "help the layman readily identify the everyday clean, safe, and reliable use of nuclear energy." Shoup suggested a symbol called "Able Atom," the appearance of which would be determined by a national contest for high school students. This, he said, would not only "enhance the empathy of the younger generations for nuclear energy" but also would "be an improvement over the iron cross radiation danger sign we now use." Although Shoup's appeal did not inspire a campaign to conceptualize and popularize "Able Atom," the importance that he attached to public acceptance of nuclear technology and the frustration he expressed with what he regarded as unwarranted public anxiety were widely shared by his colleagues.[4]

The AEC harbored the same concerns about public attitudes toward nuclear technology and fully subscribed to efforts to enhance public understanding of it. The agency conducted its own programs to familiarize the public with nuclear power and regulatory procedures, including booklets, films, exhibits, and school lectures and demonstrations. At the same time, to avoid undercutting its regulatory position, it sought to restrict its role to providing educational material and information. Duncan Clark, director of the AEC's Division of Public Information, summarized the agency's dilemma in July 1963. "To do nothing in the way of public education . . . would cause us to lose much ground and thus seriously retard the much needed public acceptance of reactors in populated areas," he wrote. "However, it would be at least equally disadvantageous to take action which could be used by critics to support charges that the Commission does not take its regulatory responsibilities as seriously as its developmental ones."[5]

The AEC sought to preserve a fine distinction between education and promotion in its public affairs activities. The purpose of its programs was more to inform than to persuade, which contrasted with the extensive advertising campaigns and lavish publications that reactor vendors and utilities sponsored. The AEC was more subdued than industry in "selling" the atom. Nevertheless, its public information programs, while acknowledging nuclear power hazards, emphasized the advantages of the technology. By presenting a view of nuclear power that invariably tilted to the favorable side, the agency crossed the line that divided education from promotion.

The AEC responded to the complaints of industry about the problem of public misunderstanding and to the protests against nuclear power by expanding its publications, speaking appearances, and exhibits in order to reach a wider audience. It also changed some of its regulatory terminology to sound more innocuous; for example, the "hazards summary report" that applicants submitted became the "safety analysis report," the "maximum credible accident" became the "design basis accident," and "engineered safeguards" became "engineered safety features." Some AEC and nuclear industry officials thought that the agency should take more aggressive action to convince the public of the safety of the technology, but most industry leaders believed that it was their own responsibility to "sell the public on nuclear power."[6]

The AEC and the industry were encouraged by signs that their new efforts to win public support were paying dividends. Articles favorable to nuclear power appeared in the *Atlantic Monthly* and *The Nation* in early 1964; in both cases the authors had attended a news briefing on reactor safety that the AEC had sponsored. *Reader's Digest*, with a worldwide circulation of about twenty-six million, ran an article in late 1965 that emphasized the advantages and the safety of nuclear power and suggested that the technology had "gone through its gawky, troubled youth and entered a promising young adulthood."[7]

The results of an opinion poll conducted in 1965 that compared public views of nuclear power in three different areas were equally gratifying. One of the locations, Buchanan, New York, was the site of an operating nuclear plant (Indian Point). The second, Philadelphia, had received some public information programs because it seemed likely that a plant would be built in the region. The third, Atlanta, had no experience with nuclear power or exposure to special educational efforts. In all three places, participants in the poll had highly favorable views of nuclear power, but the residents of Buchanan were slightly more positive in their responses than their counterparts in Philadelphia and much more so than those in Atlanta. In Buchanan, 60.5 percent were favorable (compared with 3 percent negative), in Philadelphia, 60 percent were favorable (compared to 4.5 percent negative), and in Atlanta, 45.3 percent were favorable (compared to 7.7 percent negative). The survey indicated not only that the public was generally well-disposed toward nuclear power but also that educational programs were worthwhile. It offered evidence that, as *Nuclear News* put it, "familiarity breeds confidence."[8]

Nuclear advocates remained disturbed, however, by continuing attacks on nuclear power. Utilities ordered few reactors between 1963 and 1966, and none aroused the intense opposition that the Ravenswood, Bodega Bay, and Malibu proposals encountered. Nevertheless, objections to Oyster Creek raised by a group called the New Jersey Scientists Committee for Public Information and scattered challenges to the safety of nuclear power suggested that controversies over plant siting could recur in the future. For that reason, the AEC and the industry were angered and exasperated by publications that undercut their public information programs. An article titled "Is Atomic Industry Risking Your Life?" in the June 1965 issue of *Popular Science* was especially distressing. After discussing radiation hazards and "minor operating problems" in nuclear plants, it concluded that "the accident record of the industry has been good so far." But it added: "The growth of the industry, and the continuing pressure for power reactors in populated areas, increase the likelihood of more and bigger accidents. It's also possible to question whether the industry is adequately regulated."[9]

The article elicited strong protests from the AEC, industry officials, and radiation-protection specialists, who found it an inaccurate and misleading "scare" story. The article was exceptional in its criticism of nuclear power, however, and, in general, opposition to the technology remained sporadic and localized. Nevertheless, it deeply worried the nuclear community. Chauncey Starr, president of the Atomics International Division of North American Aviation and a leading spokesman for the nuclear industry, declared in January 1966: "It is a matter of dismay to the atomic power industry to find the issue of public acceptance such a major obstacle."[10]

The obstacle became more formidable after the surge of orders for power reactors began in 1966. As the nuclear power industry expanded, so too did opposition to the construction of new plants. This was partly a result of objections to plans for specific sites, but it also was a result of concerns over the implications of building large numbers of reactors. The growth in antinuclear activism went hand in hand with the growth of the industry. There was no organized movement against nuclear power or a uniform set of goals among the protesters and intervenors. They were largely local residents who focused on issues of concern in their own communities. But the cumulative effect of their activities called attention to and increased the impact of challenges to nuclear projects beyond local areas. With varying degrees of intensity between 1966 and 1968, individuals and groups expressed well-publicized misgivings over the proposals for Indian Point II and III, Shoreham, and Bell Station in New York, Monticello in Minnesota, Calvert Cliffs in Maryland, Turkey Point in Florida, and Vermont Yankee.[11]

The most prominent leaders of the protests against reactor projects in this period reflected a wide variety of backgrounds and motivations. Some were AEC critics of long standing. Perhaps the most dogmatic was Leo Goodman, who had held various labor union posts and who was described by *Nucleonics Week* as "an elfish little man who usually has a smile on his face and one thought on his mind: to torment the U. S. Atomic Energy Commission." Goodman had first battled with the AEC in 1949 over housing for workers at Oak Ridge, and after that time had clashed repeatedly both with the AEC and the Joint Committee over the adequacy and application of radiation standards. The Greater St. Louis Committee for Nuclear Information and its succeeding organizations, founded by Washington University biologist Barry Commoner and colleagues during the fallout controversy of the 1950s, regularly questioned the benefits of nuclear power. Nuclear advocates and opponents alike took its views seriously, partly because its tone was restrained but mostly because it included respected scientists and physicians among its ranks.

Other critics emerged in the mid-1960s to protest the growing use of nuclear power. Larry Bogart, a former advertising and public relations executive, established the Anti-Pollution League and helped run a number of similarly named organizations from his home in New Jersey. Although he initially favored nuclear power as an antidote to air pollution, he became convinced that the technology was expanding too rapidly and recklessly. "We're going to have a catastrophic accident," he warned. "We'll be living in a wasteland." Bogart edited a newsletter, lobbied Congress, and spoke frequently to civic groups throughout the Northeast. Adolph J. Ackerman, an electrical and civil engineer in private practice in Madison, Wisconsin, undertook similar activities, though he directed his efforts more at professional engineers. He expressed opposition to specific reactor projects but focused his attacks on the Price-Anderson Act. Malcolm Kildale, an advertising and public relations professional, established Fact Finders, Inc. and issued a series of press releases denouncing nuclear power. He organized a protest against plans for the Shoreham plant on Long Island in which a group of women picketed AEC headquarters and the White House. The demonstrators led a gaggle of white ducks to convey their message that the AEC was "ducking" safety issues. The protesters were purportedly Long Island housewives, but at least some were models hired by Kildale. The source of funding for the antinuclear activists remained unclear; they denied nuclear industry and AEC suspicions that they were primarily supported by the coal lobby.[12]

In other cases, attacks on nuclear power unquestionably came from coal interests. A short time after being elected chairman of the National Coal Policy Conference, W. A. (Tony) Boyle, president of the United Mine Workers of America, launched a series of blistering assaults on nuclear power. To the embarrassment of other officials in coal lobbying organizations, he broke the informal truce in which coal had agreed not to attack the nuclear industry on safety issues and nuclear spokesmen had agreed not to attack coal for causing air pollution. Boyle assailed the "atomic energy monsters in Washington" for promoting "a type of powerplant whose safety is open to serious question" and that was "capable of filling both atmosphere and water with deadly radioactivity." He hailed the women who had led the ducks in demonstrating against Shoreham and expressed hope that President Johnson had heeded their message because, Boyle said, "I'm sure he doesn't want his grandchildren to be radioactive."[13]

Boyle's offensive won support in Congress from representatives of coal-producing areas, some of whom sponsored legislation to create a blue-ribbon panel to investigate the role of the federal government in atomic power programs. But reservations in Congress about the rapid growth of nuclear power were not limited to members from coal regions. A subcommittee of the House Committee on Science and Astronautics was mildly critical of nuclear power in a 1966 report on energy sources, focusing on the failure to find a way to dispose of nuclear wastes. In April 1968, Senator Edward M. Kennedy of Massachusetts introduced a bill that would impose a two-year moratorium on granting construction permits for nuclear plants.[14]

The protests against nuclear power gained momentum between 1966 and 1968, though their impact was, at best, problematical. In 1969 the antinuclear campaign gathered new strength, gained unprecedented national attention, and threatened to fulfill the worst apprehensions of nuclear proponents. In January, *Sports Illustrated* published Robert H. Boyle's unrelenting denunciation of the AEC and the nuclear industry for neglecting the problem of thermal pollution. The article increased concern about the environmental effects of nuclear plants while discrediting the AEC and the industry in ways that they found impossible to rebut effectively. John A. Harris, who had taken over as director of the AEC's Division of Public Information, advised the Commission to send a letter to the magazine explaining its position but admitted that this "would not offset the damage already done by the story."[15]

Within a month after the appearance of Boyle's broadside, nuclear proponents suffered another blow that was delivered by a book titled *The Careless Atom*. The author was Sheldon Novick, who for years had been associated with the St. Louis Committee for Nuclear Information and its successor, the Committee for Environmental Information. He was soon to assume the editorship of the organization's journal, newly named *Environment*. Novick insisted that he did not oppose nuclear power in principle, but his book was a disturbing account of the hazards of the technology and uncertainties about its safety. He highlighted the 1966 accident at the Fermi plant in Michigan, the controversy over Bodega Bay and seismic siting, the worst-case casualty and damage estimates of the original WASH–740 report and the inadequacy of the Price-Anderson Act to cover them, and the public health risks that radiation from reactors presented. Novick lamented the lack of public awareness of and debate over the dangers of nuclear power, and concluded: "Only a very much broadened base of discussion will allow us to judge wisely whether in the reactor program, as it now stands, the benefits outweigh the hazards."[16]

Despite its alarming message, the tone of Novick's book and its criticism of the AEC and the Joint Committee were muted. The appeals that the publisher, Houghton Mifflin Company, used to promote the volume were much less so. One advertisement headlined: "The Hiroshima bomb is alive—and ticking—in Indian Point." To accentuate the point, it added: "Atoms for peace can blow us to pieces!" Those assertions were misleading ballyhoo apparently written by the publisher; they did not appear in Novick's book. Such sensational statements helped *The Careless Atom* receive a great deal of notice. Supporters of nuclear power were troubled both by the arguments in the book and by the attention it commanded. They complained, in some cases bitterly, that Novick disregarded the safety record of the industry and the AEC's elaborate regulatory requirements, exaggerated the hazards of the technology, ignored its environmental benefits, and repeatedly took quotations out of context to advance his views. Yet several pronuclear reviewers, even as they took issue with many of Novick's points, conceded that he raised some valid questions and agreed that the public should learn more about nuclear power and its impact on the environment.[17]

Novick's book was followed within a short time by a much more doctrinaire antinuclear tract. Unlike Novick, the authors, Richard Curtis and Elizabeth Hogan, took an unequivocal position that the risks of nuclear power far exceeded the benefits. Hogan was a resident of New York City who had first become alarmed about atomic power after reading David Lilienthal's criticisms in a 1963 article in *McCall's* magazine during the Ravenswood controversy. She began to collect materials on the subject and became a determined foe of nuclear power. She appeared at a public hearing in 1966 to oppose a construction permit for Indian Point II and reiterated her opinions before the Joint Committee the following year. She also contacted publishers to ask if they would find a professional writer to work with her on a book. Two publishers called Curtis, a free-lance author of about 100 sports, mystery, and science fiction stories and a dozen books. He had no training or experience in science and was initially dubious about collaborating with Hogan. After examining the materials she had assembled, however, he was so astonished by what he viewed as a lack of attention to nuclear hazards that he decided to join her in writing a book.[18]

Curtis and Hogan previewed their book in an article titled "The Myth of the Peaceful Atom" that appeared in the March 1969 issue of *Natural History*, a magazine published by the American Museum of Natural History in New York. They suggested that nuclear plants were so "saturated with hazards and unknowns" that they presented the "gravest pollution threat yet to our environment." The authors emphasized that the hazards of nuclear power were twofold; not only could they be the source of a catastrophic accident but even the small amounts of radiation they released routinely were a major menace to public health. Citing the growing number of nuclear plants, the efforts to place them close to populated areas, the unsolved problem of nuclear wastes, and the laxness of the AEC's regulatory practices, Curtis and Hogan concluded that "the entire national community stands to benefit" if nuclear power were abandoned. In their view, it seemed "to be leading us toward both environmental and economic disaster." In July 1969, Curtis's and Hogan's book, *Perils of the Peaceful Atom*, which extended and elaborated on the arguments in their article, was published by Doubleday and Company. The jacket of the book set the tone by warning that "our atomic energy program involves nothing less than tampering with elemental fire." The publisher's advertisements were as sensational as those for Novick's volume. One showed a mushroom cloud under a statement in large, bold letters: "The 'peaceful uses' of atomic energy can kill you just as dead."[19]

Curtis's and Hogan's article and book polarized the debate over nuclear power more than any previous discussion of the subject. They incensed nuclear proponents, including those who had seen some merit in Novick's arguments. Supporters of nuclear power complained that Curtis and Hogan demonstrated a lack of understanding of nuclear technology and made many glaring errors in evaluating its hazards. They were even more outraged by the authors' apocalyptic tone and exaggerated depictions of nuclear risks. *Perils of the Peaceful Atom*, for example, featured chapter titles such as "Nuclear Roulette" (on the chances of a major accident) and "The Waters Ignited" (on thermal pollution). Even some friendly reviewers of the book faulted Curtis and Hogan for their shrillness and technical misconstructions. To make matters worse from the perspective of nuclear advocates, the book received wide and respectful press attention. Some newspapers ran excerpts from it and others gave it prominent reviews. *Perils of the Peaceful Atom* sold well, as did Novick's *The Careless Atom*. By January 1970 Novick's book had gone into its fourth printing and Doubleday reported that the Curtis-Hogan volume had done "quite well." Both were issued in popular-market paperback editions a short time later.[20]

Despite differences in tone and emphasis, opponents of nuclear power cited many of the same basic issues in marshalling their objections to the technology. They frequently drew comparisons between the hazards of nuclear power and nuclear weapons. The most flagrant examples were the advertisements that the publishers of the Novick and Curtis-Hogan books ran, even though the authors themselves acknowledged that a light-water reactor could not explode like an atomic bomb. It was more common to link reactors and bombs in subtler ways, such as relating the radioactivity contained in a power plant to that released by setting off a nuclear weapon. The effect was to reinforce the popular misconception that a reactor could blow up with the explosive force of an atomic bomb. Among the other events and issues that nuclear power critics emphasized were the 1966 accident at the Fermi plant, invariably described as a harrowing near-miss, and the worst-case accident consequences outlined in the 1957 WASH–740 study. They increasingly focused on thermal pollution and disposal of nuclear wastes as problems created by nuclear power generation that the AEC and the industry were neglecting.

Nuclear opponents devoted much attention to radiation hazards, which had first become a source of public concern during the fallout debate of the 1950s and early 1960s. They discussed the possibility of a catastrophic accident, stressing that it *could* happen, and rebuked the nuclear industry for pushing to build reactors in heavily populated areas. In addition, they pointed to the dangers of low levels of radioactivity that reactors released during normal operation. Curtis's and Hogan's descriptions were particularly lurid on this point; they warned at length about the "slow, but deadly, seepage of harmful products into the environment." Other critics viewed the hazards of low-level radiation as less alarming in the short-term but likely to become a more serious threat as a growing number of plants went into operation. Finally, antinuclear observers attacked the AEC, the Joint Committee, and the industry for insufficient efforts to ensure safety and undue enthusiasm for promoting nuclear technology. They emphasized the AEC's conflict of interest between its developmental and regulatory responsibilities. While faulting the agency for making promotion of the technology its first priority, they understated or ignored the role of the AEC in aborting proposals for Ravenswood, Bodega Head, and other reactor projects.[21]

The wave of attacks on nuclear power and indications that they were reaching a large audience spurred the AEC to again expand its efforts to explain its programs and counteract its critics. By the spring of 1969 it had concluded that despite the information it had provided to the public over the years through booklets, reports, films, speeches, and press conferences, "apparently the message has not penetrated to the man in the street." It continued to believe that "much of the public criticism of nuclear power has stemmed from ignorance, misinformation, or unfounded fears." John Harris, director of the Division of Public Information, advised the Commission on 17 March that although public acceptance of the technology remained "at a high level," more aggressive steps were needed to offset the recent attacks. He regarded the thermal pollution issue as the AEC's "biggest problem." He recognized, however, that other questions cited by critics were also of concern to the public and predicted that "organized attacks are likely to continue for an indefinite period."[22]

With those considerations in mind, the Commission agreed to broaden the scope of its public information programs and undertake new initiatives to answer its critics. To accomplish its goal of improved public understanding of its positions and acceptance of nuclear power, it expanded many of its existing activities. The AEC hoped, for example, to double the number of viewers of its motion pictures by making films that appealed not only to high-school and college students, its traditional audience, but also to a popular audience. It produced a film on atomic developments during 1968 that was designed for television use. The AEC also made a series of prerecorded radio programs that it offered to 6500 stations; nearly 1200 accepted. Further, it wrote regular news features that were intended for use by daily and weekly newspapers.[23]

In addition to extending existing programs, the AEC sought to win greater support from the news media by giving the commissioners more public exposure, by arranging appearances for them on news programs, and by setting up small-group sessions with officials of leading newspapers and magazines. The AEC also decided to participate more actively in meetings in the local vicinity of a reactor project. Although the agency had sent representatives to explain the AEC's regulatory procedures regarding a proposed plant on occasion in the past, it now adopted a policy of using staff experts to outline its position "in areas where problems arise." To coordinate appearances and other public relations efforts, the Commission established a task force made up of staff members from different divisions. Its mandate was to ensure "a deliberate, cohesive, and knowledgeable technical approach to the problem of public acceptance." Without creating a formal organizational unit or calling attention to its purpose, the Commission directed that the task force pursue its "public understanding activities on a priority basis." The AEC remained concerned about the impact of its public information programs on the separation of its regulatory and promotional duties. Although it insisted that it maintained a proper balance, the new initiatives further blurred the line. The irony was that the AEC stepped up its efforts to reach the public in response to the allegations of its critics, but in so doing it enhanced the credibility of charges about the conflict of interest in its dual responsibilities.[24]

One result of the AEC's decision to contest its opponents more aggressively was a series of pointed attacks on them by high-level officials. *Nucleonics Week* noted in May 1969 that the commissioners "were taking off the gloves as they jab back at critics of nuclear power." Seaborg, for example, blasted those who published one-sided accounts. "Specifically, every fact and every statement in such a story may be true," he declared, "while the article as a whole, and the conclusions it draws, may be invalid and misleading." He told the Edison Electric Institute in June 1969 that the benefits of nuclear power for the economy and the environment were too important "to allow its progress to be eroded . . . by misinformation, half-truths, and hearsay." Ramey warned against the influence of "professional stirrer-uppers" who confronted nuclear power supporters with a "stacked deck" of information that made it difficult to present a full or balanced picture to the public.[25]

At the same time that the AEC was attempting to counter the attacks on nuclear power more effectively, the Joint Committee and the nuclear industry undertook similar efforts. Holifield, who worried that "loud opposition of the kooks" prejudiced public attitudes, delivered a series of addresses that emphasized the environmental benefits of nuclear power relative to fossil-fuel technology. In June 1969 he began making plans for hearings on the environmental effects of electrical production, which got under way in October. His colleague Craig Hosmer gave a tongue-in-cheek speech to the National Coal Association the same month that poked fun at many of the objections to nuclear power. He asked his audience to imagine that the positions of the coal and nuclear industries were reversed, with coal as the new source of energy and nuclear as the established one. He detailed the problems and opportunities that might arise from the "dawn of the Coal Age." He warned that the "path to coal glory will not be easy" because the "atomic-energy people will fight back in many crafty, subtle ways." Hosmer fantasized that enemies of the new technology would "flail the industry with such terror-invoking phrases as carbon dioxide and sulphur dioxide" and that books such as "Careless Coal" would arouse public fears. In the end, he predicted that public protests would paralyze both the atomic and coal industries, that no power plants would be built, and that the nation would "plunge into darkness" for lack of electricity. The best way to avoid this disaster, he suggested, was for coal and nuclear interests to refrain from acrimonious exchanges and work together to ensure adequate supplies of energy.[26]

Like the AEC and the Joint Committee, the nuclear industry concluded that the recent attacks on nuclear power were serious enough to warrant a clearly articulated response. Both vendors and utilities feared that nuclear critics would delay the construction and operation of plants already under way and, in the long run, could threaten the development of nuclear power. General Electric, which had followed a policy of ignoring antinuclear groups, decided that it should adopt a more active program, including conducting seminars for utility officials and sponsoring television and radio commercials. Westinghouse took a similar view. James H. Wright, its leading spokesman on the environmental effects of nuclear power, warned in May 1969 that the spate of attacks on the technology presented a formidable threat. They not only interfered with the decision-making process of power companies but also could "create an untenable climate in state politics" by using "the big lie and fear tactics."[27]

Utility officials were equally concerned about the impact of antinuclear activists and sought ways to limit their appeal. In June 1969, Thomas G. Ayers, president of the Commonwealth Edison Company of Chicago, organized an *ad hoc* committee of industry representatives to discuss the problem of public acceptance and recommend action by various segments of the nuclear community. It urged that utilities, vendors, the Atomic Industrial Forum, and the AEC expand existing educational programs to inform the news media and the public about nuclear technology and to combat the arguments of nuclear critics. The burden for much of the effort would fall on individual utilities, who appeared to be in the best position to counter opposition because they could tailor their campaigns to the local population. They could make their case through advertisements, lectures, tours, exhibits, and contacts with editors, local government officials, and universities. Although the suggested activities were hardly novel, they had clearly taken on greater importance and urgency. One example of the redoubled effort on the part of utilities to win public support was an advertising campaign that Baltimore Gas and Electric, which was encountering protests against its Calvert Cliffs plants, sponsored between August and October of 1969. It ran a series of advertisements in local newspapers that attempted to explain nuclear power in lay terms and soothe public fears about radiation releases, thermal pollution, and damage to the Chesapeake Bay. One of the ads emphasized that nuclear power was so safe that President Nixon had recently purchased a home in California that was located less than three miles from the San Onofre reactor.[28]

The impact of the efforts of the AEC, the Joint Committee, and the industry to contest the arguments of their opponents and to meet a challenge of unprecedented proportions was unclear. The views of nuclear power that reports in the summer and fall of 1969 advanced were decidedly mixed. Favorable commentaries on nuclear power appeared in some prominent publications. The *Washington Star*, for example, editorialized in July 1969 that "nuclear power must be developed" and assured its readers that "the development can take place, as it surely will, without creating any serious environmental menace." The *Seattle Post-Intelligencer* ran a five-part series that presented nuclear power proponents as committed to safety and environmental protection. One of the stories in the series, headlined "Do Fission and Fishin' Mix?" found that Robert H. Boyle had exaggerated the threat of thermal pollution from nuclear plants in his *Sports Illustrated* article. Another story that placed nuclear power in a positive light came out in *Popular Science* in September 1969; it hailed the "new boom for A-power" and emphasized that an "A-power plant is attractive, silent, and clean." The AEC had provided information to the author, and it was particularly gratified that the article contrasted sharply with the critical piece the same magazine had published in 1965.[29]

The AEC's attempts to encourage the publication of favorable news stories were not always so successful. It was particularly disappointed by an article that appeared in *Life* magazine in September 1969. A *Life* reporter, Susannah McBee, had talked at length with Seaborg and other AEC officials in gathering material for the story, and they had anticipated that it would provide a positive view of the agency and of nuclear power. Instead, the article opened with a headline stating that the "promise of nuclear energy is dimmed by growing fear of contamination." It went on to declare that the industry had "skidded to an uncertain slowdown" and that the "principal villain, in many eyes, is the Atomic Energy Commission." The problem with the AEC, the story explained, was its status as the "sole nuclear regulating agency" as well as the "vigorous promoter and generous subsidizer of the peaceful atom." Although the article was more a description of the controversy over nuclear power than an attack on the technology, its tone and sympathetic treatment of antinuclear arguments came as an unpleasant surprise to the AEC. McBee called Joseph J. Fouchard, assistant director of the AEC's Division of Public Information, to tell him that she had "lost control of the story" and that the editors of the magazine had "butchered" it. The article as printed was a major setback to the AEC's efforts to gain greater public confidence.[30]

The AEC suffered other unhappy experiences in its campaign to counter nuclear critics. It had decided to depart from its previous practices and meet opponents of specific plants face-to-face in public forums, but its first appearances did not go well. In a meeting on Oyster Creek in June 1969, Andrew J. Pressesky of the AEC's Division of Reactor Development and Technology and a representative of General Electric participated on a panel that included Leo Goodman and Larry

Bogart. Faced with a hostile crowd that warmed to Goodman's and Bogart's assertions about the health hazards of nuclear power, the pronuclear spokesmen were so battered that *Nucleonics Week* called the outcome a "technical knockout."[31]

In September, the AEC stumbled again in a highly publicized conference at the University of Vermont in Burlington. Agency representatives went to Vermont at the urging of Senator George D. Aiken, a member of the Joint Committee who thought that the AEC had shown "shocking" disregard for public apprehension over the construction of the Vermont Yankee reactor. In August 1969, Aiken, joined by Vermont Governor Deane C. Davis, invited the AEC to explain nuclear technology and address citizen concerns at a public meeting. The AEC welcomed the opportunity to carry out its recent decision to increase its visibility and take its message to areas where controversies over plants had arisen. It accepted the invitation immediately and made plans for a day-long educational session that would feature appearances by Seaborg, Commissioners Ramey and Thompson, senior staff members, and well-known scientists from national laboratories. The AEC was somewhat unsettled when Governor Davis changed the original format of the meeting to add a panel that would include both AEC officials and prominent nuclear critics. But it was undeterred; it viewed the program as an important chance to educate the public and to answer its opponents.[32]

The meeting was the first occasion that AEC commissioners had sallied forth to face nuclear power critics in a public forum. It was billed as a historic confrontation, or in the words of the University of Vermont *Alumni Magazine*, "the heavyweight championship bout in the ecological world." Although the AEC did not suffer another knockout, by most accounts it did not win many points. The AEC brought a new pamphlet on "Nuclear Power and the Environment" to hand out, but its effectiveness was at least partly undercut by the antinuclear protesters who distributed copies of the recent *Life* magazine article. In accordance with its initial planning for the conference, four AEC representatives, including Ramey and Thompson, delivered prepared remarks to an audience of over 1000. They tended to be too long and too technical, and the crowd grew visibly restless. The main event pitted Ramey, Thompson, and two Oak Ridge radiation experts against Clarence A. Carlson of Cornell University (a leader of the fight against the proposed Bell Station), and Dean Abrahamson, Ernest Tsivoglou, and Arthur Tamplin, all of whom had questioned the adequacy of the AEC's radiation protection standards. The debate was low-key, inconclusive, and in some minds, dull and tedious. The best moment in the day's activities for the AEC came when Seaborg ended the program with a speech on the environmental benefits of nuclear power that was clear, balanced, and well-received.[33]

The AEC's performance in Burlington elicited mostly negative reviews. Commissioner Thompson thought the meeting was an "utter disaster." An unidentified AEC official concluded that in its first face-off with its critics, the agency "came off second best." *Nucleonics Week*, which published a special thirteen-page report on the conference, described the "general feeling" of those who attended: "It wasn't that the critics came off so well, it was just that the AEC did not." Others were more charitable in their assessments. Aiken was pleased with the AEC's appearance, Holifield was impressed with the AEC's responses to questions, and Ramey was satisfied that he and his colleagues had given creditable presentations.[34]

Despite differing views on the outcome of the Vermont meeting, AEC officials agreed that there was ample room for improvement. Assistant general manager Howard C. Brown, Jr., who coordinated the AEC's new public information program, believed that the Vermont meeting had produced "some modest gains," particularly by showing that the AEC was willing to debate its critics publicly and by demonstrating that it was concerned about preserving environmental quality. He cautioned, however, that the conference also suggested that "we need to develop effective ways of bringing facts to the public in terms they can understand." A member of Seaborg's personal staff, Stanley D. Schneider, was more blunt. He contended that the Vermont meeting indicated that "the AEC is in deep trouble with the public." Schneider argued that nuclear opponents should be challenged frequently and persistently when they aired inaccurate statistics and questionable assertions. He urged that the AEC adopt a more confrontational approach by forcing antinuclear spokesmen to spell out the costs of accepting their positions and to propose practical alternatives for producing clean electricity without nuclear power. "We are going to be fighting an uphill battle," Schneider concluded, "but let's fight it with some intelligence, verve, and imagination."[35]

The AEC continued to participate in meetings with its critics and, drawing on the lessons of its initial appearances, improved its performance. At their best, its spokesmen were knowledgeable and forceful in challenging their adversaries. Even then, they had no illusions that their arguments would win many converts, especially in cases where the audience was clearly hostile. Although both AEC and industry leaders thought that their new campaigns to reach the public had been useful, they also recognized that popular attitudes toward nuclear power remained a major problem. Despite hundreds of speeches, appearances at numerous meetings and seminars, sponsorship of exhibits and films, and other presentations by nuclear proponents, the impact of their message, focusing on the environmental advantages of and the need for nuclear power, was limited. By early 1970, the efforts of nuclear opponents were productive enough to win the notice of prominent publications. *Time* magazine, for example, commented that the "vision of clean, cheap electricity and smog-free air" from nuclear power was under attack, and it added: "The critics are vocal and active—and they are getting results." The *Chicago Tribune* echoed the same view: "The peaceful atom, heralded for a decade and a half as mankind's greatest hope for virtually unlimited power, has recently taken such a public relations beating that many experts fear America may be headed for an energy crisis."[36]

The attacks on the AEC and on nuclear power increased in frequency and intensity during 1970 and 1971. By the early part of 1970, Gofman and Tamplin were commanding a great deal of attention for their allegations that nuclear plant emissions would cause thousands of cancer deaths every year and for their denunciations of the AEC. Their charges received respectful discussion in a wide range of publications and inevitably amplified public fears of nuclear power. Articles that were critical of the AEC or gave antinuclear arguments sympathetic treatment appeared in many prominent newspapers and popular magazines, including the *New York Times*, the *Wall Street Journal*, the *Washington Post*, the *Washington Star*, the *Philadelphia Inquirer*, *Esquire*, *Newsweek*, *McCall's*, the *New Republic*, *Look*, and *Playboy*. The impressions they conveyed were reinforced by television reports. The CBS Morning News ran a five-part series in August 1970 that focused on radiation dangers and featured Gofman and Tamplin. Correspondent Joseph Benti informed his viewers that the AEC underregulated and understated the risks of "deadly radiation." He ended with a question: "Can we control our seemingly insatiable demand for electrical power long enough to determine beyond a doubt that we truly have safely harnessed the atom?" A Los Angeles station aired a documentary in May 1971 that was even less equivocal. The narrator, actor Jack Lemmon, suggested that nuclear critics had a point when they argued that "nuclear power is not only dirty and undependable . . . it's about as safe as a closetful of cobras."[37]

The basic issues that nuclear opponents cited remained the same as earlier, but between 1969 and 1971 the emphasis they placed on different problems shifted. For a time the question of primary concern was thermal pollution. It became a major source of criticism of the AEC and the industry by 1969, highlighted by Robert H. Boyle's blistering indictment in *Sports Illustrated*. By the time that the thermal pollution issue had been defused, the health effects of low-level radiation and the adequacy of existing standards had supplanted it as the foremost cause of public misgivings. The thermal pollution controversy laid the foundations for the debate over radiation standards by raising doubts about the benefits of nuclear power and about the AEC's commitment to environmental protection. Questions about radiation safety had first produced a public controversy when the state of Minnesota challenged the AEC's regulations during licensing proceedings for the Monticello plant. The risks of low-level radiation became an acute source of anxiety after the outspoken views of Gofman and Tamplin won wide attention. Their use of seemingly precise (though ever-changing) mortality statistics and claims of scientific certainty about their findings provided tangible support for the vague uneasiness that others had voiced about the health effects of nuclear plants.

The hazards of radiation released by nuclear units remained a lively issue, even after the AEC tightened its regulations. But by 1971 concern over the radiation emissions of normally operating plants had lost its primacy to even more potent apprehensions about the consequences of a major reactor accident. Seaborg noted in February 1971 that "the anti-nuclear power forces seemed to be shifting from low level radiation dangers to reactor safety." This was not a new question, but it took on increasing importance, largely as a result of uncertainties over emergency core cooling. Nuclear critics had long pointed to the WASH–740 report and the Fermi accident to support their assertions about the hazards of nuclear power. The ECCS controversy gave greater credence to those arguments and moved reactor safety to the center of the nuclear power debate.[38]

At the same time that doubts about nuclear safety were becoming a major public issue, an emerging debate over another difficult problem— disposal of high-level radioactive wastes from reactor operations—won headlines and further undercut the AEC's regulatory credibility. The focus of concern was the AEC's plan to build a permanent repository for high-level waste in an abandoned salt mine near Lyons, Kansas. The proposal prompted queries about the safety of the site from Congressman Joe Skubitz, a Kansas Republican whose district was located about 200 miles from Lyons. Skubitz grew increasingly disgruntled with the AEC's assurances that it would not proceed with building the repository unless it was convinced of the suitability of the site. He told the Joint Committee that the Lyons facility would serve as a dump for the "most dangerous garbage in the knowledge of mankind." Eventually, Skubitz gained the support of scientists, public officials, and newspapers in Kansas, and by early 1971 the issue had attracted national attention. The controversy, as the *Washington Post* observed, reflected not only the "general fear that most people have of radioactivity" but also "a basic distrust of the AEC."[39]

The suspicions and objections that each of those issues—thermal pollution, radiation standards, reactor safety, and waste disposal—generated were intensified by other problems that made headlines in the late 1960s and early 1970s. Some of them were regulatory matters, such as environmental protection and the use of mill tailings for construction purposes. Others were outgrowths of the military and promotional functions of the AEC, such as a fire at the Rocky Flats plant (which produced plutonium triggers for bombs) that released detectable amounts of plutonium to the environment, the use of nuclear blasts to free natural gas reserves, and underground weapons testing. Finally, the AEC's responsibility for both promoting and regulating nuclear power remained a frequently cited and powerful issue that antinuclear activists exploited. It was, said one, "like letting the fox guard the hen house."[40]

While most of the early critics of nuclear power remained active, they were joined, and in some cases overshadowed, by a newer group of activists who represented varying constituencies and perspectives. Some continued to be local citizens who protested the location of a nuclear plant in their communities. Others were environmentalists who faulted the AEC and the nuclear industry for failing to provide sufficient attention to protecting the environment from the effects of nuclear power generation. Some of the most active and articulate critics were practicing attorneys, such as Anthony Z. Roisman and Myron M. Cherry, who worked on behalf of intervenors in several licensing cases. Yet others were academics who contested the AEC's positions. They included

Dean Abrahamson of the University of Minnesota, Harold P. Green of George Washington University, Edward F. Radford of Johns Hopkins University, and members of the Union of Concerned Scientists.

Nuclear advocates worried more about the newer group of critics than about their predecessors. James H. Wright of Westinghouse, for example, found the newer activists to "present far more formidable opposition than the old crowd," whom he disparaged as "incompetents." The AEC's Stanley Schneider described the newer antinuclear leaders as "a group of articulate, vigorous, personable and, to a great extent, young people who have enough knowledge and a facility to use it to be extremely dangerous." Nuclear opponents, even though they recited many of the same arguments, remained largely fragmented and localized; there was no integrated antinuclear movement. The first national antinuclear group, the Committee for Nuclear Responsibility, was established in May 1971. Chaired by Gofman, writer Lenore Marshall, and former U. S. Senator Charles E. Goodell, it called for a moratorium on the construction of additional nuclear plants.[41]

The impact of nuclear critics on public opinion and on the development of nuclear power was ambiguous. In some respects, support for the technology and the prospects for its continued growth were strong. Utilities demonstrated confidence in the future of nuclear power in 1971 by announcing plans to build twenty-eight nuclear units, a number that approached the record year of 1967 and represented a considerable increase over the previous two years. Opinion polls still indicated that the public favored nuclear electricity by substantial margins. In one 1969 survey, 50 percent of the respondents favored nuclear power in their local areas while 27 percent opposed them. Another poll in the state of Washington in May 1970 showed that 70 percent of those questioned were "not opposed" to nuclear power and only 6 percent were "strongly opposed." By contrast, 22 percent were "not opposed" to fossil-fuel plants and 38 percent were "strongly opposed." In a national survey published in early 1971, 44 percent of the participants said they would support the construction of a nuclear plant in their area and 30 percent said they would object.[42]

While those signs were encouraging for nuclear proponents, they were counterbalanced by other more disquieting indications. The results of the 1971 opinion poll, although still favorable, showed a growing level of opposition to nuclear power. In a similar survey in 1967, 62 percent of the respondents had expressed support for a nuclear plant in their area. Despite the large number of nuclear power reactors for which utilities announced plans in 1971, some companies decided to buy conventional units because of their fear of public protests and licensing delays if they ordered nuclear stations.[43]

The prospects of licensing difficulties that would undermine their schedules for power availability and threaten their supplies troubled utility planners. Nuclear plants had suffered from a series of schedule slippages by 1970 that caused bitter complaints from the affected utilities. Relatively few of the delays to that time were primarily the result of public opposition. A study of the reasons for them cited construction and labor problems, backlogs in the licensing process, and the failure of pressure vessel manufacturers to keep up with demand as more decisive than public protests. During 1970 and 1971, however, intervenors increasingly contested plant applications. Several well-publicized cases where intervenors held up the issuance of construction permits or operating licenses, such as Palisades, Monticello, Calvert Cliffs, and Shoreham, provided alarming signals to utilities. In hopes of avoiding time-consuming clashes with antinuclear groups, the nuclear industry attempted to meet their objections. Utilities increasingly agreed to build cooling towers to ease concerns over thermal pollution, for example, and reactor vendors designed new measures to reduce plant effluents. The AEC tightened some of its regulations in part to placate its critics and to reassure the public about its commitment to health, safety, and environmental protection.[44]

There was additional evidence of the impact of antinuclear activities. Several members of Congress became outspoken critics. Some did not take specifically antinuclear positions but sharply questioned the AEC and the industry on particular issues of concern to them. Others proposed legislation to slow the growth of nuclear power and/or weaken the AEC's authority. Congressman Jonathan Bingham of New York sponsored a bill in October 1969 to strip the AEC of its regulatory responsibilities and assign them to the Public Health Service. Senator Mike Gravel of Alaska, who described radioactivity as "the worst conceivable pollutant and threat to life," introduced a measure in February 1971 to impose a moratorium on the construction of nuclear plants. Seven senators and two congressmen introduced new bills in May 1971 to scrutinize the government's nuclear programs. None of those proposals advanced very far, but they suggested that the Joint Committee was no longer the unchallenged guardian of atomic energy affairs in Congress.[45]

In addition to the bills introduced in Congress, elected officials and antinuclear groups in several states worked for nuclear moratoriums.

19. Demonstrators protest against Monticello nuclear plant at headquarters of Northern States Power in Minneapolis, 1971. (© 1971 *Star Tribune*, Minneapolis-St. Paul)

They were unsuccessful in campaigns in New York, California, Minnesota, Pennsylvania, and Oregon, but their efforts attracted enough attention to arouse the concern of nuclear advocates. One locality, the city of Eugene, Oregon, did vote in favor of a moratorium. In November 1968 the electorate of Eugene had overwhelmingly approved a measure that allowed its municipal utility to float bonds to build a nuclear plant. Subsequently, some citizens, impressed with the Novick and Curtis-Hogan books and other antinuclear literature, challenged the utility and urged that it suspend its plans until more information became available. In May 1970, by a narrow vote, the people of Eugene reversed themselves and supported a four-year moratorium.[46]

The decision in Eugene, like other indications of growing reservations about nuclear power, was disturbing to nuclear supporters less for what it achieved than for what it symbolized: the difficulty of winning public acceptance of the technology. The problem was made even more formidable by declining confidence in the AEC and the nuclear industry, which, in the minds of nuclear proponents, was the most serious consequence of antinuclear activism. Craig Hosmer spoke for many who shared his views when he complained in February 1971: "The scientific and political headline-seekers, the public name-callings, the anti-nuclear horror books, the sensationalized media coverage—all of these have eroded the well-earned public confidence which the outstanding safety record of nuclear power had built over the past 20 years. Until that trust is re-kindled, we can expect more legal and economic penalties which will make the job of building reactors and providing kilowatts very difficult, discouraging and costly."[47]

It was apparent that the efforts of the AEC and the nuclear industry to neutralize the impact of their critics had achieved, at best, limited success. As William C. Parler, special counsel of the Joint Committee, observed in March 1971: "It would be interesting to know (if there were a way to determine) the number of members of the public who have actually been reached since the speech, article, meeting, etc. campaigns began. My impression is that the number reached would be small." He added: "The critics, however, have no problem in reaching and influencing the public."[48]

There were several reasons that antinuclear activists were able to make a major, though not overriding, impact on public opinion and to cause great distress among nuclear supporters. One was the general disillusionment with the government, established institutions, and science that prevailed by the late 1960s, largely as a result of the Vietnam war. The AEC and the nuclear industry suffered the consequences of social trends that rejected, or at least questioned, much of what they represented. As Louis H. Roddis, Jr., president of Consolidated Edison, pointed out in early 1970: "Skepticism is endemic. And with some cause. . . . People are less ready to believe what politicians tell them. That isn't new. But they don't believe scientists either—particularly government scientists." Citing concerns over nerve gas, DDT, and x-rays from color television that turned out to be more serious than the government first indicated, he went on: "So when we wave nuclear power's fine report card in the public's face, can we reasonably expect it to be believed?" A student from Windham College in Brattleboro, Vermont put it more succinctly after listening to a debate over radiation standards between Victor Bond of Brookhaven National Laboratory and Leo Goodman: "Dr. Bond sounds good but we can't believe him. He works for the government."[49]

A second reason for the success of nuclear critics in undermining support for nuclear power was that it was easier to win the public's attention by stressing dramatic dangers than by explaining safeguards. This was a problem that nuclear advocates found particularly trying. Headline writers were especially prone to spreading alarm. Some of the most dramatic headlines announced, for example, "The Price of Nuclear Power is Death," "Invisible Death: Pollution of the Atomic Age," and "Atomic Death Factories in Your Backyard." One-sided and exaggerated depictions of the hazards of nuclear power were a staple of antinuclear literature and common in many news articles. Attempts by nuclear proponents to correct a plethora of misleading and inaccurate stories, advertisements, speeches, and other presentations inevitably failed to gain as much notice or produce the same effect. The AEC and the nuclear industry recognized that they were burdened with a severe disadvantage in trying to educate the public about a complex technology, especially one that could never be free of some risks. They found, to their dismay, that the public was usually more impressed with the risks than with their assurances about the improbability of a major accident, the slight hazards of radiation emissions from normally operating nuclear plants, and other sources of concern.[50]

A third reason that antinuclear appeals gained ground in the late 1960s and early 1970s was the cumulative effect of several related issues. The AEC and other nuclear proponents might have been less vulnerable to the arguments of their critics if a series of important controversies had not arisen within such a short time. The combination of growing concerns over thermal pollution, low-level radiation, reactor accidents, waste disposal, and a variety of other matters contributed to visibly increasing uneasiness about the technology. The cumulative effect of all those issues damaged the stature and credibility of the AEC and made effective responses to its opponents difficult. Because the technology was complex and innately frightening to many people and because uncertainties about important questions remained to be resolved, the credibility of the AEC's regulatory positions depended heavily on faith in its judgment and its commitment to safety. By 1971, the arguments of antinuclear leaders had severely undermined that faith.[51]

The AEC sabotaged its own credibility and enhanced that of its critics by consistently emphasizing the development of the nuclear industry rather than the prompt resolution of regulatory issues. Its reluctance to regulate against thermal pollution, resistance to carrying out the provisions of NEPA to the fullest extent, refusal to allow states to set radiation standards stricter than its own, and hasty publication of the ECCS interim criteria made its commitment to public health and environmental values highly suspect. The reservations that nuclear power opponents expressed were especially persuasive in an age of acute environmental awareness. Nuclear advocates acknowledged that the technology exacted some environmental costs and imposed some safety hazards, but they continued to insist that the benefits of the technology far exceeded the risks, both because of its environmental advantages and its promise of providing the means to meet growing demand for electricity. Critics offered a contrasting view that gained increasing prominence and acceptance. Faced with the prospect of scores of nuclear plants and skeptical of the AEC's ability to regulate them carefully, they expressed grave doubts that the benefits of the technology were worth the risks it involved.

The increasing strength of antinuclear sentiment was enormously frustrating to the AEC. Despite efforts to educate the public and counter the arguments of its opponents, it continued to be plagued by widely circulated reports that presented one side of nuclear issues (often in exaggerated terms), misunderstood nuclear technology, and distorted the AEC's positions. The AEC recognized that most of its critics were not irresponsible scare-mongers; it thought that they were generally ill-informed about the hazards of nuclear power and its actions to control them. For that reason, it attempted to establish a reasoned discourse with its critics, advance its own position by providing facts and correcting inaccuracies, and isolate the most frenetic and unreasonable of its opponents (the ones Ramey described as "stirrer-uppers"). It hoped to "build bridges" with those who it believed had something constructive to contribute by arranging meetings with them. The meetings that were held were cordial and mutually beneficial; they helped to ease ill-will and lack of communication between the agency and some of its most thoughtful adversaries. But they did not resolve outstanding differences or end the controversies over nuclear power.[52]

By the middle of 1971, the ambivalence that had long characterized public attitudes toward nuclear power continued to prevail. In some ways, support remained strong and the future appeared promising. For the most part, even nuclear critics did not call for an end to nuclear power or, despite the reservations they expressed, oppose it as a matter of principle. Yet there was no doubt that public confidence in the technology and its proponents had diminished since the early 1960s. The ambivalence that prevailed was increasingly weighted in the direction of an antinuclear position. This caused grave concern to supporters of nuclear power expansion. They still believed that public acceptance of the technology was vital to its long-term success. But it was apparent that winning public support had become much more uncertain. "Eventually, the fate of nuclear power will be decided by the public, and it is up to us to try to get the true facts brought to the public notice," wrote Walter H. Jordan, former assistant director of Oak Ridge National Laboratory, in October 1971. "Thus far we have not been very successful, and I see no immediate solution to this urgent need to educate as wide a sector of the public as possible."[53] Jordan's analysis of the need for public acceptance was the same as that of the editors of *Nucleonics Week* eight years earlier, but his assessment of the chances of success was considerably more pessimistic.

CHAPTER XVI

The End of an Era

On 21 July 1971, Glenn Seaborg announced that he was resigning as chairman of the AEC and returning to the University of California. When President Nixon had taken office in January 1969 he had asked Seaborg to continue in the AEC post, and in June 1970 he had reappointed him to a new five-year term. By mutual agreement, however, White House officials and Seaborg decided that he would stay for only about a year and then leave the Commission. The White House wanted time to find a suitable replacement and Seaborg thought that after nine years "the time had come . . . to leave." By the spring of 1971, the Nixon administration was actively searching for a new chairman and seeking to ensure a smooth and harmonious transition. After it settled on James Schlesinger, Seaborg asked that his tenure be extended by six months. He wanted to lead the American delegation on a long-planned official trip to Geneva for an international conference on the peaceful uses of atomic energy and to the Soviet Union for a tour of its nuclear facilities. He also hoped for additional time to relocate his family. The White House refused; it wanted Schlesinger to take over as AEC chairman without a long delay. It agreed to allow Seaborg to head the U. S. delegations to Geneva and the Soviet Union, but it insisted that he do so in a capacity as the former chairman of the AEC.[1]

During his tenure of more than a decade as AEC chairman, Seaborg presided over the tremendous growth of the nuclear power industry, which experienced a boom that exceeded the most optimistic expectations of the early 1960s. He also presided over the simultaneous growth of opposition to nuclear power, which threatened to fulfill the worst apprehensions of the early 1960s. Seaborg did not record his own feelings about departing from the AEC; he graciously welcomed his successor and quietly moved on. Despite the controversy that surrounded the AEC, he received an outpouring of acclaim after he announced his resignation. Nixon hailed his contributions "to far greater understanding and application of the miracles of the atom" and remarked that for a period of ten years, three presidents "had the benefit of [his] wisdom and counsel in making decisions which, increasingly, affect the daily lives and well-being of our fellow citizens." The *New York Times* editorialized that Seaborg deserved the "high praise President Nixon voiced." *Newsweek* reported that Seaborg was "justifiably proud of his stewardship." It cited, among his main accomplishments, the arrival of nuclear power plants as "a reality" and the beginning of "full-scale development of the nuclear breeder-reactor." *Nuclear News* expressed "congratulations and thanks for his great service in bringing to fruition so many of the objectives explicit and implicit in the Atomic Energy Act."[2]

The praise for Seaborg did not extend to the AEC's regulatory policies, and, in keeping with his focus as chairman, the favorable farewell notices made little mention of safety problems. But regulatory issues were a major concern of Schlesinger and his new colleague William Doub, who were convinced that the program was in disarray and in need of new leadership. Schlesinger quickly decided to remove Harold Price as director of regulation; at the chairman's request, Price submitted his resignation. The sixty-five-year-old Price met with his staff on 14 October 1971 to announce his departure. "Well folks, this is a once in a lifetime occasion," he said, "and you know what? I'll probably blow it." He confided that he had been thinking about retiring for a year or so, but the revision of the AEC's radiation protection regulations, the emergency core cooling issue, and the Calvert Cliffs ruling had caused him to postpone his plans. "At times," he remarked, "it seems like I have been in this job forever. . . . The job's a whole lot tougher than it used to be—the problems are bigger." He added that a colleague had urged him to find an easier job, "something like Director of Weather!" The AEC presented Price's resignation as a voluntary decision, but it was apparent that both Schlesinger and Doub wanted to replace him. As *Nucleonics Week* reported: "Price was firmly associated with the 'old' AEC which ignored the environmental effects of nuclear power, and . . . he does not fit into the 'new' commission."[3]

The AEC that Seaborg and Price left was an embattled agency. During the spring and summer of 1971, a number of vital and long-standing issues burst into headlines, aroused a new intensity of criticism, and placed enormous pressure on the AEC. The uncertainties over emergency core cooling moved reactor safety to the center of the controversy over nuclear power and made it a more tangible and plausible concern than it had been previously. The AEC's attempts to resolve the technical questions about ECCS performance, reassure the public about reactor safety, and avoid licensing delays by publishing the interim acceptance criteria in June 1971 failed on every count. The following month, the Calvert Cliffs decision sternly reproached the AEC for its circumscribed efforts to comply with NEPA. By forcing the AEC to rewrite its environmental regulations, the court's ruling effectively imposed a licensing moratorium and, ironically, caused the kind of delays that the AEC had hoped to avert when it issued the ECCS criteria.

The emergency cooling debate and the Calvert Cliffs decision were major issues in themselves that required the attention of and focused public scrutiny on the regulatory program. But they did not stand alone; a variety of other questions demanded consideration and triggered criticism at about the same time. In June 1971, the AEC published for public comment its new design objectives for emissions of radioactive materials from nuclear plants. They received a generally favorable response, but they also drew complaints from critics, some of whom argued that the revisions were excessive and some of whom found them inadequate. In either event, the new proposals failed to end the bitter controversy over radiation standards that had embroiled the AEC for over three years. By the summer of 1971 the use of mill tailings as foundation fill in Grand Junction, Colorado and the plans to bury high-level radioactive waste in Lyons, Kansas were sources of angry attacks on the AEC. There were, in addition, contested hearings over several nuclear power plants, which fueled chronic protests from both nuclear supporters and opponents about the AEC's licensing process. To deal with those matters, the AEC relied on an overburdened regulatory staff that was facing the prospect of a heavily increased workload as a result of the Calvert Cliffs decision. In that situation, Schlesinger's commitment to a new approach to regulation and greater emphasis on environmental protection was, in the words of one AEC official, "like General Patton assuming command of a beaten army."[4]

The complaints about the AEC's regulatory policies and procedures came from both sides of the debate over nuclear power. Opponents argued that the AEC failed to provide sufficient protection against radiation hazards, environmental abuse, and severe reactor accidents. They showed little confidence in the AEC's ability or willingness to regulate the nuclear industry adequately, especially in light of the agency's statutory obligation and long-standing commitment to promote the use of nuclear power. The nuclear industry, by contrast, grumbled that the AEC often was overzealous in its regulatory performance, imposing unnecessary and sometimes unreasonable demands. It was concerned that the costs of equipment to combat thermal pollution, improve ECCS, and reduce radiation emissions would undermine the competitive position of nuclear power. The capital costs of a nuclear power plant already far exceeded those of a fossil-fuel facility, and licensing delays added to the expense of construction and uncertainties about the availability of generating capacity. Regulatory requirements were not the only or even the primary source of the economic problems confronting the nuclear industry by the late 1960s. But they were a contributing cause to what Philip Sporn, in a 1970 report to the Joint Committee on Atomic Energy, called "a remarkable and ominous retrogression in the economics of our nuclear power technology."[5]

The assertions of both critics and supporters of nuclear power were frequently exaggerated and self-serving, but each side in the debate advanced arguments that contained important elements of truth. Industry representatives were correct in pointing out that the AEC's regulatory decisions and requirements conflicted, in many cases, with the views of nuclear vendors and utilities. The siting policies of the AEC were a source of chronic and bitter complaints, especially from utilities that wanted to locate close to their load centers. The AEC's ban on metropolitan siting and its actions on seismic siting, particularly in aborting the Bodega Bay and Malibu projects, generated a great deal of consternation and criticism. G. O. Wessenauer, the retired Manager of Power for TVA, approached the issue in humorous terms in a talk to an industry group in 1970, but his comments were a biting rebuke of the AEC's position on siting. "An ideal site for a nuclear plant," he declared, "is one for which there is no evidence of any seismic activity over the past millennia; is not subject to hurricanes, tornadoes, or floods. It should be in the midst of an endless expanse of unpopulated desert with an abundant supply of very cold water flowing nowhere and containing no aquatic life. Most important, it should be adjacent to a major load center."[6]

There were other instances in which the regulatory staff defied the wishes and ignored the protests of the nuclear industry. Industry representatives strongly opposed or expressed serious reservations about, for example, the more demanding requirements for pressure vessels that the staff put into effect, the emphasis on worst-case accidents in evaluating reactor safety, the focus of the AEC's research program, the more rigorous design objectives added to the radiation protection regulations, the terminology and possible impact of at least some of the general design criteria, and the creation of the Atomic Safety and Licensing Board Appeal Panel. In each of those cases, the AEC found the objections of industry unpersuasive and took action that it believed was vital to carrying out its regulatory objectives, even if its decision added to the costs of nuclear power production. The agency always weighed the impact of regulatory changes on nuclear development, but it did not always follow the recommendations or bow to the pressure of the industry. Contrary to the suggestions of antinuclear activists, the AEC did not operate in meek and heedless complicity with the interests and preferences of the industry.

Nevertheless, the critics were justified in emphasizing that the inherent conflict of interest in the AEC's dual responsibilities for promoting and regulating the nuclear industry predisposed it to treat industry concerns sympathetically. The AEC was vitally concerned with encouraging the growth of nuclear power. The government-business partnership that agency officials had long regarded as the best way to establish the nuclear industry made them acutely sensitive to actions that could discourage private development of the technology. This commitment influenced the agency's regulatory programs in many important ways. Although the regulatory staff followed an informal siting standard that prohibited the location of nuclear plants in metropolitan areas, the Commission refused to adopt the same position as official policy. Despite the regulatory staff's ban on what it defined as metropolitan siting, it approved some applications for "suburban" plants that were located reasonably close to populated areas and the load centers of utilities.

The AEC went along with the requests and shared the perspectives of the nuclear industry in other ways. It decided not to publish and then to deny the existence of the WASH–740 update because it feared that the results of the study would generate antinuclear sentiment and threaten support for nuclear development. The AEC, in a series of disputes with other federal agencies over allowable levels of radon in uranium mines, sought to make certain that the limits were not so strict that they would cause the closing of large numbers of mines and risk a shortage of fuel for the nuclear industry. The agency rushed to issue the ECCS interim acceptance criteria in large part to avoid consequences of the increasing uncertainties about emergency cooling that could discourage industrial growth. It moved slowly and grudgingly to regulate against environmental damage from nuclear plants for the same reason. Its position on thermal pollution reflected its fear that forcing nuclear plants to reduce their waste heat discharges would hurt their competitiveness with conventional sources of power. The AEC was reluctant to take an expansive view of its responsibilities under NEPA in large part because of the new delays and complications it would add to the licensing process.

In determining its regulatory policies, the AEC tried to strike a balance between necessary and excessive requirements. This was an unavoidably subjective evaluation that could not, and did not, satisfy differing opinions in an increasingly spirited controversy. The AEC was generally, though not invariably, receptive to the views of the nuclear industry because it shared the industry's conviction that the foremost problem facing the nation relating to the production of power was a shortage of electrical capacity. Like a growing number of utilities, the Joint Committee, and other nuclear proponents, the AEC believed that nuclear plants provided the best means to meet existing and future demand for power without aggravating air pollution. The agency was keenly aware of the uncertainties about the design, performance, and radiation emissions of nuclear plants, but it was confident that the technology was safe and, in comparison with other sources of power, beneficial to the environment.

Furthermore, in the AEC's estimation, nuclear power presented only a slight hazard to public health. Agency spokesmen acknowledged the possibility of a catastrophic nuclear accident, but insisted that its conservative assumptions and multiple lines of defense made the chances acceptably (and exceedingly) small. Its perspective on the advantages of nuclear power was in large part an outgrowth of its statutory mandate to promote the technology, the views of its leading officials, and the pressure of the Joint Committee. But its position was also a result of experience, investigation, and deliberation. In the collective judgment of the AEC, the benefits of nuclear power far surpassed the risks.

By the late 1960s, a growing legion of critics offered a contrasting assessment. They marshalled questions about reactor safety, many of which had first arisen within the AEC, and suggested that the risks of nuclear power exceeded its benefits.[7] The AEC and the nuclear industry took sharp exception, but they could not show empirically or demonstrate convincingly that low-level radiation emissions from nuclear plants would not cause the cancer deaths that Gofman and Tamplin claimed, that ECCS would work as designed, or that the technology was as benign as they believed. In the absence of conclusive evidence, nuclear proponents, in effect, asked the public to trust their judgment. But a growing segment of the public demurred.

The difficulties that the AEC faced and the criticism it endured reflected drastic changes in the regulatory environment that took place between 1962 and 1971. The agency actively promoted the expanding use of nuclear power, only to find that the sudden growth of the industry created complex regulatory problems and spurred increasing public opposition. The agency recognized that nuclear development depended upon safe reactors and wide public support, but the shifting technological and political environment made the achievement of those goals more formidable and more uncertain. Technologically, the rapid expansion of the nuclear industry during the 1960s, both in the number and the size of plants ordered, placed enormous pressure on the AEC's regulatory staff. Keeping up with the flood of applications was, in itself, a hardship for a staff that lacked the resources and personnel it needed. The technical problems were even more trying. The three- and four-fold increase in the size of nuclear plants raised vital new safety questions, including the potential loss of pressure vessel integrity, the effects of a core meltdown, and the performance of emergency cooling systems, that the regulatory staff had to weigh in evaluating applications. Those and other serious safety issues that accompanied the growth of nuclear units through "design by extrapolation" had not arisen in the much smaller and simpler plants that the AEC had licensed by 1962.

The dramatic transformation in the political environment immeasurably compounded the problems confronting the AEC. By the late 1960s, nuclear regulation was no longer an issue of concern only to a small number of experts, as a multiplicity of critical books, articles, interventions, and demonstrations made abundantly clear. The emergence of environmentalism as a visible and vigorous force in American politics after the mid-1960s raised questions about and generated opposition to nuclear power and the priorities of the AEC. The faith in government and confidence in the ability of science to improve standards of living that prevailed in the early 1960s fell victim to the upheavals that protests against the Vietnam war and other social movements spawned. The mounting evidence that, among other things, the government had misled the American people about the war and other issues, the heightened realization that scientific and technological developments could threaten or destroy life as well as improve it, and the growing conviction that the motives and activities of established institutions required continual scrutiny made a major impact on public attitudes toward the AEC and nuclear technology. Austin E. Penn, chairman of the board of Baltimore Gas and Electric, unhappily acknowledged the new political atmosphere in 1969. Defending his company's use of full-page newspaper advertisements that dismissed fears of nuclear power, he insisted that they were necessary "because this is an age of protest against all kinds of established institutions."[8]

A vivid example of the extent of the change in the AEC's operating environment occurred at the annual meeting of the American Association for the Advancement of Science in December 1970. Protesters accused Glenn Seaborg, whose appointment as chairman of the AEC in 1961 had received universal acclaim, of "the crime of science against the American people." Seaborg, recently elected president of the association, was prepared to deliver a paper titled "New Frontiers of the Mind" that stressed the need for cooperation between science and government. A group of about forty radical scientists had informed the press of their plans to disrupt the meeting, and at the first sign of trouble, Seaborg left the platform. A member of the dissident group then read a statement that indicted Seaborg for a litany of offenses. Among them was his leadership of the AEC, "where megadeath development and radiation pollution development are directed." The incident received wide publicity in the national press and prompted a sympathetic call to Seaborg from Nixon. The president remarked that "treating a scientist and former chancellor of a university in this fashion shows these people to be purely destroyers."[9]

In some ways, the AEC suffered consequences from events and forces over which it had little control. The disillusionment with the federal government and with science was fed by a variety of sources, most of which were divorced from the programs of the AEC. As the protest against Seaborg testified, however, dissenters denounced the AEC as a part of the military-industrial complex and viewed all of its programs with suspicion and hostility. Like the growing antipathy toward science and technology, even fear of nuclear energy was not entirely within the boundaries managed by the AEC. As Spencer R. Weart has shown, nuclear fears predated the use of the atomic bomb and the establishment of the AEC. Such anxieties were often not attributable to the agency's policies and were frequently impervious to the AEC's efforts at mollification.[10]

Some issues cited by critics that related more specifically to nuclear power and regulation were also beyond the AEC's direct control. The dual responsibilities for both promoting and regulating the nuclear industry was a part of the agency's statutory mandate, and while it tried to avoid a conflict of interest as best it could, it was committed by law and by political realities to carrying out both functions. Although the AEC welcomed the rapid development of the nuclear industry during the 1960s, it had only limited authority over the pace of growth, the design of reactors, or the size of plants. As long as the construction and operation of nuclear plants remained in the hands of private utilities, the role of the AEC in nuclear development was responsive to and dependent on decisions made by nuclear vendors and utilities—particularly since it sought to avoid taking regulatory actions that might discourage industrial growth. The federal budget stringencies of the late 1960s and early 1970s created hardships for many agencies, but they came at a particularly bad time for the AEC. They prevented the AEC from expanding its regulatory staff and from carrying out important research projects when both were urgently needed.

Yet the AEC was not merely a passive victim of forces beyond its control. Within the constraints imposed by law and by social conditions, it made judgments and reached decisions that fueled distrust of government and science, intensified concern about nuclear power, and undercut its own regulatory position. All too often it played into the hands of its critics by failing to recognize or to deal astutely with its changing environment. As a matter of emphasis, if not commitment, the AEC focused on the promotion rather than the regulation of nuclear power. This gave rise to and enhanced the credibility of charges that it was so intent on industrial development that it was incapable of effective regulation and indifferent to environmental protection. The AEC acted slowly and reluctantly to address questions about the environmental impact of nuclear power and to carry out the requirements of the National Environmental Policy Act. This not only cost it political support but also obscured the environmental advantages of nuclear technology.

The AEC refused to give light-water safety research the resources and the attention it required in the late 1960s. Although the amount of funding that the agency received was largely determined by the Bureau of the Budget, the AEC had much greater control over the *allocation* of its budget. By denying sufficient funds to light-water safety research, it forfeited its best opportunity to find answers to, or at least gather more complete information about, critical questions regarding reactor safety. The most visible result was the emergency core cooling controversy, which greatly increased already rising levels of concern about reactor safety. The AEC hoped to arrest growing opposition to nuclear power by educating the public, but its leaders did not seem to recognize the extent to which their own policies and priorities counteracted those efforts. Although nuclear fears had existed long before the controversy over nuclear power began, the AEC took actions that intensified those apprehensions. And, as Weart noted, nuclear fears "took a special place" in galvanizing the environmental movement.[11]

Both because of the changed environment in which it functioned and its own assignments of resources and priorities, the AEC's nuclear power programs were the target of sharp attacks by the late 1960s. By the summer of 1971, the criticisms had won wide attention and provoked a major debate over nuclear power. As a result, the efforts of the AEC to balance its developmental and regulatory responsibilities and the extent to which it contravened the wishes of the nuclear industry were often overlooked or discounted. The growing distrust of the AEC's reactor programs obscured the attention that the regulatory staff devoted to safety, both in writing regulations and in judging applications, and the benefits of nuclear power technology. Increasingly prevalent reservations about the AEC's performance on reactor safety issues undermined not only its regulatory credibility but its promotional objectives as well.

When Schlesinger and Doub joined the AEC, they sought to narrow the division between nuclear proponents and critics by placing greater emphasis on the AEC's regulatory commitment. Schlesinger articulated his philosophy and his intentions in a speech on 20 October 1971 to a meeting of the Atomic Industrial Forum and the American Nuclear Society in Bal Harbour, Florida. He told his listeners that although it "should be difficult to be other than bullish" about the long-term prospects for nuclear power, the pace of development would depend on two variables: "first, the provision of a safe, reliable product; second, achievement of public confidence in that product." He suggested that carrying out those objectives would be "a demanding task," but he felt confident that it could be done.

Schlesinger emphasized that fulfilling the promise of nuclear power required a major change in the role of the AEC. "From its inception the

Atomic Energy Commission has fostered and protected the nuclear industry," he declared. "Looking back one can, I think, say that this was
the right policy for that historical epoch." But, he continued, the AEC,
in recognition that the nuclear industry was "rapidly approaching mature growth," would redefine its responsibilities. "You should not expect the AEC," Schlesinger announced, "to fight the industry's political,
social, and commercial battles." Rather, he added, the AEC's role was
"primarily to perform as a referee serving the public interest."

Schlesinger made it clear that the AEC was committed to carrying
out the Calvert Cliffs decision, despite the problems it created for the
agency and the industry. "We sympathize with the difficulties that you
are facing," he declared, "but we have no intention of evading our
responsibilities under the law." He went on to affirm that even though a
number of environmentalists had displayed "bad manners," they also
had "raised many legitimate questions" that deserved careful consideration. Schlesinger concluded: "Let me reiterate: the Atomic Energy
Commission, like any government agency, exists to serve the public
interest. The public interest may overlap, but it is not coincident with
private interests. . . . The role of a government agency, designed to
achieve and enforce public goals, is distinct."[12] The message of Schlesinger's speech was unprecedented; he proclaimed a sharp break with
the AEC's history and outlined a new direction in the agency's approach
to and attitude toward its regulatory responsibilities.

Appendix 1: Glossary of Acronyms

ACRS   Advisory Committee on Reactor Safeguards

AEC   U. S. Atomic Energy Commission

ASME   American Society of Mechanical Engineers

BG&E   Baltimore Gas and Electric Company

ECCS   Emergency Core Cooling System

EPA   Environmental Protection Agency

FRC   Federal Radiation Council

GAC   General Advisory Committee of AEC

HEW   U. S. Department of Health, Education, and Welfare

ICRP   International Commission on Radiological Protection

LADWP   Los Angeles Department of Water and Power

LOFT   Loss-of-Fluid Tests

MPCA   Minnesota Pollution Control Agency

NCPC   National Coal Policy Conference

NCRP   National Committee on Radiation Protection

NEPA   National Environmental Policy Act

NUMEC   Nuclear Materials and Equipment Corporation

PG&E   Pacific Gas and Electric Company

SPERT   Special Power Excursion Reactor Tests

TVA   Tennessee Valley Authority
UCS   Union of Concerned Scientists
WL   Working Level (of radon)
WLM   Working Level Month

Appendix 2: Chronology of Regulatory History

Date	Event
Date	*Event*
1 June 1962	AEC published reactor site criteria.
20 November 1962	AEC issued "Civilian Nuclear Power—A Report to the President."
4 December 1962	Los Angeles Department of Water and Power announced plans to build Malibu plant.
10 December 1962	Consolidated Edison filed for a construction permit for the Ravenswood plant.
28 December 1962	Pacific Gas and Electric filed for a construction permit for the Bodega Bay plant.
19 February 1963	Public meeting aired opposition to Ravenswood.
20 May 1963	Secretary of the Interior Udall told AEC that Bodega Bay site was "reason for grave concern."
26 August 1963	U. S. Public Health Service announced new findings on mortality rates from lung cancer among uranium miners.
12 December 1963	Jersey Central Power and Light announced plans to build Oyster Creek plant.
3 January 1964	Consolidated Edison announced its withdrawal of Ravenswood application.
30 October 1964	Pacific Gas and Electric announced its withdrawal of Bodega Bay application.

23 March 1965	Atomic Safety and Licensing Board hearings on Malibu application opened.
18 June 1965	Chairman Seaborg sent letter to Joint Committee on probability and consequences of a major reactor accident.
14 July 1965	First Mitchell Panel report submitted.
30 July 1965	AEC's General Advisory Committee submitted its "Review of Reactor Safety Research Program."
22 November 1965	Draft general design criteria issued for public comment.
24 November 1965	ACRS sent letter expressing concern about pressure vessel reliability.
6 May 1966	Senator Muskie presided over hearing on water pollution from uranium mill tailings.
17 June 1966	TVA announced plans to build Browns Ferry plants.
5 October 1966	Accident occurred at Fermi plant.
10 March 1967	Report of the Lumb Panel on safeguarding special nuclear material submitted.
27 March 1967	Commission upheld Atomic Safety and Licensing Board decision on Malibu.
30 June 1967	Second Mitchell Panel report submitted.
27 July 1967	FRC issued a standard of 12 WLM for permissible levels of radon in uranium mines, subject to further review.
29 September 1967	Leaks revealing quality assurance problems detected at Oyster Creek.
23 October 1967	AEC released Ergen Report on emergency core cooling.
4 November 1968	AEC adopted technical specifications.
13 January 1969	U. S. Court of Appeals for the First Circuit sustained AEC position on regulating against thermal pollution.
20 January 1969	Article in *Sports Illustrated* by Robert H. Boyle criticized AEC's position on thermal pollution.
March 1969	U. S. Public Health Service published report, "Evaluation of Radon 222 Near Uranium Tailings Piles."
12 May 1969	Minnesota Pollution Control Agency required radiation limits stricter than those of the AEC in granting a waste discharge permit to Monticello plant.
18 August 1969	Atomic Safety and Licensing Board Appeal Panel established.

11 September 1969	Commissioners faced nuclear critics in a public forum for the first time at a conference on nuclear power held at the University of Vermont.
28 October 1969	Joint Committee hearings on the environmental effects of electrical production opened.
29 October 1969	Gofman and Tamplin delivered a paper arguing that radiation from nuclear plants could cause 17,000 cancer deaths annually.
1 January 1970	National Environmental Policy Act signed into law.
28 March 1970	AEC issued revised radiation-protection regulations.
31 March 1970	Backfitting regulation adopted.
2 April 1970	AEC announced plan (Appendix D) to implement NEPA.
3 April 1970	Water Quality Improvement Act signed into law.
3 June 1970	Revised Appendix D published for public comment.
17 June 1970	AEC announced plans to develop a repository for high-level nuclear wastes at a Lyons, Kansas site.
26 June 1970	Quality assurance criteria adopted.
4 December 1970	AEC published another revised version of Appendix D.
22 December 1970	United States District Court for the District of Minnesota ruled that the federal government had exclusive authority to set radiation standards.
24 December 1970	Regulation on emergency planning adopted.
19 February 1971	General design criteria adopted.
25 May 1971	EPA issued compromise ruling on radon limits in uranium mines.
4 June 1971	President Nixon delivered message on energy production and environmental protection.
7 June 1971	AEC issued revised radiation-protection regulations, including numerical design objectives, for public comment.
19 June 1971	AEC issued ECCS interim acceptance criteria.
21 July 1971	Seaborg announced resignation as chairman of AEC; Nixon announced appointments of Schlesinger and Doub.
23 July 1971	U. S. Court of Appeals for the District of Columbia Circuit rebuked the AEC in Calvert Cliffs decision.
17 August 1971	Schlesinger sworn in as chairman of the AEC and Doub sworn in as a commissioner.

26 August 1971   AEC announced that it would not appeal the Calvert
                      Cliffs decision.

14 October 1971   AEC announced resignation of Harold L. Price as direc-
                      tor of regulation.

20 October 1971   Schlesinger announced new AEC approach to regulation
                      in speech to nuclear industry groups at Bal Harbour, Flor-
                      ida.

Notes

CHAPTER I: THE CONTEXT OF NUCLEAR
REGULATION, 1946–1962

1. Glenn T. Seaborg and William R. Corliss, *Man and Atom: Building a New World through Nuclear Technology* (New York: E. P. Dutton, 1971), p. 53.

2. Richard Curtis and Elizabeth Hogan, *Perils of the Peaceful Atom: The Myth of Safe Nuclear Power Plants* (Garden City, NY: Doubleday and Co., 1969), p. xiii.

3. For background on the 1946 Atomic Energy Act and the AEC's organization, see George T. Mazuzan and J. Samuel Walker, *Controlling the Atom: The Beginnings of Nuclear Regulation, 1946–1962* (Berkeley and Los Angeles: University of California Press, 1984), chap. 1; Richard G. Hewlett and Oscar E. Anderson, Jr., *The New World, 1939/1946: Volume I of a History of the United States Atomic Energy Commission* (University Park: Pennsylvania State University Press, 1962), chaps. 13–14; Richard G. Hewlett and Francis Duncan, *Atomic Shield, 1947/1952: Volume II of a History of the United States Atomic Energy Commission* (University Park: Pennsylvania State University Press, 1969), chaps. 1–3.

4. "The Atom: The Masked Marvel," *Time*, 14 January 1952, pp. 18–22; *Nucleonics Week*, 18 December 1969, p. 1; Corbin Allardice and Edward R. Trapnell, *The Atomic Energy Commission* (New York: Praeger Publishers, 1974), pp. 15–20; Richard T. Sylves, *The Nuclear Oracles: A Political History of the General Advisory Committee of the Atomic Energy Commission, 1947–1977* (Ames: Iowa State University Press, 1987), pp. 256–270; Hewlett and Duncan, *Atomic Shield*, pp. 27–46, 581–584, 675.

5. Harold P. Green and Alan Rosenthal, *Government of the Atom: The Integration of Powers* (New York: Atherton Press, 1963), pp. 1–30, 79–103,

433

168–197; Public Law 88–72, 88th Cong., 1st Sess., 1963; Senate Report 303, 88th Cong., 1st Sess., 1963; House Report 446, 88th Cong., 1st Sess., 1963.

6. Richard Wayne Dyke, *Mr. Atomic Energy: Congressman Chet Holifield and Atomic Energy Affairs, 1945–1974* (Westport, CT: Greenwood Press, 1989), pp. 1–40, 220–239; *Congressional Record*, 91st Cong., 1st Sess., 1969, pp. 5353–5354.

7. "The Joint Committee on Atomic Energy," *Nuclear News*, 11 (February 1968): 31–42; *Nucleonics Week*, 3 December 1970, pp. 2–3, 17 December 1970, pp. 7–8.

8. Craig Hosmer speech to ANS-AIF Conference, 13 November 1968, Box 3 (AIF Discussion Notes), Office Files of James T. Ramey, Hosmer speech to Think Shop Conference, 25 September 1969, Box 7763, MH&S–11 (Industrial Hygiene), Atomic Energy Commission Records, Department of Energy, Germantown, Maryland (hereafter cited as AEC/DOE); "Joint Committee on Atomic Energy," *Nuclear News*, p. 41; Author's interview with Chet Holifield, Balboa, California, 12 July 1983.

9. *The Public Papers of Dwight D. Eisenhower, 1953* (Washington, DC: Government Printing Office, 1960), pp. 813–822; Mazuzan and Walker, *Controlling the Atom*, pp. 22–25. See also George T. Mazuzan and J. Samuel Walker, "Developing Nuclear Power in an Age of Energy Abundance, 1946–1962," *Materials and Society*, 7 (1983): 307–319, and, for a similar argument, Rebecca S. Lowen, "Entering the Atomic Power Race: Science, Industry, and Government," *Political Science Quarterly*, 102 (Fall 1987): 459–479.

10. The 1954 Atomic Energy Act is printed in U. S. Congress, Senate, *Atoms for Peace Manual: A Compilation of Official Materials on International Cooperation for Peaceful Uses of Atomic Energy*, 84th Cong., 1st Sess., 1955, pp. 209–251.

11. Joint Committee on Atomic Energy (JCAE), *Hearings on S. 3323 and H. R. 8862, To Amend the Atomic Energy Act of 1946*, 83rd Cong., 2nd Sess., 1954, pp. 334–335; *Profit Perspectives in Atomic Energy: Problems and Opportunities* (New York: American Management Association, 1957), pp. 24–53; Richard G. Hewlett and Jack M. Holl, *Atoms for Peace and War, 1953–1961: Eisenhower and the Atomic Energy Commission* (Berkeley and Los Angeles: University of California Press, 1989), pp. 201–208; Mazuzan and Walker, *Controlling the Atom*, pp. 77–78.

12. JCAE, *Hearings on Accelerating Civilian Reactor Program*, 84th Cong., 2nd Sess., 1956, pp. 1–71, 160–166, 186–204, 261–307, 447–457; Mazuzan and Walker, *Controlling the Atom*, pp. 118–121; Hewlett and Holl, *Atoms for Peace and War*, pp. 344–345, 403–429.

13. Mazuzan and Walker, *Controlling the Atom*, pp. 82–92.

14. *Ibid.*, pp. 32–58; J. Samuel Walker, "The Controversy over Radiation Safety: A Historical Overview," *JAMA: The Journal of the American Medical Association*, 262 (4 August 1989): 664–668.

15. AEC 132/64 (7 January 1964), Atomic Energy Commission Records, Nuclear Regulatory Commission, Rockville, Maryland (hereafter cited as AEC/NRC).

16. Mazuzan and Walker, *Controlling the Atom*, pp. 407–410; Hewlett

and Holl, *Atoms for Peace and War*, pp. 498–514; Elizabeth S. Rolph, *Nuclear Power and the Public Safety: A Study in Regulation* (Lexington, MA: D. C. Heath and Co., 1979), pp. 170–171.

17. George A. W. Boehm, "The AEC Gets a Different Kind of Scientist," *Fortune*, April 1961, pp. 158–160; "Seaborg's AEC: Atoms for War or Peace," *Newsweek*, 16 October 1961, pp. 63–68; "Glenn Seaborg: From Californium to the AEC," *Time*, 10 November 1961, p. 23; National Science Foundation News Release, 17 April 1988, AEC/NRC.

18. *Journal of Glenn T. Seaborg* (25 vols., Berkeley: Lawrence Berkeley Laboratory PUB–625, 1989), Vol. 1, pp. 88–89 of "Introduction," and *passim*; Seaborg speech to National Academy of Sciences, 5 May 1969, Box 25 (Pollution of the Environment), Nixon Library Materials, AEC/DOE; "This Too Is Glenn Seaborg," *Nuclear Industry*, 17 (May 1970): 44.

19. *Seaborg Journal*, Vol. 1, pp. 87–88 of "Introduction," Vol. 23, p. 552; D. Z. Robinson to D. F. Hornig, 12 April 1965, Subject File (Atomic Energy Commission, 1965), Record Group 359 (Records of the Office of Science and Technology), National Archives, Washington, DC.

20. Mazuzan and Walker, *Controlling the Atom*, pp. 409–418.

CHAPTER II: THE NUCLEAR INDUSTRY AND THE BANDWAGON MARKET

1. Arthur D. Little, Inc., "Competition in the Nuclear Power Supply Industry: Report Prepared for the U. S. Atomic Energy Commission and the Department of Justice" (Report No. C-69376), December 1968, pp. 137–145, Atomic Energy Commission Records, Nuclear Regulatory Commission, Rockville, Maryland (hereafter cited as AEC/NRC); Frank G. Dawson, *Nuclear Power: Development and Management of a Technology* (Seattle: University of Washington Press, 1976), pp. 140–143; Mark Hertsgaard, *Nuclear Inc.: The Men and Money Behind Nuclear Energy* (New York: Pantheon Books, 1983), pp. 20–39, 103–135; John F. Hogerton, "The Arrival of Nuclear Power," *Scientific American*, 218 (February 1968): 21–31.

2. Richard G. Hewlett, "Man Harnesses the Atom," in Melvin Kranzberg and Carroll W. Pursell, Jr., eds., *Technology in the Twentieth Century* (New York: Oxford University Press, 1967), p. 265; Anthony V. Nero, *A Guidebook to Nuclear Reactors* (Berkeley and Los Angeles: University of California Press, 1979), pp. 21–25.

3. Corbin Allardice and Edward R. Trapnell, *The Atomic Energy Commission* (New York: Praeger Publishers, 1974), pp. 92–93, 107; Richard G. Hewlett and Francis Duncan, *Nuclear Navy, 1946–1962* (Chicago: University of Chicago Press, 1974), pp. 39–41, 97–100, 135–142; John F. Hogerton, *The Atomic Energy Deskbook* (New York: Reinhold Publishing Corp., 1963), pp. 65–66.

4. Hewlett, "Man Harnesses the Atom," pp. 266–267; Hewlett and Duncan, *Nuclear Navy*, pp. 228–257; Nero, *Guidebook to Nuclear Reactors*, pp. 77–84.

5. Hogerton, "The Arrival of Nuclear Power," p. 24; Dawson, *Nuclear*

Power, p. 96; Little, "Competition in the Nuclear Power Supply Industry," p. 141; Hertsgaard, *Nuclear Inc.*, p. 42; W. David Montgomery and James R. Quirk, "Cost Escalation in Nuclear Power," in Lon C. Ruedisill and Morris W. Firebaugh, eds., *Perspectives on Energy: Issues, Ideas and Environmental Dilemmas* (New York: Oxford University Press, 1978), p. 343.

6. Kendall Birr, *Pioneering in Industrial Research: The Story of the General Electric Research Laboratory* (Washington, D.C.: Public Affairs Press, 1957); David F. Noble, *America by Design* (New York: Alfred A. Knopf, 1977), pp. 112–114; George Wise, "A New Role for Professional Scientists in Industry: Industrial Research at General Electric, 1900–1916," *Technology and Culture*, 21 (July 1980): 408–409; George Wise, "Science at General Electric," *Physics Today*, 37 (December 1984): 52–61; Allardice and Trapnell, *Atomic Energy Commission*, p. 20; Hewlett and Duncan, *Nuclear Navy*, pp. 38–40.

7. "Atomic Energy: The Powerhouse," *Time*, 12 January 1959, p. 75; Hogerton, *Atomic Energy Deskbook*, pp. 72–76, 592–593; R. B. Richards, "Boiling Water Reactors," October 1959, Box 349 (General Electric Co.), Papers of the Joint Committee on Atomic Energy (JCAE), Record Group 128 (Records of the Joint Committees of Congress), National Archives, Washington, D. C.

8. "Atomic Energy: The Powerhouse," *Time*, pp. 75–76, 86; Richards, "Boiling Water Reactors," JCAE Papers.

9. Harold B. Mayers, "The Great Nuclear Fizzle at Old B&W," *Fortune*, 80 (November 1969): 123–124; Walter Hamilton to Carl T. Durham, 10 March 1955, Box 97 (Atomic Power: Babcock and Wilcox Co.), JCAE Papers; Little, "Competition in the Nuclear Power Industry," pp. 138, 141–145; JCAE, *Hearings on AEC Authorizing Legislation Fiscal Year 1967*, 89th Cong., 2nd Sess., 1966, pp. 870–874; Hertsgaard, *Nuclear Inc.*, pp. 45–46. Three other companies supplied reactor systems under the AEC's Power Demonstration Reactor Program, which it established in 1955 to encourage industry development. They were: Allis-Chalmers Manufacturing Company, Atomics International, and General Atomic. Only Allis-Chalmers built a light-water reactor. Atomics International, a division of the North American Rockwell Corporation, supplied a sodium graphite reactor to a plant in Hallam, Nebraska and an organic cooled and moderated reactor to a plant in Piqua, Ohio. General Atomic, a division of General Dynamics until 1967, when it was acquired by the Gulf Oil Corporation, built high-temperature gas-cooled reactors for plants in Peach Bottom, Pennsylvania and Fort St. Vrain, Colorado.

10. JCAE, *Hearings on AEC Authorizing Legislation Fiscal Year 1967*, 1966, pp. 870–871; Little, "Competition in the Nuclear Power Supply Industry," pp. 17–30.

11. Little, "Competition in the Nuclear Power Supply Industry," pp. 176–209; Hertsgaard, *Nuclear Inc.*, 47–49.

12. Robert E. Wilson to the Commission, 10 January 1962, Wilson to James T. Ramey, 12 January 1962, Box 671, (Special Nuclear Materials, Private Ownership), John T. Conway to Chet Holifield, 10 October 1963, Box 405, (Legislation: S–2635, Private Ownership), JCAE Papers; Robert E. Wilson to Holifield, 25 January 1963, Box 42 (Energy Monopoly/Private Ownership), Chet Holifield Papers, University of Southern California, Los Angeles; JCAE, *Hearings on Pri-*

vate Ownership of Special Nuclear Materials, 88th Cong., 1st Sess., 1963, *Hearings on Private Ownership of Special Nuclear Materials*, 88th Cong., 2nd Sess., 1964; House Report 1702, 88th Cong., 2nd Sess., 1964; *Congressional Record*, 88th Cong., 2nd Sess., 1964, pp. 17851–17853, 19514–19519.

13. House Report 1702, pp. 1–3, 12–19.

14. Luther J. Carter, *Nuclear Imperatives and Public Trust: Dealing with Radioactive Waste* (Washington, D.C.: Resources for the Future, 1987), pp. 91–109; Gene I. Rochlin, Margery Held, Barbara G. Kaplan, and Lewis Kruger, "West Valley: Remnant of the AEC," *Bulletin of the Atomic Scientists*, 34 (January 1978): 17–26; Richard G. Hewlett, "Federal Policy for the Disposal of Highly Radioactive Wastes from Commercial Nuclear Power Plants: An Historical Analysis," 9 March 1978, and AEC-R 163/3 (1 March 1968), AEC/NRC.

15. Carter, *Nuclear Imperatives*, pp. 41–71; Hewlett, "Federal Policy for the Disposal of Highly Radioactive Wastes," AEC/NRC; George T. Mazuzan and J. Samuel Walker, *Controlling the Atom: The Beginnings of Nuclear Regulation, 1946–1962* (Berkeley and Los Angeles: University of California Press, 1984), pp. 345–352.

16. Little, "Competition in the Nuclear Power Supply Industry," pp. 49–51, 305–311.

17. *Ibid.*, pp. 274–290; Hertsgaard, *Nuclear Inc.*, pp. 122–123.

18. L. Silverman, D. L. Morrison, R. L. Ritzman, and T. J. Thompson, "Fission-Product Behavior and Retention in Containment Systems," and T. J. Thompson and C. R. McCullough, "The Concepts of Reactor Containment," in T. J. Thompson and J. G. Beckerley, eds., *The Technology of Nuclear Reactor Safety*, 2 vols., (Cambridge, Mass.: MIT Press, 1973), II, pp. 619–697, 755–801; Little, "Competition in the Nuclear Power Supply Industry," pp. 291–304.

19. Little, "Competition in the Nuclear Power Supply Industry," pp. 72, 119–122; John McDonald, "Westinghouse Invents a New Westinghouse," *Fortune*, 76 (October 1967): 143–147; Meyers, "The Great Nuclear Fizzle at Old B&W," p. 123.

20. Richard F. Hirsh, *Technology and Transformation in the American Electric Utility Industry* (Cambridge: Cambridge University Press, 1989), pp. 1–46; Richard F. Hirsh, "Conserving Kilowatts: The Electric Power Industry in Transition," *Materials and Society*, 7 (1983): 295–305; Irvin C. Bupp and Jean-Claude Derian, *Light Water: How the Nuclear Dream Dissolved* (New York: Basic Books, 1978), pp. 75–76.

21. Hirsh, *Technology and Transformation*, pp. 25–35, 71–82; Hirsh, "Conserving Kilowatts," pp. 296–297.

22. Little, "Competition in the Nuclear Power Supply Industry," pp. 13–16; Bruce C. Netschert, "Developing the Energy Inheritance," in Kranzberg and Pursell, eds., *Technology in the Twentieth Century*, pp. 247–250.

23. Hirsh, *Technology and Transformation*, pp. 71–81; George T. Mazuzan and J. Samuel Walker, "Developing Nuclear Power in an Age of Energy Abundance," *Materials and Society*, 7 (1983): 307–319.

24. Hogerton, "The Arrival of Nuclear Power," p. 25; Wendy Allen, *Nuclear Reactors for Generating Electricity: U. S. Development from 1946 to 1963* (Santa Monica, CA.: Rand Corporation, 1977), pp. 72–74.

25. Robert Perry, *Development and Commercialization of the Light Water Reactor, 1946–1976* (Santa Monica, CA.: Rand Corporation, 1977), pp. 26–28; JCAE, *Hearings on Development, Growth, and State of the Atomic Energy Industry*, 87th Cong., 2nd Sess., 1962, pp. 153–158.

26. Perry, *Development and Commercialization of the Light Water Reactor*, pp. 26–27.

27. Little, "Competition in the Nuclear Power Supply Industry," p. 358; JCAE, *Hearings on the Development, Growth and State of the Atomic Energy Industry*, 88th Cong., 1st Sess., 1963, pp. 561–562; JCAE, Committee Print, *Nuclear Power Economics—1962 Through 1967*, 90th Cong., 2nd Sess., 1968, p. 142.

28. Allen T. Demaree, "G.E.'s Costly Ventures into the Future," *Fortune*, 82 (October 1970); 88–93; Stephen L. Del Sesto, "The Commercialization of Civilian Nuclear Power and the Evolution of Opposition: The American Experience, 1960–1974," *Technology in Society*, 1 (1979): 300–327; "Atomic Energy: Turning the Corner," *Time*, 12 July 1963, p. 85; Perry, *Development and Commercialization of the Light Water Reactor*, p. 28; Little, "Competition in the Nuclear Power Supply Industry," pp. 360–367; Hertsgaard, *Nuclear Inc.*, pp. 43–44.

29. JV [Jack Valente], "Memo for the President," 12 June 1964 (AT: 11–23—63—7–22–66), Marguerite Owen to Douglas Cater, 25 May 1964 (AT 2), White House Central Files, Lyndon B. Johnson Papers, Lyndon B. Johnson Library, Austin, Texas; "Report on Economic Analysis for Oyster Creek Nuclear Generating Station," *Nuclear News*, 7 (January 1964): 1–46; A. F. Tegen to John O. Pastore, 17 February 1964, Box 578 (Reactors by Type or Company Name: Jersey Central), JCAE Papers; *Journal of Glenn T. Seaborg* (25 vols., Berkeley: Lawrence Berkeley Laboratory PUB–625, 1989), Vol. 7, p. 39; JCAE, *Hearings on Development . . . of Atomic Energy Industry*, 1962, pp. 152, 688–692; JCAE, *Hearings on Development . . . of Atomic Energy Industry*, 1963, pp. 479–495.

30. JCAE, *Nuclear Power Economics—1962 Through 1967*, pp. 45–46; Demaree, "G.E.'s Costly Ventures into the Future," p. 93.

31. JCAE, *Nuclear Power Economics—1962 Through 1967*, pp. 4–5, 11, 15, 32–33; Hogerton, "The Arrival of Nuclear Power," p. 27.

32. U. S. Atomic Energy Commission, WASH–1435, "Power Plant Capital Costs: Current Trends and Sensitivity to Economic Parameters," October 1974, pp. 6–7; "When the Music Stops," *Forbes*, 1 February 1969, pp. 3–4; "Neck and Neck—And Breathing Hard," *Forbes*, 1 February 1973, pp. 22–23; Del Sesto, "Commercialization of Civilian Nuclear Power," pp. 307–311; Hertsgaard, *Nuclear Inc.*, pp. 42–44; Perry, *Development and Commercialization of the Light Water Reactor*, pp. 28–37; Bupp and Derian, *Light Water*, p. 72.

33. "Nuclear Plants Turn Up the Juice," *Business Week*, 11 March 1967, p. 65; R. T. Person to Lyndon B. Johnson, 30 December 1965, Joe Califano to Johnson, 6 January 1965, Lee C. White to Johnson, 2 November 1966, Federal Power Commission, "Prevention of Power Failures," July 1967 (UT 2: Electric-

ity), White House Central File, Johnson Papers; Hogerton, "The Arrival of Nuclear Power," p. 27; Netschert, "Developing the Energy Inheritance," pp. 252–253.

34. JCAE, *Nuclear Power Economics—1962 Through 1967*, p. 12; Little, "Competition in the Nuclear Power Supply Industry," pp. 371–376; Hirsh, *Technology and Transformation*, pp. 56–70.

35. JCAE, *Hearings on AEC Authorizing Legislation Fiscal Year 1968*, 90th Cong., 1st Sess., 1967, pp. 1280–1282; JCAE, *Nuclear Power Economics—1962 Through 1967*, p. 34; Hogerton, "The Arrival of Nuclear Power," pp. 27–28; "News About Industry," *Nuclear Industry*, 12 (March 1965): 7–11.

36. Tom O'Hanlon, "An Atomic Bomb in the Land of Coal," *Fortune*, 74 (September 1966): 132–133; Little, "Competition in the Nuclear Power Supply Industry," pp. 164–172; JCAE, *Nuclear Power Economics—1962 Through 1967*, pp. 32–34.

37. Glenn T. Seaborg to Lyndon B. Johnson, 28 June 1966 (Confidential File: Agency Reports), White House Central File, Johnson Papers; Alvin M. Weinberg and Gale Young, "The Nuclear Energy Revolution—1966," *Proceedings of the National Academy of Sciences*, 57 (15 January 1967): 1–3; *Nucleonics Week*, 25 November 1965, p. 2; O'Hanlon, "Atomic Bomb in the Land of Coal," pp. 132–133.

38. JCAE, *Hearings on AEC Authorizing Legislation Fiscal Year 1970*, 91st Cong., 1st Sess., 1969, pp. 975–1017; "The Harnessed Atom: It's a Business Now," *Newsweek*, 18 April 1966, p. 84. In bi-weekly reports to President Johnson, AEC chairman Seaborg emphasized the significance of the reactor boom market. The reports are in the Confidential File (Agency Reports), White House Central File, Johnson Papers.

39. "The Early Bird," *Forbes*, 15 September 1969, p. 38; Jeremy Main, "A Peak Load of Trouble for the Utilities," *Fortune*, 80 (November 1969): 194; "JCAE Treats AEC Budget Gently But Issues Stern Warning," *Nucleonics*, 24 (June 1966): 23; *Nucleonics Week*, 20 March 1969, p. 2; JCAE, *Hearings on AEC Authorizing Legislation Fiscal Year 1970*, pp. 975–1017; JCAE, *Hearings on AEC Authorizing Legislation Fiscal Year 1971*, 91st Cong., 2nd Sess., 1970, pp. 1146–1148; JCAE, *Nuclear Power Economics—1962 Through 1967*, pp. 225, 259–260, 273–274.

40. JCAE, *Nuclear Power Economics—1962 Through 1967*, pp. 260, 274–275.

41. "The Growing Market for Nuclear KW," *Fortune*, 68 (July 1963): 173–176.

42. Mazuzan and Walker, *Controlling the Atom*, pp. 93–121, 183–213.

43. JCAE, *Hearings on Prelicensing Anti-Trust Review of Nuclear Power Plants*, 92nd Cong., 1st Sess., 1969, pp. 16–30; "New Rules of Atomic Power Plants," *Business Week*, 24 August 1968, p. 63; L. Manning Muntzing and Martin R. Hoffman to the Commission, 20 March 1972, Box 42 (Pre-Licensing Anti-Trust Review), Office Files of James T. Ramey, Atomic Energy Commission Records, Department of Energy, Germantown, Maryland.

CHAPTER III: THE REGULATORY FRAMEWORK

1. "AEC's Regulatory Division Now in Five Divisions," *Forum Memo*, 11 (April 1964): 26; *Nucleonics Week*, 29 August 1968, p. 1, 4 February 1971, p. 3, 11 November 1971, pp. 5–6; Joint Committee on Atomic Energy (JCAE), *Hearings on Licensing and Regulation of Nuclear Reactors*, 90th Cong., 1st Sess., 1967, p. 6; AEC Press Release, 17 May 1963, Atomic Energy Commission Records, Nuclear Regulatory Commission, Rockville, Maryland (hereafter cited as AEC/NRC).

2. "Remarks by Commissioner James T. Ramey at the AEC Distinguished Award Ceremony," 7 September 1967, AEC Press Releases, 25 September 1961, 7 September 1967, AEC/NRC; Craig Hosmer speech to ANS-AIF Joint Banquet, 13 November 1968, Box 3 (AIF Discussion Notes 1968), Office Files of James T. Ramey, Atomic Energy Commission Records, Department of Energy, Germantown, Maryland (hereafter cited as AEC/DOE); *Nucleonics Week*, 14 November 1968, p. 1; JCAE, *Hearings on Licensing and Regulation of Nuclear Reactors*, 1967, p. 60.

3. AEC Press Releases, 28 March 1964, 20 February 1967, Harold L. Price to the Commission, 8 February 1967, Job 8, Box 1, O&M–2 (Director of Regulation), AEC/NRC; *Nucleonics Week*, 23 February 1967, p. 1; "AEC's Regulatory Staff Now in Five Divisions," p. 26. For discussions of earlier organizational arrangements, see George T. Mazuzan and J. Samuel Walker, *Controlling the Atom: The Beginnings of Nuclear Regulation, 1946–1962* (Berkeley and Los Angeles: University of California Press, 1984), chaps. 3, 13.

4. AEC Press Release, 1 September 1966, AEC/NRC; " 'Swamped' AEC Reactor Regulators Plan Major Reorganization," *Nucleonics*, 24 (November 1966): 62–63.

5. For discussions of the AEC's regulatory process, see: Sidney G. Kingsley, "The Licensing of Nuclear Power Reactors in the United States," *Atomic Energy Law Journal*, 7 (Fall 1965): 309–352; Arthur W. Murphy, "Atomic Safety and Licensing Boards: An Experiment in Administrative Decision Making on Safety Questions," *Law and Contemporary Problems*, 33 (Summer 1968): 566–589; JCAE, *Hearings on Licensing and Regulation of Nuclear Reactors*, 1967, pp. 40–84; JCAE, *Hearings on Proposed Extension of AEC Indemnity Legislation*, 89th Cong., 1st Sess., 1965, pp. 239–244; Mazuzan and Walker, *Controlling the Atom*, pp. 60–71.

6. David Okrent, *Nuclear Reactor Safety: On the History of the Regulatory Process* (Madison: University of Wisconsin Press, 1981), pp. 8–11; S. H. Bush, "The Role of the Advisory Committee on Reactor Safeguards in the Reactor Licensing Process," *Nuclear Safety*, 13 (January-February 1972): 1–13; JCAE; *Hearings on the Licensing and Regulation of Nuclear Reactors*, 1967, pp. 33, 84–150; Mazuzan and Walker, *Controlling the Atom*, pp. 129–146, 182–199.

7. JCAE, *Hearings on the Licensing and Regulation of Nuclear Reactors*, 1967, p. 42; Kingsley, "The Licensing of Nuclear Power Reactors," p. 320.

8. Murphy, "Atomic Safety and Licensing Boards," pp. 574–577; Mazuzan and Walker, *Controlling the Atom*, pp. 60, 373–382.

9. Murphy, "Atomic Safety and Licensing Boards," pp. 574–581; Kingsley, "The Licensing of Nuclear Power Reactors," pp. 324–332; JCAE, *Hearings on the Licensing and Regulation of Nuclear Reactors,* 1967, pp. 151–177; JCAE, *Hearings on Proposed . . . Indemnity Legislation,* 1965, p. 240.

10. Kingsley, "The Licensing of Nuclear Power Reactors," pp. 332–333; JCAE, *Hearings on the Licensing and Regulation of Nuclear Reactors,* 1967, pp. 292–294.

11. *Nucleonics Week,* 28 January 1965, p. 1; "A Reform Is Proposed," *Nuclear Industry,* 12 (August 1965): 3–9.

12. Chet Holifield to Larry O'Brien, 23 January 1962, Holifield to Glenn T. Seaborg, 23 January 1962, James T. Ramey to Holifield, 15 July 1968, Box 16 (James T. Ramey), Chet Holifield Papers, University of Southern California, Los Angeles; Ramey to the Commission, 20 November 1970, Box 53 (Reactor Licensing Delays, AIF Conference), Ramey Office Files, AEC/DOE; JCAE, *Hearings on Licensing and Regulation of Nuclear Reactors,* 1967, pp. 411–412, 457; "Full Strength AEC Set to Focus on Industry Problems," *Nucleonics,* 24 (October 1966): 19–21; *Nucleonics Week,* 27 January 1966, p. 1; *Journal of Glenn T. Seaborg* (25 vols., Berkeley: Lawrence Berkeley Laboratory PUB–625, 1989), Vol. 10, p. 691; Richard G. Hewlett and Francis Duncan, *Atomic Shield, 1947–1952* (University Park: Pennsylvania State University Press, 1969), p. 256; Richard G. Hewlett and Jack M. Holl, *Atoms for Peace and War, 1953–1961: Eisenhower and the Atomic Energy Commission* (Berkeley and Los Angeles: University of California Press, 1989), pp. 426–427; Harold P. Green and Alan Rosenthal, *Government of the Atom: The Integration of Powers* (New York: Atherton Press, 1963), p. 107.

13. *Nucleonics Week,* 28 January 1965, p. 1; JCAE, *Hearings on Licensing and Regulation of Nuclear Reactors,* 1967, pp. 411–412, 432–433.

14. JCAE, *Hearings on Licensing and Regulation of Nuclear Reactors,* 1967, pp. 411–431.

15. AEC–R 4/38 (10 January 1966), AEC–R 4/42 (31 May 1966), AEC/NRC; Harold L. Price to the Commission, 24 February 1966, Box 1390, O&M–7 (ACRS), AEC/DOE; *Nucleonics Week,* 7 October 1965, p. 1, 27 January 1966, pp. 2–4, 9 June 1966, p. 1.

16. Harold L. Price to the Commission, 8 June 1967, Job 8, Box 2, O&M–7 (Regulatory Review Panel), AEC–R 4/46 (18 October 1967), AEC–R 4/52 (13 March 1968), AEC/NRC; A. O. Little to Glenn T. Seaborg, 25 March 1966, Box 80 (Director of Regulation), Office Files of Glenn T. Seaborg, AEC/DOE; "Second Mitchell Report," *Nuclear Industry,* 14 (July 1967): 6–11; JCAE, *Hearings on Licensing and Regulation of Nuclear Reactors,* 1967, pp. 43–44, 476–477.

17. *Nucleonics Week,* 30 May 1963, p. 1, 1 December 1966, pp. 1–2; "Commissioner Palfrey Answers Questions," *Nucleonics,* 22 (February 1964): 19; *Seaborg Journal,* Vol. 22, p. 51.

18. AEC 948/8 (18 April 1968), Job 8, Box 1, O&M–2 (Director of Regulation), AEC/NRC; "Olwell Report," 30 September 1968, Box 42 (Separate Agency Question), Ramey Office Files, AEC/DOE; *Nucleonics Week,* 12 September 1968, p. 2; JCAE, *Hearings on Licensing and Regulation of Nuclear Reactors,* 1967, pp. 6–7, 18, 633–635, 721–724.

19. Joseph F. Hennessey to the Commission, 31 October 1968, Box 7769, O&M–2 (Regulation), AEC/DOE; *Nucleonics Week*, 24 October 1968, p. 1.

20. AEC–R 4/55 (5 November 1968), AEC/NRC.

21. AEC–R 4/57 (9 November 1968), AEC–R 4/58 (11 December 1968), AEC–R 4/59 (18 March 1968), W. B. McCool to Joseph F. Hennessey, 8 August 1969, AEC/NRC.

22. AEC–R 4/63 (24 July 1969), AEC/NRC; Edward J. Bauser to All Committee Members, 24 December 1968, Box 872 (Memoranda from JCAE, 90th Congress), Clinton P. Anderson Papers, Library of Congress, Washington, D. C.; *Seaborg Journal*, Vol. 24, p. 140.

23. Glenn T. Seaborg to Chet Holifield, 27 November 1970, Job 9, Box 13, ID&R–14 REG (Part 50), AEC/NRC; *Nucleonics Week*, 15 December 1966, p. 1, 2 February 1967, pp. 1, 5, 16 November 1967, pp. 2–4, 27 June 1968, pp. 2–3; "Nuclear Plant Applications All Behind 'Ideal' Schedule," *Nuclear Industry*, 15 (January 1968): 30–33; G. O. Bright, "Some Effects of Public Intervention on the Reactor Licensing Process," *Nuclear Safety*, 13 (January-February 1972): 13–21.

24. R. L. Doan to Harold L. Price, 22 December 1964, MH&S–3–2 (Reactor Hazards and Control), AEC/NRC; Director of Regulation Report No. 59, 16 May 1967, Box 81 (Division of Regulation), Seaborg Office Files, Harold L. Price to the Commission, 7 January 1969, Box 7769, O&M–2 (Division of Regulation), AEC/DOE; *Nucleonics Week*, 3 August 1967, pp. 1–2, 11 January 1968, p. 3, 8 February 1968, p. 5.

25. Harold L. Price to the Commission, 7 March 1969, C. L. Henderson to Fred Schuldt, 28 April 1970, Box 7769, O&M–2 (Division of Regulation), AEC/DOE; Glenn T. Seaborg to Lee A. DuBridge, 7 April 1970, Subject File (Energy-Electric Power 1970), Record Group 359 (Records of the Office of Science and Technology Policy), National Archives, Washington, D. C.; *Seaborg Journal*, Vol. 20, pp. 714–715; *Nucleonics Week*, 18 December 1969, p. 1, 12 March 1970, p. 5, 14 January 1971, p. 8.

26. *Seaborg Journal*, Vol. 12, p. 162, Vol. 16, p. 343.

27. Earl Ewald to Glenn T. Seaborg, 19 June 1970, Box 49 (Northern States Power), Seaborg Office Files, AEC/DOE; David Okrent, *Nuclear Reactor Safety: On the History of the Regulatory Process* (Madison: University of Wisconsin Press, 1981), pp. 184–187; "Nuclear Power Is Often an Imperative Choice, Despite Pitfalls," *Nuclear Industry*, 17 (September 1970): 13–17.

28. Chet Holifield and Craig Hosmer to Glenn T. Seaborg, 24 September 1970, Seaborg to Holifield, 27 November 1970, John O. Pastore to Seaborg, 8 March 1971, Job 9, Box 13, ID&R–14 REG (Part 50), AEC/NRC; JCAE, *Hearings on AEC Authorizing Legislation Fiscal Year 1972*, 92nd Cong., 1st Sess., 1971, pp. 2496–2499.

29. Harold P. Green, " 'Reasonable Assurance' of 'No Undue Risk,' " *Scientist and Citizen*, 10 (June-July 1968): 128–140; "Complaints from Intervenors," *Nuclear Industry*, 16 (May 1969): 31; *Nucleonics Week*, 27 March 1969, p. 3.

30. Glenn T. Seaborg to Earl Ewald, 3 August 1970, Box 49 (Northern States Power), Seaborg Office Files, AEC/DOE.

CHAPTER IV: REACTORS DOWNTOWN?
THE DEBATE OVER METROPOLITAN SITING

1. George T. Mazuzan and J. Samuel Walker, *Controlling the Atom: The Beginnings of Nuclear Regulation, 1946–1962* (Berkeley and Los Angeles: University of California Press, 1984), pp. 214–219; AEC Report WASH–3 (rev.), "Summary Report of the Reactor Safeguard Committee," 31 March 1950, Atomic Energy Commission Records, Nuclear Regulatory Commission, Rockville, Maryland (hereafter cited as AEC/NRC); Charles R. Russell, *Reactor Safeguards* (New York: Pergamon Press, 1962), pp. 98–101; T. J. Thompson and C. Rogers McCullough, "The Concepts of Reactor Containment," in T. J. Thompson and J. G. Berkeley, eds., *The Technology of Nuclear Reactor Safety* (2 vols., Cambridge: MIT Press, 1964, 1973), II, p. 111.

2. Mazuzan and Walker, *Controlling the Atom*, pp. 214–245; AEC–R 2/32 (8 June 1961), AEC–R 2/39 (23 February 1962), AEC/NRC.

3. AEC–R 2/39, AEC/NRC.

4. *Ibid.*; *Federal Register*, 1 June 1962.

5. Notice of Application for Construction Permit, 10 December 1962, "Ravenswood Preliminary Hazards Report," Vol. 1, p. 1–1, "Ravenswood Site Data," pp. 1–1&2, AEC Docket 50–204, AEC/NRC; Edward J. Bauser to John T. Conway, 10 December 1962, Harold L. Price to Conway, 11 December 1962, Box 554 (Reactor by Type or Company Name: Consolidated Edison, Ravenswood), Joint Committee on Atomic Energy (JCAE) Papers, Record Group 128 (Records of the Joint Committees of Congress), National Archives; *Christian Science Monitor*, 12 December 1962.

6. Bauser to Conway, 10 December 1962, JCAE Papers; JCAE, *Hearings on the Development, Growth, and State of the Atomic Energy Industry*, 85th Cong., 1st Sess., 1963, pp. 621–622, 626–628; Minutes of Consolidated Edison (Ravenswood) Subcommittee, Advisory Committee on Reactor Safeguards (ACRS), 11 September 1963, ACRS File, AEC/NRC.

7. U.S. Congress, JCAE, *Hearings on . . . State of Atomic Energy Industry*, 1963, pp. 622–623; Minutes of Consolidated Edison Subcommittee, 11 September 1963, AEC/NRC.

8. Atomic Energy Commission, "Civilian Nuclear Power—Report to the President, 1962," printed in JCAE, *Nuclear Power Economics—1962 Through 1967*, 90th Cong., 2nd Sess., 1968, pp. 99–167; *Code of Federal Regulations (CFR)*, 10 Part 100—Reactor Site Criteria, 1962.

9. *CFR*, 10 Part 100—Reactor Site Criteria, 1962; Mazuzan and Walker, *Controlling the Atom*, pp. 240–243.

10. Robert Lowenstein to Duncan C. Clark, n.d., Job 5, Box 16, IR&A–6 REG (Consolidated Edison—Ravenswood Site), AEC/NRC; Minutes of Consolidated Edison (Ravenswood) Subcommittee, 11 September 1963, ACRS File, AEC/NRC.

11. Philip L. Cantelon, "Engineers and Alchemists: Electricity From Nuclear Energy," draft of unpublished manuscript, pp. 10–11, 32–34; Application for Construction Permit, 10 December 1962, AEC Docket 50–204, AEC/NRC; *Christian Science Monitor*, 12 December 1962; JCAE, *Hearings on . . . State of Atomic Energy Industry*, 1963, p. 624; *New York Times*, 10 May 1963.

12. Morton Kulick to Atomic Energy Commission, 14 December 1962, James Angiola to Atomic Energy Commission, 21 January, 30 January 1963, Folder No. 1 (7 December 1962—20 May 1963), AEC Docket 50–204, AEC/NRC.

13. AEC-R 102 (4 March 1963), AEC/NRC; *Nucleonics Week*, 28 February 1963, p. 6.

14. J. B. Graham to John T. Conway, 20 February 1963, Box 554 (Reactor by Type or Company Name: Consolidated Edison, Ravenswood), JCAE Papers; *Nucleonics Week*, 28 February 1963, p. 6; James V. Angiola to Jacob K. Javits, 18 March 1963, Mario J. Cariello to AEC, 15 April 1963, Harold L. Price to Javits, 2 May 1963, Folder No. 1 (7 December 1962—20 May 1963), AEC Docket 50–204, AEC/NRC.

15. Richard P. Hunt, "Atomic Question for the City," *New York Times Magazine*, 6 October 1963, p. 46; AEC/NRC; Sheldon Novick, *The Careless Atom* (Boston: Houghton Mifflin Co., 1968), p. 59; Committee for a Safe New York statement: "Hiroshima—New York," n.d.; Queens County New Frontier Regular Democratic Club, Inc., program, 25 April 1963 (Reactors by Type or Company Name: Consolidated Edison, Ravenswood), JCAE Papers; New Frontier Regular Democratic Club, "Resolution," n.d.; AEC–R 102/1 (10 May 1963), Robert K. Otterbourg to Joseph C. Clarke, 15 March 1963, AEC Docket 50–204, AEC/NRC; *New York Times*, 10 May, 15 May, 12 June 1963.

16. JCAE, *Hearings on . . . State of the Atomic Energy Industry*, 1963, p. 624.

17. *Ibid.*, pp. 704–744; David E. Lilienthal, *The Journals of David E. Lilienthal: The Harvest Years, 1959–1963* (New York: Harper and Row, 1971), pp. 460–463; *New York Times*, 5 April 1963; *Nucleonics Week*, 11 April 1963.

18. *New York Times*, 9 May, 15 May 1963.

19. Statement by Mary Hays Weik, 14 June 1963, Box 554 (Reactors by Type or Company Name: Consolidated Edison, Ravenswood), JCAE Papers; *New York Times*, 15 June 1963; AEC Press Release, 11 June 1963, AEC/NRC; *Nucleonics Week*, 20 June 1963, p. 2.

20. *Nucleonics Week*, 20 June 1963, p. 1.

21. *New York Times*, 26 August 1963.

22. Glenn T. Seaborg, Remarks to the National Convention—Sigma Delta Chi, 7 November 1963, AEC Speech File, AEC/NRC; *Nucleonics Week*, 14 November 1963, p. 6; *New York Times*, 8 November 1963.

23. *New York Times*, 8 November, 20 November 1963; *Nucleonics Week*, 14 November 1963, p. 1, 20 November 1963, p. 1; *Wall Street Journal*, 20 November 1963; Lilienthal, *Harvest Years*, pp. 521–522; John F. Hogerton, *The Atomic Energy Deskbook* (New York: Reinhold Publishing Corp. 1963), p. 24.

24. Robert Lowenstein to Consolidated Edison Co., 9 August 1963, Division of Licensing and Regulation Report No. 1 on Ravenswood to Advisory

Committee on Reactor Safeguards, 25 September 1963, AEC Docket 50–204; Minutes of Reactor Safeguards Subcommittee (Ravenswood), 10 April 1963, ACRS File, AEC/NRC.

25. Preliminary Hazards Report, chap. 2, p. 2–1, chap. 4, p. 4–1, AEC Docket 50–204, AEC/NRC.

26. *Ibid.*, chap. 1, pp. 1-4–1-5, chap. 3, pp. 2-1–2-2, 3-1–3-4, 4-3–4-4, chap. 4, pp. 3-5–3-7.

27. *Ibid.*, chap. 3, pp. 1-1–1-8; Division of Licensing and Regulation Report No. 1 on Ravenswood to Advisory Committee on Reactor Safeguards, 25 September 1963, AEC Docket 50–204, AEC/NRC.

28. Lowenstein to Con Edison, 9 August 1963, Division of Licensing and Regulation Report No. 1 on Ravenswood to Advisory Committee on Reactor Safeguards, 25 September 1963, AEC Docket 50–204, AEC/NRC.

29. Minutes of Reactor Safeguards Subcommittee (Ravenswood), ACRS, 11 September 1963, ACRS File, Division of Licensing and Regulation Report No. 1 on Ravenswood to ACRS, 25 September 1963, AEC Docket 50–204, AEC/ NRC.

30. Minutes of 50th ACRS meeting, 10–11 October 1963, ACRS File; Staff Conclusion on Ravenswood, 12 November 1963, cited in "Preapplication Meeting for Proposed Boston Edison Plant, 26 January 1966," AEC Docket 50–293, AEC/NRC.

31. F. F. Brower to R. L. Lowenstein, 14 November 1963, AEC Docket 50– 204, AEC/NRC.

32. Minutes of Reactor Safeguards Subcommittee (Ravenswood), ACRS, 10 April, 11 September, 21 October 1963, Minutes of 50th and 51st ACRS meetings, 10–11 October, 7–8 November 1963, ACRS File, AEC/NRC.

33. Harland C. Forbes to Glenn T. Seaborg, 3 January 1964, AEC Docket 50–204, AEC/NRC.

34. *New York Times,* 10 May 1963, 8 January 1964; JCAE, *Hearings on . . . State of Atomic Energy Industry,* 1963, pp. 626–627; Minutes of Reactor Safeguards Subcommittee (Ravenswood), ACRS, 11 September 1963, Minutes of 50th ACRS meeting, 10–11 October 1963, Minutes of 52nd ACRS meeting, 9–10 January 1964, ACRS File, AEC/NRC.

35. Minutes of 50th ACRS meeting, 10–11 October 1963, ACRS File, AEC/ NRC; *New York Times,* 7 January 1964.

36. David Okrent, *Nuclear Reactor Safety: On the History of the Regulatory Process* (Madison: University of Wisconsin Press, 1981), p. 67; Herbert Kouts to Glenn T. Seaborg, "Report on Engineered Safeguards," 18 November 1964, Minutes of 59th ACRS meeting, 12, 14, 15 November 1964, ACRS File, AEC/NRC.

37. AEC Press Releases, 4 March 1964, 16 December 1964, AEC/NRC; AEC Annual Report, *Major Activities in the Atomic Energy Programs, January-December 1964* (Washington, D.C.: Government Printing Office, 1965), pp. 316–317.

38. W. B. McCool to the File, 15 January 1965, Harold L. Price to Commission, 10 February 1965, Job 5, Box 19, PFC 1–1 REG (Production and Utilization Facilities), AEC/NRC.

39. *Ibid.*; Minutes of Special ACRS meeting, 6 February 1965, ACRS File, AEC/NRC; JCAE, *Hearings on Licensing and Regulation of Nuclear Reactors,* 90th Cong., 1st Sess., 1967, pp. 59, 123–130.

40. Price to the Commission, 10 February 1965, AEC/NRC.

41. Minutes of Regulatory Meeting 210, 11 February 1965, AEC/NRC.

42. *Ibid.*; W. B. McCool to Richard L. Doan, 15 February 1965, Job 5, Box 2, IR&A–6–REG (Boston Edison); James T. Ramey to the Commission, 19 February 1965, W. B. McCool to File, 4 March 1965, Job 5, Box 19, PFC–1–1 REG (Production and Utilization Facilities), AEC/NRC.

43. Minutes of special ACRS meeting, 26–27 March 1965, ACRS File, AEC/NRC.

44. *Ibid.*; W. D. Manly to Glenn T. Seaborg, 2 April 1965, ACRS File, AEC/NRC; AEC 943/19 (12 May 1965), Box 1361 (Hazards of Power Reactors), "Report on Siting Power Reactors in Metropolitan Centers," 27 March 1965, Box 1390 (O&M–7, ACRS), Atomic Energy Commission Records, Department of Energy, Germantown, Maryland.

45. Minutes of 63rd ACRS meeting, 13–15 May 1965, ACRS File, AEC/NRC.

46. Charles Avila to Glenn T. Seaborg, 5 August 1965, Job 5, Box 2, IR&A–6 REG (Boston Edison), AEC/NRC.

47. Gerald F. Tape to Charles Avila, 17 August 1965, Frances M. Staszesky to Harold L. Price, 23 September 1965, Job 5, Box 2, IR&A–6 REG (Boston Edison), Minutes of 66th ACRS meeting, 9–11 September 1965, Minutes of 67th ACRS meeting, 7–9 October 1965, ACRS File, AEC Press Release, 15 May 1964, AEC/NRC.

48. Okrent, *Nuclear Reactor Safety,* pp. 83–84; "Preapplication Meeting for Proposed Boston Edison Plant," 26 January 1966, AEC Docket 50–293, AEC/NRC.

49. "Preapplication Meetings for Proposed Boston Edison Plant," 26 January, 1 April 1966," License Application for Pilgrim Nuclear Power Station, Plymouth, Massachusetts, 23 June 1967, AEC Docket 50–293, AEC/NRC.

50. "Symposium on Locating Nuclear Power Plants in Cities," *Nuclear Industry,* 13 (April 1966): 29–33; JCAE, *Hearings on Licensing and Regulation of Nuclear Reactors,* 1967, pp. 22–23, 58–59, 66, 69–70, 75, 94–95, 123–124, 129–132, 504, 643, 649–650, 661–662, 670.

51. Notice of Application for Construction Permit, 3 December 1965, AEC Docket 50–247; AEC Press Release, 9 December 1965, Minutes of 72nd ACRS meeting, 4–6 April 1966, 73rd ACRS meeting, 5–7 May 1966, Minutes of Indian Point II Subcommittee, ACRS, 23 June 1966, ACRS File, AEC/NRC; *New York Times,* 24 November 1965.

52. Okrent, *Nuclear Reactor Safety,* pp. 85–98. The issue of pressure vessel failure and its regulatory impact is discussed in chapter 7, W. Manly to Glenn T. Seaborg, 24 November 1965, ACRS File, Consolidated Edison, Indian Point Nuclear Generating Unit No. 2, Preliminary Safety Analysis Report, Vols. 1, 2, 7 December 1965, AEC/NRC.

53. David Okrent to Glenn T. Seaborg, 16 August 1966, Initial Decision of

Atomic Safety and Licensing Board, 3 October 1966, Memorandum and Order, 20 December 1966, AEC Docket 50–247, AEC/NRC.

54. Minutes of 73rd ACRS meeting, 5–7 May 1966, 74th ACRS meeting, 8–11 June 1966, Minutes of Indian Point II Subcommittee of ACRS, 23 June 1966, Minutes of 75th ACRS meeting, 13–15 July 1966, ACRS File, AEC/NRC.

55. Notice of Application for Construction Permit, 13 December 1966, AEC Docket 50–272; AEC Press Release, 16 December 1966, Minutes of 88th ACRS Meeting, 10–12 August 1967, ACRS File, AEC/NRC.

56. Minutes of 74th ACRS meeting, 8–11 June 1966, 87th ACRS meeting, 6–8 July 1967, 88th ACRS meeting, 10–12 August 1967, ACRS File, Harold L. Price to the Commission, 17 August 1967, Job 4, Box 4, ID&R–5 REG (Public Service Electric and Gas Co.), Public Service Electric and Gas Co., Amendment to Application for Construction Permit 23 Oct. 1967, Amendment No. 3 to Application for Licenses, Salem Nuclear Generating Station, 22 January 1968, AEC Dockets 50–272 and 50–311, AEC/NRC.

57. Application for Construction Permit, 12 July 1967, Preliminary Safety Analysis Report, Vol. 1, 7 July 1967, Amendment No. 1, 15 August 1967, AEC Dockets 50–295 and 50–304, AEC Division of Reactor Licensing Report to ACRS on Zion Station Units 1 and 2, 18 March 1968, ACRS File, AEC/NRC.

58. AEC Report to ACRS, 18 March 1968, Minutes of Zion Subcommittee, ACRS, 21 March, 17 April, 29 May 1968, Minutes of 96th ACRS meeting, 4–6 April 1968, 97th ACRS meeting, 9–11 May 1968, 98th ACRS meeting, 5–8 June 1968, 99th ACRS meeting, 11–13 July 1968, ACRS File, Carroll W. Zabel to Glenn T. Seaborg, 24 July 1968, Job 9, Box 2, ID&R–5 REG (Commonwealth Edison—Zion Station), AEC/NRC.

59. CFR 10 Part 100—Reactor Site Criteria, 1962.

CHAPTER V: REACTORS AT FAULTS:
THE CONTROVERSY OVER SEISMIC SITING

1. "Big Hurdle for A-Power: Gaining Public Acceptance," *Nucleonics*, 21 (October 1963): 17–21; Kenn Sherwood Roe, "Bodega: Symbol of a National Crisis," *American Forests*, 69 (December 1963): 22–25; Joel W. Hedgpeth, "Bodega Head—A Partisan View," *Bulletin of the Atomic Scientists*, 21 (March 1965): 2–7; Sheldon Novick, *The Careless Atom* (Boston: Houghton Mifflin Co., 1969), pp. 34–36.

2. Richard L. Meehan, *The Atom and the Fault* (Cambridge: MIT Press, 1984), pp. 3–5; Hedgpeth, "Bodega Head," pp. 2–3; "Big Hurdle for A-Power," *Nucleonics*, pp. 17–20.

3. Philip Flint, "Struggle on the Seacoast," *Sierra Club Bulletin*, 46 (April 1961): 9; *Alta California*, 21 July 1961; Clem Miller to Chet Holifield, 9 December 1959, Box 543 (Reactors by Type or Company Name: Bodega Bay, Pacific Gas and Electric Co.), Papers of the Joint Committee on Atomic Energy, Record Group 128 (Records of the Joint Committees of Congress), National Archives; Rose Gaffney to Richard Smith, 26 December 1960, Mrs. Andrew

Brown to Atomic Energy Commission, 28 June 1961, Docket 50–205 (Bodega
Bay), Pacific Gas and Electric Company Press Release, 28 June 1961, Job 5, Box
7, IR&A–6 REG (PG&E—Bodega Bay Atomic Park), Atomic Energy Commission Records, Nuclear Regulatory Commission, Rockville, Maryland (hereafter
cited as AEC/NRC).

4. *San Francisco Chronicle*, 4 May, 22 May, 23 May 1962; *San Francisco
Examiner*, 22 May, 23 May, 9 June 1962; David E. Pesonen, "The Battle of
Bodega Bay," *Sierra Club Bulletin*, 47 (June 1962): 9; H. L. Price to Chairman
Seaborg, 26 September 1962, Box 8549 (Bodega Bay), Office Files of Glenn T.
Seaborg, Atomic Energy Commission Records, Department of Energy, German-
town, Maryland (hereafter cited as AEC/DOE).

5. Roe, "Bodega: Symbol of a National Crisis," pp. 23–24; Hedgpeth,
"Bodega Head—A Partisan View," pp. 4–5; Catherine Caufield, *Multiple Exposures: Chronicles of the Radiation Age* (New York: Harper and Row, 1989),
pp. 150–151; Sheldon Novick, *The Electric War: The Fight Over Nuclear
Power* (San Francisco: Sierra Club Books, 1976), p. 241; Thomas R. Wellock,
"The Battle at Bodega: The Sierra Club and Nuclear Power," paper delivered at
annual meeting of the American Society for Environmental History," 1991, pp.
13–15; "Conservation Policy Guide" Abstract of Directors' Action, 1946–
1968" (Minutes, rev. ed., July 1968), Sierra Club Records, William E. Colby
Memorial Library, Sierra Club, San Francisco.

6. AEC–R 80/4 (29 March 1962), R. S. Boyd and R. H. Wilcox to the
Files, 15 August 1962, "Preliminary Hazards Summary Report: Bodega Bay
Atomic Park Unit Number 1," 28 December 1962, Docket 50–205, AEC/NRC;
Meehan, *Atom and the Fault*, p. 9.

7. "Preliminary Hazards Summary Report," Docket 50–205, AEC/NRC;
Meehan, *Atom and the Fault*, p. 2.

8. Northern California Association to Preserve Bodega Head and Harbor,
"Memorandum of Action Concerning Late Filed Exhibit No. 48 and Related
Evidence," 6 May 1963, David E. Pesonen, "A Visit to the Atomic Park," 1962,
Rodney L. Southwick to Joe Fouchard, 28 March 1963, Docket 50–205, AEC/
NRC; Northern California Association to Preserve Bodega Head and Harbor,
Press Release, 6 May 1963, Box 543 (Reactors by Type or Company Name:
Bodega Bay, Pacific Gas and Electric Co.), JCAE Papers; "Association Fights for
Bodega," *Sierra Club Bulletin*, 48 (April-May 1963): 12.

9. Pierre Saint-Amand to Harold Gilliam, 19 April 1963, Pierre Saint-
Amand, "Geologic and Seismologic Study of Bodega Head," September 1963,
"Reply to Petition to Reopen for Further Hearing," 16 May 1963, Docket 50–
205, AEC/NRC; *Berkeley Daily Gazette*, 7 May 1963.

10. Harold L. Price to J. B. Neilands, 6 September 1962, Docket 50–205,
AEC/NRC; Meehan, *Atom and the Fault*, pp. 57–59, 78–79.

11. Eber R. Price to Frank Neumann, 3 May 1963, R. H. Bryan to the Files,
22 May 1963, Docket 50–205, AEC/NRC; James G. Terrill to Robert Lowenstein, 16 May 1963, File 780, Public Health Service Records, Division of Radiological Health, Public Health Service, Rockville, Maryland; JCAE, *Hearings on
Development, Growth, and State of the Atomic Energy Industry*, 87th Cong.,
2nd Sess., 1962, pp. 4–58; "Electrical West," 28 June 1961, copy in Box 543

(Reactors by Type or Company Name: Bodega Bay, Pacific Gas and Electric Co.), JCAE Papers; George T. Mazuzan and J. Samuel Walker, *Controlling the Atom: The Beginnings of Nuclear Regulation, 1946–1962* (Berkeley and Los Angeles: University of California Press, 1984), pp. 410–411.

12. Division of Licensing and Regulation, "Report to the Advisory Committee on Reactor Safeguards on Bodega Bay Atomic Park," 5 April 1963, Pesonen, "Visit to the Atomic Park," Docket 50–205, AEC-R 80/7 (23 April 1963), AEC/NRC; Gene Marine, "Outrage on Bodega Head," *Nation*, 196 (22 June 1963): 524–527.

13. *San Francisco Chronicle*, 28 February 1963; Stewart L. Udall, *The Quiet Crisis* (New York: Holt, Rinehart and Winston, 1963), pp. 174, 180–181.

14. Stewart L. Udall to Kermit Gordon, 18 February 1963, Glenn T. Seaborg to Udall, 8 March 1963, Job 5, Box 7, IR&A–6 REG (PG&E—Bodega Bay Atomic Park), "Summary of Discussions between Department of Interior Staff and Atomic Energy Commission Staff [on] April 9, 1963," 13 June 1963, L–4–1 (Memo of Understanding, AEC—Department of Interior and Fish and Wildlife), AEC/NRC.

15. Stewart L. Udall to Glenn T. Seaborg, 20 May 1963, W. B. McCool to Harold L. Price, 22 May 1963, Robert E. Wilson to Udall, 23 May 1963, Job 5, Box 7, IR&A–6 REG (PG&E—Bodega Bay Atomic Park), AEC/NRC; *Nucleonics Week*, 6 June 1963, p. 6.

16. *San Francisco Chronicle*, 22 May, 27 May, 31 May 1963; *San Francisco Examiner*, 22 May, 26 May 1963; *Berkeley Daily Gazette*, 22 May 1963; *Oakland Tribune*, 22 May 1963; *Santa Rosa Press Democrat*, 31 May 1963; Northern California Association to Preserve Bodega Head and Harbor, flyer, "I Like Udall," n.d., R. H. Engelken to L. Kornblith, 20 August 1963, Docket 50–205, AEC/NRC.

17. Rodney L. Southwick to Joe Fouchard, 6 June 1963, David Brower to Members of the California State Legislature, 13 June 1963, Glenn M. Anderson to Glenn T. Seaborg, 13 August 1963, Docket 50–205, AEC/NRC; *San Francisco Examiner*, 18 July 1963; Meehan, *Atom and the Fault*, pp. 7–9.

18. R. H. Bryan to the Files, 22 May 1963, Frank Neumann to Bryan, 24 May, 24 June 1963, Nathan M. Newmark to Bryan, 26 June 1963, Docket 50–205, AEC/NRC.

19. AEC–R 80/9 (3 June 1963), Robert Lowenstein to the Files, 3 June 1963, Gerald F. Hadlock to the Separated Legal Files, 7 June 1963, Hadlock to the Separated Legal Files, 13 June 1963, H. L. Price to the Files, 17 June 1963, Docket 50–205, AEC/NRC.

20. Gerald F. Hadlock to the Separated Legal Files, 21 June 1963, J. F. Newell to the Files, 23 July 1963, L. Kornblith, Jr. to Edson G. Case, 5 August 1963, Kornblith to Case, 16 September 1963, Docket 50–205, Harold L. Price to the Commission, 1 August 1963, Job 5, Box 7, IR&A–6 REG (PG&E—Bodega Bay Atomic Park), AEC/NRC; Meehan, *Atom and the Fault*, pp. 10–12.

21. *San Francisco Chronicle*, 30 August 1963, *San Francisco Examiner*, 30 August 1963, *Oakland Tribune*, 30 August 1963, *Santa Rosa Press Democrat*, 29 August 1963; Dale J. Cook to Joe Fouchard, 30 August 1963, Docket 50–205, AEC/NRC.

22. R. H. Engelken to L. Kornblith, Jr., 16 September 1963, Howard K. Shapar to H. L. Price, 26 September 1963, Docket 50–205, Price to the Commission, 13 September, 27 September 1963, Job 5, Box 7, IR&A–6 REG (PG&E— Bodega Bay Atomic Park), AEC/NRC.

23. John A. Carver to Glenn T. Seaborg, 25 September 1963, Docket 50– 205, AEC/NRC.

24. Dale J. Cook to Joe Fouchard, 7 October 1963, AEC Press Release, 4 October 1963, Docket 50–205, AEC/NRC; *San Francisco News Call Bulletin*, 4 October, 10 October 1963; *San Francisco Chronicle*, 5 October 1963; *San Francisco Examiner*, 5 October, 8 October 1963; *Oakland Tribune*, 5 October 1963; *Wall Street Journal*, 7 October 1963; *Santa Rosa Press Democrat*, 6 October 1963; Meehan, *Atom and the Fault*, pp. 12–14.

25. *San Francisco Chronicle*, 9 October, 20 October 1963; *San Francisco Examiner*, 9 October 1963; *Stockton Pathfinder*, 27 September 1963; Harold L. Price to David Pesonen, 22 October 1963, Docket 50–205, AEC/NRC.

26. Julius Schlocker and Manuel G. Bonilla, "Engineering Geology of the Proposed Nuclear Power Plant Site on Bodega Head, Sonoma County, California," December 1963, Docket 50–205, AEC/NRC.

27. Pacific Gas and Electric Company, Bodega Bay Atomic Park, Amendment No. 5, 22 January 1964, Garniss H. Curtis and Jack F. Evernden to F. F. Mautz, 26 February 1964, Docket 50–205, AEC/NRC; *San Francisco Chronicle*, 21 January, 23 March 1964; Meehan, *Atom and the Fault*, pp. 16–17.

28. R. H. Engelken to L. Kornblith, Jr., 13 January 1964, Clifford K. Beck to the Files, 7 February 1964, John F. Newell to the Files, 7 February 1964, F. N. Watson to the Files, 27 March 1964, Docket 50–205, AEC/NRC.

29. Northern California Association to Preserve Bodega Head and Harbor Press Release, 30 March 1964, Lu Watters to President Johnson, 30 March 1964, Mr. and Mrs. Robert M. Campbell to the AEC, 20 April 1964, Docket 50–205, AEC/NRC; *San Francisco Chronicle*, 14 April, 15 April 1964; *San Francisco Examiner*, 14 April 1964; L. Don Leet, Sheldon Judson, and Marvin E. Kauffman, *Physical Geology* (Englewood Cliffs, NJ: Prentice-Hall, 1978), pp. 144–153.

30. "Report to the Advisory Committee on Reactor Safeguards on . . . Bodega Bay Reactor," 30 April 1964, Harold L. Price to C. C. Whelchel, 19 May 1964, Docket 50–205, AEC/NRC; *San Francisco News Call Bulletin*, 28 May 1964.

31. Donald F. Knuth to the Files, 12 June 1964, Docket 50–205, Minutes of 55th Meeting, 6–8 May 1964, Advisory Committee on Reactor Safeguards (ACRS) File, AEC/NRC; David Okrent, *Nuclear Reactor Safety: On the History of the Regulatory Process* (Madison: University of Wisconsin Press, 1981), pp. 266–267; Meehan, *Atom and the Fault*, pp. 18–19.

32. AEC–R 80/10 (27 October 1964), "Summary Analysis: Bodega Head Nuclear Power Plant," 26 October 1964, Docket 50–205, AEC/NRC; Okrent, *Nuclear Reactor Safety*, pp. 267–273.

33. AEC Press Release, 27 October 1964, AEC/NRC; *Journal of Glenn T. Seaborg*, (25 vols., Berkeley: Lawrence Berkeley Laboratory PUB–625, 1989), Vol. 9, pp. 266, 289; PG&E Press Release, 30 October 1964, Box 543 (Reac-

tors by Type or Company Name: Bodega Bay, Pacific Gas and Electric Co.), JCAE Papers; *San Francisco Chronicle*, 28 October 1964; *Nucleonics Week*, 5 November 1964, p. 1; Meehan, *Atom and the Fault*, pp. 19–20.

34. David E. Pesenon to Harold L. Price, 5 November 1964, Pesenon to "Friend of Bodega Bay," 14 December 1964, Docket 50–205, AEC/NRC; *San Francisco Chronicle*, 22 November 1964; *Santa Rosa Press Democrat*, 7 December 1964; Author's interview with Harold L. Price, Chevy Chase, Maryland, 30 January 1989.

35. Malcolm D. Ferrier to Harold L. Price, 12 November 1964, Docket 50–205, AEC/NRC; "Demolished by Hypothesis," *Nuclear Industry*, 11 (November 1964): 3–8; "Bodega Bay Reactor Dropped by PG&E," *Nucleonics*, 22 (December 1964): 20–21; *Nucleonics Week*, 5 November 1964, pp. 1–2; *Electrical World*, 9 November 1964, p. 4.

36. Duncan Clark to the Commission, 4 December 1962, Docket 50–214 (Malibu), AEC–R 89 (10 January 1962), AEC–R 89/1 (15 March 1962), AEC–R 89/2 (6 April 1962), AEC/NRC; *Los Angeles Times*, 11 April 1965; Meehan, *Atom and the Fault*, pp. 38–39.

37. "Big Hurdle for A-Power," *Nucleonics*, p. 21.

38. *Ibid.*; Terry Waters to AEC, 20 August 1963, Malibu Citizens for Conservation, "Fact Sheet," September 1963, Docket 50–214, AEC/NRC; *Los Angeles Times*, 5 November 1963.

39. Elizabeth A. Letton to Thomas H. Kuchel, 31 December 1963, Leslie Steinmetz to Glenn T. Seaborg, 14 January 1963, Paul Siegel to the Legal Files, 24 January 1964, H. D. Hutchinson and W. C. Gregge to E. R. Price, 10 February 1964, Docket 50–214, AEC/NRC; "News About Industry," *Forum Memo*, 11 (April 1964): 9; *Nucleonics Week*, 16 April 1964, pp. 1–3, 4 June 1964, p. 1; Meehan, *Atom and the Fault*, p. 38.

40. Frank A. Morgan, "Geology of the Proposed Corral Canyon Nuclear Reactor Site," 11 February 1964, Thomas L. Bailey, "Geology of Malibu Coastal Belt in the Vicinity of Proposed Nuclear Reactor Site in Corral Canyon," 12 February 1964, Clifford K. Beck to the Files, 19 February, 26 February 1964, Docket 50–214, AEC/NRC; "Memorandum for the Chairman," 1 February 1964, Box 8549 (Bodega Bay), Seaborg Office Files, AEC/DOE.

41. Richard H. Jahns, "Geologic Conditions at Proposed Site for a Nuclear Power Station at Corral Canyon," 9 July 1964, U. S. Coast and Geodetic Survey, "Report on the Seismicity of the Malibu, California Area," 4 June 1964, Geologic Survey, "Engineering Geology Summary of the Proposed Nuclear Power Plant, Malibu Site," Draft, 5 June 1964, R. F. Yerkes and C. M. Wentworth, "Geologic Report on the Proposed Corral Canyon Nuclear Power Plant Site," December 1964, Los Angeles Department of Water and Power Press Release, 12 January 1965, Docket 50–214, AEC/NRC; LADWP, "Condensation of the Public Hearing for the Malibu Nuclear Plant," 16 December 1965, Box 10 (Malibu), Office Files of James T. Ramey, AEC/DOE.

42. Division of Reactor Licensing, "Report to the Advisory Committee on Reactor Safeguards . . . on Malibu Nuclear Plant," 1 July 1964, M. Keith Woodard to the Files, 14 July, 5 August, 9 November 1964, Herbert Kouts to Glenn T. Seaborg, 15 July 1964, Edson G. Case to the Files, 25 September 1964,

D. F. Knuth to E. G. Case, 1 December 1964, R. L. Doan to Harold L. Price, 9 December 1964, W. D. Manly to Seaborg, 25 January 1965, "Hazards Analysis by the Division of Reactor Licensing in the Matter of . . . Malibu Nuclear Plant," 15 February 1965, Docket 50–214, AEC/NRC; *Los Angeles Times*, 4 February 1965.

43. Barclay Kamb, "Geology of the Corral Canyon Site in Relation to the Proposed Nuclear Reactor," 20 January 1965, Docket 50–214, AEC/NRC.

44. Gerald F. Hadlock to the Separated Legal Files, 25 January 1965, Verl R. Wilmarth to Edson G. Case, 25 February 1965, Docket 50–214, AEC/NRC; *Los Angeles Times*, 31 March 1965; *Nucleonics Week*, 1 April 1965, p. 1; "Four Intervenors to Oppose LADWP," *Nuclear Industry*, 12 (March 1965): 11.

45. Thomas B. Nolan to Harold L. Price, 19 March 1965, Gerald F. Hadlock to the Separated Legal Files, 29 April 1965, LADWP Press Release, 21 April 1965, Price to the Commission, 4 May 1965, L. Kornblith to Edson G. Case, 18 May 1965, R. F. Yerkes and C. M. Wentworth, "Structure, Quaternary History, and General Geology of the Corral Canyon Area," July 1965, "Proposed Findings of Fact and Conclusions of Law by Applicant," 2 February 1966, Docket 50–214, F. T. Hobbs to the Commissioners, 6 May 1965, Job 5, Box 5, IR&A–6 (Los Angeles Dept. of W & P), AEC/NRC; *Santa Monica Evening Outlook*, 17 July, 23 July, 27 July, 28 July, 29 July, 4 August, 20 October 1965.

46. Harold L. Price to the Commission, 21 April, 17 November 1965, Docket 50–214, "Circumstances Regarding Changes in Wording of U. S. Geological Survey Draft Reports," n. d., Job 5, Box 5, IR&A–6 (Los Angeles Dept. of W & P), AEC/NRC; John T. Conway to Chet Holifield, 28 April 1966, Box 177 (Members: Holifield, Chet), JCAE Papers.

47. Gerald F. Hadlock to Separated Legal Files, 12 May 1966, Docket 50–214, AEC-R 89/30 (15 July 1966), AEC/NRC; *Santa Monica Evening Outlook*, 11 May 1966.

48. Gerald F. Hadlock to the Files, 15 August 1966, Docket 50–214, "Exceptions of Intervenor Marblehead Land Company," 31 August 1966, Job 4, Box 2, ID&R–5 REG (Los Angeles, City of), "Exceptions of the Regulatory Staff," 1 September 1966, Docket 50–214, AEC/NRC; "Commission Extends Deadline For Malibu Decision Appeals," *Nuclear Industry*, 13 (August 1966): 14–16; *Nucleonics Week*, 21 July 1966, pp. 1–2; *Santa Monica Outlook*, 1 September 1966.

49. "Exceptions of the Regulatory Staff," 1 September 1966, Marshall K. Ray, "Status of Malibu Project," 21 July 1970, Docket 50–214, AEC/NRC; JCAE, *Hearings on Licensing and Regulation of Nuclear Reactors*, 90th Cong., 1st Sess., 1967, pp. 19–25, 395–403; *Santa Monica Outlook*, 7 September 1966; *Nucleonics Week*, 29 September 1966, p. 4, 30 March 1967, p. 1; "Malibu Case Back to Board," *Nuclear Industry*, 14 (April 1967): 45–48; "LADWP Drafts Design Criteria on Malibu Seismic Problem," *ibid.*, 15 (March 1968): 19–21; "LADWP Reconciled to 'Waiting Out' AEC on Malibu Site," *ibid.*, (October 1968): 14.

50. AEC–R 108 (16 September 1963), AEC–R 108/12 (30 January 1964), AEC/NRC; "Big Hurdle for A-Power," *Nucleonics*, pp. 21–22.

51. AEC–R 108, AEC–R 108/12, R. A. Williamson to Edson Case, 10 August 1964, Docket 50–214, AEC/NRC; "Big Hurdle for A-Power," *Nucleonics*, pp. 21–22.

52. AEC–R 2/45 (26 February 1964), AEC–R 2/47 (1 April 1965), AEC/NRC.

53. "Criteria Relating to Seismic Factors in Design and Location of Reactors," draft, 9 November 1966, Job 4, Box 4, ID&R–6 (Siting), AEC/NRC.

54. "Seismic and Geologic Siting and Design Criteria for Nuclear Power Plants," drafts, 21 March, 13 April, 11 May 1967, Harold L. Price to the Commission, 21 March, 14 April, 24 May 1967, Job 4, Box 4, ID&R–6 (Siting), AEC/NRC.

55. Minutes of Seismic Activity/Tsunamis Subcommittee Meeting, 5 February 1969, ACRS File, AEC/NRC; *Nucleonics Week*, 11 June 1970, p. 3; Richard G. Hewlett and Francis Duncan, *Nuclear Navy, 1946–1962* (Chicago: University of Chicago Press, 1974), p. 211.

56. AEC Press Release, 13 January 1968, Minutes of Seismic Activity/Tsunamis Subcommittee, 5 February 1969, AEC/NRC; JCAE, *Hearings on Use of Nuclear Power for the Production of Fresh Water from Salt Water*, 88th Cong., 2nd Sess., 1964, and *Hearings on Proposed Large-Scale Combination Nuclear Power-Desalting Project*, 89th Cong., 2nd Sess., 1966; Okrent, *Nuclear Reactor Safety*, 189–190, 277–282.

57. "Seismic and Geologic Siting Criteria for Nuclear Power Plants," draft, 19 March 1969, SECY–R 311 (27 August 1971), Box 43 (Seismic Matters), Ramey Office Files, AEC/DOE; *Seaborg Journal*, Vol. 19, pp. 358, 365; Minutes of Seismic Activity/Tsunamis Subcommittee, 5 February 1969, AEC/NRC; *Nucleonics Week*, 17 July 1969, p. 1.

58. *Atomic Energy Commission Reports: Opinions and Decisions of the Atomic Energy Commission with Selected Orders*, Vol. 4, pp. 89–99, 447–466, Vol. 6, pp. 929–969; Harold L. Price to the Commission, 30 April 1971, Box 43 (Seismic Matters), Ramey Office Files, AEC/DOE; James B. Graham to George F. Murphy, Jr., 10 June 1971, Box 50 (San Onofre, 1971–1973), Chet Holifield Papers, University of Southern California, Los Angeles; *New York Times*, 12 April 1971; *Nucleonics Week*, 22 July 1971, p. 2; Okrent, *Nuclear Reactor Safety*, p. 190.

CHAPTER VI: DILEMMA OVER DISASTERS:
THE EXTENSION OF INDEMNITY LEGISLATION

1. George T. Mazuzan and J. Samuel Walker, *Controlling the Atom: The Beginnings of Nuclear Regulation, 1946–1962* (Berkeley and Los Angeles: University of California Press, 1984), pp. 93–121, 199–213; John W. Johnson, *Insuring against Disaster: The Nuclear Industry on Trial* (Macon, GA: Mercer University Press, 1986), pp. 41–56; Richard T. Sylves, *The Nuclear Oracles: A Political History of the General Advisory Committee of the Atomic Energy Commission, 1947–1977* (Ames: Iowa State University Press, 1987), p. 263; United States Atomic Energy Commission, "Theoretical Possibilities and Consequences of Major Accidents in Large Nuclear Power Plants" (WASH–740),

March 1957, Atomic Energy Commission Records, U. S. Nuclear Regulatory Commission, Rockville, Maryland (hereafter cited as AEC/NRC).

2. General Public Utilities Corporation and Jersey Central Power and Light Company, "Statement to Atomic Energy Commission on Indemnity Problems," 23 October 1963, Box 148 (Price-Anderson Indemnity), Office Files of Glenn T. Seaborg, Records of the Atomic Energy Commission, Department of Energy, Germantown, Maryland (hereafter cited as AEC/DOE).

3. Joint Committee on Atomic Energy (JCAE), *Hearings on AEC Omnibus Bills for 1963 and 1964*, 88th Cong., 2nd Sess., 1964, pp. 173–179; John O. Pastore and Chet Holifield to Glenn T. Seaborg, 5 December 1963, Job 5, Box 1, BAF–12 REG (Insurance), AEC/NRC; Seaborg to Pastore, 3 January 1964, Seaborg to Pastore, 28 April 1964, John T. Conway to All Committee Members, 29 April 1964, Sherman R. Knapp to Pastore, 8 May 1964, Box 382 (Indemnity), Joint Committee on Atomic Energy Papers, Record Group 128 (Records of the Joint Committees of Congress), National Archives.

4. Richard H. K. Vietor, *Environmental Politics and the Coal Coalition* (College Station, TX: Texas A & M University Press, 1980), pp. 13–57.

5. JCAE, *Hearings on Development, Growth, and State of the Atomic Energy Industry*, 88th Cong., 1st Sess., 1963, pp. 547–560, 923–927; *Nucleonics Week*, 11 April 1963, p. 3, 5 September 1963, p. 1; "Nuclear Industry Takes Off Kid Gloves, Counterattacks Coal," *Nucleonics*, 22 (January 1964): 17–20.

6. *Congressional Record*, 88th Cong., 2nd Sess., 1964, pp. 5504–5505; JCAE, *Hearings on AEC Authorizing Legislation Fiscal Year 1965*, 88th Cong., 2nd Sess., 1964, pp. 649–699.

7. JCAE, *Hearings on AEC Omnibus Bills*, pp. 119–126.

8. W. B. McCool to the File, 8 June 1964, Job 5, Box 1, C&R–2 REG (Topics and Discussions w/ REG Staff), Clifford K. Beck to Maurice Goldhaber, 29 June 1964, MH&S–3–2 (WASH–740), AEC/NRC.

9. "Clifford Keith Beck," *Physics Today*, 40 (March 1987): 86–87; "'First Temple of the Atom': The Story of the Raleigh Research Reactor on the North Carolina State College Campus," pamphlet, c. 1953, AEC/NRC.

10. V. R. O'Leary to Clarke Williams, 27 July 1964, Williams to Clifford K. Beck, 31 July 1964, MH&S–3–2 (WASH–740), AEC/NRC.

11. Joseph A. Lieberman to Those Listed Below, 23 July 1964, Minutes of Ad Hoc Committee on Revision of WASH–740, Executive Session, 5 August 1964, Stanley A. Szawlewicz to U. M. Staebler, 13 August 1964, MH&S–3–2 (WASH–740), AEC/NRC.

12. Lieberman to Those Listed Below, 23 July 1964, Minutes of Ad Hoc Committee, 5 August 1964, WASH–740 (1957), AEC/NRC.

13. F. P. Cowan to J. B. H. Kuper, 6 August 1964, Minutes of Steering Committee on Revision of WASH–740, 21 October 1964, 28 January 1965, MH&S–3–2 (WASH–740), AEC Press Release, 25 June 1973, AEC/NRC.

14. Stanley A. Szawlewicz to U. M. Staebler, 13 November 1964, MH&S–3–2 (WASH–740), AEC/NRC.

15. William B. Cottrell to Kenneth Downes, 18 November 1964, Summary of Discussions on WASH–740 Rewrite, 24 November 1964, MH&S–3–2 (WASH–740), AEC/NRC.

16. S. A. Szawlewicz to U. M. Staebler, 21 December 1964, "Minutes of Steering Committee on Revision of WASH–740, December 16, 1964," 6 January 1965, MH&S–3–2 (WASH–740), AEC/NRC.

17. Stuart A. Krieger to Clifford K. Beck, 30 November 1964, Robert J. Mulvihill to Beck, 5 January 1965, MH&S–3–2 (WASH–740), AEC/NRC.

18. C. K. Beck, "Probability of Catastrophic Accidents," working drafts, 12 January 1965, 21 January 1965, MH&S–3–2 (WASH–740), AEC/NRC.

19. Brookhaven National Laboratory, "Upper Limit Accident Calculations for Large Nuclear Power Plants," working draft, 22 January 1965, Stanley A. Szawlewicz to U. M. Staebler, 28 January 1965, MH&S–3–2 (WASH–740), AEC/NRC.

20. H. G. Hembree to U. M. Staebler, 18 November 1964, S. A. Szawlewicz to U. M. Staebler, 27 November 1964, 17 February 1965, John G. Palfrey speech, 3 December 1964, MH&S–3–2 (WASH–740), AEC/NRC.

21. Szawlewicz to Staebler, 17 February 1965, "Minutes of the Steering Committee on the Revision of WASH–740," 28 January 1965, MH&S–3–2 (WASH–740), AEC/NRC.

22. Harold E. Vann to Clifford K. Beck, 25 February 1965, MH&S–3–2 (WASH–740), AEC/NRC; *Nucleonics Week*, 25 February 1965, p. 2.

23. Clifford K. Beck, draft letter to Harold E. Vann with handwritten note to Mr. Palfrey on p. 1, 24 March 1965, MH&S–3–2 (WASH–740), AEC/NRC; Beck to the Commission, 17 March 1965, Box 16 (Indemnification—Price-Anderson), Office Files of James T. Ramey, Minutes of Commission Meeting 2093 (17 March 1965), AEC Meeting Minutes, AEC/DOE.

24. John G. Palfrey to the Commission, 25 March 1965, Box 16 (Indemnification—Price-Anderson), Ramey Office Files, AEC/DOE; *Nucleonics Week*, 9 August 1962, p. 1; Mazuzan and Walker, *Controlling the Atom*, pp. 104–106.

25. Clifford K. Beck to Members of the Steering Committee for Brookhaven Report (WASH–740), 21 April 1965, MH&S–3–2 (WASH–740), AEC/NRC.

26. W. B. McCool to the File, 8 June 1965, 21 June 1965, Glenn T. Seaborg to Chet Holifield, 18 June 1965, Box 1361, MH&S–3 (Hazards from Power Reactors), Minutes of Information Meetings, 10 June, 14 June 1965, Box 5667, OGM Files (Information Meeting Minutes 1965), AEC/DOE.

27. Minutes of Information Meetings, 10 June, 18 June 1965, Box 5667, OGM Files (Information Meeting Minutes 1965), AEC/DOE.

28. Clifford K. Beck to Members of the Steering Committee, WASH–740, 23 June 1965, MH&S–3–2 (WASH–740), AEC/NRC.

29. David E. Pesonen to John Palfrey, 14 August 1965, Palfrey to Pesonen, 27 August 1965, Pesonen to Palfrey, 13 September 1965, Palfrey to Pesonen, 8 October 1965, MH&S–3–2 (WASH–740), AEC/NRC.

30. David E. Pesonen, "Atomic Insurance: The Ticklish Statistics," *Nation*, 201 (18 October 1965): 242–245; Duncan Clark to the Commission, 26 October 1965, Harold L. Price to Abraham Ribicoff, 29 November 1965, Philippe G. Jacques to the Commission, 27 January 1966, Box 1361 (MH&S–3, Hazards from Power Reactors, Vol. 3), AEC/DOE.

31. Ernest B. Tremmel to the Files, 12 March 1965, Tremmel to Arnold R. Fritsch, 15 March 1965, Tremmel to R. E. Hollingsworth, 12 May 1965, 3 June

1965, Box 49 (National Coal Association), Seaborg Office Files, AEC/DOE; *Journal of Glenn T. Seaborg* (25 vols., Berkeley: Lawrence Berkeley Laboratory PUB–625, 1989), Vol. 10, pp. 284, 496; *Nucleonics Week*, 13 May 1965, p. 6.

32. JCAE, Committee Print, *Selected Materials on Atomic Energy Indemnity Legislation*, 89th Cong., 1st Sess., 1965, pp. 1–40.

33. JCAE, *Hearings on Proposed Extension of AEC Indemnity Legislation*, 89th Cong., 1st Sess., 1965, pp. 3–39, 46, 50–51, 56–57, 60–61, 71, 94, 97, 154.

34. *Ibid.*, pp. 176–211, 261; Leonard M. Trosten to the Files, 1 February 1965, J. H. Merritt to Trosten, 10 February 1965, Box 382 (Indemnity), JCAE Papers.

35. JCAE, *Hearings on . . . Indemnity Legislation*, pp. 8–9, 118–131, 441–449.

36. *Ibid.*, pp. 131–150; *Nucleonics Week*, 2 September 1965, p. 1; John W. Johnson, "The 'Coal Boys' Attempt to Split the Atomic Power Lobby: A Tale of Two Technologies and Government Policy in the 1960s," paper delivered at the joint meeting of the History of Science Society and the Society for the History of Technology, 1987, pp. 19–21.

37. *Nucleonics Week*, 2 September 1965; John T. Conway to Glenn T. Seaborg, 26 November 1965, Box 382 (Indemnity), JCAE Papers.

38. Leonard M. Trosten and William T. England, "Waiving Defenses: A New Approach to Protecting the Public Against Financial Loss from Use of Atomic Energy," *Federal Bar Journal*, 27 (Winter 1967): 27–56; David F. Cavers, "Improving Financial Protection of the Public Against the Hazards of Nuclear Power," *Harvard Law Review*, February 1964, reprinted in JCAE, *Hearings on . . . Indemnity Legislation*, pp. 396–440; JCAE, *Selected Materials on . . . Indemnity Legislation*, pp. 31–40.

39. AEC 785/91 (10 February 1966), Box 16 (Indemnification—Price-Anderson), Ramey Office File, AEC/DOE; Leonard M. Trosten to John T. Conway, 23 February 1966, Bertram Schur to Trosten, 26 May 1966, Box 382 (Indemnity), JCAE Papers; *Nucleonics Week*, 7 April 1966, p. 1.

40. William T. England to the Files, 21 February 1966, England to John Conway, 27 April 1966, Box 382 (Indemnity), JCAE Papers; *Nucleonics Week*, 21 July 1966, p. 4; JCAE, *Hearings on Proposed Amendments to Price-Anderson Act Relating to Waiver of Defenses*, 89th Cong., 2nd Sess., 1966, pp. 32–35, 78–79; Trosten and England, "Waiving Defenses," pp. 47–48.

41. AEC 785/98 (9 February 1968), AEC 785/102 (14 October 1968), AEC Press Release, 30 October 1968, AEC/NRC; William T. England, "The 1966 Waiver-of-Defenses Amendment (P. L. 89–645) to the Price-Anderson Act," 6 September 1968, Box 382 (Indemnity), JCAE Papers.

CHAPTER VII: REACTOR SAFETY: GROWING CONCERN OVER LARGER REACTORS

1. For a lucid introduction to nuclear technology, written for nonspecialists, see Richard Wolfson, *Nuclear Choices: A Citizen's Guide to Nuclear Tech-*

nology (Cambridge, MA: MIT Press, 1991). For a clear explanation of the design, operation, and hazards of power reactors, see Anthony V. Nero, Jr., *A Guidebook to Nuclear Reactors* (Berkeley and Los Angeles: University of California Press, 1979). For a compilation of important documents and a retrospective view of reactor safety issues from a prominent participant, see David Okrent, *Nuclear Reactor Safety: On the History of the Regulatory Process* (Madison: University of Wisconsin Press, 1981). For discussions of reactor safety systems and issues as viewed from the perspective of the mid-1960s, see: Paul Schmitz and Roger Griffin, "Power Reactor Containment," and Henry Gitterman, "Reactor Safeguard Systems," in *Nucleonics*, 23 (October 1965): 50–64; U. S. Atomic Energy Commission, *Atomic Power Safety* (one of a series of booklets on "Understanding the Atom"), 1964, Atomic Energy Commission Records, Nuclear Regulatory Commission, Rockville, Maryland (hereafter cited as AEC/NRC); AEC Press Release, "How Safe Is a Nuclear Reactor?" 26 January 1964, Herbert Kouts to Glenn T. Seaborg, 18 November 1964, Box 629 (Reactor Safeguards and Site Criteria), Papers of the Joint Committee on Atomic Energy (JCAE), Record Group 128 (Records of the Joint Committees of Congress), National Archives, Washington, D. C.

2. "Minutes of Steering Committee on Revision of WASH–740, December 16, 1964," 6 January 1965, MH&S–3–2 (WASH–740), AEC/NRC; J. N. Inglima, "Engineering Safeguards," *Nuclear News*, 8 (January 1965): 42; JCAE, *Hearings on Licensing and Regulation of Nuclear Reactors*, 90th Cong., 1st Sess., 1967, p. 62.

3. USAEC, *Atomic Power Safety*, pp. 32–33; ACRS draft letter to Glenn T. Seaborg, 16 January 1965, Draft Minutes of Engineered Safeguards Subcommittee Meeting, 22 April 1965, ACRS File, AEC/NRC.

4. Marvin M. Mann to A. R. Leudecke, 24 January 1963, MH&S 3–2 (Reactor Hazards and Control), U. S. Atomic Energy Commission, *Nuclear Reactors* (one of a series of booklets on "Understanding the Atom"), 1970, AEC/NRC; AEC 949/19 (12 May 1965), Box 1361, MH&S–3 (Hazards of Power Reactors), Atomic Energy Commission Records, Department of Energy, Germantown, Maryland (hereafter cited as AEC/DOE); "AEC Plans Reactor-Safety Engineering Test Programs," *Nucleonics*, 21 (February 1963): 19; JCAE, *Hearings on AEC Authorizing Legislation Fiscal Year 1965*, 88th Cong., 2nd Sess., 1964, pp. 398–409.

5. Stanley A. Szawlewicz, "Current Trends in Nuclear Safety and Development," *Nuclear News*, 6 (August 1963): 3–6; "AEC Plans Reactor-Safety Engineering Test Programs," *Nucleonics*, p. 19.

6. AEC 943/10 (10 January 1963), AEC/NRC; A. R. Leudecke to David B. Hall, 14 February 1963, Box 1361 (MH&S–3, Hazards from Power Reactors), AEC/DOE.

7. R. H. Wilcox to H. J. C. Kouts, 7 June 1963, T. J. Thompson to Safety Research Subcommittee, 24 July 1963, ACRS File, AEC/NRC; Szawlewicz, "Current Trends," p. 5; "AEC Plans Reactor-Safety Engineering Test Programs," p. 19.

8. AEC 943/11 (2 August 1963), AEC 943/12 (20 November 1963), Box 1361 (MH&S–3, Hazards from Power Reactors), AEC/DOE.

9. Dwight A. Ink to David B. Hall, 14 January 1964, Box 1390 (O&M–7, ACRS), AEC/DOE; AEC Press Release, 19 November 1963, AEC/NRC.

10. Minutes of Safety Research Subcommittee Meetings, 27 January 1964, 23 October 1964, Meeting of ACRS Safety Research Subcommittee with GAC Reactor Subcommittee, 26 May 1965, ACRS File, AEC/NRC; AEC 943/19 (12 May 1965), Box 1361 (MH&S–3, Hazards from Power Reactors), AEC/ DOE.

11. Meeting of ACRS Safety Research Subcommittee with GAC Reactor Subcommittee, 26 May 1965, ACRS File, AEC/NRC; General Advisory Committee, Reactors Subcommittee, "Review of Reactor Safety Research Program," 30 July 1965, Box 1361 (MH&S–3, Hazards from Power Reactors), AEC/ DOE; Richard T. Sylves, *The Nuclear Oracles: A Political History of the General Advisory Committee of the Atomic Energy Commission, 1947–1977* (Ames: Iowa State University Press, 1987).

12. Harold L. Price to the Commission, 14 February 1964, Box 80 (Director of Regulation), Office Files of Glenn T. Seaborg, A. J. Vander Weyden to Clifford K. Beck, 4 February 1964, Box 17 (Nuclear Safety Liaison), Office Files of James T. Ramey, AEC/DOE; Beck to R. E. Hollingsworth, 28 April 1964, MH&S 3–2 (Reactor Hazards and Control), Meeting of ACRS Safety Research Subcommittee with GAC Reactor Subcommittee, 26 May 1965, ACRS File, AEC/NRC; "The AEC's New Budget," *Nuclear Industry*, 12 (February 1965): 4–9.

13. GAC Reactors Subcommittee, "Review of Reactor Safety Research Program," AEC/DOE; "Report to the Atomic Energy Commission by the Regulatory Review Panel (Mitchell Panel)," 14 July 1965, printed in JCAE, *Hearings on Licensing and Regulation of Nuclear Reactors*, pp. 410–433.

14. AEC 943/23 (22 July 1965), Edward J. Bloch and Harold L. Price to the Commission, 12 August 1965, Box 1361 (MH&S–3, Hazards from Power Reactors), F. T. Hobbs to the File, 17 August 1965, Box 7729 (O&M–7, Steering Committee on Reactor Safety Research), AEC/DOE.

15. Minutes of Meeting of the Steering Committee on Reactor Safety Research, 7 September 1965, Box 7729 (O&M–7, Steering Committee on Reactor Safety Research), AEC/DOE; James Graham to John T. Conway, 22 April 1966, Box 629 (Reactor Safety), JCAE Papers; JCAE, *Hearings on AEC Authorizing Legislation Fiscal Year 1966*, 89th Cong., 1st Sess., 1965, p. 872; *Forum Memo*, 11 (February 1964): 6.

16. Minutes of Meeting of the Steering Committee on Reactor Safety Research, 29 September 1965, Box 7729 (O&M–7, Steering Committee on Reactor Safety Research), AEC/DOE; JCAE, *AEC Authorizing Legislation*, p. 872.

17. AEC 1204 (1 December 1965), Box 7729 (O&M–7, Steering Committee on Reactor Safety Research), AEC/DOE.

18. N. J. Palladino, "Mechanical Design of Components for Reactor Systems," in *The Technology of Nuclear Reactor Safety*, eds. T. J. Thompson and J. G. Beckerley, 2 vols. (Cambridge, MA: MIT Press, 1964, 1973), II, pp. 107–146; *Nucleonics Week*, 16 February 1967, p. 2; Okrent, *Nuclear Reactor Safety*, p. 85.

19. C. Rogers McCullough, "Some Philosophical Comments on Accident Models and Hazard Evaluation," April 1965, Box 1361 (MH&S–3, Hazards from Power Reactors), AEC/DOE; Okrent, *Nuclear Reactor Safety*, pp. 85–86.

20. Minutes of Meeting of ACRS Subcommittee on Reactor Pressure Vessels, 23–24 April 1965, ACRS File, AEC/NRC; Okrent, *Nuclear Reactor Safety*, pp. 85–86; "ACRS Qualms on Possible Vessel Failure Startle Industry," *Nucleonics*, 24 (January 1966): 17–18.

21. AEC Press Release, 11 January 1966, AEC/NRC; Okrent, *Nuclear Reactor Safety*, p. 86.

22. Minutes of Meeting of Dresden 2 Subcommittee, 1 September 1965, ACRS File, AEC/NRC; Okrent, *Nuclear Reactor Safety*, p. 87.

23. Draft Statement by N. J. Palladino regarding Pressure Vessel Failures, 15 November 1965, Statement by T. J. Thompson regarding Pressure Vessel Failures, 16 November 1965, ACRS File, AEC/NRC; Okrent, *Nuclear Reactor Safety*, pp. 87–88.

24. W. D. Manly to Glenn T. Seaborg, 24 November 1965, in AEC Press Release, 3 December 1965, AEC/NRC.

25. "Statement by AEC on Safety Research Program," in AEC Press Release, 3 December 1965, AEC/NRC.

26. *Nucleonics Week*, 9 December 1965, pp. 1–2; "ACRS Qualms on Possible Pressure Vessel Failure," *Nucleonics*, pp. 17–18.

27. "Summary Notes of Meeting with Advisory Committee on Reactor Safeguards," 9 February 1966, Box 1390 (O&M–7, ACRS), AEC/DOE; David Okrent, "On the History of the Evolution of Light Water Safety in the United States," typescript, n.d., pp. 2–130—2–139, AEC/NRC. This is a somewhat longer version of Okrent's *Nuclear Reactor Safety*.

28. M. W. Libarkin to ACRS Members, 6 December 1966, ACRS File, Harold L. Price to the Commission, 2 August 1967, Job 4, Box 11 (ID&R–6 REG—Reactor Pressure Vessels), AEC/NRC; Price to the Commission, 10 January 1967, Box 7729 (O&M–7, ACRS), AEC/DOE; Okrent, *Nuclear Reactor Safety*, pp. 221–222.

29. M. W. Libarkin to ACRS Members, 26 July 1967, ACRS File, AEC/NRC; Okrent, *Nuclear Reactor Safety*, p. 221.

30. "Tentative Regulatory Supplementary Criteria for ASME Code-Constructed Nuclear Pressure Vessels," 23 August 1967, Box 602 (Reactor by Type or Company Name, Pressure Vessels), JCAE Papers; "AEC Subjects Plant Applicants to 'Tentative' Vessel Criteria," *Nuclear Industry*, 15 (February 1968): 41–42.

31. "AEC Pressure Vessel Criteria Get Adverse Forum Comments," *Nuclear Industry*, 14 (December 1967): 19–21; *Nucleonics Week*, 7 March 1968, pp. 2–3; AEC 943/45 (12 August 1968), AEC–R 69/15 (24 September 1968), AEC/NRC.

32. R. F. Fraley to ACRS Members, 13 June 1968, Carroll W. Zabel to Glenn T. Seaborg, draft, 15 July 1968, ACRS File, AEC/NRC; Okrent, *Nuclear Reactor Safety*, pp. 154–161.

33. William J. Lanouette to Jack Riley, 11 December 1972, Box 36 (Miscellaneous), Ramey Office Files, AEC/DOE; "Current Status and Future Technical and Economic Potential of Light Water Reactors" (WASH–1082), March 1968, AEC/NRC; JCAE, *Hearings on Licensing and Regulation of Nuclear Reactors*, pp. 75–76; Joel Primack and Frank Von Hippel, "Nuclear Reactor Safety," *Bulletin of the Atomic Scientists*, 30 (October 1974): 5–7; E. E. Lewis, *Nuclear Power Reactor Safety* (New York: John Wiley and Sons, 1977), pp. 437–444; Okrent, *Nuclear Reactor Safety*, pp. 103–107.

34. T. J. Thompson to ACRS Members, 23 May 1966, R. H. Wilcox to T. J. Thompson and H. Etherington, 28 December 1965, ACRS File, AEC/NRC; Okrent, *Nuclear Reactor Safety*, pp. 111–112.

35. Okrent, *Nuclear Reactor Safety*, pp. 112–122.

36. David Okrent to Glenn T. Seaborg, 16 August 1966, draft, ACRS File, AEC/NRC; Okrent, *Nuclear Reactor Safety*, pp. 122–124, 131–133.

37. Draft Minutes of Indian Point II Subcommittee Meeting, 23 June 1966, Draft Minutes of Dresden III Subcommittee Meeting, 7 July 1966, Minutes of Reactor Safety Research Subcommittee Meeting, 10 August 1966, ACRS File, Okrent, "On the History of the Evolution of Light Water Safety in the United States," pp. 2–205—2–206, AEC/NRC; Harold L. Price to the Commission, 30 August 1966, Box 17 (ACRS), Seaborg Office Files, AEC/DOE; Okrent, *Nuclear Reactor Safety*, p. 125.

38. Harold L. Price to the Chairman, 30 August 1966, Box 17 (ACRS), Seaborg Office Files, AEC/DOE; Minutes of ACRS Meeting, 8–10 September 1966, ACRS File, AEC/NRC; Okrent, *Nuclear Reactor Safety*, p. 126.

39. Minutes of ACRS Meeting, September 8–10, 1966, ACRS File, David Okrent to Glenn T. Seaborg, 12 October 1966, AEC Press Release, 25 October 1966, AEC/NRC.

40. Okrent, *Nuclear Reactor Safety*, pp. 122–123.

41. Minutes of Reactor Safety Research Subcommittee Meeting, 10 August 1966, ACRS File, AEC/NRC.

42. George T. Mazuzan and J. Samuel Walker, *Controlling the Atom: The Beginnings of Nuclear Regulation, 1946–1962* (Berkeley and Los Angeles: University of California Press, 1984), pp. 122–182; Octave J. DuTemple, "A Report on Fermi," *Nuclear News*, 7 (January 1964): 28–33; John G. Fuller, *We Almost Lost Detroit* (New York: Reader's Digest Press, 1975), pp. 123–127, 182–195.

43. Earl M. Page, "The Fuel Melting Incident," in E. Pauline Alexanderson, ed., *Fermi-I: New Age for Nuclear Power* (LaGrange Park, IL: American Nuclear Society, 1979), pp. 225–254; Power Reactor Development Company, "Report on the Fuel Melting Incident in the Enrico Fermi Atomic Power Plant," 15 December 1968, U. S. Atomic Energy Commission, Division of Reactor Licensing, "Report to the ACRS: The Enrico Fermi Atomic Power Plant," 21 November 1969, ACRS File, AEC/NRC; "Fermi Mishap's 'Actual' Cause Draws Congressional Criticism," *Nuclear Industry*, 15 (February 1968): 17–19. Fuller gives a dramatic but exaggerated account of the hazards of the Fermi accident in *We Almost Lost Detroit*.

CHAPTER VIII: ACCIDENT PREVENTION:
THE EMERGENCY COOLING IMBROGLIO

1. R. F. Fraley to the Files, 1 December 1966, R. H. Wilcox to the Files, 27 February 1967, ACRS Files, James F. Young to Glenn T. Seaborg, 7 October 1966, Job 4, Box 4, ID&R–6 REG (Hazard Evaluation), Atomic Energy Commission Records, Nuclear Regulatory Commission, Rockville, Maryland (hereafter cited as AEC/NRC); *Journal of Glenn T. Seaborg*, (25 vols., Berkeley: Lawrence Berkeley Laboratory PUB–625, 1989), Vol. 13, p. 461.

2. Fraley to the Files, 1 December 1966, "Draft Minutes of Engineered Safeguards Subcommittee Meeting Held on January 11, 1967," 27 January 1967, ACRS File, Clifford K. Beck to the Commission, 28 October 1966, Job 4, Box 4, ID&R–6 REG (Hazard Evaluation), AEC/NRC.

3. Wilcox to the Files, 27 February 1967, ACRS File, AEC/NRC; Peter A. Morris to Harold L. Price, 6 January 1967, Box 62 (Nuclear Power Reactors), Office Files of James T. Ramey, Atomic Energy Commission Records, Department of Energy, Germantown, Maryland (hereafter cited as AEC/DOE); "Licensed and New Reactors Given Safeguards Review," *Nuclear Industry*, 13 (November 1966): 29–31.

4. Harold L. Price to the Commission, 20 October 1966, Job 4, Box 4, ID&R–6 REG (Hazard Evaluation), AEC Press Release, 27 October 1966, AEC/NRC; David Okrent, *Nuclear Reactor Safety: On the History of the Regulatory Process* (Madison: University of Wisconsin Press, 1981), pp. 164–166.

5. M. W. Libarkin to ACRS Members, 6 December 1966 and 23 May 1966, R. H. Wilcox to H. O. Monson, 17 February 1967, ACRS File, Harold L. Price to the Commission, 27 April 1967, Job 4, Box 4, ID&R–6 REG (Hazard Evaluation), AEC/NRC; Minutes of Fifth Meeting of the Industry Advisory Committee on Power Reactor Emergency Core Cooling Systems, 16–20 January 1967, Box 3364, RDT Files (Task Force-Emergency Core Cooling), AEC/DOE.

6. R. J. Impara to Members of the Advisory Committee on Power Reactor Emergency Core Cooling Systems, 3 May 1967, Box 3364 (Task Force-Emergency Core Cooling), AEC/DOE; Harold L. Price to the Commission, 18 July 1967, Job 4, Box 4, ID&R–6 REG (Hazard Evaluation), AEC/NRC; *Seaborg Journal*, Vol. 14, p. 519.

7. *Emergency Core Cooling: Report of Advisory Task Force on Power Reactor Emergency Cooling*, October 1967, Job 4, Box 4, ID&R–6 REG (Hazard Evaluation), AEC/NRC.

8. Edson G. Case to Harold L. Price, 6 October 1967, RD–8–3 (ECCS), Harold L. Price to the Commission, 18 October 1967, Job 4, Box 4, ID&R–6 REG (Hazard Evaluation), AEC Press Release, 23 October 1967, AEC/NRC; W. K. Ergen to Task Force, 29 September 1967, Box 3364, RDT Files (Task Force-Emergency Core Cooling), AEC/DOE.

9. P. A. Morris to D. J. Skovholt, L. D. Low, and others, 31 October 1967, Division of Reactor Standards, "Report to the Advisory Committee on Reactor Safeguards in the Matter of the Advisory Task Force Report on Emergency Core Cooling," 4 December 1967, RD–8–3 (ECCS), AEC/NRC.

10. Minutes of Meeting of ACRS Subcommittee on Emergency Core Cooling, 23 October 1967, ACRS File, AEC/NRC; Okrent, *Nuclear Reactor Safety*, pp. 169–170, 174–176.

11. *Emergency Core Cooling: Report of the Advisory Task Force on Power Reactor Emergency Cooling*, pp. 49–50, Joseph A. Lieberman to M. Shaw and others, 20 February 1968, RD–8–3 (ECCS), ARC/NRC; Okrent, *Nuclear Reactor Safety*, pp. 105–107, 170.

12. R. F. Fraley to the File, 29 February 1968, J. C. McKinley to D. Okrent, 6 March 1968, McKinley to ACRS Members, 19 June 1968, ACRS File, S. Levine to Harold L. Price and others, 5 March 1968, RD–8–3 (ECCS), AEC/NRC.

13. "Water-Reactor Safety Program: Summary Description," draft, January 1967, printed in U. S. Congress, Joint Committee on Atomic Energy (JCAE), *Hearings on AEC Authorizing Legislation Fiscal Year 1968*, 90th Cong., 1st Sess., 1967, pp. 1351–1423; AEC 943/46 (30 August 1968), Box 7751, ID&R–6 (Hazards, Vol. 1), AEC/DOE; Frank Schroeder, "Overview of the Water-Reactor Safety Program," 19 September 1967, O&M–7 (ASLB), AEC/NRC.

14. "Water-Reactor Safety Program" in JCAE, *AEC Authorizing Legislation 1968*, pp. 1365–1367, 1387–1391, 1395, 1401–1402, 1406, 1412–1416; AEC 943/46, AEC/DOE.

15. AEC 943/46, AEC/DOE; "LOFT Experiment Redirected to Emergency Core Cooling," *Nuclear Industry*, 15 (January 1968): 33–34.

16. "Water-Reactor Safety Program" in JCAE, *AEC Authorizing Legislation 1968*, pp. 1385–1387; T. R. Wilson, "Engineered Safeguards Systems Investigations," 19 September 1967, O&M–7 (ASLB), AEC/NRC.

17. JCAE, *Hearings on Licensing and Regulation of Nuclear Reactors*, 90th Cong., 1st Sess., 1967, pp. 435–438.

18. *Ibid*, pp. 641–643; McKinley to Okrent, 6 March 1968, ACRS File, AEC/NRC; Notes on ACRS-AEC-Industry Meeting on Safety Research, 27–28 February 1968, Box 3364, RDT Files (ACRS R&D Subcommittee), AEC/DOE.

19. J. C. McKinley to David Okrent, 6 March 1968, 5 November 1968, R. F. Fraley to H. S. Isbin, 6 March 1967, ACRS File, AEC/NRC; N. J. Palladino to Glenn T. Seaborg, 14 April 1967, Box 7729 (ACRS), AEC/DOE.

20. Edson G. Case to John Ryan, 24 March 1967, P. A. Morris to H. L. Price and others, 25 June 1968, O&M–6 (AIF), AEC/NRC.

21. Edwin A. Wiggin to George M. Kavanagh, 6 June 1967, O&M–6 (AIF), McKinley to Okrent, 6 March 1968, ACRS File, AEC/NRC; Milton Shaw to R. E. Hollingsworth, 24 January 1967, Box 3364, RDT Files (General Manager), AEC/DOE; JCAE, *Hearings on Licensing and Regulation of Nuclear Reactors*, pp. 209–210, 252–253, 262.

22. Peter A. Morris to Edson G. Case, 18 April 1968, Clifford K. Beck to Milton Shaw, 24 April 1968, RD–8–3 (ECCS), AEC/NRC; Case to J. A. Lieberman, 19 January 1968, Box 3364, RDT Files (ACRS R&D Subcommittee), AEC/DOE.

23. JCAE, *AEC Authorizing Legislation 1968*, pp. 1330–1333.

24. Summary of 118th ACRS Meeting (5–7 February 1970), 20 February 1970, ACRS File, AEC/NRC; AEC 671/24 (1 July 1968), AEC 671/25 (24 July 1968), Box 7776, P&C–9 (Idaho Nuclear Corporation), AEC/DOE.

25. AEC 671/25, AEC 671/27 (18 December 1968), AEC 671/29 (4 March 1969), Box 7776, P&C–9 (Idaho Nuclear Corporation), M. A. Rosen to Milton Shaw, 23 June 1969, Job 1145, RDT Files, Box 12 (Davis-Bacon), AEC/DOE; AEC Press Release, 2 February 1966, AEC/NRC.

26. Phillips Petroleum Company, "LOFT Program Evaluation," 18 January 1967, Milton Shaw to W. L. Ginkel, 31 May 1967, Shaw to Harold L. Price, 26 July 1967, Job 1145, RDT Files, Box 8 (Loss of Fluid Test), S. G. Forbes to C. W. Bills, 27 January 1969, RDT Files, Box 7 (LOFT—Semi-Scale Blowdown), H. L. Coplen to R. E. Swanson, 9 June 1970, RDT Files, Box 11 (Contractor to ID Correspondence), AEC/DOE; JCAE, *Hearings on AEC Authorizing Legislation Fiscal Year 1970*, 91st Cong., 1st Sess., 1969, p. 948.

27. Milton Shaw, "RDT Standards Program," n. d., S. G. Forbes to D. E. Williams, 4 June 1968, Job 1145, RDT Files, Box 8 (Standards and Codes), R. E. Swanson to W. H. Layman, 31 December 1970, RDT Files, Box 6 (RDT to ID Correspondence), AEC/DOE; Summary of 118th ACRS Meeting, AEC/NRC.

28. AEC 943/49 (3 October 1969), Box 7751, ID&R–6 (Hazards, Vol. 1), AEC/DOE; "Late News," *Nuclear Industry*, 15 (January 1968): 35.

29. AEC 943/49, AEC/DOE.

30. *Ibid.*, Glenn T. Seaborg to Frank Church, 17 November 1970, Box 64 (Reactor Development), Ramey Office Files, AEC/DOE; JCAE, *Hearings on AEC Authorizing Legislation Fiscal Year 1972*, 92nd Cong., 1st Cong., 1971, pp. 852–862.

31. Anthony V. Nero, Jr., *A Guidebook to Nuclear Reactors* (Berkeley and Los Angeles: University of California Press, 1979), pp. 203–219; George T. Mazuzan and J. Samuel Walker, *Controlling the Atom: The Beginnings of Nuclear Regulation, 1946–1962* (Berkeley and Los Angeles: University of California Press, 1984), p. 124.

32. Speeches by James T. Ramey and Milton Shaw before the American Power Conference, 23 April 1968, Box 566 (Reactors by Type or Company Name-Fast Ceramic Reactor), JCAE Papers, Record Group 128 (Records of the Joint Committees of Congress), National Archives; Glenn T. Seaborg to Paul W. McCracken, 30 October 1970, White House Central File-SMOF Edward David (White House-Energy, Vol. II), Richard M. Nixon Papers, Nixon Presidential Materials Project, Alexandria, Virginia; JCAE, *Authorizing Legislation 1972*, pp. 538–542; *Nucleonics Week*, 18 December 1969, p. 3; Mazuzan and Walker, *Controlling the Atom*, pp. 413–415.

33. Seaborg to Church, 17 November 1970, AEC/DOE; JCAE, *Authorizing Legislation 1972*, p. 862.

34. R. E. Swanson to Don E. Williams, 21 August 1968, Job 1145, RDT Files, Box 12 (LOFT-Rpt. Correspondence), "Blowdown Meeting: Opening Remarks by H. G. Hembree," 17 December 1968, RDT Files, Box 7 (LOFT Semi-Scale Blowdown), "1969 INC Annual Appraisal," n. d., RDT Files, Box 8 (LOFT Supporting Tasks, General), J. C. Haire to W. B. Cottrell, 15 October 1970, Haire to D. R. deBoisblanc, 14 December 1970, RDT Files, Box 11 (Contractor to ID Correspondence), C. M. Rice to W. L. Ginkel, 5 February 1971, RDT Files, Box 13 (LOFT Costs, Schedule, and Budget), AEC/DOE; JCAE, *Authorizing Legislation 1972*, pp. 861–862.

35. AEC 943/49, AEC/DOE; "Report of DRD&T Meeting with Water Reactor Vendors, May 26, 1970," 2 June 1970, ACRS File, AEC/NRC.

36. "Summary of AEC-AIF Meeting on Water-Reactor Safety Research Program, November 17–18, 1969," n. d., Box 629 (Reactor Safety), JCAE Papers; "A New Call for 'Realism' in AEC Reactor Safety R&D," *Nuclear Industry*, 16 (October 1969): 22–26; *Nucleonics Week*, 27 November 1969, pp. 3–4.

37. R. C. DeYoung to Saul Levine, 6 March 1969, Milton Shaw to W. L. Ginkel, 18 April 1969, O&M–7 (ACRS), "Minutes of PBF/LOFT Experimental Programs," 26–27 September 1969, ACRS File, AEC/NRC; AEC 943/50 (5 November 1969), Box 7751, ID&R–6 (Hazards, Vol. 1), AEC/DOE; *Nucleonics Week*, 4 June 1970, pp. 3–4; Joseph M. Hendrie to Glenn T. Seaborg, 12 November 1969, printed in JCAE, *Hearings on AEC Licensing Procedure and Related Legislation*, 92nd Cong., 1st Sess., 1971, pp. 1115–1118.

38. Morris Rosen to R. C. DeYoung, 2 April 1969 (Exhibit 801), Edson G. Case to P. A. Morris, 28 May 1970 (Exhibit 802), J. C. Haire to C. W. Bills, 1 June 1970 (Exhibit 803), DeYoung to Morris, 2 June 1970 (Exhibit 804), Morris to Harold L. Price (Exhibit 805), 20 July 1970, Docket RM 50–1 (In the Matter of Acceptance Criteria for Emergency Core Cooling Systems), AEC/NRC; E. H. Davidson to the Files, 20 August 1970, and enclosures, Job 1145, RDT Files, Box 16 (SEFET), AEC/DOE.

39. Idaho Nuclear Corporation, Nuclear Safety Program Division, *Monthly Report*, January 1971 (Exhibit 34), "Evaluation of Semiscale—ECC Tests for Applicability to Pressurized Water Reactors," September 1971 (Exhibit 758), Docket RM 50–1, AEC/NRC; *Nucleonics Week*, 13 May 1971, p. 5; Robert Gillette, "Nuclear Reactor Safety: A Skeleton at the Feast?" *Science*, 172 (28 May 1971): 918–919.

40. Fred C. Finlayson, "A View from the Outside," *Bulletin of the Atomic Scientists*, 31 (September 1975): 20–25; *Nucleonics Week*, 13 May 1971, p. 5.

41. David Okrent, "On the History of the Evolution of Light Water Reactor Safety in the United States," typescript, n. d., pp. 6–64 to 6–65, AEC/NRC; "Denwood F. Ross, Jr., Personal Qualifications," in Insert following p. 474, Docket RM 50–1, AEC/NRC; Author's interview with Stephen H. Hanauer, Rockville, Maryland, 25 June 1990.

42. Okrent, "Evolution of Light Water Reactor Safety," pp. 6–62 to 6–63; JCAE, *AEC Licensing Procedure*, 1971, pp. 484–486.

43. Edson G. Case to Harold L. Price and others, 5 March 1971 (Exhibit 707), Case to Steve Hanauer, 8 March 1971 (Exhibit 708), Docket RM 50–1, AEC/NRC; Handwritten Notes of "Meeting in Bethesda," 22 March 1971, Job 1145, RDT Files, Box 13 (LOFT Supporting Tasks-General), AEC/DOE.

44. William B. Cottrell to A. J. Pressesky, 30 June 1970 (Exhibit 1029), P. L. Rittenhouse, "Progress in Zircaloy Cladding Failure Modes Research," December 1970 (Exhibit 101), D. O. Hobson and P. L. Rittenhouse to Morris Rosen, 1 March 1971 (Exhibit 251), Cottrell to Pressesky, 4 March 1971 (Exhibit 1030), Docket RM 50–1, AEC/NRC; P. L. Rittenhouse, "Fuel-Rod Failure and Its Effects in Light-Water-Reactor Accidents," *Nuclear Safety*, 12 (September-October 1971): 487–495.

45. *Seaborg Journal*, Vol. 25, p. 686; James B. Graham to Edward J. Bauser,

4 May 1971, Box 629 (Reactor Safety), JCAE Papers; Glenn T. Seaborg to John O. Pastore, 27 April 1971, O&M–7 (ACRS), AEC/NRC; *Nucleonics Week*, 6 May 1971, p. 1.

46. A. J. Pressesky to Milton Shaw, 14 May 1971, Job 1145, RDT Files, Box 13 (LOFT Supporting Tasks-General), AEC/DOE; W. B. McCool to the File, 23 April 1971, "Proposed Questions and Answers," 23 April 1971, Job 9, Box 13, ID&R–14 REG (Part 50), AEC Press Release, 27 May 1971, AEC/NRC; R. E. Lapp to Chet Holifield, 11 May 1971, Box 48 (Nuclear Power Plants 1971), Chet Holifield Papers, University of Southern California, Los Angeles; John A. Harris to Edward J. Bauser, 3 May 1971, Bauser to All Committee Members, 11 May 1971, "el" memorandum, 13 May 1971, Box 566 (ECCS), JCAE Papers; "Regulators Apparently Satisfied on Emergency Core Cooling Test Data," *Nuclear Industry*, 18 (May 1971): 25; *Nucleonics Week*, 20 May 1971, p. 2; *Washington Post*, 26 May 1971; Gillette, "Skeleton at the Feast," p. 918.

47. David Dinsmore Comey to Glenn T. Seaborg, 3 June 1971, Box 7809, ID&R–6 (Hazards, Vol. 2), AEC/DOE; Edward E. David to Seaborg, 25 May 1971, White House Central File–SMOF Edward David (OST White House Administration, EED Chron.), Nixon Papers; *Nucleonics Week*, 10 June 1971, p. 5; *New York Times*, 6 June 1971; Gillette, "Skeleton at the Feast," pp. 918–919.

48. *Seaborg Journal*, Vol. 25, pp. 1–2, 12; *Nucleonics Week*, 20 May 1971, p. 2.

49. Edson G. Case to Milton Shaw, 16 April 1971, Job 1145, RDT Files, Box 16 (Liaison–REG), C. M. Rice to W. L. Ginkel, 14 May 1971, RDT Files, Box 13 (LOFT Costs, Schedule, and Budget), AEC/DOE; Robert J. Colmar and Morris Rosen to Task Force Members, 1 June 1971 (Exhibit 715), "Status Report by ECCS Task Force," 16 June 1971 (Exhibit 757), "Emergency Core Cooling in Water-Cooled Power Reactors," draft, November 1971 (Exhibit 740B), Docket RM 50–1, AEC/NRC.

50. W. B. McCool to the File, 7 July 1971, Box 7809, ID&R–6 (Hazards, Vol. 2), AEC/DOE; "Interim Acceptance Criteria for Emergency Core Cooling Systems for Light-Water Power Reactors," 25 June 1971, Job 9, ID&R–6 REG (Hazard Evaluation), AEC Press Release, 19 June 1971, "Status Report by ECCS Task Force," 16 June 1971, AEC/NRC; Hanauer interview, 25 June 1990.

51. *Seaborg Journal*, Vol. 25, pp. 206, 223, 226–230, 275–276; "Excerpt of Regulatory Information Meeting 486 Discussion," 9 June 1971, W. B. McCool to the File, 11 June 1971, Job 9, ID&R–6 (Hazard Evaluation), "Edited Transcript of Press Briefing on Interim Acceptance Criteria," 19 June 1971, RD–8–3 (ECCS), Okrent, "Evolution of Light Water Reactor Safety," pp. 6–62 to 6–76, AEC/NRC; "Emergency Core Cooling," *Nuclear Industry*, 18 (June 1971): 13–15; "Nuclear News Briefs," *Nuclear News*, 14 (July 1971): 17; *Nucleonics Week*, 27 May 1971, p. 1.

52. *Seaborg Journal*, Vol. 25, p. 230; Stephen H. Hanauer to Commissioner Ramey, 29 June 1971 (Exhibit 718), Docket RM 50–1, AEC/NRC; *Nucleonics Week*, 3 June 1971, p. 4.

53. Okrent, "Evolution of Light Water Reactor Safety," pp. 6–68 to 6–69, AEC/NRC; Gillette, "Skeleton at the Feast," p. 919.

54. Glenn T. Seaborg to the President, 2 March 1971, White House Special

File, White House Central File, FG 78 (Atomic Energy Commission), John C. Whitaker to John D. Ehrlichman, 8 April 1971, White House Central File—Subject File, AT 2 (Industrial); R. K. Price, Jr., "Notes for the President's File: Meeting of the Cabinet," 13 April 1971, White House Special File, President's Office File, Box 84 (Memoranda for the President), Nixon Papers.

55. Will Kriegsman to John Ehrlichman, 16 April 1971, White House Special File, White House Central File—Subject File Confidential, UT (Utilities), EAC to Searl, 9 June 1971, Edward E. David, Jr. to Peter Flanigan, 10 June 1971, White House Central File—SMOF Edward David (White House—Energy, Vol. 2), Nixon Papers; JCAE, *AEC Licensing Procedure*, pp. 519–528; W. Henry Lambright and Dianne Rahm, "Presidential Management of Technology" in Michael E. Kraft and Norman J. Vig, eds., *Technology and Politics* (Durham: Duke University Press, 1988), pp. 81–97.

56. "A Cooling Threat to Nuclear Power," *Business Week*, 5 June 1971, p. 48; Gillette, "Skeleton at the Feast," p. 918.

57. Harold L. Price to Edward J. Bauser, 7 October 1971, Job 9, ID&R–6 REG (Hazard Evaluation), "Emergency Core Cooling in Water-Cooled Power Reactors," draft, November 1971 (Exhibit 740B), "Testimony of the AEC Regulatory Staff," 28 December 1971 (Exhibit 1001), Docket RM 50–1, AEC/NRC; U. S. Congress, House, Committee on Interstate and Foreign Commerce, Subcommittee on Communications and Power, *Hearings on Powerplant Siting and Environmental Protection*, 92nd Cong., 1st Sess., 1971, pp. 939–957; JCAE, *AEC Licensing Procedure*, pp. 13–20; Robert Gillette, "Nuclear Reactor Safety: A New Dilemma for the AEC," *Science*, 173 (9 July 1971): 126–130.

58. Union of Concerned Scientists, "Nuclear Reactor Safety: An Evaluation of New Evidence," July 1971, Box 7809, ID&R–6 (Hazards, Vol. 2), AEC/DOE; Union of Concerned Scientists, "A Critique of the New A.E.C. Design Criteria for Reactor Safety Systems," October 1971, Job 9, ID&R–6 REG (Hazard Evaluation), AEC/NRC; Ian A. Forbes and others, "Nuclear Reactor Safety: An Evaluation of New Evidence," *Nuclear News*, 14 (September 1971): 32–40; *New York Times*, 27 July 1971; Joel Primack and Frank von Hippel, *Advice and Dissent: Scientists in the Political Arena* (New York: Basic Books, 1974), pp. 209–214.

59. Stephen H. Hanauer to ECCS Task Force, 2 December 1971 (Exhibit 810), William B. Cottrell to L. Manning Muntzing, 6 December 1971 (Exhibit 1020), Docket RM 50–1, AEC/NRC; *Nucleonics Week*, 16 December 1971, p. 3.

60. For discussions of later ECCS experiments, see William B. Cottrell, "The ECCS Rule-Making Hearing," *Nuclear Safety*, 15 (January-February 1974): 30–55; Gary L. Downey, "Reproducing Cultural Identity in Negotiating Nuclear Power: The Union of Concerned Scientists and Emergency Core Cooling," *Social Studies of Science*, 18 (May 1988): 231–264.

CHAPTER IX: THE FIRST LINE OF DEFENSE—
AND OTHER SAFETY MEASURES

1. "Report to the Atomic Energy Commission on the Reactor Licensing Program by the Internal Study Group," June 1969, printed in Joint Committee

on Atomic Energy (JCAE), *Hearings on AEC Licensing Procedure and Related Legislation*, 92nd Cong., 1st Sess., 1971, p. 1048.

2. John T. Conway to All Committee Members, 5 December 1964, Box 577 (Reactor by Type or Company Name—Oyster Creek), Papers of the Joint Committee on Atomic Energy, Record Group 128 (Records of the Joint Committees of Congress), National Archives; "A Precedent-Setting Decision," *Nuclear Industry*, 11 (December 1964): 4–9; *Nucleonics Week*, 10 December 1964, pp. 1–3.

3. Harold L. Price to the Commission, 10 December 1971, Job 5, Box 5, IR&A–6 REG (Jersey Central Power and Light Co.), Troy B. Conner to Separated Legal Files, 10 December 1964, Docket 50–219, Atomic Energy Commission Records, Nuclear Regulatory Commission, Rockville, Maryland (hereafter cited as AEC/NRC); "Staff Asks AEC to Review Jersey Central Decision," *Nuclear Industry*, 12 (January 1965): 19–21.

4. John T. Conway to All Committee Members, 5 January 1965, Box 577 (Reactor by Type or Company Name—Oyster Creek), JCAE Papers; AEC–R 114/4 (6 January 1965), AEC–R 114/12 (6 May 1965), Job 5, Box 5, IR&A–6 REG (Jersey Central Power and Light Co.), AEC/NRC.

5. Draft Minutes of ACRS Subcommittee Meeting on Reactor Design and Operating Criteria, 15 July 1965, ACRS File, Minutes of AEC Regulatory Meeting 210, 11 February 1965, AEC/NRC; James T. Ramey to the Commission, 20 May 1965, Harold L. Price to Commissioners Bunting and Ramey, 24 May 1965, Box 26 (Reactor Siting Criteria), Office Files of James T. Ramey, Atomic Energy Commission Records, Department of Energy, Germantown, Maryland (hereafter cited as AEC/DOE); David Okrent, *Nuclear Reactor Safety: On the History of the Regulatory Process* (Madison: University of Wisconsin Press, 1981), pp. 214–216.

6. "Report to the Atomic Energy Commission by the Regulatory Review Panel," 14 July 1965, printed in JCAE, *Hearings on Licensing and Regulation of Nuclear Reactors*, 90th Cong., 1st Sess., 1967, pp. 429–431.

7. "Criteria for Nuclear Power Plant Construction Permits," draft, 20 May 1965, Box 26 (Reactor Siting Criteria), Ramey Office Files, AEC/DOE.

8. Draft Minutes of Subcommittee Meetings on Reactor Design and Operating Criteria, 9 June, 30 July, 3 September 1965, W. D. Manly to Glenn T. Seaborg, 9 August 1965, ACRS File, AEC/NRC; Okrent, *Nuclear Reactor Safety*, pp. 216–217.

9. AEC–R 2/49 (5 November 1965), Minutes of Regulatory Meetings 223 (10 November 1965) and 224 (17 November 1965), AEC Press Release, 22 November 1965, AEC/NRC.

10. Edwin A. Wiggin to Clifford Beck, 28 February 1966, O&M–6 (AIF, 1965–1968), AEC/NRC; A. O. Little to Glenn T. Seaborg, 16 November 1965, Box 80 (Director of Regulation), Office Files of Glenn T. Seaborg, "Summary Notes of Briefing on Safety Analysis Procedures," 3 August 1966, Box 3734 (AEC Minutes), AEC/DOE.

11. AEC–R 2/57 (16 June 1967), AEC/NRC; Okrent, *Nuclear Reactor Safety*, p. 218.

12. AEC–R 2/57, "Comparison of Drafts . . . for General Design Criteria

for Nuclear Power Plant Construction Permits," 6 February 1967, L–4–1 Part 50 (General Design Criteria), Draft Minutes of Reactor Design and Operating Criteria Subcommittee, 10 August 1966, R. F. Fraley to the File, 7 April 1967, ACRS File, AEC/NRC; Okrent, *Nuclear Reactor Safety*, pp. 218–219.

13. Comments on Draft General Design Criteria, L–4–1 Part 50 (General Design Criteria), S. H. Hanauer to H. E. Etherington, 20 February 1967, ACRS File, AEC Press Release, 10 July 1967, AEC/NRC; "AEC Nuclear Plant Criteria," and Edson G. Case, "Regulatory Functions-Division of Reactor Standards," *Nuclear News*, 10 (November 1967): 11, 44–47; "Forum Group Recommends Design Criteria Changes," *Nuclear Industry*, 14 (October 1967): 21–25.

14. Harold L. Price to the Commission, 6 July 1970, Edwin A. Wiggin to Edson G. Case, 6 October 1970, Job 9, Box 13, ID&R–14 REG (Part 50), "General Design Criteria for Nuclear Power Units," 15 July 1969, L–4–1 Part 50 (General Design Criteria), Minutes of Meetings of Reactor Design and Operating Criteria Subcommittee, 26 May 1969, 9 July 1969, Minutes of Meeting of Codes, Standards, and Criteria Subcommittee, 4 December 1970, ACRS File, David Okrent, "On the History of the Evolution of Light Water Reactor Safety in the United States," typescript, n.d., pp. 3–89 to 3–91, AEC/NRC.

15. SECY–R 143 (28 January 1971), SECY–R 257 (11 June 1971), Job 9, Box 13, ID&R–14 REG (Part 50), AEC/NRC; "AEC Issues New Version of Controversial General Design Criteria," *Nuclear Industry*, 18 (March 1971): 32–33.

16. Marvin M. Mann, speech, "The Establishment of Standards for Safety Evaluation of Reactors," February 1965, L–4–1 Part 50 (Technical Specifications), AEC–R 2/50 (30 June 1966), Job 4, Box 6, ID&R–14 REG (Part 50), AEC/NRC.

17. AEC–R 2/50, AEC–R 2/67 (16 October 1968), Job 4, Box 12, ID&R–14 REG (Part 50), Minutes of Regulatory Meetings 239 (3 August 1966) and 270 (4 November 1968), AEC/NRC; "New Tech Specs Regulation Is Published for Comment," *Nuclear Industry*, 13 (September 1966): 38–40; *Nucleonics Week*, 18 August 1966, pp. 2–3.

18. AEC–R 69/17 (22 November 1968), Job 4, Box 11, ID&R–6 REG (Reactor Pressure Vessels), AEC/NRC; *Nucleonics Week*, 27 November 1969, pp. 1–3.

19. "Emergency Core Cooling Evaluation Guidelines," January 1968, RD–8–3 (ECCS), Harold L. Price to the Commission, 25 June 1970, Job 4, Box 12, ID&R–14 REG (Part 50), AEC Press Release, 13 November 1970, AEC/NRC; JCAE, *AEC Licensing Procedure*, 1971, pp. 45–47.

20. Speeches by James T. Ramey and Milton Shaw, printed in JCAE, *Hearings on AEC Authorizing Legislation Fiscal Year 1968*, 90th Cong., 1st Sess., 1967, pp. 1287–1303; Francis Duncan, *Rickover and the Nuclear Navy: The Discipline of Technology* (Annapolis, MD: Naval Institute Press, 1990), pp. 71–73.

21. *Hearings on AEC Authorizing Legislation 1968*, pp. 762, 1291–1292, 1296–1297.

22. *Ibid.*, p. 1303.

23. R. L. Tedesco to the Files, 3 January 1967, Victor Stello to Roger S.

Boyd, 10 August 1967, Tedesco to Boyd, 22 September 1967, William H. McElwain to Harold L. Price, 12 October 1967, Gerald F. Tape to McElwain, 19 October 1967, Docket 50–219, AEC/NRC; Peter A. Morris to John E. Logan, 16 October 1967, Box 577 (Reactors by Type or Company Name—Oyster Creek), JCAE Papers; *Nucleonics Week,* 9 November 1967, p. 1.

24. JCAE, *Hearings on Licensing and Regulation of Nuclear Reactors,* pp. 216–236; Division of Compliance, Report of Reactor Pressure Vessel Repair Program, 6 January 1969, Docket 50–219, AEC/NRC.

25. Arnold R. Fritsch to Glenn T. Seaborg, 12 March 1968, Box 187 (Jersey Central P & L Co.), Seaborg Office Files, AEC/DOE; Edward J. Bauser to John T. Conway, 8 November 1967, Conway to All Committee Members, 8 November 1967, Box 578 (Reactors by Type or Company Name—Oyster Creek), JCAE Papers; Richard L. Doan speech, "Quality Assurance in the Design, Construction and Operation of Nuclear Power Plants," July 1969, L–4–1 Part 50 (Quality Assurance), AEC/NRC; *Nucleonics Week,* 23 November 1967, pp. 1–2.

26. Saul Levine to Harold L. Price and others, with attached statement of ACRS, 10 June 1968, Docket 50–219, AEC/NRC; John T. Conway to All Committee Members, 25 November 1967, Box 578 (Reactors by Type or Company Name—Oyster Creek), JCAE Papers; Philip H. Abelson, "Nuclear Power—Rosy Optimism and Harsh Reality," *Science,* 161 (12 July 1968): 113; *Nucleonics Week,* 23 November 1967, p. 1.

27. Richard L. Doan, Comments on Jersey Central Power and Light Company's Report on Oyster Creek Pressure Vessel Repair Program, 28 March 1968, "Quality Control Inspection of the Oyster Creek Reactor Facility," 6 May 1968, Docket 50–219, AEC/NRC.

28. *Nucleonics Week,* 15 February 1968, p. 1; *Wall Street Journal,* 16 July 1969; *Washington Post,* 24 August 1970.

29. R. H. Engelken to Roger S. Boyd, 5 February 1968, Docket 50–219, AEC/NRC; Arnold R. Fritsch to Glenn T. Seaborg, 12 March 1968 (n. 25 above), AEC/DOE; *Nucleonics Week,* 9 November 1967, p. 1, 25 January 1968, p. 1, 2 May 1968, pp. 4–5; "A Landmark Nuclear Plant Makes Tardy But Hopeful Debut," *Nuclear Industry,* 16 (April 1969): 12–14.

30. Peter A. Morris to John E. Logan, 22 March 1968, Victor Stello, Jr. to Roger S. Boyd, 31 January 1968, Docket 50–219, AEC/NRC; *Nucleonics Week,* 30 November 1967, p. 1, 4 January 1968, p. 4, 11 January 1968, pp. 2–3, 2 May 1968, pp. 4–5.

31. Arnold R. Fritsch to Glenn T. Seaborg, 20 May 1968, Harold L. Price to John T. Conway, 6 June 1968, Box 187 (Jersey Central P & L Co.), Seaborg Office Files, AEC/DOE; P. A. Morris to H. L. Price, 20 May 1968, L. Porse to the Files, 16 August 1968, V. Stello and L. Porse to Roger S. Boyd, 27 August 1968, J. P. O'Reilly to Boyd, 3 October 1968, Docket 50–219, AEC-R 114/15 (17 December 1968), Job 4, Box 9, ID&R–5 REG (Jersey Central), AEC Press Releases, 27 December 1968, 10 April 1969, AEC/NRC.

32. N. C. Moseley to J. P. O'Reilly, 28 February 1969, O'Reilly to Peter A. Morris, 8 April 1969, O'Reilly to R. S. Boyd, with attached Compliance Investigation Report, 2 May 1969, O'Reilly to Boyd, with attached Compliance Investigation Report, 2 June 1969, Compliance Investigation Report, 9 June 1969,

Docket 50–219, "Notification of Alleged Deficiencies: Report to the Director of Regulation," 20 March 1969, W. B. McCool to the File, 14 April 1969, Job 4, Box 9, ID&R–5 REG (Jersey Central), AEC/NRC.

33. Glenn T. Seaborg to Harrison A. Williams, Jr., 29 August 1969, Docket 50–219, AEC/NRC; *Journal of Glenn T. Seaborg*, (25 vols., Berkeley: Lawrence Berkeley Laboratory PUB–625, 1989), Vol. 19, pp. 358, 383, 395, 431; "New Oyster Creek Setback," *Nuclear Industry*, 16 (July 1969): 10–12; *Nucleonics Week*, 7 August 1969, p. 1.

34. James T. Ramey speech to the American Power Conference, 23 April 1968, Box 45 (Backup for APC Remarks), Ramey Office Files, AEC/DOE; *Nucleonics Week*, 7 November 1968, pp. 4–5.

35. Atomic Safety and Licensing Board, "Order Reopening Proceeding to Receive Additional Evidence," 17 October 1968, OGM Files, Box 5625 (Quality Assurance—RDT 1969), AEC/DOE; *Nucleonics Week*, 13 June 1968, p. 2, 24 October 1968, p. 1; Abelson, "Nuclear Power—Rosy Optimism and Harsh Reality," p. 113.

36. Minutes of Atomic Safety and Licensing Board Panel Meeting, 8 July 1969, Box 7771, O&M–7 (AS&LB), AEC/DOE; Minutes of Quality Assurance Subcommittee Meeting, 5 February 1969, ACRS File, AEC/NRC.

37. AEC–R 2/76 (24 March 1969), Job 4, Box 12, ID&R–14 REG (Part 50), Harold L. Price to Edward J. Bauser, 14 April 1969, L–4–1 Part 50 (Quality Assurance), AEC/NRC.

38. AEC–R 2/76, AEC/NRC.

39. C. L. Henderson to Harold L. Price and others, 2 May 1969, L–4–1 Part 50 (Quality Assurance), Comments on Proposed Rule, Docket PR–50 (Quality Assurance, Appendix B), AEC Press Release, 17 April 1969, AEC/NRC.

40. AEC–R 2/85 (28 April 1970), Job 4, Box 12, ID&R–14 REG (Part 50), Minutes of Regulatory Meeting 285, 16 June 1970, AEC Press Release, 26 June 1970, AEC/NRC.

41. *Nucleonics Week*, 10 July 1969, p. 4.

42. Julius H. Rubin, "Memorandum of Appointment," 16 April 1968, Box 46 (General Electric), Seaborg Office Files, AEC/DOE.

43. AEC–R 2/69 (6 January 1969), AEC–R 2/71 (29 January 1969), Minutes of Regulatory Meeting 271, 4 February 1969, AEC/NRC; Okrent, *Nuclear Reactor Safety*, pp. 231–232.

44. AEC–R 2/74 (7 March 1969), AEC–R 2/80 (12 March 1970), AEC/NRC; *Nucleonics Week*, 14 August 1969, p. 5; Okrent, *Nuclear Reactor Safety*, p. 232.

45. AEC 664/1 (7 April 1958), AEC 907/7 (19 May 1958), AEC/NRC; Kenneth A. Dunbar to G. Mennen Williams, 6 June 1958, "AEC Recommendations to Local Authorities for Dealing with Incidents Involving Radioactive Materials," n. d., Box 236 (Atomic Energy Study Committee), G. Mennen Williams Papers, Michigan Historical Collections, Bentley Historical Library, University of Michigan, Ann Arbor; Philip L. Cantelon and Robert C. Williams, *Crisis Contained: The Department of Energy at Three Mile Island* (Carbondale: Southern Illinois University Press, 1981), pp. 20–21.

46. Marvin M. Mann to the Commission, 10 March 1967, Box 7729,

O&M–7 (ACRS), AEC/DOE; W. B. McCool to George M. Kavanagh and C. L. Henderson, 17 March 1967, Job 4, Box 4, ID&R–6 REG (Hazard Regulation), Draft Minutes of Subcommittee Meeting on Reactor Design Criteria, 22 December 1966, ACRS File, Okrent, "On the History of the Evolution of Light Water Reactor Safety," p. 2–286, AEC/NRC.

47. Harold L. Price to the Commission, 26 December 1967, Job 4, Box 4, ID&R–6 REG (Hazard Regulation), Price to the Commission, 5 November 1969, W. B. McCool to Price, 14 November 1969, Job 4, Box 12, ID&R–14 REG (Part 50), AEC/NRC; Price to the Commission, 10 March 1970, Box 36 (Siting and Licensing, Safety), Nixon Library Materials, AEC/DOE; JCAE, *Licensing and Regulation of Nuclear Reactors*, 1967, p. 215; *Nucleonics Week*, 25 January 1968, p. 3.

48. AEC–R 2/84 (17 April 1970), Job 4, Box 12, ID&R–14 REG (Part 50), AEC/NRC.

49. SECY–R 80 (6 November 1970), Harold L. Price to Edward J. Bauser, 18 December 1970, Job 9, Box 13, ID&R–14 REG (Part 50), AEC Press Release, 20 May 1970, AEC/NRC.

50. W. Altman, J. Hockert, and E. Quinn, *A Safeguards Case Study of the Nuclear Materials and Equipment Corporation Uranium Processing Plant: Main Report* (NUREG–0627), March 1980, AEC/NRC; Leonard M. Brenner and Samuel C. T. McDowell, "U. S. Safeguards History and the Evolution of Safeguards Research and Development," *Journal of the Institute of Nuclear Materials Management*, 17 (July 1989): 4–7; George T. Mazuzan and J. Samuel Walker, *Controlling the Atom: The Beginnings of Nuclear Regulation, 1946–1962* (Berkeley and Los Angeles: University of California Press, 1984), pp. 82–84.

51. AEC–R 123/1 (15 February 1966), AEC–R 123/2 (18 February 1966), AEC/NRC; "Dialogue Intensifies over Proliferation Treaty Issues," *Nucleonics*, 24 (January 1966): 19–20; Glenn T. Seaborg with Benjamin S. Loeb, *Stemming the Tide: Arms Control in the Johnson Years* (Lexington, Mass.: Lexington Books, 1987), pp. 111–118.

52. "Summary Notes of [14 February 1966] Briefing on Safeguards and Domestic Material Accountability," 25 March 1966, O&M–6 (Briefings), Glenn T. Seaborg to Chet Holifield, 14 February 1966, Box 1359 (Materials–9), AEC/DOE; *New York Times*, 18 September 1966; *Wall Street Journal*, 13 June 1968; Seaborg, *Stemming the Tide*, p. 259.

53. AEC 213/107 (28 February 1966), AEC–R 38/11 (3 May 1966), AEC/NRC; "Summary Notes of Briefing on Safeguards and Domestic Material Accountability," AEC/DOE.

54. AEC 1230 (23 June 1966), AEC 1230/1 (7 July 1966), AEC 675/31 (8 July 1966), AEC/NRC; Glenn T. Seaborg to Chet Holifield, 20 July 1966, Box 7729, O&M–7 (Safeguarding SNM, Adv. Panel on), AEC/DOE.

55. "Report to the Atomic Energy Commission by the Ad Hoc Advisory Panel on Safeguarding Special Nuclear Material," 10 March 1967, Box 7730, O&M–7 (Safeguarding SNM, Adv. Panel on), AEC/DOE.

56. AEC 132/112 (16 January 1967), AEC 132/113 (24 January 1967), AEC 132/115 (3 February 1967), AEC/NRC.

57. AEC–R 38/18 (5 December 1969), AEC–R 38/19 (17 March 1970), Job 4, Box 12, ID&R–14 REG (Part 73), "Proposed Guide for Preparation of Fundamental Material Controls and Nuclear Materials Safeguards Procedures, Revision A," April 1968, Job 4, Box 6, ID&R–14 REG (Part 70), R. P. Wischow to J. A. Waters, 27 September 1967, O&M–3 (Division of Nuclear Materials Safeguards), AEC/NRC; "Proposed Guide for Preparation of Fundamental Material Controls and Nuclear Materials Safeguards Procedures," December 1967, Box 648 (Safeguards: Nuclear Materials), JCAE Papers.

58. AEC 213/145 (3 September 1969), Discussion Paper, "Resident Inspection on Domestic Nuclear Materials Safeguards," 2 May 1969, Delmar L. Crowson to Walter F. Mondale, 26 May 1968, O&M–3 (Division of Nuclear Materials Safeguards), AEC/NRC; "Proceedings of the Symposium on Safeguards Research and Development (WASH–1076)," 1967, Box 7730, O&M–7 (Safeguarding SNM, Adv. Panel on), AEC/DOE; E. J. Bloch to John T. Conway, 1 June 1967, Box 647 (Safeguards of Nuclear Materials), JCAE Papers.

59. "From the Editor," *Nuclear News*, 12 (December 1969): 15; *Nucleonics Week*, 1 July 1971, pp. 2–3.

CHAPTER X: REGULATING MINES AND MILLS:
HEALTH, ENVIRONMENTAL, AND
BUREAUCRATIC HAZARDS

1. George T. Mazuzan and J. Samuel Walker, *Controlling the Atom: The Beginnings of Nuclear Regulation, 1946–1962* (Berkeley and Los Angeles: University of California Press, 1984), pp. 304–324.

2. George Vranesh, "Radiation—The Miner and the Law," *The Mines Magazine*, July 1961, copy in Box 14 (Correspondence Forwarded from White House), Casper A. Nelson to George D. Clyde, 18 May 1964, Box 16 (Comments on Workmen's Compensation and Radiation Injury), G. A. Franz, Jr. to Paul C. Tompkins, 19 August 1966, Anthony Bennett to Tompkins, 22 August 1966, Carlyle F. Gronning to Tompkins, 26 September 1966, Box 1 (Correspondence 1965–1966–1967), Papers of the Federal Radiation Council, Record Group 412 (Records of the Environmental Protection Agency), National Archives (hereafter cited as FRC Papers); William J. Bair, "Inhaled Radioactive Particles and Gases," *Science*, 146 (16 October 1964): 440–444.

3. Joseph K. Wagoner et al., "Mortality Patterns among United States Uranium Miners and Millers, 1950 through 1962," *Atomic Energy Law Journal*, 6 (Spring 1964): 1–13; Joseph K. Wagoner et al., "Cancer Mortality Patterns Among U. S. Uranium Miners and Millers, 1950 through 1962," *Journal of the National Cancer Institute*, 32 (April 1964): 787–801; U. S. Congress, House, Committee on Education and Labor, Select Subcommittee on Labor, *Hearings on Federal Metal and Nonmetallic Mine Safety Act*, 89th Cong., 1st Sess., 1965; Department of Health, Education, and Welfare Press Release, 26 August 1963, Box 660 (Source Materials—Uranium), Papers of the Joint Committee on Atomic Energy, Record Group 128 (Records of the Joint Committees of Congress), National Archives (hereafter cited as JCAE Papers).

4. Federal Radiation Council Press Release, 4 October 1963, Box 522

(Radiation: Federal Radiation Council), JCAE Papers; Mazuzan and Walker, *Controlling the Atom*, pp. 256–276.

5. Anthony J. Celebrezze, Memorandum for the President, 26 May 1964, White House Central Files, EX FG 655/A (FG 662, Federal Radiation Council), Lyndon B. Johnson Papers, Lyndon B. Johnson Library, Austin, Texas; Federal Radiation Council Action Paper, "Radiation Protection Policy," 12 August 1965, Box 28 (Federal Radiation Council), Office Files of James T. Ramey, Atomic Energy Commission Records, Department of Energy, Germantown, Maryland (hereafter cited as AEC/DOE).

6. Technical Advisory Committee on Uranium Miners Study, Minutes of Sixth Meeting, 23 July 1965, Box 15 (Technical Advisory Committee on Uranium Miners Study), E. C. Van Blaroom to DRM Files, 12 August 1966, Box 13 (Archer's First Draft on Phase II), FRC Papers; AEC 503/8 (10 September 1965), Box 7717 (MH&S–11, Uranium Miners), AEC/DOE; Mazuzan and Walker, *Controlling the Atom*, pp. 310–311.

7. Technical Advisory Committee on Uranium Miners Study, Minutes of Seventh Meeting, 5 December 1966, Box 15 (Technical Advisory Committee on Uranium Miners Study), FRC Papers; Jim Graham to John T. Conway, 30 June 1966, Box 520 (Radiation: General), JCAE Papers; Raye C. Ringholz, *Uranium Frenzy: Boom and Bust on the Colorado Plateau* (New York: W. W. Norton, 1989), pp. 86–89, 171–173.

8. AEC 544/35 (21 December 1966), AEC 544/36 (29 December 1966), Box 7717, MH&S–11 (Uranium Miners), AEC/DOE.

9. Clayton M. Phelan, "Environmental Status of Uranium Miners for Calendar Year 1966," December 1966, Box 12 (PHS Report), Federal Radiation Council Staff Meeting on Radiation Protection Guidance, Transcript, 9 March 1967, Box 15 (no folder), FRC Papers; Keith Schiager, "Inhalation Hazards of Radon Decay Products in Uranium Mines," *Nuclear News*, 9 (June 1966): 21–23; "The Mining Industry Edges Toward a Full-Scale Effort," *Nucleonics*, 24 (September 1966): 23–24.

10. *Washington Post*, 9 March, 10 March 1967; Ringholz, *Uranium Frenzy*, pp. 239–240; Interview with Leo Goodman, Washington, D. C., 15 March 1982.

11. Federal Radiation Council, Draft Minutes of Working Group Meeting, 14 March 1967, Box 1 (Correspondence 1965–1966–1967), FRC Papers; J. V. Reistrup to W. Willard Wirtz, 9 March 1967, Box 76 (Occupational Health and Safety), Records of the Office of the Secretary of Labor—W. Willard Wirtz, Record Group 174 (General Records of the Department of Labor), National Archives; *Washington Post*, 13 March 1967; Select Subcommittee on Labor, *Hearings on . . . Mine Safety Act*, pp. 57–58; Mazuzan and Walker, *Controlling the Atom*, p. 320.

12. Esther Peterson to W. Willard Wirtz, 31 March 1967, Wirtz to Stewart L. Udall, 14 April 1967, David A. Swankin to Frank W. Erwin, 20 April 1967, Box 76 (Occupational Health and Safety), Records of the Office of the Secretary of Labor; Ringholz, *Uranium Frenzy*, pp. 236–237.

13. Andrew J. Biemiller to Paul Tompkins, 15 March 1967, Box 13 (Uranium Study Panel Meeting), FRC Papers; Nelson M. Bortz to W. Willard Wirtz,

21 April 1967, Box 77 (Radiation in Uranium Mines), Records of the Office of the Secretary of Labor; AEC 544/41 (17 March 1967), Box 7717, MH&S–11 (Uranium Miners), AEC/DOE.

14. *Washington Post*, 14 April, 18 April 1967; Ringholz, *Uranium Frenzy*, p. 241.

15. *Journal of Glenn T. Seaborg*, (25 vols., Berkeley: Lawrence Berkeley Laboratory PUB–625, 1989), Vol. 14, pp. 462, 519; John A. Erlewine to Dr. Fritsch, 27 April 1967, Box 28 (Federal Radiation Council), Office Files of Glenn T. Seaborg, Minutes of Information Meetings, 19 April, 21 April, 26 April 1967, Box 5667, OGM Files (Information Meeting Minutes), AEC/DOE.

16. Federal Radiation Council Action Paper, "Radiation Protection Policy," 21 April 1967, Box 28 (Federal Radiation Council), Seaborg Office Files, AEC/DOE; Federal Radiation Council, Minutes and Record of Actions, 4 May 1967, Box 3 (Council Meeting of May 4, 1967), FRC Papers; Melvin Price to W. Willard Wirtz, 21 April 1967, Box 85 (Joint Committee on Atomic Energy), Records of the Office of the Secretary of Labor; Robert Austin Milch to Dr. Hornig, 3 May 1967, Subject File (Life Sciences—Uranium Miners 1967), Record Group 359 (Records of the Office of Science and Technology), National Archives; *Seaborg Journal*, Vol. 14, pp. 523, 529, 534, 535, 558.

17. Department of Labor Press Release, 5 May 1967, Box 76 (Occupational Health and Safety), Records of the Office of the Secretary of Labor; *Seaborg Journal*, Vol. 14, p. 567; *Washington Post*, 6 May 1967.

18. JCAE, *Hearings on Radiation Exposure of Uranium Miners*, 90th Cong., 1st Sess., 1967, pp. 3–56; *Washington Post*, 9 May 1967.

19. JCAE, *Hearings on . . . Uranium Miners*, pp. 60–154.

20. *Ibid.*, pp. 169–228, 323–324, 353.

21. *Ibid.*, pp. 318–529; Summary of Federal Radiation Council Staff Meeting on Radiation Protection Guidance Applicable to Uranium Mining, 9 March 1967, Box 13 (Uranium Study Panel Meeting, March 9, 1967), FRC Papers.

22. Department of Labor Press Release, 10 June 1967, Box 47 (Council-Federal Radiation), Records of the Office of the Secretary of Labor; Statement by Congressmen Holifield and Price, 12 June 1967, Box 12 (Workmen's Compensation, Uranium Miners), Ramey Office Files, AEC/DOE; John T. Conway to All Committee Members, 12 June 1967, Box 45 (Uranium Mine Radiation), Chet Holifield Papers, University of Southern California, Los Angeles; "Mr. Wirtz Reconsiders," *Nuclear Industry*, 14 (June 1967): 3–9.

23. Esther Peterson to W. Willard Wirtz, 10 July 1967, Box 47 (Council-Federal Radiation), David A. Swankin to Frank W. Erwin, 13 July 1967, Box 77 (Radiation in Uranium Mines), Records of the Office of the Secretary of Labor; Federal Radiation Council Action Paper, 1 August 1967, Box 28 (Federal Radiation Council), Seaborg Office Files, AEC/DOE; John T. Conway to All Committee Members, 28 July 1967, Box 45 (Uranium Mine Radiation), Holifield Papers.

24. *Washington Post*, 31 March 1967; *Wall Street Journal*, 6 October 1967; Milton Viorst and J. V. Reistrup, "Radon Daughters and the Federal Government," *Bulletin of the Atomic Scientists*, 23 (October 1967): 25–29; W. A. Boyle to Craig Hosmer, 8 November 1967, Box 530 (Radiation: Radiation Exposure, Uranium Miners), JCAE Papers.

25. Glenn T. Seaborg to W. Willard Wirtz, 6 June 1967, Box 53 (Interdepartmental Committee on Uranium Mine Radiation Exposure Problem), Esther Peterson to Wirtz, 16 June 1967, David Swankin to Frank W. Erwin, 23 June 1967, Box 77 (Radiation in Uranium Mines), Records of the Office of the Secretary of Labor; *Seaborg Journal*, Vol. 14, pp. 739–741.

26. E. L. Newman, "Activities by States—Uranium Mining," 1 August 1967, Box 85 (Testimony on Uranium Mines), Records of the Office of the Secretary of Labor; Esther Peterson to Robert Milch, 16 November 1967, Subject File (Life Sciences—Uranium Miners 1967), Records of the Office of Science and Technology; "Surveys Show Improvement in Mine Radiation Levels," *Nuclear Industry*, 14 (December 1967): 32–34; "Miner Exposures Drop Sharply; Protective Measures Costly," *ibid.*, 15 (February 1968): 29–31; JCAE, *Hearings on Radiation Standards for Uranium Mining*, 91st Cong., 1st Sess., 1969, pp. 6–7.

27. Tim McGinley to W. Willard Wirtz, 3 August 1967, Box 77 (Radiation in Uranium Mines), Records of the Office of the Secretary of Labor; J. Cordell Moore to Glenn T. Seaborg, 20 December 1967, Box 7717, MH&S–11 (Uranium Miners), AEC/DOE.

28. Glenn T. Seaborg to Melvin Price, 24 August 1967, Seaborg to Michael J. Kirwan, 24 August 1967, Kirwan to Seaborg, 9 November 1967, Seaborg to John T. Conway, 11 April 1967, Box 7717, MH&S–11 (Uranium Miners), AEC 544/80 (16 April 1968), Box 13 (Workmen's Compensation, Uranium Miners), Ramey Office Files, Federal Radiation Council Information Paper, "Radiation Protection Policy," 31 January 1968, Box 28 (Federal Radiation Council), Seaborg Office Files, AEC/DOE.

29. "Status of State Workmen's Compensation Laws in Meeting the Eleven Standards Adopted by the U. S. Atomic Energy Commission," c. March 1967, Box 12 (Guide for Uranium Miners), Ramey Office Files, John A. Love to James T. Ramey, 7 March 1967, Box 7717, MH&S–11 (Uranium Miners), AEC/ DOE; Esther Peterson to W. Willard Wirtz, 31 March 1967, Box 77 (Radiation in Uranium Mines), Records of the Office of the Secretary of Labor; *Congressional Record*, 90th Cong., 1st Sess., 1967, pp. 33600–33610; Mazuzan and Walker, *Controlling the Atom*, pp. 335–341.

30. "Report of the Steering Committee on Health and Social Responsibilities of the Federal Government to Uranium Miners," 27 February 1968, Box 13 (Workmen's Compensation, Uranium Miners), C. L. Henderson to John Ryan, 22 May 1967, Box 12 (Guide for Uranium Miners), Ramey Office Files, AEC/ DOE; Charles F. Eason to Those Listed Below, 14 May 1968, MH&S–3–7 (Mill Survey and Tailings), Atomic Energy Commission Records, Nuclear Regulatory Commission, Rockville, Maryland (hereafter cited as AEC/NRC); Esther Peterson to W. Willard Wirtz, 31 March 1967 (cited in note 29), James T. Ramey to Wirtz, 3 August 1967, Box 53 (Committee—Interdepartmental Committee on Uranium Mine Radiation Exposure Problem), Records of the Office of the Secretary of Labor; William T. England to Edward J. Bauser, 28 February 1969, Box 530 (Radiation: Radiation Exposure, Uranium Miners), JCAE Papers; "Compensation for Radiation-Injured Miners," *Nuclear Industry*, 15 (February 1968): 30.

31. C. Elkins and F. Schuldt to Sam Hughes, 5 June 1968, Subject File (Life Sciences—Uranium Miners), Records of the Office of Science and Technology.

32. Paul C. Tompkins to Charles Dunham, 6 October 1967, *ibid.*; John B. Radcliffe to the Files, 5 September 1968, Box 530 (Radiation: Radiation Exposure, Uranium Miners), JCAE Papers; National Academy of Sciences, *Radiation Exposure of Uranium Miners: A Report of an Advisory Committee for the Division of Medical Sciences*, August 1968, printed in JCAE, *Hearings on Radiation Standards for Uranium Mining*, pp. 187–221.

33. Resource Management Corporation, "Control of Radiation Exposure in Uranium Mines: A Cost and Economic Analysis," 4 November 1968, printed in JCAE, *Hearings on Radiation Standards for Uranium Mining*, pp. 223–299; Federal Radiation Council Information Paper, "Radiation Protection Policy," 25 July 1968, Box 12 (Workmen's Compensation, Uranium Miners), Ramey Office Files, B. R. Buck to Melvin Price, 20 September 1968, Box 5574, OGM Files (FRC—Uranium Mine Radiation Standard), AEC/DOE.

34. Paul C. Tompkins to Philip R. Lee, 3 October 1968, Box 1 (1968 Correspondence by Date), FRC Papers; Department of Labor Press Release, 16 October 1968, Box 64 (Occupational Health and Safety), Records of the Office of the Secretary of Labor; Federal Radiation Council Action Paper, "Radiation Protection Policy," 18 October 1968, Box 7764, MH&S–11 (Uranium Miners), AEC/DOE.

35. Esther Peterson to W. Willard Wirtz, 24 October 1968, Box 64 (Occupational Health and Safety), Records of the Office of the Secretary of Labor; AEC 544/88 (14 November 1968), Box 7764, MH&S–11 (Uranium Miners), AEC/DOE.

36. Esther Peterson to W. Willard Wirtz, 2 December 1968, Box 45 (Council—Federal Radiation), Records of the Office of the Secretary of Labor; Federal Radiation Council Action Paper, "Draft Minutes and Record of Actions," 9 December 1968, Box 28 (Federal Radiation Council), Seaborg Office Files, J. C. Ryan Memo and attached paper by John A. Erlewine, 22 November 1968, Box 13 (Workmen's Compensation, Uranium Miners), Ramey Office Files, AEC/DOE; *Nucleonics Week*, 28 November 1968, p. 3.

37. Esther Peterson to W. Willard Wirtz, 2 December 1968, Box 88 (Peterson, Esther), Records of the Office of the Secretary of Labor; AEC 544/91 (29 January 1969), AEC 544/94 (4 February 1969) Box 7764, MH&S–11 (Uranium Miners), AEC/DOE; Edward J. Bauser to All Committee Members, 14 January 1969, Box 530 (Radiation: Radiation Exposure, Uranium Miners), JCAE Papers; "Present Mine Radiation Exposure Standard Readopted for 1969–70," *Nuclear Industry*, 16 (January 1969): 35–37; "Radon and Its Daughters—Politics and the Mines," *Nuclear News*, 12 (May 1969): 27–28.

38. JCAE, *Hearings on Radiation Standards for Uranium Mining*, pp. 6–8.

39. *Ibid.*, pp. 55–111; Robert H. Finch to Wayne Aspinall, 20 February 1969, Finch to Melvin Price, 24 April 1969, Box 3 (Hearings on Uranium Mining, 1967–1969), FRC Papers; *Nucleonics Week*, 6 February 1969, p. 1; *Washington Post*, 18 March 1969; "Joint Committee Finds Outlook on Mine Safety Standards Still Unclear," *Nuclear Industry*, 16 (March 1969): 43–49.

40. AEC 544/95 (20 February 1969), W. B. McCool to the File, 26 Febru-

ary 1969, Box 7764, MH&S–11 (Uranium Miners), AEC/DOE; "Improvements in Mine Radiation Exposures Tapering Off," *Nuclear Industry*, 17 (February 1970): 33–34; JCAE, *Hearings on Radiation Standards for Uranium Mining*, pp. 144–145.

41. Elliot Richardson to Chet Holifield, 2 December 1970, James B. Graham to Edward J. Bauser, 10 March, 18 March 1971, Box 530 (Radiation: Radiation Exposure, Uranium Miners), JCAE Papers; SECY–1191 (19 March 1971), SECY–1227 (25 March 1971), SECY–1232 (25 March 1971), Box 7822, MH&S–11 (Uranium Miners), AEC/DOE; *Nucleonics Week*, 5 November 1970, p. 3, 24 December 1970, pp. 6–8, 1 April 1971, p. 2; "Stricter Mine Standards Postponed," *Nuclear News*, 14 (February 1971): 33.

42. Glenn T. Seaborg to John O. Pastore, 30 April 1971, Telephone Call from William D. Ruckelshaus, 8 May 1971, Box 184 (Uranium Miners), Seaborg Office Files, AEC/DOE; John A. Griffin to James B. Graham, 28 May 1971, Box 530 (Radiation: Radiation Exposure, Uranium Miners), JCAE Papers; Edward J. Bauser to All Members, 14 May 1971, George F. Murphy to All Members, 1 June 1971, Box 48 (Uranium Mines/Miners), Holifield Papers; *Nucleonics Week*, 10 June 1971, p. 4; "How Much Time to Conform to 4 WLM Mine Radiation Standard?" *Nuclear Industry*, 18 (June 1971): 40–41.

43. Edward J. Bauser to All Committee Members, 28 June 1971, Box 48 (Staff Memoranda 1971), Holifield Papers; *Nucleonics Week*, 27 May 1971, p. 1, 3 June 1971, p. 3, 1 July 1971, p. 1, 15 July 1971, pp. 7–8; "Four WLM 'Guidance' Confirmed; Variance Rules Published," *Nuclear Industry*, 18 (July 1971): 51–52.

44. Department of Labor Press Release, 9 May 1967, Box 76 (Occupational Health and Safety), Records of the Office of the Secretary of Labor.

45. Luther J. Carter, "Uranium Mill Tailings: Congress Addresses a Long-Neglected Problem," *Science*, 202 (13 October 1978): 191–195; Mazuzan and Walker, *Controlling the Atom*, pp. 314–315.

46. AEC–R 18/2 (22 July 1960), AEC-R 18/3 (10 December 1960), AEC/NRC; Program Director, RH to Chief, State Assistance Branch, DRH, 11 March 1964, File 47, U. S. Public Health Service Records, Division of Radiological Health, U. S. Public Health Service, Rockville, Maryland; *Wall Street Journal*, 14 June 1968.

47. Program Director, RH to Chief, State Assistance Branch, DRH, 11 March 1964, Public Health Service Records; AEC–R 18/17 (20 November 1963), Robert F. Barker to the Files, 24 October 1961, Donald I. Walker to Leo Dubinski, 1 December 1965, MH&S–3–7 (Mill Survey and Tailings), AEC/NRC; Edward J. Bauser to the Files, 1 October 1963, Box 842 (Joint Committee on Atomic Energy—Fallout, Radioactivity), Clinton P. Anderson Papers, Library of Congress; Carter, "Uranium Mill Tailings," p. 191.

48. James G. Terrill, Jr. to Robert Lowenstein, 9 August 1963, File 47, Public Health Service Records; AEC–R 18/17, AEC–R 18/28 (11 October 1965), Donald A. Nussbaumer to the Files, 14 November 1963, MH&S–3–7 (Mill Survey and Tailings), AEC/NRC; W. B. McCool to the File, 7 January 1966, AEC 544/16 (21 October 1965), Box 7717, MH&S–11 (Mill Tailings), AEC/DOE.

49. U. S. Public Health Service, "Disposition and Control of Uranium Mill Tailings Piles in the Colorado River Basin," Advanced Copy, October 1965, Harold L. Price to John T. Conway, Box 184 (Uranium Mill Tailings), Seaborg Office Files, AEC/DOE; AEC–R 18/32 (3 December 1965), Price to W. B. McCool, 26 November 1965, Donald A. Nussbaumer to the Files, 18 March 1964, 7 January 1966, MH&S–3–7 (Mill Survey and Tailings), AEC/NRC; Program Director, RH to Chief, State Assistance Branch, DRH, 11 March 1964, Public Health Service Records.

50. Comments on the Article, "Uranium Mystery in the Colorado Basin," n. d., Donald A. Nussbaumer to the Files, 22 March 1966, J. A. McBride to Harold L. Price, 13 April 1966, AEC 544/17 (18 April 1966), Transcripts from CBS Television News and CBS Radio News, 20 April 1966, MH&S–3–7 (Mill Survey and Tailings), AEC/NRC; A. O. Little to Glenn T. Seaborg, 2 March 1966, Box 184 (Uranium Mill Tailings), Philippe G. Jacques to Judith Eckerson, 21 April 1966, Box 7717, MH&S–11 (Mill Tailings), AEC/DOE; *New Republic*, 5 March 1966, p. 6, 16 April 1966, pp. 36–37.

51. U. S. Congress, Senate, Subcommittee on Air and Water Pollution, Committee on Public Works, *Hearings on Radioactive Water Pollution in the Colorado River Basin*, 89th Cong., 2nd Sess., 1966, pp. 1–49; *Rocky Mountain News*, 29 April 1966, 7 May 1966; *Denver Post*, 8 May 1966, 10 May 1966; "Radioactive Mining Wastes," *Scientist and Citizen*, 8 (August 1966): 10–12.

52. G. D. Carlyle Thompson to Glenn T. Seaborg, 20 April 1966, Roy L. Cleere to Seaborg, 25 April 1966, 21 September 1966, 6 January 1967, G. J. Van Heuvelen to Seaborg, 26 April 1966, Robert Alberts to Seaborg, 12 May 1966, Edwin O. Wicks to Seaborg, 3 June 1966, Gerald F. Tape to Thompson, 9 June 1966, AEC 544/29 (21 October 1966), R. E. Hollingsworth to Frank C. DiLuzio, 8 December 1966, Box 7717, MH&S–11 (Mill Tailings), AEC/DOE; Donald A. Nussbaumer to Wayne Kerr, 10 November 1966, J. A. McBride to Harold L. Price, 1 December 1966, Frank E. McGinley to Elton A. Youngberg, 16 November 1971, MH&S–3–7 (Mill Survey and Tailings), AEC/NRC.

53. AEC 544/33 (21 December 1966), AEC 544/40 (13 March 1967), AEC 544/64 (2 August 1967), AEC 544/100 (11 June 1969), Box 7717, MH&S–11 (Mill Tailings), AEC/DOE; U. S. Public Health Service, "Evaluation of Radon 222 Near Uranium Tailings Piles," March 1969, Rafford L. Faulkner to the Files, 3 April 1970, MH&S–3–7 (Mill Survey and Tailings), AEC/NRC; Edward J. Bauser to Chet Holifield, 10 July 1969, Box 46 (JCAE Outgoing Correspondence), Holifield Papers; *Seaborg Journal*, Vol. 14, p. 221; *New York Times*, 20 September 1969.

54. Stephen H. Greenleigh, "Potential AEC Liability from Use as Construction Fill of Sand Tailings," 18 December 1970, MH&S–3–7 (Mill Survey and Tailings), AEC/NRC; JCAE, *Hearings on Use of Uranium Mill Tailings for Construction Purposes*, 92nd Cong., 1st Sess., 1971, pp. 24–25, 395–396; *Washington Daily News*, 1 September 1966; *Denver Post*, 19 December 1969; "Hot Town," *Time*, 20 December 1971, p. 56.

55. J. A. McBride to Harold L. Price, 11 March, 27 March 1969, Martin B. Biles to John A. Erlewine, 17 June 1970, MH&S–3–7 (Mill Survey and Tailings), AEC/NRC; AEC 544/106 (14 November 1969), Box 7764, MH&S–11

(Uranium Miners), AEC/DOE; *Denver Post*, 19 December 1969; *Rocky Mountain News*, 19 December 1969.

56. Transcript from ABC Evening News, 20 January 1970, MH&S–3–7 (Mill Survey and Tailings), AEC/NRC; Edward J. Burger to David Freeman and Gordon Moe, 24 March 1970, Subject File (Environment T.F. 1970), Records of the Office of Science and Technology; *Denver Post*, 19 December 1969, 9 January 1970; *New York Times*, 1 March 1970; Steve Gascoyne, "How I Learned to Live with Radioactivity and Love It—in Colorado," *Commonweal*, 13 March 1970, pp. 7–9.

57. Donald A. Nussbaumer to Harold L. Price, 27 February 1970, Nussbaumer to Stanley T. Robinson, 6 March 1970, Lyall Johnson to Price, 24 August 1970, Martin B. Biles to J. A. Erlewine and others, 23 July 1971, MH&S–3–7 (Mill Survey and Tailings), AEC/NRC; *Denver Post*, 12 December 1969; H. Peter Metzger, " 'Dear Sir: Your House Is Built on Radioactive Uranium Waste,' " *New York Times Magazine*, 31 October 1971, pp. 14–15, 59–65.

58. C. L. Henderson to Harold L. Price, 5 August 1971, Stella A. [Teets] to Mr. Price, 17 August 1971, Frank E. McGinley to Elton A. Youngberg, 31 August 1971, Martin B. Biles to Edward J. Bauser, 28 September 1971, MH&S–3–7 (Mill Survey, Uranium Mills and Tailings), AEC/NRC; JCAE, *Hearings on Use of Uranium Mill Tailings*, pp. 33–36.

59. John A. Erlewine to George F. Murphy, 5 October 1970, SECY–1900 (19 July 1971), Box 7822, MH&S–11 (Uranium Mines), AEC/DOE; Colorado Indoor Radon Study, First Meeting of Joint Working Committee, n.d., MH&S–3–7 (Mill Survey and Tailings), AEC/NRC.

60. C. E. Larson to the Commission, 18 November 1970, Martin B. Biles to the Commission, 27 July 1971, SECY–1806 (29 July 1971), Box 7822, MH&S–11 (Uranium Miners), Julius H. Rubin to Donald B. Rice, 9 December 1970, Theos J. Thompson to the Commission, 20 August 1970, Box 184 (Uranium Mill Tailings), Seaborg Office Files, AEC/DOE; Rafford L. Faulkner to Robert E. Hollingsworth, 24 September 1970, MH&S–3–7 (Mill Survey and Tailings), AEC/NRC; *Seaborg Journal*, Vol. 23, p. 475, Vol. 25, p. 356.

61. John A. Erlewine to the Commission, 29 September 1971, James R. Schlesinger to John O. Pastore, 16 May 1972, Box 7822, MH&S–11 (Uranium Miners), AEC/DOE; JCAE, *Hearings on Use of Uranium Mill Tailings*, p. 121; "JCAE Inquiry into Mill Tailings Controversy Is Inconclusive," *Nuclear Industry*, 18 (October-November 1971): 31–39; *Denver Post*, 23 September 1971; *New York Times*, 27 September 1971; *Washington Star*, 7 December 1971.

CHAPTER XI: THE AEC AND THE
ENVIRONMENT: THE THERMAL POLLUTION
CONTROVERSY

1. For useful overviews of environmental issues, see Samuel P. Hays, *Beauty, Health, and Permanence: Environmental Politics in the United States, 1955–1985* (New York: Cambridge University Press, 1987); Martin V. Melosi, *Coping with Abundance: Energy and Environment in Industrial America* (Phila-

delphia: Temple University Press, 1985); Martin V. Melosi, "Lyndon Johnson and Environmental Policy" in Robert A. Divine, ed., *The Johnson Years, Volume Two: Vietnam, the Environment, and Science* (Lawrence: University Press of Kansas, 1987); Richard H. K. Vietor, *Environmental Politics and the Coal Coalition* (College Station, TX: Texas A&M University Press, 1980); Roderick Nash, *Wilderness and the American Mind*, 3d ed. (New Haven: Yale University Press, 1982); and Edward W. Lawless, *Technology and Social Shock* (New Brunswick: Rutgers University Press, 1977).

2. *Considerations Affecting Steam Power Plant Site Selection* (Washington, D. C.: Executive Office of the President, Office of Science and Technology, 1968); *Nucleonics Week*, 12 December 1968, pp. 4–5; Jeremy Main, "A Peak Load of Trouble for the Utilities," *Fortune*, 80 (November 1969): 116–119ff; "Why Utilities Can't Meet Demand," *Business Week*, 29 November 1969, pp. 48–62; *Wall Street Journal*, 23 March 1970; "Danger of More Power 'Blackouts,' " *U. S. News and World Report*, 20 April 1970, pp. 48–50: Richard F. Hirsh, *Technology and Transformation in the American Electric Utility Industry* (Cambridge: Cambridge University Press, 1989), pp. 56–62, 101.

3. *Considerations Affecting Steam Power Plant Site Selection*, chap. 4; James G. Terrill, Jr., E. D. Harward, and I. Paul Leggett, Jr., "Environmental Aspects of Nuclear and Conventional Powerplants," *Journal of Industrial Medicine and Surgery*, 36 (July 1967): 412–419, reprinted in Joint Committee on Atomic Energy (JCAE), *Selected Materials on Environmental Effects of Producing Electric Power*, 91st Cong., 1st Sess., 1969, pp. 121–133; *Nucleonics Week*, 11 November 1965, p. 2; National Rural Electric Cooperative Association, *The Electric Power Crisis: Its Impact on Workers and Consumers*, 1971, copy in Office Files of James T. Ramey, Box 35 (Reaction of Utilities, JCAE, etc.), Atomic Energy Commission Records, Department of Energy, Germantown, Maryland (hereafter cited as AEC/DOE); President's Science Advisory Committee, *Restoring the Quality of Our Environment: Report of the Environmental Pollution Panel*, 1965.

4. Federal Power Commission Press Release, 12 November 1968, copy in Box 7763, MH&S–11, (Industrial Hygiene), AEC/DOE; *Considerations Affecting Steam Power Plant Site Selection*, chap. 4; *The Electric Power Crisis*, pp. 18–26; "A Crisis in Fossil Fuels," *Nuclear Industry*, 17 (June 1970): 9–15; Melosi, *Coping with Abundance*, pp. 209–215, 266–269.

5. Conservation Foundation, *CF Letter*, March 1970, copy in Box 181 (Pollution), Office Files of Glenn T. Seaborg, AEC/DOE; Main, "A Peak Load of Trouble," p. 205.

6. *Nucleonics Week*, 25 February 1965, p. 1; J. H. Wright, "Nuclear Power and the Environment," *Atomic Power Digest*, published by Westinghouse Nuclear Energy Systems, 1969, copy in Box 53 (Westinghouse Electric Co.), Seaborg Office Files, AEC/DOE; JCAE, *Environmental Effects of Producing Electric Power*, 91st Cong., 2nd Sess., 1970, pp. 1512–1526; *New York Times*, 29 September 1970; AI–Public Information to AI Supervision, 18 January 1971, Box 194 (Anti-Nuclear Organizations), Craig Hosmer Papers, University of Southern California, Los Angeles.

7. "News About Industry," *Nuclear Industry*, 12 (March 1965): 7–11;

"Northern States Power Puts the Accent on Environment," *ibid.*, 14 (November 1967): 32–33.

8. "Conservation Policy Guide," Abstract of Directors' Actions, 1946–1968, (Minutes), rev. ed., July 1968, Sierra Club Records, William E. Colby Memorial Library, Sierra Club, San Francisco; Thomas E. Dustin to Glenn T. Seaborg, 17 February 1967, Box 7717, MH&S–11 (Industrial Hygiene), Thomas L. Kimball to Seaborg, 6 November 1970, Box 181 (Pollution), Seaborg Office Files, AEC/DOE; Malcolm L. Peterson, "Environmental Contamination from Nuclear Reactors," *Scientist and Citizen*, 8 (November 1965): 1–11; Norman Cousins, "Breakfast with Dr. Teller," *Saturday Review*, 19 March 1966, pp. 26, 54.

9. "Are Nuclear Plants Winning Acceptance?" *Electrical World*, 24 January 1966, pp. 115–117; "Nuclear Power and the Community: Familiarity Breeds Confidence," *Nuclear News*, 9 (May 1966): 15–16.

10. U.S. Atomic Energy Commission, *Nuclear Power and the Environment* (one of a series of booklets on "Understanding the Atom"), 1969, Atomic Energy Commission Records, Nuclear Regulatory Commission, Washington, D. C. (hereafter cited as AEC/NRC); Glenn T. Seaborg speeches, 13 December 1966, Box 7717, MH&S–11 (Industrial Hygiene), 5 May 1969, Box 25 (Pollution of the Environment), and 20 April 1970, Box 35 (Regulatory-General), Nixon Library Materials, James T. Ramey speech, 13 August 1970, Box 35 (H. Peter Metzger), Ramey Office Files, AEC/DOE; *Washington Evening Star*, 16 June 1969; *Journal of Glenn T. Seaborg* (25 vols., Berkeley: Lawrence Berkeley Laboratory PUB–625, 1989), Vol. 17, p. 139.

11. John G. Palfrey to Donald F. Hornig, 5 September 1964, Box 168 (Office of Science and Technology), Seaborg Office Files, Glenn T. Seaborg to Edward Wenk, Jr., 19 January 1966, Box 1362, MH&S–3–3 (Contamination and Decontamination), AEC/DOE; John R. Totter speech, 3 December 1969, Box 194 (Environment-General), Hosmer Papers; U.S. Atomic Energy Commission, *Atoms, Nature, and Man* (one of a series on "Understanding the Atom"), 1966, and USAEC, *Nuclear Power and the Environment*, AEC/NRC.

12. Lawrence D. Low to P. A. Morris and others, 22 August 1967, U.S. Department of Health, Education, and Welfare, *Radiological Surveillance at a Boiling Water Nuclear Power Reactor*, n.d., MH&S–17 (HEW), AEC/NRC; U.S. Senate, Subcommittee on Air and Water Pollution, *Hearings on Clean Air*, 88th Cong., 2nd Sess., 1964, pp. 1069–1097, and *Hearings on Thermal Pollution*–1968, pp. 1248–1262.

13. Lester R. Rogers to Harold L. Price and others, 13 June 1963, Glenn T. Seaborg to Stewart L. Udall, 27 March 1964, John F. Newell to the Files, 9 June 1964, Troy Conner to Harold L. Price, 25 May 1965, Harold L. Price to Commissioner Ramey, 2 February 1968, L–4–1 (Memo of Understanding, AEC-Dept. of Interior and Fish and Wildlife), AEC/NRC.

14. Troy B. Conner, Jr. to the Separated Legal Files, 9 July 1964, Harold L. Price to Lewis A. Sigler, 6 July 1965, L–4–1 (Memo of Understanding, AEC-Dept. of Interior and Fish and Wildlife), AEC/NRC.

15. Federal Power Commission, *Problems in Disposal of Waste Heat from Steam-Electric Plants*, 1969, Box 7763 (MH&S–11, Bulky Package), AEC/

DOE; U.S. Atomic Energy Commission, *Thermal Effects and U.S. Nuclear Power Stations*, WASH–1169, 1971, MH&S–3–1 (Thermal Effects, Nov. 1970–), AEC/NRC; Ralph E. Lapp, "Power and Hot Water," *New Republic*, 6 February 1971, pp. 20–23.

16. Federal Power Commission, *Problems in Disposal of Waste Heat*, pp. 1–2, 17–22; Roger Don Shull, "Thermal Discharges to Aquatic Environments," 9 June 1965, attachment to Clifford K. Beck to James T. Ramey, 6 November 1967, Legal–4–1 (Federal Water Pollution Agency), AEC/NRC; CW [Charles Weaver] memorandum, 27 April 1967, File 23–1–10 (Vermont Yankee Nuclear Power Plant Material, 1967), George D. Aiken Papers, University of Vermont, Burlington; John R. Clark, "Thermal Pollution and Aquatic Life," *Scientific American*, 220 (March 1969): 19–26; Wolfgang Langewiesche, "Can Our Rivers Stand the Heat?" *Reader's Digest*, 96 (April 1970): 76–80.

17. Federal Power Commission, *Problems in Disposal of Waste Heat*, pp. 18–21; Clark, "Thermal Pollution and Aquatic Life," pp. 19–22; JCAE, *Hearings on Participation by Small Electrical* Utilities *in Nuclear* Power, 90th Cong., 2nd Sess., 1968, pp. 89–92; *New York Times*, 1 July 1966; Robert H. Boyle, "A Stink of Dead Stripers," *Sports Illustrated*, 26 April 1965, pp. 81–84; Allan R. Talbot, *Power Along the Hudson: The Storm King Case and the Birth of Environmentalism* (New York: E. P. Dutton, 1972), pp. 112–114.

18. William M. Holden, "Hot Water: Menace and Resource," *Science News*, 94 (17 August 1968): 164–166; John Cairns, Jr., "We're In Hot Water," *Scientist and Citizen*, 10 (October 1968): 187–198; Frank Graham, Jr., "Tempest in a Nuclear Teapot," *Audubon*, 72 (March 1970): 13–19; Federal Power Commission, *Problems in Disposal of Waste Heat*, pp. 21–22; Atomic Energy Commission, *Thermal Effects and U.S. Nuclear Power Stations*, pp. 18–23.

19. Cairns, "We're In Hot Water," p. 187; Clark, "Thermal Pollution and Aquatic Life," p. 19; *Philadelphia Inquirer*, 30 September 1970; "The Problem of Thermal Pollution," NBC Television Broadcast, 24 May 1970, Box 37 (Environmental Effects, Including Thermal), Nixon Library Materials, AEC/DOE.

20. Holden, "Hot Water: Menace and Resource," pp. 164–165; "Problems Associated with U.S. Thermal Effects Standards Examined," *Nuclear Industry*, 17 (August 1970): 32–36; Atomic Industrial Forum, *Info: Thermal Effects*, 20 February 1970, copy in Box 18 (AIF Meeting, Thermal Considerations), Ramey Office Files, AEC/DOE; John A. Harris to the Commission, April 8, 1970, MH&S–3–1 (Thermal Effects), AEC/NRC; Glenn T. Seaborg and William R. Corliss, *Man and Atom: Building A New World Through Nuclear Technology* (New York: E. P. Dutton, 1971), pp. 81–83, 117–120.

21. "The Effects and Control of Heated Water Discharges: A Report to the Federal Council for Science and Technology by the Committee on Water Resources Research," November 1970, copy in Box 169 (Office of Science and Technology), Seaborg Office Files, "Thermal Effects Studies by Nuclear Power Plant Licensees and Applicants," 1969, Box 5625 (Environmental Pollution), OGM Files, AEC/DOE.

22. R. E. Hollingsworth to Joseph D. Tydings, 23 January 1970, MH&S–3–1 (Thermal Effects), SECY–812 (28 December 1970), AEC/NRC; Milton

Shaw to Commissioner Larson, 20 February 1970, Box 7751, ID&R–6 (Hazards, Vol. 2), AEC/DOE; *Nucleonics Week*, 4 May 1967, pp. 1–2.

23. Steve Elonka, "Cooling Towers: A Special Report," *Power*, March 1963, copy in Box 16 (Reactors-General), Office Files of Wilfrid E. Johnson, P. N. Ross, "Presentation to President's Water Pollution Control Advisory Board," 6 December 1968, Box 3 (Water Pollution-Muskie Bill), Ramey Office Files, AEC/DOE; *Considerations Affecting Steam Power Plant Site Selection*, chap. 4.

24. Elonka, "Cooling Towers," Ross, "Presentation to . . . Advisory Board," Frank H. Rainwater, "Thermal Waste Treatment and Control," 29 June 1970, Box 18 (AIF Meeting, Thermal Considerations), Ramey Office Files, Glenn T. Seaborg to Sally Morrison, 3 February 1970, Box 7763, MH&S–11 (Environmental Studies, Vol. 5), AEC/DOE; *Nucleonics Week*, 4 May 1967, pp. 1–2; "Outlook for Cooling Towers," *Nuclear Industry*, 14 (October 1967): 8–14; "Wealth of New Data on Nuclear Plant Cooling Methods, Costs," *ibid.*, 17 (July 1970): 7–15; "Utilities Burn Over Cooling Towers," *Business Week*, 3 April 1971, pp. 52–54; *Considerations Affecting Steam Power Plant Site Selection*, chap. 5.

25. Shaw to Larson, 20 February 1970, AEC/DOE; "Priorities for Review of Environmental Matters," 27 July 1971, Job 9, Box 19 (Legal–11 REG Litigation), Hollingsworth to Tydings, 23 January 1970, SECY–812, AEC/NRC; "Outlook for Cooling Towers," *Nuclear Industry*, pp. 8–12.

26. Troy Conner to Harold L. Price, 25 May 1965, L–4–1 (Memo of Understanding, AEC-Department of Interior and Fish and Wildlife), AEC/NRC; "A Jurisdictional 'No Man's Land,' " *Nuclear Industry*, 15 (March 1968): 3–5; Ellen Thro, "The Controversy Over Thermal Effects," *Nuclear News*, 11 (December 1968): 49–53.

27. "Preliminary Evaluation of Possible Effects on Fish and Shellfish of the Proposed Millstone Nuclear Reactor," 15 December 1965, Box 587 (Reactors: Millstone Point), Papers of the Joint Committee on Atomic Energy, Record Group 128 (Records of the Joint Committees of Congress), National Archives, Washington, D. C. (hereafter cited as JCAE Papers). This document is printed in U. S. Congress, House, Committee on Merchant Marine and Fisheries, Subcommittee on Fisheries and Wildlife Conservation, *Hearings on Miscellaneous Fisheries Legislation*, 89th Cong., 2nd Sess., 1966, pp. 191–194.

28. Clarence F. Pautzke to Harold L. Price, 23 March 1966, Box 587 (Reactors: Millstone Point), JCAE Papers. See also Subcommittee on Fisheries and Wildlife Conservation, *Hearings on Miscellaneous Fisheries Legislation*, pp. 191–192.

29. Harold L. Price to the Commission, 28 March 1966, L–4–1 (Memo of Understanding, AEC-Department of Interior and Fish and Wildlife), AEC/NRC; Chet Holifield to Glenn T. Seaborg, 21 March 1966, Box 202 (Regulatory Matters-General Files), Seaborg Office Files, AEC/DOE; John D. Dingell to Chet Holifield, 17 March 1966, John T. Conway to All Committee Members, 2 April 1966, Box 512 (Pollution: Thermal Pollution), JCAE Papers; John D. Dingell to Clinton P. Anderson, 18 March 1966, Box 845 (Joint Committee on Atomic Energy—General 1966), Clinton P. Anderson Papers, Library of Congress, Washington, D. C.

30. Howard K. Shapar to the Files, 18 April 1966, Box 512 (Pollution: Thermal Pollution), JCAE Papers; Joseph F. Hennessey to the Commission, 21 April 1966, L–4–1 (Memo of Understanding, AEC-Department of Interior and Fish and Wildlife), AEC/NRC.

31. Howard K. Shapar to Mr. Trosten, 9 May 1966, John T. Conway to All Committee Members, 10 May 1966, William T. England to John T. Conway, 17 May 1966, Box 512 (Pollution: Thermal Pollution), JCAE Papers; *Nucleonics Week*, 19 May 1966, p. 5; Subcommittee on Fisheries and Wildlife Conservation, *Hearings on Miscellaneous Fisheries Legislation*, pp. 112–113.

32. Subcommittee on Fisheries and Wildlife Conservation, *Hearings on Miscellaneous Fisheries Legislation*, pp. 97, 207–222; William T. England to John T. Conway, 17 May 1966, Box 512 (Pollution: Thermal Pollution), JCAE Papers.

33. James T. Ramey to Donald F. Hornig, 14 December 1966, Box 62 (Nuclear Power Reactors), Clarence F. Pautzke to Harold L. Price, 8 February 1968, James T. Ramey to David Black, 9 February 1968, Draft Memorandum, "Legislation on Thermal Effects," 12 March 1968, Box 64 (Thermal Effects or Pollution), Ramey Office Files, AEC/DOE; R. E. Baker to John F. Newell, 13 March 1967, Legal–4–1 (Federal Water Pollution Agency), Price to Ramey, 8 October 1969, Legal–4–1 (Memo of Understanding), AEC/NRC.

34. Richard M. Klein, "Bananas in Vermont," *Natural History*, 79 (February 1970): 11–18; John Walsh, "Vermont: A Power Deficit Raises Pressure for New Plants," *Science*, 173 (17 September 1971): 1110–1115; John Walsh, "Vermont: Forced to Figure in Big Power Picture," *Science*, 174 (1 October 1971): 44–47; Steven Ebbin and Raphael Kasper, *Citizen Groups and the Nuclear Power Controversy* (Cambridge: MIT Press, 1974), pp. 90–94.

35. *Rutland Herald*, 25 July, 28 July 1967; *Burlington Free Press*, 20 October 1967; *Boston* Globe, 10 November 1967; *Nucleonics Week*, 7 September 1967, p. 1, 14 September 1967, p. 5; "Thermal Effects: An Acute Issue," *Nuclear Industry*, 14 (September 1967): 8–14.

36. *Rutland Herald*, 25 July 1967, 31 October 1967, 9 December 1967; *Nucleonics Week*, 23 November 1967, p. 6, 14 December 1967, p. 5; "Thermal Effects: An Acute Issue," *Nuclear Industry*, pp. 8–11.

37. Edmund S. Muskie to Glenn T. Seaborg, 20 September 1967, Harold L. Price to Muskie, 23 October 1967, Box 512 (Pollution: Thermal Pollution), JCAE Papers.

38. Edmund S. Muskie to Glenn T. Seaborg, 25 October 1967, John T. Conway to All Committee Members, 31 October 1967, Seaborg to Muskie, 4 November 1967, Box 512 (Pollution: Thermal Pollution), JCAE Papers; *Nucleonics Week*, 2 November 1967, p. 2; Bryce Nelson, "Thermal Pollution: Senator Muskie Tells AEC to Cool It," *Science*, 158 (10 November 1967): 755–756.

39. U. S. Senate, Subcommittee on Air and Water Pollution, *Hearings on Thermal Pollution—1968*, pp. 311–346.

40. Joseph F. Hennessey to Frank M. Wozencraft, 16 November 1967, Box 7717, MH&S–11 (Industrial Hygiene), AEC/DOE; Frank M. Wozencraft to Joseph F. Hennessey, 25 April 1968, Box 512 (Pollution: Thermal Pollution), JCAE Papers.

41. AEC–R 141/34 (9 April 1968), AEC–R 141/36 (23 April 1968), AEC/
NRC; *Nucleonics Week*, 18 July 1968, p. 6; "AEC Holds the Line on Jurisdictional Contention," *Nuclear News*, 12 (January 1969): 8–9; *Bennington Banner*, 16 June 1969; *State of New Hampshire v. Atomic Energy Commission*, 406
F.2d 170 (1969).

42. Joseph F. Hennessey to the Commission, 29 April 1968, Glenn T.
Seaborg to James F. C. Hyde, Jr., 12 June 1968, Seaborg to Edmund S. Muskie,
15 October 1968, Legal–4–1 (Federal Water Pollution Agency), AEC/NRC;
Seaborg to Charles Schultze, 11 December 1967, AEC 783/98 (21 August 1968),
Box 64 (Thermal Effects or Pollution), Ramey Office Files, AEC/DOE; John T.
Conway to John O. Pastore, 27 August 1968, Pastore to Muskie, 5 September
1968, General Files—Atomic Energy (Pastore Outgoing Mail), John O. Pastore
Papers, Providence College, Providence, Rhode Island; *Nucleonics Week*, 4 July
1968, p. 1, 26 September 1968, pp. 3–5, 10 October 1968, p. 6; "Thermal
Effects Jurisdiction Stirs Further Controversy," *Nuclear Industry*, 15 (Septem-
ber 1968): 36–38.

43. "How to Stop the Pillage of America," *Sports Illustrated*, 6 December
1967, pp. 40–48; "Letter from the Publisher," *ibid.*, 20 January 1969, p. 4;
Harold Price to Sidney L. James, 17 January 1968, Arthur Brawley to Price, 11
February 1968, Box 98 I&P 4–6 (Articles), AEC/NRC; Talbot, *Power Along
the Hudson*, pp. 112–114.

44. Robert H. Boyle, "The Nukes Are in Hot Water," *Sports Illustrated*, 20
January 1969, pp. 24–28.

45. "19th Hole: The Readers Take Over," *Sports Illustrated*, 3 February
1969, p. 62, 10 February 1969, p. 76, 17 February 1969, p. 72; *Congressional
Record*, 91st Cong., 1st Sess., 1969, pp. 1700–1703, 5353–5354.

46. "Are the 'Nukes' in Hot Water?" *Nuclear News*, 12 (March 1969): 12–
15; AEC 688/62 (17 March 1969), Box 7751, I&P–4 (Public Information),
AEC/DOE; Paul Turner, "The Radiation Controversy," *Vital Speeches of the
Day*, 37 (1 September 1971): 697.

47. Luther J. Carter, "Thermal Pollution: A Threat to Cayuga's Waters?"
Science, 162 (8 November 1968): 649–650; Dorothy Nelkin, *Nuclear Power
and Its Critics: The Cayuga Lake Controversy* (Ithaca: Cornell University Press,
1971).

48. Claude R. Kirk, Jr. to Glenn T. Seaborg, 12 December 1967, Box 180
(Pollution), Seaborg Office Files, Seaborg to Dante Fascell, 1 February 1968,
AEC 544/79 (5 April 1968), Box 7717, MH&S–11 (Industrial Hygiene), AEC/
DOE; Harold L. Price to the Commission, 27 February 1970, MH&S–3–1
(Thermal Effects), AEC/NRC; *Nucleonics Week*, 10 July 1969, p. 3, 17 July
1969, p. 3, 26 February 1970, p. 3, 5 March 1970, pp. 2–3, 1 April 1971, p. 5,
8 July 1971, p. 5, 2 September 1971, p. 4, 2 December 1971, p. 4; *Christian
Science Monitor*, 2 February 1970; "Florida Utility Wins a Round in Thermal
Effects Court Action," *Nuclear Industry*, 17 (April 1970): 29–30; "Cloudy
Sunshine State," *Time*, 13 April 1970, pp. 48–49.

49. *Chicago Daily News*, 6 October 1969; *New York Times*, 28 April, 30
October 1970, 24 March, 26 March 1971; *Wall Street Journal*, 11 May 1970;
Nucleonics Week, 2 October 1969, p. 3, 9 October 1969, p. 4, 14 May 1970,

pp. 3–4, 15 April 1971, pp. 6–8; "Proposed One Degree Rise Limit for Lake Michigan Causes Alarm," *Nuclear Industry*, 17 (May 1970): 8; "New Basis Is Proposed for Measuring Heat Discharge into Lake Michigan," *ibid.*, 17 (October 1970): 20–25; Philip F. Gustafson, "Nuclear Power and Thermal Pollution: Zion, Illinois," *Bulletin of the Atomic Scientists*, 26 (March 1970): 17–23; "Utilities Burn Over Cooling Towers," p. 52; John A. Harris to the Commission, 14 January 1970, Box 7763, MH&S–11 (Environmental Studies), AEC/ DOE; Harold L. Price to the Commission, 15 May 1970, Box 98, MH&S–11 REG (Thermal Effects), AEC/NRC.

50. *New York Times*, 17 March 1971; *Nucleonics Week*, 2 July 1970, p. 5, 25 February 1971, p. 1, 18 March 1971, p. 3; "In Government," *Nuclear Industry*, 17 (July 1970): 33–39; "Cooling Tower Concession Seen as Aim of Palisades Intervenors," *ibid.*, 17 (August 1970): 22–26; Frances Gendlin, "The Palisades Protest: A Pattern of Citizen Intervention," *Bulletin of the Atomic Scientists*, 27 (November 1971): 53–56; James B. Graham to Edward J. Bauser, 29 September 1970, Box 512 (Pollution: Thermal Pollution), JCAE Papers; Bauser to All Committee Members, 30 June 1970, Box 181 (Reactors— Palisades), Hosmer Papers.

51. Julius H. Rubin, Memorandum of Meeting, 18 January 1969, Box 180 (Pollution), Seaborg Office Files, AEC/DOE; AEC 1318/1 (4 September 1969), AEC/NRC; *Seaborg Journal*, Vol. 18, p. 167.

52. Walter G. Belter to Frank Nowak, 26 March 1969, Box 64 (Thermal Effects), Ramey Office Files, Harold L. Price to George D. Aiken, 17 April 1969, Box 53 (Congressional), Seaborg Office Files, Belter to Milton Shaw and G. M. Kavanagh, 24 July 1969, Box 5626 (Environmental Pollution, 1969), OGM Files, John A. Erlewine to the Commission, 9 February 1970, Box 7763, MH&S–11 (Environmental Studies), AEC/DOE; R. E. Hollingsworth to Joseph D. Tydings, 23 January 1970, MH&S–3–1 (Thermal Effects), AEC/NRC; USAEC, *Nuclear Power and the Environment*, pp. 14–22.

53. Daniel Merriman, "The Calefaction of a River," *Scientific American*, 222 (May 1970): 2–12; *Washington Post*, 25 August 1970; *Nucleonics Week*, 27 August 1970, pp. 7–8; USAEC, *Thermal Effects and U.S. Nuclear Power Stations*, pp. 15–36.

54. "Muskie Thermal Effects Bill Provides State Certification," *Nuclear Industry*, 16 (January 1969): 54–57; *Washington Post*, 29 January 1969.

55. Chet Holifield to Edmund S. Muskie, 26 February 1969, Box 46 (JCAE General Correspondence), Edward J. Bauser to All Committee Members, 6 April 1970, Box 47 (Bauser Memos), Chet Holifield Papers, University of Southern California, Los Angeles; AEC 783/106 (13 February 1969), AEC 1318/6 (12 September 1969), AEC/DOE; U. S. Congress, House, Committee on Public Works, *Hearings on Federal Water Pollution Control Act Amendments—1969*, 91st Cong., 1st Sess., 1969, pp. 407–423; "In Government," *Nuclear Industry*, 16 (March 1969): 37–42; "Muskie Bill for Thermal Effects State Control Gains Vital Support," *Nuclear News*, 12 (April 1969): 22; *New York Times*, 13 March 1970; *Nucleonics Week*, 24 April 1969, pp. 4–5, 19 March 1970, pp. 2–3.

56. Louis Harris and Associates, *A Survey of Public and Leadership Atti-*

tudes toward Nuclear Power Development in the United States (New York: Ebasco Services, Inc., 1975), pp. 39, 56.

57. JCAE, *Hearings on Participation by Small Utilities in Nuclear Power*, p. 43.

CHAPTER XII: RADIATION STANDARDS:
DEBATE WITHIN AND CHALLENGES FROM
WITHOUT

1. George T. Mazuzan and J. Samuel Walker, *Controlling the Atom: The Beginnings of Nuclear Regulation, 1946–1962* (Berkeley and Los Angeles: University of California Press, 1984), pp. 32–58; Catherine Caufield, *Multiple Exposures: Chronicles of the Radiation Age* (New York: Harper and Row, 1989), pp. 3–74.

2. Jack Schubert and Ralph E. Lapp, *Radiation: What It Is and How It Affects You* (New York: Viking Press, 1957), pp. 65–87, 98–136; Merril Eisenbud, *Environmental Radioactivity* (New York: McGraw-Hill Book Company, 1963), pp. 24–32; National Committee on Radiation Protection, *Maximum Permissible Amounts of Radioisotopes in the Human Body and Maximum Permissible Concentrations in Air and Water*, Handbook 52 (Washington, D. C.: National Bureau of Standards, 1953).

3. Mazuzan and Walker, *Controlling the Atom*, pp. 38–46; Schubert and Lapp, *Radiation*, pp. 108–136, 181–201; Isaac Asimov and Theodore Dobzhansky, *The Genetic Effects of Radiation* (Washington, D. C.: U. S. Atomic Energy Commission, 1966, a part of the "Understanding the Atom" series).

4. Schubert and Lapp, *Radiation*, pp. 21–25, 37–38; Anthony V. Nero, Jr., *A Guidebook to Nuclear Reactors* (Berkeley and Los Angeles: University of California Press, 1979), pp. 32–39.

5. "Radiation Exposure and Biological Effects," *Nucleonics*, 21 (March 1963): 46–47; L. D. Hamilton, "Somatic Effects," *ibid.*, pp. 48–53; Eugene P. Cronkite, "Diagnosis and Treatment of Radiation Injury in Man," *Nuclear News*, 7 (March 1964): 8–15; Eisenbud, *Environmental Radioactivity*, p. 15; Schubert and Lapp, *Radiation*, p. 74; Patricia J. Lindop and J. Rotblat, "Radiation Pollution of the Environment," *Bulletin of the Atomic Scientists*, 27 (September 1971): 17–24; Mazuzan and Walker, *Controlling the Atom*, pp. 49–57.

6. National Committee on Radiation Protection, *Permissible Dose from External Sources of Ionizing Radiation*, Handbook 59 (Washington, D. C.: National Bureau of Standards, 1954); Lauriston S. Taylor, "Radiation Protection Standards," *Nucleonics*, 21 (March 1963): 58–60; "Radiation Exposure and Biological Effects," *ibid.*, p. 46. Asimov and Dobzhansky, *Genetic Effects of Radiation*, pp. 27–29; Eisenbud, *Environmental Radioactivity*, pp. 25–37, 170.

7. Karl Z. Morgan, "Permissible Exposure to Ionizing Radiation," *Science*, 139 (15 February 1963): 565–571; International Commission on Radiological Protection, *Radiation Protection*, ICRP Publication 9 (Oxford: Pergamon Press, 1966), pp. 6–8; Mazuzan and Walker, *Controlling the Atom*, p. 55.

8. Mazuzan and Walker, *Controlling the Atom*, pp. 54–56, 251–260,

340. For a fascinating, detailed study of the NCRP, see Gilbert F. Whittemore, "The National Committee on Radiation Protection, 1928–1960: From Professional Guidelines to Government Regulation." Ph.D. diss., Harvard University, 1986.

9. Glenn T. Seaborg to Abraham Ribicoff, 7 April 1969, Box 7763, MH&S–6 (Health and Safety Standards), Atomic Energy Commission Records, Department of Energy, Germantown, Maryland (hereafter cited as AEC/DOE); Ellen Thro, "Current Trends in Biological Research," *Nuclear News*, 12 (August 1969): 39–42; J. J. Davis, "Radiation Effects in Man's Environment," *ibid.*, pp. 42–43; Asimov and Dobzhansky, *Genetic Effects of Radiation*, pp. 40–43.

10. U. S. Congress, House, Committee on Energy and Commerce, Subcommittee on Energy Conservation and Power, *American Nuclear Guinea Pigs: Three Decades of Radiation Experiments on U. S. Citizens*, 99th Cong., 2nd Sess., 1986; George W. Farwell to John R. Totter, 16 July 1969, Carl G. Heller, et al., "Protection of the Rights and Welfare of Prison Volunteers: Policies Followed Throughout a 17-Year Medical Research Program," n. d., AEC/DOE; *Willamette Week*, 16 February 1976; *Washington Post*, 28 February 1976; *Seattle Post-Intelligencer*, 3 March 1976.

11. J. Samuel Walker, "Evaluating Fallout Hazards: The Public Health Service, the Atomic Energy Commission, and the Weiss Reports," paper delivered at the annual meeting of the Organization of American Historians, Reno, Nevada, 1988; AEC 636/8 (12 September 1962), AEC 636/10 (2 July 1964), Robert E. Jones to Glenn T. Seaborg, 10 November 1964, Box 1362 (MH&S 3–3, Contamination and Decontamination), AEC/DOE; U. S. Congress, House, Committee on Government Operations, Subcommittee on Natural Resources and Power, *Hearings on Water Pollution Control and Abatement (Part 5, Pacific Northwest)*, 88th Cong., 1st Sess., 1963, pp. 2845–2848; Robert D. O'Neill to John T. Conway, 5 September 1968, Box 512 (Pollution: Radioactivity in Columbia River), Papers of the Joint Committee on Atomic Energy, Record Group 128 (Records of the Joint Committees of Congress), National Archives.

12. Edward W. Lawless, *Technology and Social Shock* (New Brunswick: Rutgers University Press, 1977), pp. 199–207.

13. Mazuzan and Walker, *Controlling the Atom*, pp. 324–335.

14. Clifford K. Beck to William T. England, 25 June 1969, Harold L. Price to the Commission, 22 April 1970, Job 4, Box 11, ID&R–14 REG (Part 20), Lester R. Rogers to Price, 12 May 1969, Legal–4–1 (Part 20), Atomic Energy Commission Records, Nuclear Regulatory Commission, Rockville, Maryland (hereafter cited as AEC/NRC).

15. Saul Levine to the Files, 31 March 1966, Forrest Western to Harold L. Price, 12 July 1968, Legal–4–1 (Part 20), AEC/NRC; "Notes on Conversation between D.D.C. [David D. Comey] and Dr. Peter Morris," 7 January 1969, Box 440 (SIPI—New York Chapter, Cayuga Lake), Barry Commoner Papers, Library of Congress.

16. Western to Price, 12 July 1968, AEC Press Release, 28 March 1964, AEC/NRC.

17. R. Waterfield to the Files, 16 August 1968, Saul Levine to Forrest Western, 23 August 1968, Edson G. Case to Western, 9 December 1968, Milton Shaw to Martin B. Biles, 19 December 1968, R. B. Chitwood to J. A. McBride, 23 December 1968, Legal 4–1 (Part 20), AEC/NRC.

18. "Chronology of Significant Events Involving the Northern States Power Company's Monticello Nuclear Generating Plant," n. d., Box 182 (Minnesota Pollution Problems), Office Files of Glenn T. Seaborg, AEC/DOE; *Nucleonics Week*, 14 April 1966, p. 1; "Scientists Question Reactor Effects," *Scientist and Citizen*, 10 (August 1968): 154–157.

19. "Dr. Dean E. Abrahamson," Box 35 (Dean Abrahamson), Ramey Office Files, AEC/DOE; Dean E. Abrahamson to Edward J. Bauser, 10 July 1969, Box 195 (Monticello), Craig Hosmer Papers, University of Southern California, Los Angeles; *Nucleonics Week*, 24 September 1970, p. 6; *Minneapolis Star*, 14 February 1968; *Minneapolis Tribune*, 14 February 1968; *St. Paul Pioneer Press*, 14 February 1968.

20. "Chronology of Significant Events," AEC/DOE; John P. Badalich to Harold L. Price, 29 July 1968, Legal–4–2 (State Legislation-Minnesota), AEC/NRC; *Minneapolis Tribune*, 14 February 1968; *St. Paul Pioneer Press*, 15 February, 21 February 1968.

21. Ernest C. Tsivoglou, "Radioactive Pollution Control in Minnesota: Final Report," 31 January 1969, Box 182 (Minnesota Pollution Problems), Seaborg Office Files, AEC/DOE; "Statement of E. C. Tsivoglou," 8 April 1969, Legal–4–2 (State Legislation-Minnesota), AEC/NRC.

22. *Minneapolis Star*, 13 February, 20 February 1969; *St. Paul Dispatch*, 26 February 1969; Philip M. Boffey, "Radioactive Pollution: Minnesota Finds AEC Standards Too Lax," *Science*, 163 (7 March 1969): 1043–1045; "Cooling It in Minnesota," *Environment*, 11 (March 1969): 21–25; Minnesota Committee for Environmental Information, *M.C.E.I. Bulletin*, March 1969, Box 512 (Pollution: Thermal Pollution), JCAE Papers.

23. Boffey, "Radioactive Pollution," p. 1043.

24. *Ibid.*, p. 1046; *Minneapolis Star*, 11 February 1969; Mazuzan and Walker, *Controlling the Atom*, chap. 10.

25. *Minneapolis Tribune*, 13 May 1969; *Nucleonics Week*, 15 May 1969, pp. 3–4.

26. *Journal of Glenn T. Seaborg* (25 vols., Berkeley: Lawrence Berkeley Laboratory PUB–625, 1989), Vol. 16, p. 442; Harold LeVander to Seaborg, 12 May 1969, Box 181 (Minnesota Pollution Problems), Seaborg Office Files, AEC/DOE; Seaborg to LeVander, 2 June 1969, Box 195 (Monticello), Hosmer Papers; William T. England to Edward J. Bauser, 26 May 1969, Box 512 (Pollution, Environmental), JCAE Papers; Chet Holifield to Robert C. Tuveson, 3 May 1969, Box 46 (JCAE Outgoing Correspondence), Chet Holifield Papers, University of Southern California, Los Angeles.

27. W. B. McCool to the File, 28 May 1969, Job 4, Box 13, MH&S–11 (Thermal Effects, Vol. 2), AEC/NRC; *Seaborg Journal*, Vol. 19, p. 268; Joseph F. Hennessey to Edward J. Bauser, 5 September 1969, Box 182 (Minnesota Pollution Problems), Seaborg Office Files, AEC/DOE; Clay T. Whitehead to Nils Boe, 29 May 1969, White House Central File, Subject File (ATZ-Industrial), Richard

M. Nixon Papers, Nixon Presidential Materials Project, Alexandria, Virginia; S. David Freeman to Lee A. DuBridge, 6 June 1969, 18 June 1969, Box 862, Subject File (Atomic Energy Commission, 1969), Record Group 359 (Records of the Office of Science and Technology), National Archives.

28. "Who Gets Last Radioactive Word?" *Business Week*, 31 May 1969, p. 28; *Wall Street Journal*, 28 August 1969; *Nucleonics Week*, 28 August 1969, p. 2; *Minneapolis Star*, 8 September 1969.

29. S. David Freeman to Tom Whitehead, 9 September, 16 September, 15 December 1969, Box 862, Subject File (Atomic Energy Commission, 1969), Office of Science and Technology Records; Chet Holifield to William D. Ruckelshaus, 8 October 1969, Freeman to Whitehead, 10 October 1969, Whitehead to Freeman, 11 October 1969, White House Central File, Subject File (AT Atomic Energy), Nixon Papers; Craig Hosmer to Richard M. Nixon, 9 October 1969, Box 184 (Administration Officials), Hosmer Papers; William D. Ruckelshaus to Holifield, 23 October 1969, Box 46 (JCAE Incoming Correspondence 1969), Holifield Papers; *Seaborg Journal*, Vol. 20, pp. 348–349, 661, Vol. 21, p. 236; *Nucleonics Week*, 2 October 1969, p. 1, 23 October 1969, p. 3, 26 March 1970, p. 6; *Electrical World*, 6 October 1969, p. 27.

30. AEC 1318/20 (10 November 1969), AEC/NRC; James T. Ramey to Earl Ewald, 12 November 1969, Julius H. Rubin to Glenn T. Seaborg, 20 November 1969, Box 182 (Minnesota Pollution Problems), Ewald to Seaborg, 28 November 1969, Box 49 (Northern States Power), Seaborg Office Files, AEC/DOE; *Nucleonics Week*, 5 February 1970, pp. 7–8, 7 May 1970, p. 2; *Electrical World*, 9 March 1970, p. 23.

31. AEC 1318/62 (3 March 1970), AEC/NRC; Edward J. Bauser to All Committee Members, 16 September 1969, File 23–5–3 (Atomic Energy General), George D. Aiken Papers, University of Vermont, Burlington; "Utility Puts to Court Test Minnesota Challenge to AEC Authority," *Nuclear Industry*, 16 (September 1969): 38–39; "The Environment: A Challenge for the Seventies," *Nuclear News*, 13 (April 1970): 18–26; *New York Times*, 16 February 1970.

32. *Nucleonics Week*, 5 March 1970, p. 7; Harry Foreman, ed., *Nuclear Power and the Public* (Minneapolis: University of Minnesota Press, 1970), pp. 124, 237.

33. *Northern States Power Company v. State of Minnesota*, 320 F.Supp. 172 (1970) and 447 F.2d 1143 (1971).

34. Lester R. Rogers to Leon G. Billings, 19 July 1969, Job 4, Box 11, ID&R–14 REG (Part 20), AEC/NRC; U. S. Congress, Joint Committee on Atomic Energy, *Hearings on Environmental Effects of Producing Electric Power*, 91st Cong., 1st Sess., 1969, p. 213.

35. Lester Rogers to Clifford K. Beck et al., 14 October 1969, Legal–4–1 (Part 20), AEC/NRC; Department of the Interior, Prepared Statement for Meeting with Minnesota-Wisconsin Delegation, 9 October 1969, Box 182 (Minnesota Pollution Problems), Seaborg Office Files, AEC/DOE; James B. Graham to Edward J. Bauser, 9 October 1969, Box 512 (Pollution, Environmental), JCAE Papers; Thomas R. Dunlap, *DDT: Scientists, Citizens, and Public Policy* (Princeton: Princeton University Press, 1981).

36. Lester R. Rogers to Harold L. Price, 27 March 1969, Forrest Western to

Price, 3 April 1969, Legal–4–1 (Part 20), Price to the Commission, 24 July 1969, Job 4, Box 11, ID&R–14 REG (Part 20), Minutes of Regulatory Meeting 274, 9 April 1969, AEC/NRC.

37. Rogers to Price, 27 March 1969, Price to the Commission, 24 July 1969, AEC/NRC.

38. *Nucleonics Week*, 1 May 1969; "AEC Commissioner Thompson and Aide, Jack Rosen, Killed in Plane Crash," *Nuclear Industry*, 17 (November-December 1970): 70; "Theos J. Thompson, 1918–70," *Nuclear News*, 14 (January 1971): 23; *Seaborg Journal*, Vol. 20, p. 51, Vol. 23, pp. 524–33; Minutes of Regulatory Meeting 274, 9 April 1969, AEC/NRC. Thompson's term on the Commission was cut short by a tragedy; he was killed in a plane crash in November 1970 at the age of fifty-two.

39. Theos J. Thompson, "Social Responsibility and Nuclear Engineering Education," *Nuclear News*, 12 (August 1969): 46–49; Thompson Notes from Aspen Conference, n. d., Box 8 (Seminar on Progress in a Living Environment), Office Files of Theos J. Thompson, AEC/DOE; *Nucleonics Week*, 1 May 1969, p. 1.

40. Harold L. Price to the Commission, 8 September 1969, Theos J. Thompson to the Commission, 3 October 1969, Box 13 (Part 20), Thompson Office Files, AEC/DOE.

41. Thompson to the Commission, 3 October 1969, AEC/DOE; "Peaceful Uses of Nuclear Explosives," *Nuclear News*, 11 (March 1968): 24–29.

42. AEC 180/66 (24 October 1969), AEC/NRC.

43. *Seaborg Journal*, Vol. 20, pp. 317, 349.

44. *Seaborg Journal*, Vol. 19, p. 205; James B. Graham to Edward J. Bauser, 15 September 1969, Box 512 (Pollution, Environmental), JCAE Papers; *Nucleonics Week*, 19 June 1969, p. 1; JCAE, *Hearings on Environmental Effects of Producing Electric Power*, pp. 489–490.

45. *Nucleonics Week*, 30 October 1969, p. 2, 6 November 1969, p. 4; CW [Charles Weaver] to George Aiken, 1 December 1969, File 23–5–7 (Atomic Energy General), Aiken Papers; JCAE, *Hearings on Environmental Effects of Producing Electric Power*, pp. 89, 204–207.

46. *Seaborg Journal*, Vol. 20, p. 463; W. B. McCool to Harold L. Price, 14 November 1969, Theos J. Thompson to the Commission, 4 November 1969, Job 4, Box 11, ID&R–14 REG (Parts 20 and 50), AEC/NRC.

47. Harold L. Price to the Commission, 12 December 1969, Job 4, Box 11, ID&R–14 REG (Parts 20 and 50), AEC/NRC.

48. *Seaborg Journal*, Vol. 20, pp. 652, 661; Vol. 21, pp. 174, 236, 258–260; W. E. Johnson to the Commission, 3 February 1970, Johnson to the Files, 6 February 1970, Job 4, Box 11, ID&R–14 REG (Parts 20 and 50), AEC/NRC; "Commissioner Clarence E. Larson," *Nuclear News*, 12 (October 1969): 25–26; *Nucleonics Week*, 14 July 1966, p. 1.

49. *Seaborg Journal*, Vol. 21, pp. 284, 322; Harold L. Price to the Commission, 19 February 1970, with Draft Proposed Amendments to Parts 20 and 50, Job 4, Box 11, ID&R–14 REG (Parts 20 and 50), AEC/NRC; R. M. Ketchel to the File, 9 February 1970, Henry E. Bliss to Dennis Spurgeon, 5 February 1970, Box 13 (Part 20), Thompson Office Files, AEC/DOE; Flyer of Anti-Pollution

League, 6 February 1970, Series 2, Box 1 (Anti-Pollution League), Chesapeake Bay Foundation Papers, University of Maryland, College Park.

50. *Seaborg Journal*, Vol. 21, pp. 322, 349, 357, 362; Harold L. Price to the Commission, 25 February 1970, Job 4, Box 11, ID&R–14 REG (Parts 20 and 50), AEC/NRC; James B. Graham to Edward J. Bauser, 23 February 1970, Box 531 (Radiation Standards for Protection), JCAE Papers.

51. *Seaborg Journal*, Vol. 21, p. 432; "Statement of Commissioner James T. Ramey," 27 March 1970, Box 41 (Effluent Releases), Ramey Office Files, AEC/ DOE; *Nucleonics Week*, 2 April 1970, p. 1; "Down to What's Practicable," *Science News*, 97 (4 April 1970): 341; "AEC Acts to Strengthen Radioactive Discharge Standards," *Nuclear Industry*, 17 (April 1970): 36–38; *St. Paul Pioneer Press*, 29 March 1970.

52. Westinghouse Electric Corporation Press Release, 6 May 1970, Box 47 (Environmental Problems), Holifield Papers; *Seaborg Journal*, Vol. 21, pp. 321–322; *Nucleonics Week*, 7 May 1970, p. 1, 19 November 1970, p. 2.

53. JCAE, *Environmental Effects of Producing Electric Power*, p. 1185.

CHAPTER XIII: FALLOUT OVER RADIATION

1. Philip M. Boffey, "Ernest J. Sternglass: Controversial Prophet of Doom," *Science*, 166 (10 October 1969): 195–200; Leslie J. Freeman, *Nuclear Witnesses: Insiders Speak Out* (New York: W. W. Norton, 1981), pp. 57–59.

2. Alice Stewart, Josefine Webb, and David Hewitt, "A Survey of Childhood Malignancies," *British Medical Journal*, 1 (28 June 1958): 1495–1508; Brian MacMahon, "Pre-Natal X-Ray Exposure and Childhood Cancer," *Journal of the National Cancer Institute*, 28 (May 1962): 1173–1191; Ralph E. Lapp, "Nevada Test Fallout and Radioiodine in Milk," *Science*, 137 (7 September 1962): 756–757; Boffey, "Ernest J. Sternglass," p. 197.

3. Ernest J. Sternglass, "Cancer: Relation of Prenatal Radiation to Development of the Disease in Childhood," *Science*, 140 (7 June 1963): 1102–1104.

4. U. S. Congress, Joint Committee on Atomic Energy (JCAE), *Hearings on Fallout, Radiation Standards, and Countermeasures*, 88th Cong., 1st Sess., 1963, pp. 417–419, 600–601, 1088; A. B. Brill, "Comments on Recent Article by Sternglass in *Science*," 25 June 1963, File 1462, U. S. Public Health Service Records, Division of Radiological Health, U. S. Public Health Service, Rockville, Maryland; Boffey, "Ernest J. Sternglass," p. 197.

5. Ernest J. Sternglass, "Leukemia: Evidence for Induction of the Disease by Fallout Radiation at Low Dose-Rates," 18 May 1968, File 1581, Public Health Service Records; Sternglass to John O. Pastore, 24 June 1968, "DBM Comments on the Sternglass Paper," 25 July 1968, Box 311 (Fallout), Papers of the Joint Committee on Atomic Energy, Record Group 128 (Records of the Joint Committees of Congress), National Archives; *New York Times*, 21 June 1968; Boffey, "Ernest J. Sternglass," p. 197.

6. Ernest J. Sternglass, *Secret Fallout: Low-Level Radiation from Hiroshima to Three-Mile Island* (New York: McGraw-Hill, 1981), pp. 55–83; *Pittsburgh Post-Gazette*, 24 October 1968; Boffey, "Ernest J. Sternglass," p. 198.

7. Ernest J. Sternglass, "Infant Mortality and Nuclear Tests," *Bulletin of*

the Atomic Scientists, 25 (April 1969): 18–20; Sternglass, "Evidence for Low-Level Radiation Effects on the Human Embryo and Fetus," 7 May 1969, File 1581, Sternglass, "Strontium–90: Evidence for a Possible Genetic Effect in Man," 8 June 1969, File 1583, Public Health Service Records.

8. *New York Times*, 21 June 1969; Ernest J. Sternglass, "Can the Infants Survive?" *Bulletin of the Atomic Scientists*, 25 (June 1969): 26–27; Sternglass to Craig Hosmer, 11 July 1969, Box 190 (ABM—Anti-Ballistic Missile System), Craig Hosmer Papers, University of Southern California, Los Angeles; Gregg Herken, *Counsels of War* (New York: Alfred A. Knopf, 1985), pp. 187–200, 229–238.

9. Ernest J. Sternglass, "The Death of All Children: A Footnote to the A.B.M. Controversy," *Esquire*, September 1969, pp. 1a–1d; *Washington Post*, 29 July 1969; *New York Times*, 29 July 1969; Boffey, "Ernest J. Sternglass," p. 198.

10. *Toronto Star*, 9 July 1969; *New York Post*, 8 July 1969; Sternglass interviews, "The Today Show," 24 July 1969, "Panorama," 29 July 1969, "Martin Agronsky's Washington," 29 July 1969, "Newsfront," 5 August 1969, "Huntley-Brinkley Report," 6 August 1969, "CBS Morning News," 6 August 1969, Energy History Collection (Sternglass Papers and Interviews), Atomic Energy Commission Records, U. S. Department of Energy, Germantown, Maryland (hereafter cited as AEC/DOE); Boffey, "Ernest J. Sternglass," p. 198.

11. John B. Storer to John S. Kelley, 24 March 1969, Bernard Shore to Richard Hamburger, 24 March 1969, Energy History Collection (Sternglass—Internal Info Only), AEC/DOE.

12. *Journal of Glenn T. Seaborg* (25 vols., Berkeley: Lawrence Berkeley Laboratory PUB–625, 1989), Vol. 19, pp. 480, 505; John A. Harris to the General Manager, 25 July 1969, Box 182 (Sternglass), "Doctors Storer and Sagan Interviewed," 28 July 1969, Box 136 (Gofman-Tamplin Statements), Office Files of Glenn T. Seaborg, AEC/DOE; AEC 688/96 (30 December 1969), Atomic Energy Commission Records, Nuclear Regulatory Commission, Rockville, Maryland (hereafter cited as AEC/NRC); "Infant Mortality Controversy: Sternglass and His Critics," *Bulletin of the Atomic Scientists*, 25 (October 1969): 26–33; *New York Times*, 28 July 1969.

13. Ernest J. Sternglass to Robert Finch, 9 June 1969, File 1581, Chris A. Hansen to the Surgeon General, 14 July 1969, "Notes on Sternglass for Dr. Telles' Use," n. d., "Statement on the Assertions of E. J. Sternglass," n. d., File 1583, Public Health Service Records; *New York Times*, 13 September 1969.

14. J. Rotblat to Saul J. Harris, 17 November 1969, File 1581, Public Health Service Records; Michael W. Friedlander and Joseph Klarmann, "How Many Children?" *Environment*, 11 (December 1969): 3–8; Patricia J. Lindop and J. Rotblat, "Strontium–90 and Infant Mortality," *Nature*, 224 (27 December 1969): 1257–1260; *Washington Post*, 29 January 1969; Sternglass, *Secret Fallout*, pp. 23–26, 104, 116–117.

15. Alice Stewart, "The Pitfalls of Extrapolation," *New Scientist*, 43 (24 July 1969): 181; Amembassy Stockholm to Department of State, "Swedish Press on Sternglass Theory on Nuclear Fallout," 22 August 1969, File 1583, Public Health Service Records.

16. Freeman J. Dyson, "Comment on Sternglass Thesis," *Bulletin of the Atomic Scientists*, 25 (June 1969): 27; "Professor Sternglass, Fallout, and Infant Mortality," *American Journal of Public Health*, 59 (December 1969): 2129.

17. Richard G. Hewlett and Francis Duncan, *Atomic Shield, 1947/1952: Vol. II of a History of the United States Atomic Energy Commission*, (University Park: Pennsylvania State University Press, 1969), pp. 581–584; Richard T. Sylves, *The Nuclear Oracles: A Political History of the General Advisory Committee of the Atomic Energy Commission, 1947–1977* (Ames: Iowa State University Press, 1987), pp. 259–261.

18. AEC 859/12 (4 May 1963), Box 35 (Gofman and Tamplin), Office Files of James T. Ramey, AEC/DOE; A. R. Leudecke to John O. Pastore, 23 May 1963, Box 310 (Fallout), JCAE Papers.

19. AEC 859/12, "Excerpt from Minutes of Commission Meeting 1929," 6 May 1963, "Comment on the Atherosclerogenic Index and Dr. John W. Gofman," by Hardin B. Jones, 11 May 1970, Box 35 (Gofman and Tamplin), Ramey Office Files, AEC/DOE; Freeman, *Nuclear Witnesses*, pp. 81–89.

20. John W. Gofman, "The Hazards to Man from Radioactivity," 1964, Energy History Collection (Gofman-Tamplin), AEC/DOE; John B. Radcliffe to John T. Conway, 18 December 1967, Michael M. May to Conway, 13 September 1968, Box 397 (Lawrence Radiation Laboratory, Livermore, California), JCAE Papers; "Low-Level Radiation Effects on Man: Good, Bad, or Both?" *Nucleonics*, 22 (June 1964): 18; *New York Times*, 23 April 1964.

21. AEC 604/97 (17 August 1966), Box 7716, MH&S–3 (Radiation), John R. Totter to the Commission, 20 April 1970, Box 136 (Gofman-Tamplin), Seaborg Office Files, AEC/DOE; Clifford E. Nelson to Compliance and Control Program, 19 March 1969, File 1583, Public Health Service Records; Pete Winslow, "Fallout from the 'Peaceful Atom'," *Nation*, 212 (3 May 1971): 557–561.

22. AEC 688/92 (10 September 1969), Energy History Collection (Gofman-Tamplin Staff Papers), AEC/DOE; Arthur R. Tamplin, Yvonne Ricker, and Marguerite F. Longmate, "A Criticism of the Sternglass Article on Fetal and Infant Mortality," 22 July 1969, File 1583, Public Health Service Records.

23. AEC 688/92, Arthur R. Tamplin to Eugene Rabinowitch, 29 July 1969, Richard S. Lewis to Tamplin, 15 August 1969, Energy History Collection (Gofman-Tamplin Staff Papers), Sheldon Novick to Tamplin, 7 August 1969, Energy History Collection (Sternglass—Internal Info Only), AEC/DOE.

24. AEC 688/92, Appendix B to SECY–103 (16 July 1970), Energy History Collection, no folder, AEC/DOE.

25. John R. Totter to Arthur R. Tamplin, 10 September 1969, Box 7752, I&P 4–6 (Articles), AEC/DOE.

26. Arthur R. Tamplin, "Fetal and Infant Mortality and the Environment," *Bulletin of the Atomic Scientists*, 25 (December 1969): 23–29; Sheldon Novick to CEI Board of Directors, 15 October 1969, Box 441 (CEI-Scientific Division), Barry Commoner Papers, Library of Congress.

27. *Seaborg Journal*, Vol. 20, p. 294; *Nuclear Power and the Environment: Proceedings*, 11 September 1969, Box 7764, MH&S–11 (Environmental Studies), AEC/DOE; Arthur R. Tamplin and John W. Gofman, *'Population Control'*

through Nuclear Pollution (Chicago: Nelson-Hall Company, 1970), pp. 142–143; Harry Foreman, ed., *Nuclear Power and the Public* (Minneapolis: University of Minnesota Press, 1970), pp. 45–51.

28. John W. Gofman and Arthur R. Tamplin, "Low Dose Radiation, Chromosomes, and Cancer," 29 October 1969, printed in JCAE, *Hearings on Environmental Effects of Producing Electric Power*, 91st Cong., 1st Sess., 1969, pp. 640–652; *San Francisco Chronicle*, 30 October 1969.

29. U. S. Congress, Senate, Committee on Public Works, Subcommittee on Air and Water Pollution, *Hearings on Underground Uses of Nuclear Energy*, 91st Cong., 1st Sess., 1969, pp. 58–99; *Seaborg Journal*, Vol. 20, pp. 581, 589.

30. John W. Gofman to Glenn T. Seaborg, 28 November 1969, Box 136 (Lawrence Radiation Laboratory), Seaborg Office Files, AEC/DOE.

31. AEC 1318/27 (5 December 1969), Box 7763, MH&S–11 (Environmental Studies), AEC/DOE; JCAE, *Hearings on Environmental Effects of Producing Electric Power*, pp. 685–693.

32. *Seaborg Journal*, Vol. 20, pp. 632, 666, 669, 673; Tamplin and Gofman, *'Population Control' through Nuclear Pollution*, pp. 153, 224–225.

33. John R. Totter to the Commission, 29 December 1969, Arthur R. Tamplin, "Nuclear Reactors and the Public Health and Safety," December 1969, Box 7763, MH&S–11 (Environmental Studies), James T. Ramey to Roger E. Batzel, 5 January 1970, Box 35 (Gofman and Tamplin), Ramey Office Files, AEC/DOE; Tamplin and Gofman, *'Population Control' through Nuclear Pollution*, pp. 153, 224; *Nucleonics Week*, 8 January 1970, p. 3.

34. John A. Harris to Edward J. Bloch, 23 December 1969, Glenn T. Seaborg to Edmund S. Muskie, 24 December 1969, John A. Harris to the Commission, 31 December 1969, W. B. McCool to the File, 6 January 1970, Box 7763, MH&S–11 (Environmental Studies), AEC/DOE; *Washington Star*, 30 December 1969; *Philadelphia Inquirer*, 30 December 1969; *Baltimore Sun*, 30 December 1969; *Minneapolis Star*, 1 January 1970; *Los Angeles Times*, 27 January 1970.

35. John W. Gofman to Theos J. Thompson and Clarence E. Larson, 11 February 1970, Box 512 (Pollution, Environmental), JCAE Papers; Gofman to Charles W. Mays, 18 February 1970, Energy History Collection (Gofman-Tamplin II), AEC/DOE; "The Atom as Pollutant," *Newsweek*, 16 February 1970, p. 57.

36. "Public Hearing of the Special Committee for the Environmental Protection of the City Council of the City of New York," transcript, 4 March 1970, Energy History Collection (Gofman-Tamplin: New York Council), "Unedited Transcript of Remarks of Dr. John W. Gofman," 5 March 1970, Energy History Collection (Gofman-Tamplin II), AEC/DOE; Subcommittee on Air and Water Pollution, *Hearings on Underground Uses of Nuclear Energy*, pp. 281–290.

37. Robert W. Holcomb, "Radiation Risk: A Scientific Problem?" *Science*, 167 (6 February 1970): 853–855; Harold L. Davis, "Clean Air Misunderstanding," *Physics Today*, 23 (May 1970): 104; Federal Radiation Council Information Paper, 1 May 1970, File 203, Public Health Service Records.

38. JCAE, *Hearings on Environmental Effects of Producing Electric Power*, pp. 1110–1420; *Nucleonics Week*, 8 January 1970, pp. 2–3, 5 February 1970, pp. 3–4; *Washington Evening Star*, 18 December 1969; *New York Times*, 16 March 1969; Transcripts, "Huntley-Brinkley Report," 4 March 1970, "CBS Morning News," 4 March 1970, Energy History Collection (Gofman-Tamplin II), AEC/DOE; Paul J. Bailey to Jess W. Malcolm, 20 December 1969, Series 2, Box 1 (Calvert Cliffs Nuclear Power Plant), Chesapeake Bay Foundation Papers, University of Maryland, College Park.

39. JCAE, *Hearings on Environmental Effects of Producing Electric Power*, p. 1397.

40. Robert H. Finch to Edmund S. Muskie, 23 January 1970, Federal Radiation Council, Minutes and Record of Actions, 3 June 1970, File 203, Public Health Service Records; Paul C. Tompkins, "Draft Action Staff Paper," 20 March 1970, Federal Radiation Council Action Paper, "Radiation Protection Policy," 6 April 1970, Box 3 (Council Meeting of May 8, 1970), Federal Radiation Council Papers, Record Group 412 (Records of the Environmental Protection Agency), National Archives; *Nucleonics Week*, 26 February 1970, pp. 1–2; *Baltimore News American*, 24 March 1970.

41. "Statement of Commissioner James T. Ramey," 27 March 1970, Box 41 (Effluent Releases), Ramey Office Files, AEC/DOE; *Nucleonics Week*, 2 April 1970, p. 1; *New York Times*, 28 March 1970.

42. John W. Gofman and Arthur R. Tamplin to the Members of the International Commission on Radiological Protection, n.d., Energy History Collection (Gofman-Tamplin II), AEC/DOE; *Washington Sunday Star*, 22 March 1970; *San Antonio Light*, 22 March 1970; *San Francisco Sunday Examiner and Chronicle*, 22 March 1970; *Washington Post*, 8 April 1970.

43. Transcript, "The Big News," 13 April 1970, Box 136 (Gofman-Tamplin), Seaborg Office Files, AEC/DOE; John W. Gofman to Atomic Industrial Forum, 14 May 1970, Box 50 (Gofman-Tamplin 1965–1971), Chet Holifield Papers, University of Southern California, Los Angeles; "AEC Commissioner Criticizes Industry on Radwaste Claims," *Nuclear Industry*, 17 (June 1970): 19–20; *Nucleonics Week*, 16 April 1970, pp. 6–7, 30 April 1970, p. 6.

44. Donald C. Kull to the Commission, 14 April 1970, Box 7763, MH&S–1 (Environmental Studies), AEC/DOE; *Seaborg Journal*, Vol. 21, pp. 705, 709, 730, 737, 755, Vol. 22, p. 234; Michael M. May to Paul Turner, 12 June 1970, May to Glenn T. Seaborg, 8 April 1970, Box 512 (Pollution, Environmental), JCAE Papers.

45. *Washington Post*, 5 July 1970; *Oakland Tribune*, 6 July 1970; *San Francisco Examiner*, 6 July 1970; *San Francisco Chronicle*, 6 July 1970; Ralph Nader to Edmund S. Muskie, 5 July 1970, Box 136 (Gofman-Tamplin), Seaborg Office Files, AEC/DOE.

46. "Statement by the Atomic Energy Commission," 7 July 1970, "Statement of Dr. Roger E. Batzel," n.d., Box 398 (Lawrence Livermore Laboratory), JCAE Papers.

47. "Staff Report on Allegations Made by Drs. Gofman and Tamplin," 21 July 1970, Energy History Collection (no folder), Edmund S. Muskie to Glenn

T. Seaborg, 5 August 1970, Energy History Collection (Gofman-Tamplin III), AEC/DOE; Arthur R. Tamplin to Chet Holifield, 27 July 1970, John W. Gofman to Holifield, 29 July 1970, Box 50 (Drs. Gofman and Tamplin), Holifield Papers; William Bevan to Muskie, 17 December 1970, Box 1840, File 500 Public Works (Gofman-Tamplin Controversy Correspondence), Edmund S. Muskie Papers, Edmund S. Muskie Archives, Bates College, Lewiston, Maine; *Washington Post*, 6 August 1970; *Washington Evening Star*, 6 August 1970.

48. Glenn T. Seaborg to Edmund S. Muskie, 23 July 1970, Box 41 (Effects of Radiation), Nixon Library Materials, John W. Gofman to Seaborg, 29 July 1970, Box 136 (Gofman-Tamplin), Seaborg Office Files, AEC/DOE.

49. *Wall Street Journal*, 20 May 1970; *New York Times*, 16 July, 17 September 1970; *Washington Evening Star*, 23 July 1970; *Washington Post*, 11 September 1970; *National Enquirer*, 27 September 1970; "AEC Attacked on Licensing and Attitude Toward Nuclear Critics," *Nuclear Industry*, 17 (September 1970): 32–33; *Nucleonics Week*, 9 July 1970, p. 1, 27 August 1970, p. 3, 1 October 1970, p. 5; Philip M. Boffey, "Gofman and Tamplin: Harassment Charges against AEC, Livermore," *Science*, 169 (28 August 1970): 838–843; Joshua Lederberg to Chairman, Select Committee on Nuclear Electricity, 16 October 1970, Box 512 (Pollution, Environmental), JCAE Papers.

50. C. G. Stewart to Chet Holifield, 3 July 1970, Box 50 (Gofman-Tamplin), Holifield Papers; AEC 1318/63 (5 March 1970), AEC/NRC; *Nucleonics Week*, 29 May 1970, p. 8.

51. Edward J. Bauser to All Members, 26 January 1971, Box 48 (Basic Radiation Protection Criteria), Lauriston S. Taylor to Chet Holifield, 5 February 1971, Box 48 (National Council on Radiation Protection), Holifield Papers; "NCRP to Take 'As Low as Practicable' Approach to Radiation Guides," *Nuclear Industry*, 18 (February 1971): 73–75; *Nucleonics Week*, 28 January 1971, p. 3; *Washington Post*, 27 January 1971; *New York Times*, 27 January 1971.

52. SECY–R 52 (30 September 1970), W. B. McCool to Lester R. Rogers, 16 November 1970, Harold L. Price to Chet Holifield, 30 November 1970, Job 9, Box 12, ID&R–14 REG, (Parts 20 and 50), Minutes of Regulatory Meeting 288 (2 October 1970), AEC/NRC; "Forum Committee Comments on AEC Proposed Radwaste Rule Changes," *Nuclear Industry*, Vol. 17 (June 1970): 43–44; "AEC Receives Comments on Proposed New Ruling," *Nuclear News*, 13 (August 1970): 57–59; *Nucleonics Week*, 16 July 1970, pp. 5–6, 10 December 1970, p. 2.

53. Harold L. Price to the Commission, 22 December 1970, Job 9, Box 12, ID&R–14 REG (Parts 20 and 50), AEC/NRC; *Seaborg Journal*, Vol. 24, pp. 289, 498, 504; Price and R. E. Hollingsworth to the Commission, 4 March 1971, Box 35 (J. Lederberg), Ramey Office Files, AEC/DOE; *Washington Post*, 2 May 1971; *Nucleonics Week*, 25 February 1971, p. 1a.

54. Harold L. Price to the Commission, 30 March 1971, Job 9, Box 12, ID&R–14 REG (Parts 20 and 50), SECY–R 210 (8 April 1971), AEC/NRC; *Nucleonics Week*, 27 May 1971, p. 2.

55. *Seaborg Journal*, Vol. 24, pp. 512, 571, 582, 694, Vol. 25, pp. 23, 32, 294; James B. Graham to George F. Murphy, 4 June 1971, Box 48 (Staff

Memoranda 1971), Holifield Papers; *Nucleonics Week*, 27 May 1971, p. 2, 10 June 1971, p. 3.

56. AEC Press Release, 7 June 1971, AEC/NRC; *New York Times*, 8 June 1971; *Nucleonics Week*, 27 May 1971, pp. 2–3; "Radiation Discharge Limits," *Nuclear Industry*, 18 (June 1971): 17–19.

57. *Seaborg Journal*, Vol. 25, pp. 123–124, 126; *New York Times*, 10 July 1970; "AEC Radioactive Effluent Standards-Setting, FRC to New Agency," *Nuclear Industry*, 17 (June 1970): 16–19; "The Goals and Responsibilities of EPA's Office of Radiation," *ibid.*, 18 (February 1971): 52–53.

58. *Seaborg Journal*, Vol. 25, pp. 169, 170, 194; Robert Gillette, "Reactor Emissions: AEC Guidelines Move Toward Critics' Position," *Science*, 172 (18 June 1971): 1215–1216.

59. "Comment by the Standards Committee, Health Physics Society," 15 September 1971, Box 48 (JCAE Incoming Correspondence), Holifield Papers; *New York Times*, 13 June 1971; "Radiation Discharge Limits," *Nuclear Industry*, pp. 18–19; "Reactor Emissions," *Science*, p. 1216; *Nucleonics Week*, 17 June 1971, pp. 5–6.

CHAPTER XIV: ENVIRONMENTAL LAW AND
LITIGATION: FROM NEPA TO THE CALVERT
CLIFFS DECISION

1. For the legislative history of NEPA, see Richard A. Liroff, *A National Policy for the Environment: NEPA and Its Aftermath* (Bloomington: Indiana University Press, 1976), and Richard N. L. Andrews, *Environmental Policy and Administrative Change* (Lexington, MA: D. C. Heath, 1976).

2. The text of the law is printed in Liroff, *National Policy*, pp. 217–223, and Andrews, *Environmental Policy*, pp. 167–172.

3. George F. Trowbridge speech, "Environmental Issues in Reactor Licensing," September 1970, Box 53 (Reactor Licensing Delays, AIF), Office Files of James T. Ramey, Atomic Energy Commission Records, Department of Energy, Germantown, Maryland (hereafter cited as AEC/DOE).

4. *Congressional Record*, 91st Cong., 1st Sess., 1969, pp. 29053–29056; *Washington Post*, 8 October 1969; Edward J. Bauser to the File, 15 August 1969, Box 413 (Legislation, S. 414 . . . Practical Value, 91st Cong.), Papers of the Joint Committee on Atomic Energy, Record Group 128 (Records of the Joint Committees of Congress), National Archives (hereafter cited as JCAE Papers). For an excellent brief discussion of the AEC and NEPA, see Roger R. Trask, "The Atomic Energy Commission and the Environment: NEPA and the Calvert Cliffs Decision, 1969–1972," draft, Historical Office, Nuclear Regulatory Commission, Rockville, Maryland.

5. Joseph F. Hennessey to the Commission, 14 October, 23 December 1969, C. E. Larson to Wilfred H. Rommel, 29 December 1969, Box 7763, MH&S–11 (Environmental Studies), AEC/DOE; AEC 1318/23 (17 November 1969), Atomic Energy Commission Records, Nuclear Regulatory Commission (hereafter cited as AEC/NRC).

6. Harold L. Price to Thomas J. Meskill, 6 February 1970, C. L. Henderson to Price, 2 February 1970, MH&S–3–1 (Thermal Effects), AEC/NRC.

7. AEC 783/150 (19 February 1970), AEC/NRC; Trask, "AEC and the Environment," pp. 18–20.

8. AEC 783/150, AEC/NRC.

9. *Journal of Glenn T. Seaborg* (25 vols., Berkeley: Lawrence Berkeley Laboratory PUB–625, 1989), Vol. 21, pp. 322, 432.

10. John A. Harris to the Commission, 31 March 1970, MH&S–3–2–2 (NEPA), Harold L. Price to Edward J. Bauser, 31 March 1970, MH&S–3–2–2 (Hazards and Control), AEC/NRC; Bauser to All Committee Members, 10 March 1970, Box 47 (Environmental Problems), Chet Holifield Papers, University of Southern California, Los Angeles; "AEC Sets Procedures for Implementing Environmental Policy Act," *Nuclear Industry*, 17 (April 1970): 42–44; *Nucleonics Week*, 9 April 1970, p. 4.

11. "Environmental Policy Guides Issued: Industry Asks AEC Clarification," *Nuclear Industry*, 17 (May 1970): 38–40; *Nucleonics Week*, 12 March 1970, p. 1, 21 May 1970, pp. 1–2.

12. Joseph F. Hennessey to the Commission, 5 June 1970, Box 182 (National Environmental Policy Act), Office Files of Glenn T. Seaborg, AEC/DOE.

13. Glenn T. Seaborg to Russell E. Train, 28 May 1970, Job 4, Box 11, ID&R–14 REG (Parts 20 and 50, 170), Harold L. Price to the Commission, 21 May 1970, "Appendix D—Statement of General Policy and Procedure," 1 June 1970, Job 4, Box 12, ID&R–14 REG (Part 50), AEC/NRC.

14. *Nucleonics Week*, 18 May 1970, pp. 3–4, 18 June 1970, pp. 10–11, 16 July 1970, pp. 3–4.

15. *Ibid.*, 28 May 1970, p. 4; Trask, "AEC and the Environment," p. 21; Liroff, *National Policy*, pp. 89–97.

16. AEC–R 178/3 (18 June 1970), AEC/NRC.

17. Howard C. Brown to Harold L. Price and others, 23 May 1969, MH&S–3–2–1 (Environmental Meetings), AEC/NRC; Glenn T. Seaborg to Robert P. Mayo, 7 April 1970, Box 7769 (O&M–2, Regulation), AEC/DOE.

18. Harold L. Price to the Commission, 11 May 1970, MH&S–3–2 (Hazards and Control), AEC Press Release, 16 June 1970, AEC/NRC.

19. James B. Graham to Edward J. Bauser, 12 September 1970, Box 194 (Licensing and Regulation), Craig Hosmer Papers, University of Southern California, Los Angeles; "Ramey Proposes Drastic Revision of AEC Licensing Process," *Nuclear Industry*, 17 (November-December 1970): 7–10.

20. AEC Press Release, 3 August 1970, Harold L. Price to the Commission, 24 August 1970, Job 9, Box 13, ID&R–14 (Part 50), AEC/NRC; *Newsday*, 15 September 1970.

21. Joseph E. Karth to Glenn T. Seaborg, 30 September 1970, Box 192 (National Environmental Policy Act), Seaborg Office Files, AEC/DOE; Philip A. Hart to Seaborg, 3 August 1970, MH&S–3–2–2 (National Environmental Policy Act), AEC/NRC; James B. Graham to Edward J. Bauser, 29 September 1970, Box 512 (Pollution: Thermal Pollution), JCAE Papers; *New York Times*, 30 September 1970.

22. C. L. Henderson to Harold L. Price, 14 September 1970, Howard K. Shapar to H. L. Price and others, 23 September 1970, SECY–R 56 (1 October 1970), MH&S–3–2–2 (National Environmental Policy Act), AEC/NRC.

23. SECY–R 56, AEC/NRC.

24. *Seaborg Journal*, Vol. 23, p. 204; Minutes of AEC Regulatory Meeting 289, 14 October 1970, AEC/NRC.

25. SECY–R 72 (2 November 1970), Box 35 (NEPA), Ramey Office Files, AEC/DOE.

26. Joseph F. Hennessey to the Commission, 5 November, 10 November 1970, MH&S–3–2–2 (National Environmental Policy Act), SECY–R 89 (18 November 1970), Minutes of AEC Regulatory Meeting 292, 23 November 1970, AEC/NRC; *Seaborg Journal*, Vol. 23, p. 367.

27. U. S. Congress, House, Committee on Merchant Marine and Fisheries, Subcommittee on Fisheries and Wildlife Conservation, *Appendixes to Hearings on Administration of the National Environmental Policy Act*, 91st Cong., 2nd Sess., 1970, pp. 194–211; "NEPA's Role Clarified," *Nuclear Industry*, 17 (November-December 1970): 48–49.

28. SECY–R 98 (3 December 1970), AEC/NRC; *Nucleonics Week*, 10 December 1970, pp. 3–4.

29. Subcommittee on Fisheries and Wildlife Conservation, *Hearings on Administration of the National Environmental Policy Act*, 91st Cong., 2nd Sess., 1970, pp. 158–214; "Environmental Group Leaders and AEC Regulators Compare Notes," *Nuclear Industry*, 18 (January 1971): 26–28; *Nucleonics Week*, 14 January 1971, p. 4; Marvin M. Mann to Edward J. Bauser, 1 March 1971, Box 512 (Pollution, Environmental), JCAE Papers.

30. Edward L. Strohbehn to Howard Shapar, 5 February 1970, MH&S–3–2–2 (National Environmental Policy Act), AEC/NRC.

31. Edward P. Radford and others, "Statement of Concern," *Environment*, 11 (September 1969): 18–27.

32. *Ibid.*; Arthur W. Sherwood to George W. Geppart, 12 December 1967, Sherwood to Joseph Tydings, 10 May 1968, Jess W. Malcolm to Paul J. Bailey, 11 December 1968, Series 2, Box 1 (Calvert Cliffs Nuclear Power Plant Correspondence), Chesapeake Bay Foundation Papers, University of Maryland, College Park; *Washington Post*, 13 May, 14 May, 15 May 1969; *Washington Evening Star*, 14 May 1969.

33. Radford, "Statement of Concern," pp. 22–24; *Congressional Record*, 91st Cong., 1st Sess., 1969, pp. 26115–26121; *Washington Evening Star*, 14 May 1969; *Washington Post*, 14 May 1969.

34. Clifford K. Beck to the Commission, 15 July 1969, Job 4, Box 13, MH&S–11 REG (Thermal Effects), AEC 1318/38 (19 January 1970), AEC/NRC; "State vs. Federal Authority: Growth vs. Environment," *Nuclear Industry*, 17 (March 1970): 42–44.

35. "Nuclear Power Plants and Our Environment: A Report to the Maryland Academy of Sciences," January 1970, Box 181 (Pollution), Seaborg Office Files, Joseph J. Fouchard to the Commission, 26 January 1970, Box 7763, MH&S–11 (Environmental Studies), AEC/DOE; *Nucleonics Week*, 4 June 1970, pp. 1–2; *Baltimore Sun*, 14 October 1970; *Washington Post*, 20 October 1970.

36. AEC–R 192/9 (30 June 1970), AEC/NRC; Anthony Z. Roisman interview, 17 June 1980; Claude E. Barfield, "Energy Report: Calvert Cliffs Decision Requires Agencies to Get Tough with Environmental Laws," *National Journal*, 18 September 1971, p. 1930.

37. AEC–R 192/9, AEC Press Release, 3 August 1970, AEC/NRC.

38. Joseph F. Hennessey to the Commission, 4 December 1970, MH&S–3–2–2 (National Environmental Policy Act), AEC/NRC; *Nucleonics Week*, 17 December 1970, p. 1; "Federal Appeals Court to Rule on Whether AEC Properly Meets Its Responsibilities under NEPA," *Nuclear Industry*, 18 (March 1971): 35–37.

39. "Federal Appeals Court to Rule on . . . NEPA," pp. 35–36; Reply Brief for Petitioners, n. d., Hearing and Opposition File, AEC Dockets 50–317 and 50–318 (Calvert Cliffs), AEC/NRC.

40. Brief for the Respondents, n. d., ID&R–5 REG (Baltimore-Calvert Cliffs), Job 9, Box 1, AEC/NRC.

41. Marcus A. Rowden interview, 20 August 1980; Roisman interview, 17 June 1980; Martin G. Malsch interview, 22 June 1988; Trask, "AEC and the Environment," pp. 34–39.

42. Roisman interview, 17 June 1980; William C. Parler interview, 22 June 1988; Trask, "AEC and the Environment," p. 109; *Nucleonics Week*, 13 May 1971, pp. 5–6.

43. *Calvert Cliffs Coordinating Committee, Inc. et al. v. United States Atomic Energy Commission*, 449 F.2d 1109 (1971), printed in JCAE, *Selected Materials on the Calvert Cliffs Decision: Its Origin and Aftermath*, 92nd Cong., 1st Sess., 1972, pp. 125–146.

44. Edward J. Bauser to All Members, 24 July 1971, Box 49 (Calvert Cliffs vs. AEC), Holifield Papers; M. A. Rowden to Commissioner Ramey, 23 July 1971, Box 35 (Calvert Cliffs Decision and Analysis), Ramey Office Files, AEC/DOE; *Seaborg Journal*, Vol. 25, p. 511; *Nucleonics Week*, 29 July 1971, p. 1; Trask, "AEC and the Environment," pp. 52–54.

45. *Seaborg Journal*, Vol. 25, pp. 534–535; AEC Press Release, 26 August 1971, AEC/NRC; Trask, "AEC and the Environment," pp. 48–55.

46. Harold L. Price to the Commission, 27 July 1971, Job 9, Box 19, Legal–11 REG (Litigation), L. Rogers to E. J. Bloch, draft, 4 August 1971, MH&S–3–2–2 (NEPA), AEC/NRC; *Nucleonics Week*, 12 August 1971, pp. 1–2; "The Calvert Cliffs Case," *Nuclear News*, 14 (September 1971): 28–29.

47. SECY–R 297 (30 July 1971), "Interim Guidance on Modification in Applicant's Environmental Reports and AEC Statements under NEPA," 4 August 1971, Job 9, Box 1, ID&R–5 REG (Baltimore Gas and Electric-Calvert Cliffs), Harold L. Price to the Commission, 10 August 1971, Job 9, Box 13, ID&R–14 REG (Part 50), Stephen H. Hanauer to Lester Rogers, 10 August 1971, MH&S–3–2–2 (NEPA), AEC/NRC; "Period of Adjustment," *Nuclear Industry*, 18 (September 1971): 17–24.

48. Harold L. Price to the Commission, 29 August 1971, Box 35 (Revision of Appendix D-Drafts), Price to the Commission, 30 August 1971, Box 35 (Reaction of Utilities, JCAE, etc.), Ramey Office Files, AEC/DOE; "Appendix D—Interim Statement of General Policy and Procedure," draft, 29 August

1971, MH&S–3–2–2 (NEPA), AEC/NRC; *Nucleonics Week*, 2 September 1971, pp. 1–4; "Period of Adjustment," *Nuclear Industry*, pp. 17–19.

49. On the Corps of Engineers, see, for example, Jeffrey K. Stine, "Environmental Politics in the American South: The Fight over the Tennessee-Tombigbee Waterway," *Environmental History Review*, 15 (Spring 1991): 1–24; Michael C. Robinson, "The Relationship between the Army Corps of Engineers and the Environmental Community," *Environmental Review*, 13 (Spring 1989): 1–41; and Daniel A. Mazmanian and Jeanne Nienaber, *Can Organizations Change? Environmental Protection, Citizen Participation, and the Corps of Engineers* (Washington, D. C.: Brookings Institution, 1979).

CHAPTER XV: THE PUBLIC AND NUCLEAR POWER

1. Hazel Gaudet Erskine, "The Polls: Atomic Weapons and Nuclear Energy," *Public Opinion Quarterly*, 27 (Summer 1963): 164; Charles B. Delafield, "How Con Ed Informs the Public," *Forum Memo*, 7 (December 1960): 10–13; Allan C. Fisher, Jr., "You and the Obedient Atom," *National Geographic Magazine*, 64 (September 1958): 303–352; "Big Hurdle for A-Power: Gaining Public Acceptance," *Nucleonics*, 21 (October 1963): 17–24; Stanley M. Nealy, Barbara D. Melber, and William L. Rankin, *Public Opinion and Nuclear Energy* (Lexington, Mass.: D. C. Heath and Co., 1983), p. 4; Michael Smith, "Advertising the Atom," in Michael J. Lacey, ed., *Government and Environmental Politics* (Washington, D. C.: Wilson Center Press, 1989), pp. 233–262.

2. *Nucleonics Week*, 20 June 1963, p. 1.

3. Minutes of Steering Group Meeting, Public Information Committee, Atomic Industrial Forum, 31 July 1963, Script for "Atomic Power Today: Safety and Service," 5 August 1964, File 532, U. S. Public Health Service Records, Division of Radiological Health, U. S. Public Health Service, Rockville, Maryland; *Nucleonics Week*, 28 November 1963, pp. 2–3.

4. William E. Shoup, "The Atom, the Public, and You," *Nuclear News*, 8 (August 1965): 13–17. For other comments on public views of nuclear power, see, for example, G. C. Laurence, "The Atom's Public Relations," *Nuclear News*, 6 (December 1963): 3–5; "What Does the Public Think: Telling the Atomic Story," *ibid.*, 7 (January 1964): 4–14; "Public Education Most Urgent, Reactor-Site Experts Agree," *Nucleonics*, 23 (March 1965): 22–23.

5. Duncan Clark to Those Listed Below, 11 July 1963, MH&S–3–2 (Reactor Hazards and Control), Atomic Energy Commission Records, Nuclear Regulatory Commission, Rockville, Maryland (hereafter cited as AEC/NRC); Walter A. Kee, "Information Systems for Tomorrow: Role of the USAEC," *Nuclear News*, 9 (December 1966): 22–24; Glenn T. Seaborg and Daniel M. Wilkes, *Education and the Atom* (New York: McGraw-Hill Book Co., 1964), pp. 94–110.

6. AEC–R 2/48 (27 April 1965), Jack Vanderryn to Duncan C. Clark, 23 July 1963, MH&S–3–2 (Reactor Hazards and Control), W. B. McCool to the File, 2 February 1967, Job 4, Box 4, ID&R–6 REG (Hazard Evaluation), AEC/

NRC; Jerome D. Luntz to Glenn T. Seaborg, 2 August 1963, Box 1341, I&P–6 (Articles, Vol. 1), Atomic Energy Commission Records, Department of Energy, Germantown, Maryland (hereafter cited as AEC/DOE); "Educating the Public about Nuclear Power—the AEC View," *Nucleonics*, 21 (September 1963): 6; "Industry Views the Commission," *ibid.*, 24 (October 1966): 20.

7. Duncan Clark to the Commission, 28 February, 24 March 1964, Box 1341, I&P–6 (Articles, Vol. 1), AEC/DOE; "The Atlantic Report: Science and Industry," *Atlantic Monthly*, 213 (March 1964): 22, 25; Carl Dreher, "Atomic Power: The Fear and the Promise," *Nation*, 198 (23 March 1964): 289–293; Joseph P. Blank, "Atomic Power Comes of Age," *Reader's Digest*, 87 (December 1965): 109–112.

8. "Are Nuclear Plants Winning Acceptance?" *Electrical World*, 24 January 1966, pp. 115–118; "Nuclear Power and the Community: Familiarity Breeds Confidence," *Nuclear News*, 9 (May 1966): 15–16.

9. Seymour T. Zenchelsky and Vernon Bryson to Glenn T. Seaborg, 5 January 1965, Box 187 (Jersey Central P&L), Office Files of Glenn T. Seaborg, AEC/DOE; Mary Hays Weik, "The Story Nobody Prints," 1965, Job 5, Box 19, IR&A–6 REG (Jersey Central P&L Co.), AEC/NRC; *Washington Post*, 29 May 1965; Wallace Cloud, "Is Atomic Industry Risking Your Life?" *Popular Science*, 186 (June 1965): 45–49, 176–178.

10. "*Popular Science* Article on Atomic Industry Draws Strong Protests," *Popular Science*, 187 (September 1985): 38–42; Chauncey Starr, "The Social Value of Nuclear Power," *Nuclear News*, 9 (April 1966): 28–31.

11. Leonard M. Trosten to George D. Aiken, 28 September 1967, File 23–1–10 (Vermont Yankee Nuclear Power Plant Material, 1967), George D. Aiken Papers, University of Vermont, Burlington; Anti-Pollution League, "Bulletin," June 1969, Box 1541, File 800, Atomic Energy Commission (1969 Resource/Projects), Edmund S. Muskie Papers, Edmund S. Muskie Archives, Bates College, Lewiston, Maine; John A. Harris to R. E. Hollingsworth, 19 October 1967, I&P–12–3 (Complaints and Criticisms), AEC/NRC; *Burlington Free Press*, 30 August 1967.

12. Larry Bogart to Clifford Case, 3 October 1967, Box 577 (Reactor Type or Company Name: Jersey Central, Oyster Creek), Papers of the Joint Committee on Atomic Energy, Record Group 128 (Records of the Joint Committees of Congress), National Archives; Bogart to Glenn T. Seaborg, 10 November 1965, Box 200 (Con Ed), Seaborg Office Files, AEC/DOE; Helen Matheson, "Atomic Power and the Problem of Public Safety," *The Rotarian*, December 1966, copy in Box 720, File 800, Atomic Energy Commission (1967 Correspondence), Muskie Papers; *Los Angeles Times*, 5 February 1968; *Nucleonics Week*, 19 October 1967, p. 7, 21 December 1967, p. 6, 18 January 1968, p. 6.

13. *Journal of Glenn T. Seaborg* (25 vols., Berkeley: Lawrence Berkeley Laboratory PUB–625, 1989), Vol. 15, p. 596; *Congressional Record*, 90th Cong., 1st Sess., 1967, pp. 33260–33262; "Coal's Atomic Energy 'Problem,' " *Nuclear Industry*, 14 (March 1967): 6; "Morton Calls for Review of Nuclear Program," *ibid.*, 15 (March 1968): 36–38.

14. "Siting Bill Would Ban AEC Licensing for Two Years," *Nuclear Indus-*

try, 15 (May 1968): 4; *Nucleonics Week*, 27 October 1966, p. 3, 25 April 1968, pp. 3–4.

15. Robert H. Boyle, "The Nukes Are in Hot Water," *Sports Illustrated*, 20 January 1969, pp. 24–28; John A. Harris to the Commission, 28 January 1969, Box 7751, I&P–4–6 (Articles), AEC/DOE.

16. Sheldon Novick, *The Careless Atom* (Boston: Houghton Mifflin Co., 1969); *Nucleonics Week*, 13 February 1969, p. 4.

17. *New York Times Book Review*, 2 March, 9 March 1969; Ellen Thro, "Independent Indictment," *Nuclear News*, 12 (May 1969): 26; "In Pursuit of Public Awareness," *ibid.*, 12 (June 1969): 2–23; *Nucleonics Week*, 13 February 1969, p. 4; Atomic Industrial Forum, *Info*, March 1969, copy in Box 3 (Atomic Industrial Forum), Office Files of James T. Ramey, AEC/DOE.

18. AEC 688/82 (30 July 1969), Box 5626, OGM Files (Environmental— Pollution), AEC/DOE; *Nucleonics Week*, 22 September 1966, pp. 3–4.

19. Richard Curtis and Elizabeth Hogan, "The Myth of the Peaceful Atom," *Natural History*, 78 (March 1969): 6–15, 71–76; Curtis and Hogan, *Perils of the Peaceful Atom* (Garden City, NY: Doubleday and Co., 1969); *Publishers' Weekly*, 19 May 1969, p. 18.

20. Joseph J. Fouchard to Howard Brown, 1 July 1969, RD 25–1 (Myth of the Peaceful Atom), AEC/NRC; Harold L. Price to Glenn T. Seaborg, 15 May 1969, Box 81 (Director of Regulation), Seaborg to Richard L. Ottinger, 28 May 1969, Box 181 ("Myth of the Peaceful Atom"), Seaborg Office Files, AEC/ DOE; Edward J. Bauser to Chet Holifield, 14 July 1969, Box 46 (JCAE Outgoing Correspondence), Chet Holifield Papers, University of Southern California, Los Angeles; James G. Beckerley and Norman Hilberry, "The Perils of 'Declaratory Truth,'" *Nuclear News*, 12 (September 1969): 59–64; Ivan Bloch, "Attack on Atomic Power," *Environment*, 11 (October 1969): 28; David Rittenhouse Inglis, "The Hazardous Industrial Atom," *Bulletin of the Atomic Scientists*, 26 (February 1970): 50–54; *Washington Post*, 23 July 1969; *Philadelphia Inquirer*, 27 July 1969; *Electrical World*, 21 July 1969; *Nucleonics Week*, 3 July 1969, p. 2, 8 January 1970, p. 5.

21. The summary is drawn from the literature cited above. The Curtis-Hogan quote is in "The Myth of the Peaceful Atom," p. 8.

22. AEC 688/62 (17 March 1969), Box 7751, I&P–4 (Public Information), AEC/DOE; Howard C. Brown, Jr., "AEC Goes Public—A Case History," *Nuclear Safety*, 11 (September-October 1970): 365–369.

23. AEC 688/62, AEC 688/68 (29 April 1969), Box 7751, I&P–4 (Public Information), AEC/DOE.

24. "Draft Talking Paper," 28 April 1969, Box 7751, I&P–4 (Public Information), Information Meeting Minutes, 9 May 1969, Box 5667, OGM Files (Information Meeting Minutes), "Summary—Environmental Group Organization," n.d., Box 5586 (Environment—Minutes), AEC/DOE; *Seaborg Journal*, Vol. 20, p. 314; JCAE, *Hearings on Environmental Effects of Producing Electric Power*, 91st Cong., 1st Sess., 1969, pp. 111–116.

25. *Nucleonics Week*, 8 May 1969, p. 4, 12 June 1969, p. 2, 31 July 1969, p. 5.

26. Information Meeting Minutes, 21 May 1969, Box 5667, OGM Files

(Information Meeting Minutes), AEC/DOE; William T. England to Edward J. Bauser, 26 May 1969, Box 511 (Pollution—Environmental), JCAE Papers; Chet Holifield to William E. Warne, 12 September 1969, Box 46 (JCAE Outgoing Correspondence), Holifield Papers; *Nucleonics Week*, 19 June 1969, pp. 4–5.

27. J. H. Wright, Memorandum on "Anti-Nuclear Activities in New Hampshire," 13 May 1969, Box 182 (Westinghouse Electric Corporation), Craig Hosmer Papers, University of Southern California, Los Angeles; W. E. Johnson to the Files, 1 April 1969, Box 8395 (General Electric), Seaborg Office Files, AEC/DOE; *Nucleonics Week*, 11 December 1969, pp. 3–4.

28. W. B. Behnke to Thomas G. Ayers, 24 April 1969, Ayers to James T. Ramey, 1 May 1969, and attached "Memorandum on Nuclear Power," Ad Hoc Committee on Nuclear Power Agenda, 19 June 1969, Box 35 (Ayers Group), John W. Gore to Ramey, 8 October 1969, Box 18 (Baltimore Chamber of Commerce), Ramey Office Files, AEC/DOE; *Washington Star*, 19 August 1969; *Baltimore Evening Sun*, 19 August, 26 September 1969; *Baltimore Morning Sun*, 29 August, 2 September, 12 September, 7 October 1969.

29. *Washington Star*, 15 July 1969; *Seattle Post-Intelligencer*, 17–22 July 1969; Alden P. Armagnac, "World's Biggest Atom-Power Plant," *Popular Science*, 195 (September 1969): 94–97, 207–209; John A. Harris to the Commission, 4 September 1969, Box 7752, I&P–4–6 (Articles), AEC/DOE.

30. AEC 1318/10 (22 September 1969), Box 7763, MH&S–11 (Environmental Studies), "Miscellaneous Notes," n.d., Box 5586, OGM Files (Environment, Janet's File), AEC/DOE; *Seaborg Journal*, Vol. 19, p. 162, Vol. 20, p. 51; "'Peaceful Atom' Sparks a War," *Life*, 12 September 1969, pp. 26–33.

31. *Nucleonics Week*, 3 July 1969, pp. 3–4.

32. Deane C. Davis to Glenn T. Seaborg, 14 August 1969, Howard C. Brown, Jr. to the Commission, 3 September 1969, Box 182 (Program on Nuclear Power, Burlington, Vt.), Seaborg Office Files, AEC/DOE; John E. McEwem, Jr. to C. L. Henderson, 26 August 1969, MH&S–3–2–1 (Environmental Meetings), AEC/NRC; Deane C. Davis Press Release, 26 August 1969, File 23–5–20 (Meeting in Vermont, Correspondence), Aiken Papers; George D. Aiken to Chet Holifield, 7 July 1969, Box 511 (Pollution-Environmental), JCAE Papers.

33. University of Vermont *Alumni Magazine*, 50 (November 1969): 4–5, copy in Box 182 (Program on Nuclear Power, Burlington, Vt.), Seaborg Office Files, AEC/DOE; *New York Times*, 12 September 1969; *Burlington Free Press*, 12 September 1969; *Rutland Herald*, 12 September, 13 September 1969; *Barre-Montpelier Times-Argus*, 12 September 1969; *Nucleonics Week*, 18 September 1969, pp. 4–12; "AEC Goes Public on Nuclear Safety," *Business Week*, 20 September 1969, pp. 52–53; "AEC Holds First Public Meeting on Nuclear Power," *Nuclear News*, 12 (October 1969): 14–16.

34. CW [Charles Weaver] Memorandum on "AEC Meeting in Vermont," 15 September 1969, File 23–5–21 (Meeting in Vermont, Memos), George D. Aiken to James T. Ramey, 12 September 1969, File 23–5–20 (Meeting in Vermont, Correspondence), Aiken Papers; W. B. McCool to H. L. Price, 13 October 1969, Job 4, Box 11, ID&R–6 (Hazards Evaluation), AEC/NRC;

Nucleonics Week, 18 September 1969, p. 13, 25 September 1969, pp. 3–4; "AEC Goes Public," *Business Week,* p. 53.

35. AEC 1318/11 (25 September 1969), Box 7763, MH&S–11 (Environmental Studies), Stanley D. Schneider, "Reactions to Trip to Burlington, Vermont . . . And its Implications," 16 September 1969, Box 182 (Program on Nuclear Power, Burlington, Vermont), Seaborg Office Files, AEC/DOE.

36. Memorandum (author unidentified) to Commissioner Ramey, 24 March 1970, Joseph J. DiNunno to Ramey, 16 April 1970, Box 35 (Ayers Group), Ramey Office Files, AEC/DOE; "The Peaceful Atom: Friend or Foe?" *Time,* 19 January 1970, pp. 42–43; *Chicago Tribune,* 18 January 1970; *Nucleonics Week,* 16 October 1969, pp. 1–2, 30 October 1969, pp. 3–4.

37. See, for example: *New York Times,* 16 July 1970; *Washington Post,* 15 June 1970; *Philadelphia Inquirer,* 19 July 1970; *Washington Star,* 22–25 July 1970; *Wall Street Journal,* 25 January 1971; Roy Bongartz, "Muckuppery along the Potomac," *Esquire,* 73 (June 1970): 75–78; "The Controversial Atomic Energy Commission," *Newsweek,* 4 January 1971, pp. 37–40; Nancy Wood, "America's Most Radioactive City," *McCall's,* 97 (September 1970): 46–50; Howard G. Paster, "Water and Air," *New Republic,* 163 (31 October 1970): 24–26; Jack Shepherd, "The Nuclear Threat Inside America," *Look,* 15 December 1970, pp. 21–27; "Playboy Interview: Dr. Paul Ehrlich," *Playboy,* (August 1970), Transcripts of CBS-TV Morning News, 10–14 August 1970, MH&S–3–2 (Hazards and Control), AEC/NRC; Transcript of "Powers That Be: A KNBC Public Affairs Documentary," 17 May 1971, Box 7809, I&P–4–4 (TV), AEC/DOE.

38. *Seaborg Journal,* Vol. 24, p. 202.

39. *New York Times,* 17 February 1971; *Washington Post,* 19 March 1971; "A Nuclear Graveyard," *Newsweek,* 29 March 1971, p. 60; Constance Holden, "Nuclear Waste: Kansans Riled by AEC Plans for Atom Dump," *Science,* 172 (16 April 1971): 249–250; Randy Brown, "The AEC Has Something for Kansas," *Nation,* 212 (7 June 1971): 712–716; U. S. Atomic Energy Commission, *Environmental Statement: Radioactive Waste Repository, Lyons, Kansas* (WASH–1503), June 1971, AEC/NRC.

40. Shepherd, "Nuclear Threat Inside America," p. 21.

41. J. H. Wright, Memorandum on "Anti-Nuclear Activities in New Hampshire," 13 May 1969, Hosmer Papers; Stanley D. Schneider, "Reactions to Trip to Burlington, Vermont," 16 September 1969, John A. Harris to the Commission, 24 May 1971, Box 35 (Miscellaneous), Ramey Office Files, AEC/DOE; *Nucleonics Week,* 3 June 1971, pp. 5–6.

42. *San Jose Mercury News,* 10 January 1971; H. G. Slater, "Public Opposition to Nuclear Power: An Industry Overview," *Nuclear Safety,* 12 (September-October 1971): 448–456; *Nucleonics Week,* 6 May 1971, p. 4; Elizabeth S. Rolph, *Nuclear Power and the Public Safety: A Study in Regulation* (Lexington, Mass.: D. C. Heath and Co., 1979), pp. 170–182.

43. *San Jose Mercury News,* 10 January 1971; *Nucleonics Week,* 8 May 1969, p. 5, 13 August 1970, p. 3.

44. Lee A. DuBridge to Edmund S. Muskie, 4 May 1970, White House

Central File, Subject Files, [GEN] UT 2 (DuBridge, Lee A.), Richard M. Nixon Papers, Nixon Presidential Materials Project, Alexandria, Virginia; "Something to Cheer About," *Nuclear Industry*, 17 (February 1970): 20–22; "Westinghouse Offers Systems for 'Essentially Zero Release' of Radwaste," *ibid.*, 17 (May 1970): 40; JCAE, *Hearings on Environmental Effects of Producing Electric Power*, 91st Cong., 2nd Sess., 1970, pp. 1692–1697; *Nucleonics Week*, 19 February 1970, pp. 3–4; G. O. Bright, "Some Effects of Public Intervention on the Reactor Licensing Process," *Nuclear Safety*, 13 (January-February 1972): 13–21.

45. James B. Graham to Chet Holifield, 12 February 1970, Box 47 (Bauser Memos), Holifield Papers; *New York Times*, 11 January, 17 February 1971; "Probing the Atom," *Nuclear News*, 14 (May 1971): 30; *Nucleonics Week*, 30 October 1969, p. 1, 15 January 1970, p. 5, 11 February 1971, p. 3, 29 April 1971, p. 6.

46. "Antinuclear Forces Push for Moratoriums," *Nuclear News*, 14 (July 1971): 29–32; *Nucleonics Week*, 10 December 1970, p. 5, 27 May 1971, p. 3; Daniel Pope, "'We Can Wait. We Should Wait.' Eugene's Nuclear Power Controversy, 1968–1970," *Pacific Historical Review*, 59 (August 1990): 349–373.

47. *Congressional Record*, 92nd Cong., 1st Sess., 1971, pp. 5267–5269; "Something to Cheer About," *Nuclear Industry*, p. 21.

48. William C. Parler to Edward J. Bauser, 5 March 1971, Box 194 (Public Acceptance of Nuclear Power), Hosmer Papers.

49. "The Environment: A Challenge for the Seventies," *Nuclear News*, 13 (April 1970): 18–26; *Nucleonics Week*, 30 October 1969, p. 3; Spencer R. Weart, *Nuclear Fear: A History of Images* (Cambridge, MA: Harvard University Press, 1988), pp. 323–327; Daniel J. Kevles, *The Physicists: The History of a Scientific Community in America* (New York: Alfred A. Knopf, 1977), pp. 398–405.

50. Robert W. Newlin to James Hanchett, 24 March 1970, Box 193 (Anti-Nuclear Organizations), Hosmer Papers; *Barre-Montpelier Times-Argus*, 12 September 1970; *Philadelphia Inquirer*, 28 September 1970; JCAE, *Hearings on Nuclear Reactor Safety*, 93rd Cong., 1st Sess., 1973, pp. 8–10.

51. For discussions of the difficulty of winning public confidence on safety issues that were matters of uncertainty, see Joseph G. Morone and Edward J. Woodhouse, *The Demise of Nuclear Energy? Lessons for Democratic Control of Technology* (New Haven, Conn.: Yale University Press, 1989), pp. 88–95; and Morone and Woodhouse, *Averting Catastrophe: Strategies for Regulating Risky Technologies* (Berkeley and Los Angeles: University of California Press, 1986), pp. 74–75.

52. Howard C. Brown, Jr. to the Commission, 23 April 1971, Box 7809, I&P-4 (Public Information), AEC/DOE; Glenn T. Seaborg, "On Misunderstanding the Atom," *Bulletin of the Atomic Scientists*, 27 (September 1971): 46–53; *Nucleonics Week*, 14 January 1971, p. 4, 21 January 1971, p. 6, 4 March 1971, pp. 1–2.

53. Walter H. Jordan, "The Issues Concerning Nuclear Power," *Nuclear News*, 14 (October 1971): 43–49.

CHAPTER XVI: THE END OF AN ERA

1. *Journal of Glenn T. Seaborg* (25 vols., Berkeley: Lawrence Berkeley Laboratory PUB–625, 1989), Vol. 21, pp. 753–754, Vol. 22, pp. 299, 334, Vol. 25, pp. 420, 439; Will Kriegsman to John Ehrlichman, 22 March 1971, Edward E. David, Jr. to Fred Malek, 10 June 1971, Malek to Peter Flanigan, 11 June 1971, Malek to Flanigan, 17 June 1971, White House Special File, White House Central File, FG 78 (Atomic Energy Commission), Richard M. Nixon Papers, Nixon Presidential Papers Materials Project, Alexandria, Virginia; *New York Times*, 22 July 1971.

2. *New York Times*, 30 July 1971; "New Man at the AEC," *Newsweek*, 2 August 1971, p. 49; *Washington News*, 22 July 1971; "Commission Appointments," *Nuclear Industry*, 18 (July 1971): 3–4; "Memorandum from the Editor," *Nuclear News*, 14 (September 1971): 27.

3. W. B. McCool, "Secretary's Meeting Report," 21 October 1971, Box 3321 (Secretary's Meeting Reports), Atomic Energy Commission Records, Department of Energy, Germantown, Maryland; AEC Press Release, 21 October 1971, Atomic Energy Commission Records, Nuclear Regulatory Commission, Rockville, Maryland; Harold L. Price, "Meeting with Regulatory Staff," 14 October 1971, Harold L. Price Papers, Herbert Hoover Library, West Branch, Iowa; *Nucleonics Week*, 18 October 1971, p. 2; Interview with Harold L. Price, Chevy Chase, Maryland, 30 January 1989.

4. "There's a *Bird Watcher* Running the Atomic Energy Commission," *National Wildlife*, 10 (August-September 1972): 17.

5. "Nuclear Power: Setback?" *Forbes*, 15 November 1968, pp. 55, 58; "Economics of Nuclear Power Badly Eroded—Sporn," *Nuclear News*, 13 (March 1970): 25–28; *Nucleonics Week*, 6 February 1969, pp. 3–4.

6. "The 'Ideal' Nuclear Plant Site," *Nuclear Industry*, 17 (September 1970): 16.

7. For an insightful discussion of how issues debated within the AEC eventually became key elements of the public controversy, see Brian Balogh, *Chain Reaction: Expert Debate and Public Participation in American Commercial Nuclear Power, 1945–1975* (New York: Cambridge University Press, 1991).

8. Jeffrey K. Stine, *A History of Science Policy in the United States, 1940–1985*, Report Prepared for the Task Force on Science Policy, Committee on Science and Technology, House of Representatives, 99th Cong., 2nd Sess., 1986; Bruce L. R. Smith, *American Science Policy since World War II* (Washington, D. C.: Brookings Institution, 1990); *Baltimore Morning Sun*, 3 October 1969.

9. *Seaborg Journal*, Vol. 23, pp. 678–686; *Washington Post*, 31 December 1970; *Washington Star*, 31 December 1970; *New York Times*, 31 December 1970; *Los Angeles Times*, 31 December 1970.

10. Spencer R. Weart, *Nuclear Fear: A History of Images* (Cambridge, MA: Harvard University Press, 1988).

11. *Ibid.*, p. 325.

12. Schlesinger's speech is printed in U. S. Congress, Joint Committee on Atomic Energy, *Selected Materials on the Calvert Cliffs Decision, Its Origin and Aftermath*, 92nd Cong., 1st Sess., 1972, pp. 392–398.

Essay on Sources

The basic documentary sources for this book are the records of the Atomic Energy Commission. When the AEC was abolished in 1975, its records were divided along functional lines between the agencies that succeeded it, the Energy Research and Development Administration, which later became a part of the Department of Energy, and the Nuclear Regulatory Commission. Files that clearly related to the AEC's regulatory programs went to the NRC; the remainder went to ERDA, and then DOE. AEC records at both agencies contain important materials on the history of regulatory policies and actions. The records at the NRC cited in the notes (as AEC/NRC) are available for research at the NRC's Public Document Room, 2120 "L" Street NW, Washington, D. C. For information about the documents at DOE cited in the notes (as AEC/DOE), researchers should contact the DOE History Division.

Detailed information about the nature of AEC records is provided in George T. Mazuzan and J. Samuel Walker, *Controlling the Atom: The Beginnings of Nuclear Regulation, 1946–1962* (Berkeley and Los Angeles: University of California Press, 1984), pp. 499–501, and Richard G. Hewlett and Jack M. Holl, *Atoms for Peace and War, 1953–1961: Eisenhower and the Atomic Energy Commission* (Berkeley and Los Angeles: University of California Press, 1989), pp. 660–661.

Other collections of government records and personal papers also provide valuable information. The papers of the Joint Committee on Atomic Energy, included in Record Group 128 (Records of the Joint

Committees of Congress) at the National Archives in Washington, D. C., are a rich source. They are especially useful in documenting the Joint Committee's role in regulatory issues, but shed light on other matters as well. Joint Committee records become available for research according to a twenty-year rule established by the U. S. Senate.

Several other collections at the National Archives contain important materials. The papers of the Federal Radiation Council, a part of Record Group 412 (Records of the Environmental Protection Agency), are helpful in understanding a number of issues in which the agency was involved. For the purposes of this volume, they were particularly valuable on the radiation controversy and on radon standards for uranium mines. The records of the Office of the Secretary of Labor—W. Willard Wirtz, a part of Record Group 174 (General Records of the Department of Labor), are equally indispensable on the controversy surrounding uranium mine safety. The records of the Office of Science and Technology (Record Group 359) and the records of the Bureau of the Budget (Record Group 51) include a few significant items relating to nuclear regulation.

The Division of Radiological Health of the U. S. Public Health Service is a key source of documents relating to radiation protection and safety. The records are available for research on microfilm at the division's library in Rockville, Maryland.

In addition to the records of government agencies, several collections of personal papers contain documents of vital and singular importance. The presidential papers of Richard M. Nixon, housed under the custody of the National Archives in the Richard M. Nixon Presidential Materials Project in Alexandria, Virginia, include some interesting correspondence, especially on the Monticello and emergency core cooling issues. The papers of Lyndon B. Johnson at the Lyndon B. Johnson Library in Austin, Texas, are less helpful but have a few noteworthy items. The papers of Harold L. Price, the AEC's director of regulation, have been accessioned by the Herbert Hoover Library in West Branch, Iowa. Although the volume of the collection is not large, it contains important documents relating to a wide range of regulatory issues, especially in the late 1960s and early 1970s.

The papers of several members of the Joint Committee on Atomic Energy include useful materials. The papers of Chet Holifield and Craig Hosmer, both at the University of Southern California in Los Angeles, are rich collections that feature documents on a wide variety of subjects over a long period of time. Both should be consulted by any researcher interested in the history of nuclear energy. Other members of the Joint Committee left papers that include useful materials. The papers of Clinton P. Anderson at the Library of Congress in Washington, D. C. are better for earlier years but still have some helpful materials for the 1963–1971 period. The George D. Aiken papers at the University of Vermont in Burlington are especially valuable on issues in which Aiken took a deep personal interest—the controversy over Vermont Yankee and the antitrust aspects of nuclear licensing. The John O. Pastore papers at Providence College in Providence, Rhode Island are generally not as rich as the Holifield or Hosmer collections but still contain many useful documents from his tenure on the Joint Committee.

Manuscript collections that document the activities and views of nuclear critics are rare and frequently nonexistent. There are a few sources, however, that were useful for this volume. The Edmund S. Muskie papers, housed in the Edmund S. Muskie Archives at Bates College in Lewiston, Maine, contain a fair amount of material relating to nuclear power and regulation. In addition, there are items of interest in the Barry Commoner papers at the Library of Congress, the Chesapeake Bay Foundation papers at the University of Maryland, College Park, and Sierra Club records in San Francisco.

A source of unique and extraordinary importance for any scholar working on nuclear energy issues during the 1960s and early 1970s is the *Journal of Glenn T. Seaborg*. The *Journal* for the years of Seaborg's chairmanship of the AEC runs twenty-five volumes; it was published by Lawrence Berkeley Laboratory of the University of California in 1989 (PUB–625). The *Journal* includes Seaborg's meticulous notes on each day's activities, ranging from accounts of meetings and phone conversations to the results of baseball games. It also includes many documents relating to the events described in the daily notes. In all cases, the *Journal* is an invaluable source; in some cases, such as the AEC's deliberations over tightening radiation standards, it provides information and insight that is unavailable elsewhere. Seaborg gave copies of his *Journal* to several archives and manuscript libraries, where it is open for research. He donated complete copies to the libraries of three University of California campuses—Berkeley, Santa Barbara, and UCLA. In addition, Seaborg sent portions of the *Journal* covering the administrations of the three presidents under whom he served to the appropriate presidential library: volumes 1–6 to the John F. Kennedy Library, Boston, Massachusetts, volumes 7–17 to the Lyndon B. Johnson Library, and volumes 18–25 to the Richard M. Nixon Presidential Materials Project.

The Library of Congress has accessioned Seaborg's personal papers for the years 1931–1988; when they are opened for research they will include a copy of the complete *Journal*.

This book also drew on many published primary sources, such as government reports and congressional hearings, and a wide variety of newspaper and magazine stories. Nuclear industry trade journals, especially *Nucleonics Week*, are an indispensable source for studying any topic relating to the history of nuclear power. They followed industry developments and regulatory policies with scrupulous care and generally reported them with accuracy and frankness.

The secondary literature on nuclear power and regulation has traditionally been dominated by partisan accounts. In recent years, however, a few books and articles have taken a scholarly approach and provided a balanced treatment of a complex topic. Those that I have found most useful are cited in the notes. Much work remains to be done to gain a full understanding of the history of nuclear regulation. Indeed, I think that practically every reactor built and many that were not built are worthy of careful study; each reflects important trends in political, technological, scientific, environmental, and business history, and in the areas in which plants were built or planned, local and social history.

I am grateful to many participants in the events recounted in this book who shared their experiences and views with me (or in a few cases, with my predecessors at the NRC). They provided important perspectives and information that was not always available from documentary sources. I benefited from the contributions of: Edson G. Case, Robert Colmar, Joseph J. Fouchard, Stephen H. Hanauer, Christopher L. Henderson, Albert P. Kenneke, Ralph E. Lapp, Norman Lauben, Morton W. Libarkin, Martin G. Malsch, Woodford B. McCool, Peter A. Morris, David Okrent, William C. Parler, Harold L. Price, James T. Ramey, Anthony Z. Roisman, Marcus A. Rowden, Glenn T. Seaborg, Howard K. Shapar, and Lauriston S. Taylor.

Index

Abelson, Philip H., 219
Abrahamson, Dean E., 310, 342, 403, 408
Accidents at nuclear facilities:
  Brookhaven reports on, 114–115,
  117–121; consequences, 115, 120,
  121, 127; and containment integrity,
  79–80, 160; core meltdown, 160,
  161, 162; economic costs of, 2, 114;
  engineered safety features and, 120–
  121, 126, 140, 142; environmental
  considerations under NEPA, 384–
  385; extraordinary nuclear occurrence defined, 136–138; Fermi plant
  fuel melt, 166–168, 395, 397, 406;
  first-line-of-defense concept, 203,
  216, 232; internally generated missiles, 156, 206; minor incidents, 125;
  model for analysis, 123, 124–125,
  129; population density and, 75; pressure vessel failure, 152–153, 155,
  156; prevention, 169–202, 216; public fears of, 406; radiation releases
  and dispersion, 123, 125, 129, 137;
  risk of, 10, 17, 113, 115, 123, 124–
  125, 156, 161, 170, 392, 420; safety
  systems for mitigating, 69, 73; severity classification, 385; size of plants
  and, 120, 121, 139; testing effects of,
  142, 144, 147; worst-case, 121, 122–
  123, 127, 139, 141, 142, 172, 173,
  200, 397–398. *See also* Loss-of-
  coolant accidents; WASH-740 update
Ackerman, Adolph J., 95, 393

Advanced Test Reactor, 213
Advisory Committee on Reactor Safeguards: Bodega Bay site endorsement,
  89–90, 97, 98; core melt concerns,
  161–164; criticisms of, 55, 164; design criteria review, 206–207, 208;
  ECCS position, 170, 174, 195–196;
  emergency planning views, 224; engineered safeguards, 141; Environmental Subcommittee, 323; members, 42,
  141, 153, 155, 156, 162, 170, 224;
  metropolitan siting of reactors, 58,
  76, 77, 81, 141, 165–166; pressure vessel concerns, 153–160; purpose of, 42, 164, 165; radiation protection standards, 323, 325; Ravenswood site review, 68; regulatory staff
  relationship with, 98, 100, 143; role
  in licensing process, 42–43, 48; safety
  research program concerns, 144–145,
  148–149, 150, 152, 179, 188; seismic
  siting views, 89–90, 97, 98, 104,
  108–109; views on nuclear power,
  43; WASH-740 update, 121, 126–
  127
Advisory Committee on X-Ray and Radium Protection, 297–298
Advisory Panel on Safeguarding Special
  Nuclear Material, 228–230
Aerojet-General Corporation, 182
Aiken, George D., 35–36, 403, 404
Air pollution, from fossil fuel plants, 32–
  33, 61, 102, 116–117, 268–269

Designer: U.C. Press Staff
Compositor: Huron Valley Graphics
Text: 10/13 Sabon
Display: Sabon
Printer: BookCrafters
Binder: BookCrafters